Immunotherapy in Hepatocellular Carcinoma

Immunotherapy in Hepatocellular Carcinoma

Editors

Naoshi Nishida
Masatoshi Kudo

MDPI • Basel • Beijing • Wuhan • Barcelona • Belgrade • Manchester • Tokyo • Cluj • Tianjin

Editors
Naoshi Nishida
Department of
Gastroenterology
and Hepatology
Faculty of Medicine
Kindai University
Osaka
Japan

Masatoshi Kudo
Department of
Gastroenterology
and Hepatology
Faculty of Medicine
Kindai University
Osaka
Japan

Editorial Office
MDPI
St. Alban-Anlage 66
4052 Basel, Switzerland

This is a reprint of articles from the Special Issue published online in the open access journal *Cancers* (ISSN 2072-6694) (available at: www.mdpi.com/journal/cancers/special_issues/Immuno_HCC).

For citation purposes, cite each article independently as indicated on the article page online and as indicated below:

LastName, A.A.; LastName, B.B.; LastName, C.C. Article Title. *Journal Name* **Year**, *Volume Number*, Page Range.

ISBN 978-3-0365-6691-7 (Hbk)
ISBN 978-3-0365-6690-0 (PDF)

© 2023 by the authors. Articles in this book are Open Access and distributed under the Creative Commons Attribution (CC BY) license, which allows users to download, copy and build upon published articles, as long as the author and publisher are properly credited, which ensures maximum dissemination and a wider impact of our publications.

The book as a whole is distributed by MDPI under the terms and conditions of the Creative Commons license CC BY-NC-ND.

Contents

Ying-Chun Shen, Ching-Ping Yeh, Yung-Ming Jeng, Chiun Hsu, Chih-Hung Hsu and Zhong-Zhe Lin et al.
Limited Predictive or Prognostic Role of Tumor-Infiltrating Tissue-Resident Memory CD8 T Cells in Patients with Hepatocellular Carcinoma Receiving Immunotherapy
Reprinted from: *Cancers* 2021, 13, 5142, doi:10.3390/cancers13205142 1

Piera Federico, Angelica Petrillo, Pasqualina Giordano, Davide Bosso, Antonietta Fabbrocini and Margaret Ottaviano et al.
Immune Checkpoint Inhibitors in Hepatocellular Carcinoma: Current Status and Novel Perspectives
Reprinted from: *Cancers* 2020, 12, 3025, doi:10.3390/cancers12103025 13

Sophia Heinrich, Darko Castven, Peter R. Galle and Jens U. Marquardt
Translational Considerations to Improve Response and Overcome Therapy Resistance in Immunotherapy for Hepatocellular Carcinoma
Reprinted from: *Cancers* 2020, 12, 2495, doi:10.3390/cancers12092495 33

Akira Asai, Hidetaka Yasuoka, Masahiro Matsui, Yusuke Tsuchimoto, Shinya Fukunishi and Kazuhide Higuchi
Programmed Death 1 Ligand Expression in the Monocytes of Patients with Hepatocellular Carcinoma Depends on Tumor Progression
Reprinted from: *Cancers* 2020, 12, 2286, doi:10.3390/cancers12082286 63

Naoshi Odagiri, Hoang Hai, Le Thi Thanh Thuy, Minh Phuong Dong, Maito Suoh and Kohei Kotani et al.
Early Change in the Plasma Levels of Circulating Soluble Immune Checkpoint Proteins in Patients with Unresectable Hepatocellular Carcinoma Treated by Lenvatinib or Transcatheter Arterial Chemoembolization
Reprinted from: *Cancers* 2020, 12, 2045, doi:10.3390/cancers12082045 77

Won-Mook Choi, Danbi Lee, Ju Hyun Shim, Kang Mo Kim, Young-Suk Lim and Han Chu Lee et al.
Effectiveness and Safety of Nivolumab in Child–Pugh B Patients with Hepatocellular Carcinoma: A Real-World Cohort Study
Reprinted from: *Cancers* 2020, 12, 1968, doi:10.3390/cancers12071968 93

Naoshi Nishida and Masatoshi Kudo
Immune Phenotype and Immune Checkpoint Inhibitors for the Treatment of Human Hepatocellular Carcinoma
Reprinted from: *Cancers* 2020, 12, 1274, doi:10.3390/cancers12051274 107

Masatoshi Kudo
Scientific Rationale for Combined Immunotherapy with PD-1/PD-L1 Antibodies and VEGF Inhibitors in Advanced Hepatocellular Carcinoma
Reprinted from: *Cancers* 2020, 12, 1089, doi:10.3390/cancers12051089 125

Shigeharu Nakano, Yuji Eso, Hirokazu Okada, Atsushi Takai, Ken Takahashi and Hiroshi Seno
Recent Advances in Immunotherapy for Hepatocellular Carcinoma
Reprinted from: *Cancers* 2020, 12, 775, doi:10.3390/cancers12040775 137

Sooyeon Oh, YoungJoon Park, Hyun-Jung Lee, Jooho Lee, Soo-Hyeon Lee and Young-Seok Baek et al.
A Disintegrin and Metalloproteinase 9 (ADAM9) in Advanced Hepatocellular Carcinoma and Their Role as a Biomarker During Hepatocellular Carcinoma Immunotherapy
Reprinted from: *Cancers* **2020**, *12*, 745, doi:10.3390/cancers12030745 . . . . **153**

Pei-Chang Lee, Yee Chao, Ming-Huang Chen, Keng-Hsin Lan, Chieh-Ju Lee and I-Cheng Lee et al.
Predictors of Response and Survival in Immune Checkpoint Inhibitor-Treated Unresectable Hepatocellular Carcinoma
Reprinted from: *Cancers* **2020**, *12*, 182, doi:10.3390/cancers12010182 . . . . **169**

David Tai, Su Pin Choo and Valerie Chew
Rationale of Immunotherapy in Hepatocellular Carcinoma and Its Potential Biomarkers
Reprinted from: *Cancers* **2019**, *11*, 1926, doi:10.3390/cancers11121926 . . . . **183**

Edoardo G. Giannini, Andrea Aglitti, Mauro Borzio, Martina Gambato, Maria Guarino and Massimo Iavarone et al.
Overview of Immune Checkpoint Inhibitors Therapy for Hepatocellular Carcinoma, and The ITA.LI.CA Cohort Derived Estimate of Amenability Rate to Immune Checkpoint Inhibitors in Clinical Practice
Reprinted from: *Cancers* **2019**, *11*, 1689, doi:10.3390/cancers11111689 . . . . **211**

Zuzana Macek Jilkova, Caroline Aspord and Thomas Decaens
Predictive Factors for Response to PD-1/PD-L1 Checkpoint Inhibition in the Field of Hepatocellular Carcinoma: Current Status and Challenges
Reprinted from: *Cancers* **2019**, *11*, 1554, doi:10.3390/cancers11101554 . . . . **233**

Wei Tse Li, Angela E. Zou, Christine O. Honda, Hao Zheng, Xiao Qi Wang and Tatiana Kisseleva et al.
Etiology-Specific Analysis of Hepatocellular Carcinoma Transcriptome Reveals Genetic Dysregulation in Pathways Implicated in Immunotherapy Efficacy
Reprinted from: *Cancers* **2019**, *11*, 1273, doi:10.3390/cancers11091273 . . . . **247**

Article

Limited Predictive or Prognostic Role of Tumor-Infiltrating Tissue-Resident Memory CD8 T Cells in Patients with Hepatocellular Carcinoma Receiving Immunotherapy

Ying-Chun Shen [1,2,3], Ching-Ping Yeh [2], Yung-Ming Jeng [4], Chiun Hsu [1,2,3], Chih-Hung Hsu [2,3], Zhong-Zhe Lin [1,2,5], Yu-Yun Shao [2,3], Li-Chun Lu [2,3], Tsung-Hao Liu [2,3], Chien-Hung Chen [5,6] and Ann-Lii Cheng [1,2,3,5,*]

1. Department of Medical Oncology, National Taiwan University Cancer Center, Taipei 10672, Taiwan; yingchunshen@ntu.edu.tw (Y.-C.S.); hsuchiun@ntu.edu.tw (C.H.); zzlin7460@ntu.edu.tw (Z.-Z.L.)
2. Department of Oncology, National Taiwan University Hospital, Taipei 10002, Taiwan; 113297@ntuh.gov.tw (C.-P.Y.); chihhunghsu@ntu.edu.tw (C.-H.H.); yuyunshao@ntu.edu.tw (Y.-Y.S.); lichun@ntuh.gov.tw (L.-C.L.); 017027@ntuh.gov.tw (T.-H.L.)
3. Graduate Institute of Oncology, College of Medicine, National Taiwan University, Taipei 10055, Taiwan
4. Department of Pathology, National Taiwan University Hospital, Taipei 10002, Taiwan; chengym@ntu.edu.tw
5. Department of Internal Medicine, National Taiwan University Hospital, Taipei 10002, Taiwan; chenhcc@ntuh.gov.tw
6. Department of Internal Medicine, National Taiwan University Hospital Yunlin Branch, Yunlin 64041, Taiwan
* Correspondence: alcheng@ntu.edu.tw

Simple Summary: Total tumor-infiltrating CD8 T cells inconsistently correlate with the efficacy of immune checkpoint blockade (ICB) in hepatocellular carcinoma. Tumor-infiltrating CD8 tissue-resident memory T cells (T_{RM}) are considered a surrogate of tumor-specific T cells and correlated better with survival in patients with melanoma, non–small-cell lung cancer, head and neck cancer or bladder cancer who received ICB. However, in this study, compared with total tumor-infiltrating CD8 T cells, tumor-infiltrating CD8 T_{RM} cells failed to provide additional advantages in predicting the efficacy of ICB-based immunotherapy in patients with hepatocellular carcinoma.

Abstract: Purpose: Tumor-infiltrating tissue-resident memory CD8 T cells (CD8 T_{RM}; CD103+ CD8+) are considered tumor-specific and may correlate better with the tumor response to immune checkpoint blockade (ICB). This study evaluated the association of tumor-infiltrating CD8 T_{RM} and their subsets with the efficacy of immunotherapy in patients with advanced hepatocellular carcinoma (HCC). Experimental Design: Consecutive HCC patients who received ICB in prospective trials were analyzed. Formalin-fixed paraffin-embedded tumor sections were stained for DAPI, CD8, CD103, CD39, programmed cell death-1 (PD-1), and programmed cell death ligand 1 (PD-L1) using a multiplex immunohistochemical method. The densities of CD8 T cells, CD8 T_{RM}, and CD39+ or PD-L1+ subsets of CD8 T_{RM} were correlated with tumor response and overall survival (OS). Results: A total of 73 patients were identified, and 48 patients with adequate pretreatment tumor specimens and complete follow-up were analyzed. A median of 32.7% (range: 0–92.6%) of tumor-infiltrating CD8 T cells were T_{RM}. In subset analyses, 66.6% ± 34.2%, 69.8% ± 33.4%, and 0% of CD8 T_{RM} cells coexpressed CD39, PD-L1, and PD-1, respectively. The objective response rates for CD8 T cell-high, CD8 T_{RM}-high, CD39+ CD8 T_{RM}-high, and PD-L1+ CD8 T_{RM}-high groups were 41.7%, 37.5%, 37.5%, and 29.2%, respectively. Patients with CD8 T cell-high, but not those with CD8 T_{RM}-high, CD39+ CD8 T_{RM}-high, or PD-L1+ CD8 T_{RM}-high, tumors, had significantly prolonged OS ($p = 0.0429$). Conclusions: Compared with total tumor-infiltrating CD8 T cells, tumor-infiltrating CD8 T_{RM} or their subsets failed to provide additional advantages in predicting the efficacy of immunotherapy for HCC.

Keywords: tissue-resident memory CD8 T cells; hepatocellular carcinoma; immune checkpoint blockade; immunotherapy

Citation: Shen, Y.-C.; Yeh, C.-P.; Jeng, Y.-M.; Hsu, C.; Hsu, C.-H.; Lin, Z.-Z.; Shao, Y.-Y.; Lu, L.-C.; Liu, T.-H.; Chen, C.-H.; et al. Limited Predictive or Prognostic Role of Tumor-Infiltrating Tissue-Resident Memory CD8 T Cells in Patients with Hepatocellular Carcinoma Receiving Immunotherapy. *Cancers* 2021, 13, 5142. https://doi.org/10.3390/cancers13205142

Academic Editors: Tim Kendall and Masaru Enomoto

Received: 27 September 2021
Accepted: 8 October 2021
Published: 14 October 2021

Publisher's Note: MDPI stays neutral with regard to jurisdictional claims in published maps and institutional affiliations.

Copyright: © 2021 by the authors. Licensee MDPI, Basel, Switzerland. This article is an open access article distributed under the terms and conditions of the Creative Commons Attribution (CC BY) license (https://creativecommons.org/licenses/by/4.0/).

1. Introduction

CD8 T cell-infiltrated tumors are generally considered to be more immunogenic and more likely to respond to immune checkpoint blockade (ICB) [1]. However, total tumor-infiltrating CD8 T cells did not correlate well with the objective tumor response to nivolumab in patients with advanced hepatocellular carcinoma (HCC; CheckMate 040 study) [2]. Recent studies have demonstrated that bystander CD8 T cells targeting tumor-irrelevant antigens are abundant among tumor-infiltrating CD8 T cells in multiple cancer types [3,4]. This may partly explain why total tumor-infiltrating CD8 T cells did not correlate well with the response to ICB in patients with HCC and highlights the need to analyze tumor-specific T cells selectively.

Tumor-infiltrating tissue-resident memory CD8 T cells (T_{RM}; expressing the tissue residency marker CD103) are considered to be highly tumor-specific and are correlated better with survival in patients with various types of cancers [5–9]. This subpopulation of tumor-infiltrating CD8 T cells is retained in the tumor microenvironment following initial activation and expansion and plays an essential role in tumor-immune equilibrium [10] and tumor surveillance [11,12]. Tumor-infiltrating CD8 T_{RM} are characterized by the higher clonality of T-cell receptor repertoires [7,13] and can efficiently kill autologous tumor cells in a major histocompatibility complex class I-dependent manner [6,14]. By contrast, compared with their non-T_{RM} counterparts, tumor-infiltrating CD8 T_{RM} cells more frequently express inhibitory molecules such as CD39 (the rate-limiting enzyme in the conversion of ATP to immunosuppressive adenosine) [15], programmed cell death-1 (PD-1), cytotoxic T lymphocyte antigen-4, lymphocyte activation gene-3, and T-cell immunoglobulin and mucin domain-3 while maintaining effector functions [7,16]. Tumor-infiltrating CD8 T_{RM} cells were increased in patients with melanoma or non-small cell lung cancer who responded to ICB, but not in those nonresponders [5,17]. These findings suggest that tumor-infiltrating CD8 T_{RM} cells are responsive to ICB-invoked immune regulation. Moreover, the CD39 coexpression of tumor-infiltrating CD8 T_{RM} cells has been linked to higher tumor specificity and reactivity [3,14]. Therefore, tumor-infiltrating CD8 T_{RM} cells or their subsets, instead of total tumor-infiltrating CD8 T cells, may exhibit a better correlation with the efficacy of ICB-based immunotherapy in patients with HCC.

The current study characterized CD8 T_{RM} cells in the tumor microenvironment of HCC and investigated the association between CD8 T_{RM} cells and their subsets expressing CD39 or PD-1/PD-L1 signaling and the efficacy of ICB-based immunotherapy in patients with advanced HCC.

2. Materials and Methods

2.1. Patients

Patients with advanced HCC who met the following criteria were included in this study: (1) received ICB-based immunotherapy in prospective clinical trials from August 2015 to March 2019; (2) had high-quality pre-immunotherapy archived tumor tissues with viable tumor parts, as assessed by a senior independent pathologist; and (3) had complete clinical follow-up information and an evaluable tumor response to ICB-based immunotherapy according to Response Evaluation Criteria in Solid Tumors (RECIST; version 1.1) [18]. Clinical information including patients' characteristics and their tumors, immunotherapy regimens, prior systemic therapy, date of immunotherapy initiation, the best response according to RECIST (version 1.1), and date of death was obtained from electronic medical records. Objective responses included complete response (CR) and partial response (PR). Overall survival (OS) was defined as the time from the initiation of immunotherapy to death due to any cause or the last follow-up. This study was approved by the Research Ethics Committee of National Taiwan University Hospital (202001070RIND) and conducted in compliance with the Declaration of Helsinki and other ethical guidelines.

2.2. Multiplex Fluorescent Immunohistochemical Staining

Hematoxylin/eosin (H/E)-stained slides of formalin-fixed paraffin-embedded (FFPE) tumor blocks were evaluated by an independent pathologist (YMJ). The block with largest area of viable tumors was selected for sectioning at a thickness of 5 μm. The viable tumor parts were marked on H/E slides. Selected FFPE sections were deparaffinized, rehydrated, antigen retrieved, and stained using a customized multiplex fluorescent immunohistochemical (IHC) panel (Opal 7-color manual IHC kit; Akoya, Marlborough, MA, USA) according to the manufacturer's instructions. The primary antibodies used were CD8 (clone: C8/144B; 1:400) from DAKO (Santa Clara, CA, USA), CD39 (clone: polyclonal; 1:100) from Sigma (St. Louise, MO, USA), CD103 (clone: EPR4166 [2]; 1:100) from Abcam (Cambridge, UK), and PD-1 (clone: EH33; 1:100) and PD-L1 (clone: E1L3N; 1:200) both from Cell Signaling (Danvers, MA, USA). Spectral 4′,6-diamidino-2-phenylindole was used for nuclear counterstaining. FFPE sections from a tonsillectomy specimen and a PD-L1-high non–small-cell lung cancer specimen were used for the optimization of the staining protocol and as positive controls for PD-1 and PD-L1 staining. FFPE sections from a known CD8 T_{RM}-rich HCC tumors were included in each staining batch to detect any batch effects as a quality control measure.

2.3. Multispectral Fluorescent Imaging and Analysis

Visualization of multiplex fluorescent imaging was performed using Vectra Polaris Automated Quantitative Pathology Imaging Systems (Perkin Elmer, Hopkinton, MA, USA). Color separation, tissue and cell segmentation, and cell phenotyping were performed using inForm Software v2.4.2 (Perkin Elmer, Hopkinton, MA, USA). Multispectral regions of interest (200× magnification field) were randomly selected from the viable tumor part of each slide—as many as possible. The densities (number/mm^2) of CD8 T cells (CD8+), CD8 T_{RM} (CD103+ and CD8+), CD39+ CD8 T_{RM} (CD39+, CD103+, and CD8+), PD-1+ CD8 T_{RM} cells (PD-1+, CD103+, and CD8+), and PD-L1+ CD8 T_{RM} cells (PD-L1+, CD103+, and CD8+) in each area of interest were automatically quantitated under the supervision of a skilled researcher (CPY) and a senior pathologist (YMJ), who were blinded to the response status. The average density of each cell type for each tumor was calculated. The median value of immune cell density of interest among all tumors was used to divide tumors into "high (infiltration)" and "low (infiltration)" groups.

2.4. Statistical Analyses

The Mann–Whitney test was performed to compare binary outcome variables. Fisher's exact test was used when proportions were compared between binary variables. Non-parametric Spearman correlation was performed to measure the degree of association between two variables. The log-rank (Mantel–Cox) test was used to compare OS. Above analyses were conducted in GraphPad Prism (GraphPad Software, La Jolla, CA, USA). Cox regression analyses were performed to evaluate the risk factors for death and were conducted in IBM SPSS Statistics version 28.0.0.0 (New York, NY, USA).

3. Results

3.1. Baseline Characteristics and Treatment Outcomes of Enrolled Patients

A total of 73 patients with advanced HCC who received ICB-based immunotherapy in global open-label clinical trials were identified. Of these 73 patients, 25 were subsequently excluded due to the following reasons: no archived tumor specimens in 18; scant tumor cells in the FFPE slide in 5; no tumor part in the FFPE slide in 1; and death before tumor assessment in 1 (Figure S1). Finally, 48 patients were included in this study, and their baseline characteristics are shown in Table 1. Most of them were men (42, 87.5%) and HBV carriers (36, 75%). All of them had a Child–Pugh Classification A liver function and Eastern Cooperative Oncology Group (ECOG) performance status of 0–1 according to the eligibility criteria of clinical trials (Table S1). A total of 41 (85.4%) and 20 (41.7%) patients had extrahepatic metastasis and vascular invasion, respectively. Half of them had never received first-line sorafenib for advanced HCC. Most (31, 64.6%) of them received ICB-based combination therapy.

Table 1. Characteristics of enrolled patients and their archived tumors.

Variable	All (N = 48)	CR/PR (N = 15)	SD/PD (N = 33)	p-Value *
Age (years-old)				NS
Median	63	63.9	61.9	
Range	25.2–76.9	50.2–75.9	25.2–76.9	
Gender				NS
Male	42	12	30	
Female	6	3	3	
Viral status				NS
HBV	36	12	24	
HCV	10	3	7	
Non-HBV and non-HCV	2	0	2	
Vascular invasion				0.0401
No	28	12	16	
Yes	20	3	17	
Extrahepatic spread				NS
No	7	3	4	
Yes	41	12	29	
Lung	24	6	18	
Lymph node	21	6	15	
Bone	8	3	5	
Peritoneum/pleura	5	1	4	
Adrenal gland	2	1	1	
AFP level				
≤400 ng/mL	30	10	20	NS
>400 ng/mL	18	5	13	
Prior sorafenib				NS
No	24	7	17	
Yes	24	8	16	
Regimen of immunotherapy				NS
Anti-PD-1 or anti-PD-L1 monotherapy	16	4	12	
Nivolumab	12	3	9	
Tislelizumab	1	0	1	
Atezolizumab	1	0	1	
Durvalumab	2	1	1	
Anti-CTLA-4 monotherapy	1	1	0	
Tremelimumab	1	1	0	
Anti-PD-1 plus anti-CTLA-4	14	5	9	
Nivolumab plus ipilimumab	8	4	4	
Durvalumab plus tremelimumab	6	1	5	
Anti-PD-L1 + Anti-glypican-3V	2	1	1	
Atezolizumab plus codrituzumab	2	1	1	
Anti-PD-L1 + Anti-VEGF	15	4	11	
Atezolizumab plus bevacizumab	15	4	11	
Archived tumor				NS
Surgical specimen	26	9	17	
Biopsy specimen	22	6	16	
Time from tumor sampling to immunotherapy (month)				NS
Median	7.2	17.2	6.4	
Range	0.2–144.4	0.4–144.4	0.2–74.8	
Organ of specimen				NS
Liver	40	13	27	
Lung	3	1	3	
Lymph node	2	1	1	
Bone	2	0	2	

* comparison between CR/PR and SD/PD; CR, complete response; PR, partial response; SD, stable disease; PD, progressive disease; NS, not statistically significant; HBV, hepatitis B virus; HCV, hepatitis C virus; AFP, alpha-fetoprotein; VEGF, vascular endothelial growth factor.

Fifteen (31.3%) of these patients were responders (4 showed CR and 11 showed PR). The only difference in baseline features between responders and nonresponders (patients with stable disease or progressive disease) was vascular invasion (20% vs. 51.5%; $p = 0.0401$; Table 1). During a median follow-up of 30.7 months, the median OS for all patients, responders, and nonresponders was 35, not reached, and 16 months, respectively (Figure S2).

3.2. Characteristics of Archived Tumors and Their Multispectral Image Acquisition

The characteristics of archived tumors and their multispectral image acquisition are shown in Table S2. Most (40, 83.3%) of the archived tumors were primary hepatic tumors, and 21 (52.5%) of them were obtained from previous hepatectomy. The median ages of surgical and biopsy specimens were 17.2 and 0.9 months, respectively ($p = 0.002$). On average, 28.4 and 7.3 multispectral regions of interest were selected from each surgical and biopsy specimen, respectively. The representative figures are shown in Figure S3.

3.3. Tumor-Infiltrating CD8 T_{RM} Cells

The density of tumor-infiltrating CD8 T_{RM} cells correlated well with that of tumor-infiltrating CD8 T cells (Spearman $r = 0.8770$; $p < 0.0001$; Figure 1A). A median of 32.7% (range: 0%–92.6%) of tumor-infiltrating CD8 T cells were T_{RM}. Compared with non-T_{RM} CD8 T cell counterparts, CD8 T_{RM} cells more frequently coexpressed CD39 or PD-L1, but not PD-1 (Figure 1B). On average, 66.6% ± 34.2%, 69.8% ± 33.4% and 0% of tumor-infiltrating CD8 T_{RM} cells coexpressed CD39, PD-L1, and PD-1, respectively. The representative figures are shown in Figure 2. The density of tumor-infiltrating CD8 T_{RM} cells or their CD39+ or PD-L1+ subsets did not significantly correlate with the etiologies of HCC (Figure 1C).

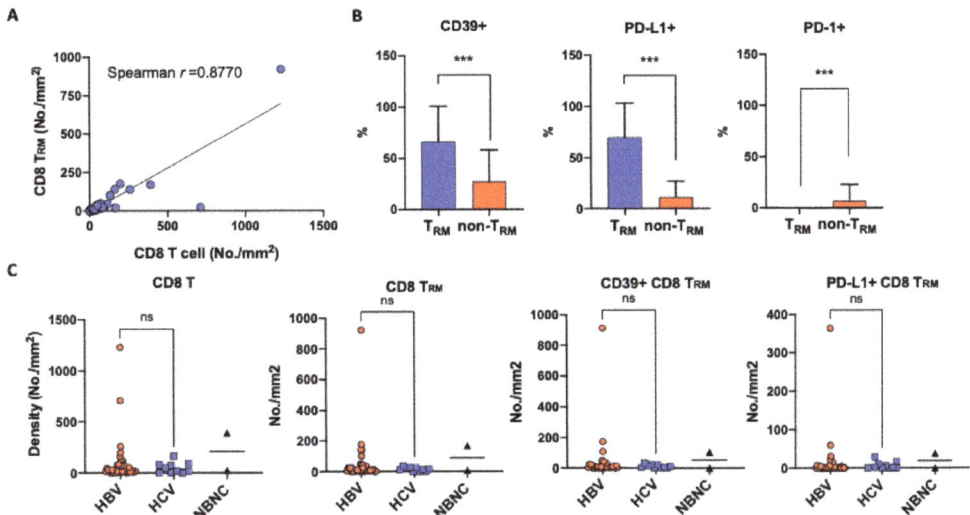

Figure 1. CD8 T_{RM} cells in the tumor microenvironment of HCC. (**A**) Correlation between the densities of tumor-infiltrating CD8 T cells and densities of tumor-infiltrating CD8 T_{RM} cells; (**B**) Coexpression of CD39, PD-L1, and PD-1 on tumor-infiltrating CD8 T_{RM} cells and CD8 non-T_{RM} cells; ***: $p < 0.001$; (**C**) Correlation between densities of tumor-infiltrating CD8 T cells, CD8 T_{RM} cells, CD39+ CD8 T_{RM} cells, or PD-L1+ CD8 T_{RM} cells and the etiologies of HCC. Statistical analyses were performed only for comparisons between HBV and HCV; ns: not statistically significant.

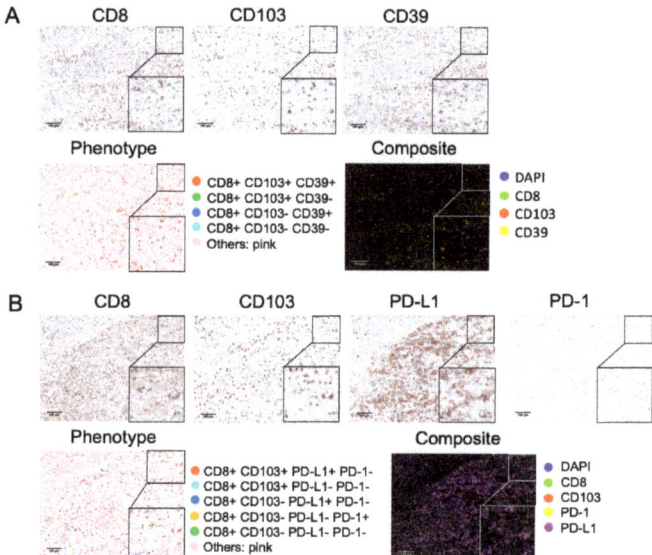

Figure 2. Representative images of tumor-infiltrating CD8 T_{RM}, CD39+ CD8 T_{RM}, and PD-L1+ CD8 T_{RM} cells. (**A**) CD8, CD103, and CD39 staining images and the composite image; (**B**) CD8, CD103, PD-L1 and PD-1 staining images and the composite image. PD-1 was primarily expressed on non-CD8 T cells. Each is shown at 200× magnification field (400× magnification field for the right lower corner square).

3.4. Correlations with Efficacy of Immunotherapy

The densities of tumor-infiltrating CD8 T, CD8 T_{RM}, CD39+ CD8 T_{RM}, and PD-L1+ CD8 T_{RM} cells are shown by the best response in Figure 3A. Objective responses were associated with higher densities of tumor-infiltrating CD39+ CD8 T_{RM} cells ($p = 0.04$ for both CR/PR vs. stable disease and CR/PR vs. progressive disease comparisons). The objective response rates of patients with CD8 T cell-high, CD8 T_{RM}-high, CD39+ CD8 T_{RM}-high, and PD-L1+ CD8 T_{RM}-high tumors were 41.7%, 37.5%, 37.5%, and 29.2%, respectively (Figure 3B). Patients with CD8 T cell-high tumors were associated with significantly prolonged OS ($p = 0.0429$). Patients with CD8 T_{RM}-high, CD39+ CD8 T_{RM}-high, and PD-L1+ CD8 T_{RM}-high tumors showed a trend of better survival; however, the finding was not statistically significant (Figure 4). The baseline characteristics were not different between patients with CD8 T cell-high tumors and those with CD8 T cell-low tumors, likewise between patients with CD8 T_{RM}-high tumors and those with CD8 T_{RM} low tumors.

3.5. Prognostic Factors in HCC Patients Receiving Immunotherapy

Univariate Cox regression analysis revealed that vascular invasion posed a higher risk for death (hazard ratio: 2.934; $p = 0.020$) while objective response and high CD8 T cell density posed lower risks for death (hazard ratio: 0.084, and 0.398, respectively; $p = 0.002$ and 0.048, respectively). However, high CD8 T cell density was no longer an independent prognostic factor after controlling all variables in multivariate Cox regression analysis (Table 2).

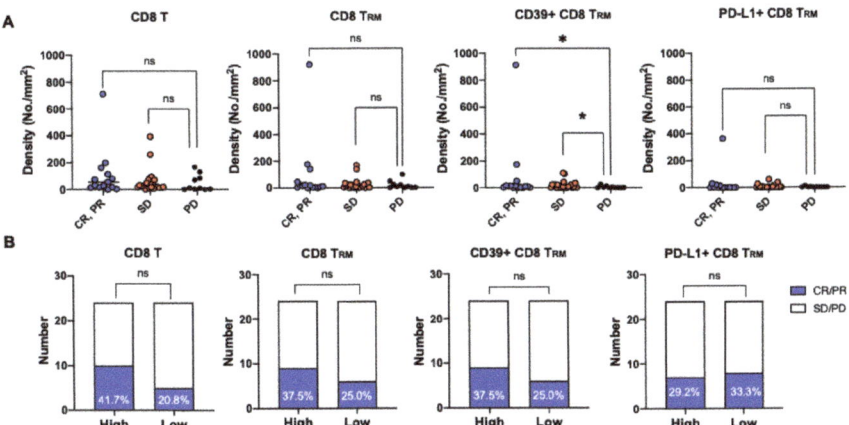

Figure 3. Correlations of CD8 T$_{RM}$ cells or their subsets with response to ICB-based immunotherapy. (A) Correlations among densities of tumor-infiltrating CD8 T cells, CD8 T$_{RM}$ cells, CD39+ CD8 T$_{RM}$ cells, or PD-L1+ CD8 T$_{RM}$ cells and the best response according to RECIST version 1.1; (B) Numbers of responders (CR/PR) and nonresponders (SD/PD) according to the infiltration levels of CD8 T cells, CD8 T$_{RM}$ cells, CD39+ T$_{RM}$ cells, and PD-L1+ T$_{RM}$ cells (high vs. low). CR, complete response; PR, partial response, SD, stable disease; PD, progressive disease; ns, not statistically significant; *, $p < 0.05$ but > 0.01.

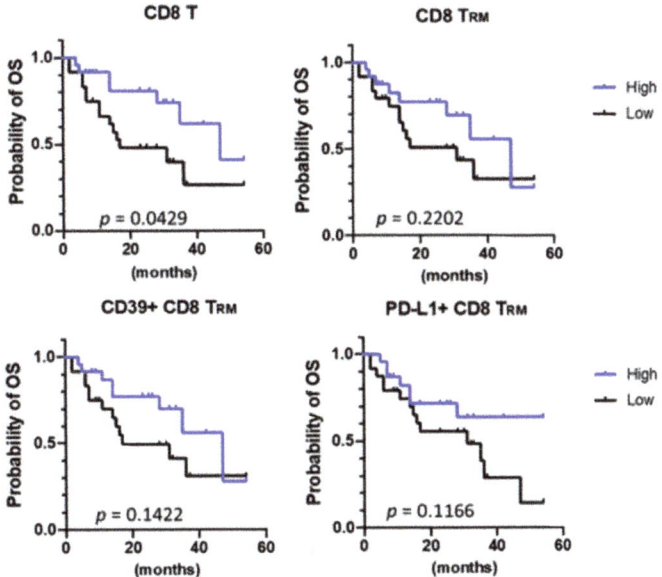

Figure 4. Correlations of CD8 T$_{RM}$ cells or their subsets with overall survival. Overall survival (OS) curves according to the infiltration levels of CD8 T cells, CD8 T$_{RM}$ cells, CD39+ CD8 T$_{RM}$ cells, and PD-L1+ CD8 T$_{RM}$ cells are shown (Kaplan-Meier method). Significance was tested using log-rank (Mantel-Cox) test.

Table 2. Cox regression analysis of risk factors for death.

Variable	Univariate Analysis				Multivariate Analysis			
	HR	95% IC (Lower)	95% IC (Upper)	p	HR	95% IC (Lower)	95% IC (Upper)	p
≥63 years-old	0.559	0.231	1.353	0.197				
Male	1.206	0.348	4.176	0.767				
HBV-related	0.935	0.341	2.565	0.897				
Vascular invasion	2.943	1.184	7.312	0.020 *	6.766	1.631	28.067	0.008 *
Extrahepatic spread	1.394	0.320	6.084	0.658				
AFP >400 ng/mL	1.047	0.433	2.531	0.919				
Prior sorafenib use	0.914	0.368	2.266	0.846				
ICI-based combination	0.454	0.187	1.104	0.082				
Objective response (CR/PR)	0.084	0.018	0.393	0.002 *	0.059	0.007	0.500	0.009 *
High CD8 T cell density	0.398	0.159	0.994	0.048 *	0.974	0.193	4.922	0.975
High CD8 TRM cell density	0.574	0.238	1.387	0.218				
High CD39+ CD8 TRM cell density	0.521	0.215	1.259	0.147				
High PD-L1+ CD8 TRM cell density	0.492	0.198	1.220	0.126				

HR: hazard ratio; IC: interval of confidence; p: p-value; HBV: hepatitis B virus; AFP: alfa-fetal protein; ICI: immune checkpoint inhibitor; CR: complete response; PR: partial response; TRM: tissue-resident memory T cells; *, statistically significant ($p < 0.05$).

4. Discussion

The role of tumor-infiltrating CD8 T_{RM} cells, a potential surrogate of tumor-specific CD8 T cells, in predicting the efficacy of immunotherapy in cancer patients remains elusive. The current study correlated the densities of tumor-infiltrating CD8 T_{RM} cells, their subsets, and total CD8 T cells with the efficacy of ICB-based immunotherapy in clinical trial patients with advanced HCC. Our data revealed that neither tumor-infiltrating CD8 T_{RM} nor its CD39+ or PD-L1+ subset provided additional advantages over total tumor-infiltrating CD8 T cells in predicting the efficacy of ICB-based immunotherapy in patients with advanced HCC. However, neither total tumor-infiltrating CD8 T cells nor tumor-infiltrating CD8 T_{RM} cells are independent prognostic factors in HCC patients receiving ICB-based immunotherapy.

In contrast to the current study, three previous studies have demonstrated that tumor-infiltrating CD8 T_{RM} cells correlated with prolonged survival in cancer patients treated with ICB. Two of them were conducted using bulk RNA-sequencing data obtained from clinical trials or published studies of melanoma, non–small-cell lung cancer, and bladder cancer [19,20]. Another study used a multiplex fluorescent IHC method in patients with non–small-cell lung cancer [17]. Banchereau et al. [19] quantified tumor-infiltrating CD8 T_{RM} cells by using *ITGAE* (encoding CD103) gene expression, whereas Zhang et al. [20] used a CD8 T_{RM} signature consisting of *CXCR6*, *ZNF683*, and *ITGAE* genes. The *ZNF683* gene encodes a T_{RM}-specific transcriptional factor; however, *CXCR6*- and *ITGAE*-encoded proteins are expressed in a wide range of immune cells such as regulator T cells, natural killer cells, and dendritic cells. Thus, the abundance of CD8 T_{RM} cells determined by either *ITGAE* gene expression or T_{RM} signature may not be specific enough for CD8 T_{RM}. Corgnac et al. [17] and the current study used the multiplex fluorescent IHC method, which is superior to the gene expression approach for its simultaneous colocalization and visualization [21]. However, our results did not show any significant correlation of tumor-infiltrating CD8 T_{RM} cells with OS in patients with HCC patients who were treated with ICB-based immunotherapy. These findings indicated that the roles of tumor-infiltrating CD8 T_{RM} cells in antitumor immunity may vary by cancer type.

CD8 T_{RM} is a heterogenous population [22]. CD39+ CD8 T_{RM} cells have been well characterized as a highly tumor-reactive subset of CD8 T_{RM} cells in non–small-cell lung cancer [6], head and neck cancer [14], and endometrial cancer [23], but they have never been investigated in HCC. A recent study reported that the frequency of CD39+ CD8 T cells well correlated with tumor mutation burden as well as high-affinity neoantigen burden in HCC [24]. Moreover, sorted tumor-infiltrating CD39+ CD8 T cells from human HCC, but not CD39− CD8 T cells, elicited high-affinity neoantigen-specific T-cell responses upon ex vivo neoantigen peptide stimulation. This finding strongly suggests that neoantigen-specific CD8 T cells are enriched in CD39+ CD8 T cells. According to our data, 66.2% ± 33.1% of tumor-infiltrating CD39+ CD8 T cells were T_{RM}. Therefore, tumor-infiltrating CD39+ CD8 T_{RM} cells in HCC are considered highly tumor-specific and responsive. The densities of CD39+ CD8 T_{RM} cells were significantly higher in responders (Figure 3A); however, high infiltration of CD39+ CD8 T_{RM} cells failed to predict an objective response or prolonged overall survival in our patients. We hypothesized that the effector functions of CD39+ CD8 T_{RM} cells in patients with HCC may be limited by local cytokine milieu [25] or metabolic fitness [26]. Therefore, future studies should focus on the functional characterization of CD39+ CD8 T_{RM} cells to better understand their role in anti-tumor immunity in HCC.

Aside from the heterogeneity of CD8 T_{RM} cells, two other reasons may also explain the limited predictive or prognostic value of tumor-infiltrating CD8 T_{RM} cells in HCC patients receiving immunotherapy. First and most importantly, tumor-infiltrating CD8 T_{RM} cells are not a good surrogate of tumor-specific CD8 T cells in HCC. The theory of differential tumor specificity between tumor-infiltrating CD8 T_{RM} cells and CD8 non- T_{RM} cells was initially established in lung [6] and breast [7] cancers, which were not virus-associated cancers. However, approximately 80% of HCC arise from virus-infected liver. Virus-specific CD8 T cells may be coincidently present in the tumor microenvironment of HCC. This argument is supported by a recent study, in which the investigators identified not only tumor-specific CD8 T_{RM} cells but also HBV-specific CD8 T_{RM} cells from HCC-infiltrating T cells using peptide-major histocompatibility complex tetramers and single cell RNA sequencing [27]. It indicates that tumor-infiltrating CD8 T_{RM} cells are not highly tumor-specific in HCC, especially in virus-associated HCC. Second, a significant overlap between CD8 T cell-high tumors and CD8 T_{RM}-high tumors was noted in our HBV-related HCC-predominant cohort. Twenty (83.3%) out of 24 CD8 T cell-high tumors were also characterized as CD8 T_{RM}-high tumors. Therefore, CD8 T_{RM} cells are less likely to provide additional advantages than total CD8 T cells in predicting the outcome of immunotherapy in our patient cohort. It is necessary to recruit more patients with non-HBV-related HCC for further validation.

Our study has several limitations. First, the sample size was relatively small, and the treatment regimens were heterogenous; however, this insufficiency may be partly alleviated by the stringency of conducting global prospective trials. Ideally, such a study should be conducted under a single large-scale clinical trial; however, pretreatment archived tumor samples are usually not absolutely required for recruitment in large-scale clinical trials. Therefore, obtaining an adequate number of archived tumors from a single clinical trial would be considerably difficult. Second, the lack of functional characterization of CD8 T_{RM} may limit the implication of the results. Third, approximately 46% of archived tumors analyzed in this study were obtained through core biopsies that were often small pieces of tissues. Thus, whether intratumor heterogeneity of CD8 T_{RM} cells and their subsets may affect the reliability of estimating the immune composition of the whole tumor by measuring such a small piece of tissue remains unclear. Nevertheless, we recently indicated that the intratumor heterogeneity of the immune tumor microenvironment may not be a major concern in HCC [28].

5. Conclusions

We demonstrated that tumor-infiltrating CD8 T_{RM} cells or their subsets may not have significant predictive or prognostic value in patients with advanced HCC who received ICB-based immunotherapy. Further studies are required to elucidate the contradictive roles of tumor-infiltrating CD8 T_{RM} cells in various cancer types.

Supplementary Materials: The following are available online at https://www.mdpi.com/article/10.3390/cancers13205142/s1, Figure S1: Patient selection flowchart. Figure S2: Overall survival of all patients, responders (CR/PR) and nonresponders (SD/PD). Figure S3: Selection of multispectral regions of interest. Table S1: Clinical trials of immune checkpoint blockade-based immunotherapy for advanced hepatocellular carcinoma during 2015–2019. Table S2: Characteristics of archived tumors and their multispectral imaging acquisition.

Author Contributions: Conceptualization, Y.-C.S.; Methodology, Y.-C.S., C.-P.Y. and Y.-M.J.; Software, C.-P.Y.; Validation, Y.-M.J.; Formal Analysis, Y.-C.S.; Investigation, Y.-C.S., C.-P.Y., C.H., C.-H.H., Z.-Z.L., Y.-Y.S., L.-C.L., T.-H.L. and C.-H.C.; Resources, Y.-C.S. and A.-L.C.; Data Curation, Y.-C.S.; Writing—Original Draft Preparation, Y.-C.S.; Writing—Review & Editing, A.-L.C.; Visualization, C.-P.Y.; Supervision, A.-L.C.; Project Administration, C.-P.Y.; Funding Acquisition, Y.-C.S. All authors have read and agreed to the published version of the manuscript.

Funding: NTU-109L901403 from Center of Precision Medicine from The Featured Areas Research Center Program within the framework of the Higher Education Sprout Project by the Ministry of Education, Taiwan.

Institutional Review Board Statement: The study was conducted according to the guidelines of the Declaration of Helsinki and approved by the Review Ethics Committee of National Taiwan University Hospital (202001070RIND; approved on 11 March 2020).

Informed Consent Statement: Patient informed consents were waived by Review Ethics Committee of National Taiwan University Hospital.

Data Availability Statement: The data presented in this study are available in this article (and supplementary material).

Conflicts of Interest: All authors declare no conflict of interest.

Abbreviations

HCC	hepatocellular carcinoma
ICB	immune checkpoint blockade
T_{RM}	tissue-resident memory T cells
FFPE	formalin-fixed parafilm-embedded
IHC	immunohistochemical
OS	overall survival
MHC	major histocompatibility complex
RECIST	Response Evaluation Criteria in Solid Tumors
CR	complete response
PR	partial response
H/E	hematoxylin/eosin
ECOG	Eastern Cooperative Oncology Group

References

1. Chen, D.S.; Mellman, I. Elements of cancer immunity and the cancer-immune set point. *Nature* **2017**, *541*, 321–330. [CrossRef]
2. Sangro, B.; Melero, I.; Wadhawan, S.; Finn, R.S.; Abou-Alfa, G.K.; Cheng, A.L.; Yau, T.; Furuse, J.; Park, J.-W.; Boyd, Z.; et al. Association of inflammatory biomarkers with clinical outcomes in nivolumab-treated patients with advanced hepatocellular carcinoma. *J. Hepatol.* **2020**, *73*, 1460–1469. [CrossRef]
3. Simoni, Y.; Becht, E.; Fehlings, M.; Loh, C.Y.; Koo, S.L.; Teng, K.W.W.; Yeong, J.P.S.; Nahar, R.; Zhang, T.; Kared, H.; et al. Bystander CD8(+) T cells are abundant and phenotypically distinct in human tumour infiltrates. *Nature* **2018**, *557*, 575–579. [CrossRef]
4. Gokuldass, A.; Draghi, A.; Papp, K.; Borch, T.H.; Nielsen, M.; Westergaard, M.C.W.; Andersen, R.; Schina, A.; Bol, K.F.; Chamberlain, C.A.; et al. Qualitative Analysis of Tumor-Infiltrating Lymphocytes across Human Tumor Types Reveals a Higher Proportion of Bystander CD8(+) T Cells in Non-Melanoma Cancers Compared to Melanoma. *Cancers* **2020**, *12*, 3344. [CrossRef]
5. Edwards, J.; Wilmott, J.S.; Madore, J.; Gide, T.N.; Quek, C.; Tasker, A.; Ferguson, A.; Chen, J.; Hewavisenti, R.; Hersey, P.; et al. CD103(+) Tumor-Resident CD8(+) T Cells Are Associated with Improved Survival in Immunotherapy-Naïve Melanoma Patients and Expand Significantly During Anti-PD-1 Treatment. *Clin. Cancer Res.* **2018**, *24*, 3036–3045. [CrossRef]
6. Djenidi, F.; Adam, J.; Goubar, A.; Durgeau, A.; Meurice, G.; de Montpréville, V.; Validire, P.; Besse, B.; Mami-Chouaib, F. CD8+CD103+ tumor-infiltrating lymphocytes are tumor-specific tissue-resident memory T cells and a prognostic factor for survival in lung cancer patients. *J. Immunol.* **2015**, *194*, 3475–3486. [CrossRef] [PubMed]
7. Savas, P.; Virassamy, B.; Ye, C.; Salim, A.; Mintoff, C.P.; Caramia, F.; Salgado, R.; Byrne, D.J.; Teo, Z.L.; Dushyanthen, S.; et al. Single-cell profiling of breast cancer T cells reveals a tissue-resident memory subset associated with improved prognosis. *Nat. Med.* **2018**, *24*, 986–993. [CrossRef] [PubMed]

8. Wang, B.; Wu, S.; Zeng, H.; Liu, Z.; Dong, W.; He, W.; Chen, X.; Dong, X.; Zheng, L.; Lin, T.; et al. CD103+ Tumor Infiltrating Lymphocytes Predict a Favorable Prognosis in Urothelial Cell Carcinoma of the Bladder. *J. Urol.* **2015**, *194*, 556–562. [CrossRef] [PubMed]
9. Lim, C.J.; Lee, Y.H.; Pan, L.; Lai, L.; Chua, C.; Wasser, M.; Lim, T.K.H.; Yeong, J.; Toh, H.C.; Lee, S.Y.; et al. Multidimensional analyses reveal distinct immune microenvironment in hepatitis B virus-related hepatocellular carcinoma. *Gut* **2019**, *68*, 916–927. [CrossRef]
10. Park, S.L.; Buzzai, A.; Rautela, J.; Hor, J.L.; Hochheiser, K.; Effern, M.; McBain, N.; Wagner, T.; Edwards, J.; McConville, R.; et al. Tissue-resident memory CD8(+) T cells promote melanoma-immune equilibrium in skin. *Nature* **2019**, *565*, 366–371. [CrossRef]
11. Amsen, D.; van Gisbergen, K.; Hombrink, P.; van Lier, R.A.W. Tissue-resident memory T cells at the center of immunity to solid tumors. *Nat. Immunol.* **2018**, *19*, 538–546. [CrossRef]
12. Park, S.L.; Gebhardt, T.; Mackay, L.K. Tissue-Resident Memory T Cells in Cancer Immunosurveillance. *Trends Immunol.* **2019**, *40*, 735–747. [CrossRef]
13. Boddupalli, C.S.; Bar, N.; Kadaveru, K.; Krauthammer, M.; Pornputtapong, N.; Mai, Z.; Ariyan, S.; Narayan, D.; Kluger, H.; Deng, Y.; et al. Interlesional diversity of T cell receptors in melanoma with immune checkpoints enriched in tissue-resident memory T cells. *JCI Insight* **2016**, *1*, 88955. [CrossRef]
14. Duhen, T.; Duhen, R.; Montler, R.; Moses, J.; Moudgil, T.; de Miranda, N.F.; Goodall, C.P.; Blair, T.C.; Fox, B.A.; McDermott, J.E.; et al. Co-expression of CD39 and CD103 identifies tumor-reactive CD8 T cells in human solid tumors. *Nat. Commun.* **2018**, *9*, 2724. [CrossRef]
15. Moesta, A.K.; Li, X.Y.; Smyth, M.J. Targeting CD39 in cancer. *Nat. Rev. Immunol.* **2020**, *20*, 739–755. [CrossRef] [PubMed]
16. Ganesan, A.P.; Clarke, J.; Wood, O.; Garrido-Martin, E.M.; Chee, S.J.; Mellows, T.; Samaniego-Castruita, D.; Singh, D.; Seumois, G.; Alzetani, A.; et al. Tissue-resident memory features are linked to the magnitude of cytotoxic T cell responses in human lung cancer. *Nat. Immunol.* **2017**, *18*, 940–950. [CrossRef]
17. Corgnac, S.; Malenica, I.; Mezquita, L.; Auclin, E.; Voilin, E.; Kacher, J.; Halse, H.; Grynszpan, L.; Signolle, N.; Dayris, T.; et al. CD103(+)CD8(+) T(RM) Cells Accumulate in Tumors of Anti-PD-1-Responder Lung Cancer Patients and Are Tumor-Reactive Lymphocytes Enriched with Tc. *Cell Rep. Med.* **2020**, *1*, 100127. [CrossRef] [PubMed]
18. Eisenhauer, E.A.; Therasse, P.; Bogaerts, J.; Schwartz, L.H.; Sargent, D.; Ford, R.; Dancey, J.; Arbuck, S.; Gwyther, S.; Mooney, M.; et al. New response evaluation criteria in solid tumours: Revised RECIST guideline (version 1.1). *Eur. J. Cancer* **2009**, *45*, 228–247. [CrossRef] [PubMed]
19. Banchereau, R.; Chitre, A.S.; Scherl, A.; Wu, T.D.; Patil, N.S.; de Almeida, P.; Kadel, E.E., III; Madireddi, S.; Au-Yeung, A.; Takahashi, C.; et al. Intratumoral CD103+ CD8+ T cells predict response to PD-L1 blockade. *J. Immunother. Cancer* **2021**, *9*, e002231. [CrossRef] [PubMed]
20. Zhang, C.; Yin, K.; Liu, S.Y.; Yan, L.X.; Su, J.; Wu, Y.L.; Zhang, X.C.; Yang, X.-N. Multiomics analysis reveals a distinct response mechanism in multiple primary lung adenocarcinoma after neoadjuvant immunotherapy. *J. Immunother. Cancer* **2021**, *9*, e002312. [CrossRef] [PubMed]
21. Mehnert, J.M.; Monjazeb, A.M.; Beerthuijzen, J.M.T.; Collyar, D.; Rubinstein, L.; Harris, L.N. The Challenge for Development of Valuable Immuno-oncology Biomarkers. *Clin. Cancer Res.* **2017**, *23*, 4970–4979. [CrossRef] [PubMed]
22. Milner, J.J.; Toma, C.; He, Z.; Kurd, N.S.; Nguyen, Q.P.; McDonald, B.; Quezada, L.; Widjaja, C.E.; Witherden, D.A.; Crowl, J.T.; et al. Heterogenous Populations of Tissue-Resident CD8(+) T Cells Are Generated in Response to Infection and Malignancy. *Immunity* **2020**, *52*, 808–824. [CrossRef]
23. Workel, H.H.; van Rooij, N.; Plat, A.; Spierings, D.C.J.; Fehrmann, R.S.N.; Nijman, H.W.; de Bruyn, M. Transcriptional Activity and Stability of CD39+CD103+CD8+ T Cells in Human High-Grade Endometrial Cancer. *Int. J. Mol. Sci.* **2020**, *21*, 3770. [CrossRef] [PubMed]
24. Liu, T.; Tan, J.; Wu, M.; Fan, W.; Wei, J.; Zhu, B.; Guo, J.; Wang, S.; Zhou, P.; Zhang, H.; et al. High-affinity neoantigens correlate with better prognosis and trigger potent antihepatocellular carcinoma (HCC) activity by activating CD39(+)CD8(+) T cells. *Gut* **2020**, *70*, 1965–1977. [CrossRef] [PubMed]
25. Mackay, L.K.; Wynne-Jones, E.; Freestone, D.; Pellicci, D.G.; Mielke, L.A.; Newman, D.M.; Braun, A.; Masson, F.; Kallies, A.; Belz, G.; et al. T-box Transcription Factors Combine with the Cytokines TGF-β and IL-15 to Control Tissue-Resident Memory T Cell Fate. *Immunity* **2015**, *43*, 1101–1111. [CrossRef] [PubMed]
26. Lin, R.; Zhang, H.; Yuan, Y.; He, Q.; Zhou, J.; Li, S.; Sun, Y.; Li, D.Y.; Qiu, H.-B.; Wang, W.; et al. Fatty Acid Oxidation Controls CD8(+) Tissue-Resident Memory T-cell Survival in Gastric Adenocarcinoma. *Cancer Immunol. Res.* **2020**, *8*, 479–492. [CrossRef] [PubMed]
27. Cheng, Y.; Gunasegaran, B.; Singh, H.D.; Dutertre, C.A.; Loh, C.Y.; Lim, J.Q.; Crawford, J.C.; Lee, H.K.; Zhang, X.; Lee, B.; et al. Non-terminally exhausted tumor-resident memory HBV-specific T cell responses correlate with relapse-free survival in hepatocellular carcinoma. *Immunity* **2021**, *54*, 1825–1840. [CrossRef] [PubMed]
28. Shen, Y.C.; Hsu, C.L.; Jeng, Y.M.; Ho, M.C.; Ho, C.M.; Yeh, C.P.; Yeh, C.Y.; Hsu, M.-C.; Hu, R.-H.; Cheng, A.-L. Reliability of a single-region sample to evaluate tumor immune microenvironment in hepatocellular carcinoma. *J. Hepatol.* **2020**, *72*, 489–497. [CrossRef]

Review

Immune Checkpoint Inhibitors in Hepatocellular Carcinoma: Current Status and Novel Perspectives

Piera Federico [1,*], Angelica Petrillo [1,2], Pasqualina Giordano [1], Davide Bosso [1], Antonietta Fabbrocini [1], Margaret Ottaviano [1,3], Mario Rosanova [1], Antonia Silvestri [1], Andrea Tufo [4], Antonio Cozzolino [5] and Bruno Daniele [1]

[1] Medical Oncology Unit, Ospedale del Mare, 80147 Napoli, Italy; angelic.petrillo@gmail.com (A.P.); giopas@email.it (P.G.); davidebosso84@gmail.com (D.B.); antonietta.fabbrocini@gmail.com (A.F.); margaretottaviano@gmail.com (M.O.); rosanovamario@hotmail.com (M.R.); antonia.silv@libero.it (A.S.); b.daniele@libero.it (B.D.)
[2] Division of Medical Oncology, Department of Precision Medicine, School of Medicine, University of Study of Campania "L. Vanvitelli", 80131 Napoli, Italy
[3] Department of Clinical Medicine and Surgery, University of Naples "Federico II", 80131 Naples, Italy
[4] Surgical Unit, Ospedale del Mare, 80147 Napoli, Italy; tufo.andrea@gmail.com
[5] Gastroenterology Unit, Ospedale del Mare, 80147 Napoli, Italy; ancozzolino@libero.it
* Correspondence: pierafederico@yahoo.it; Tel.: +39-081-1877-5339

Received: 23 September 2020; Accepted: 15 October 2020; Published: 18 October 2020

Simple Summary: Immune checkpoint inhibitors represent a promising treatment choice in many kind of tumours, including hepatocellular carcinoma (HCC). In this review, we provide an overview of the role of these new agents in the management of HCC according to the Barcelona staging system, alongside with a critical evaluation of the current status and future directions. Several clinical trials are focusing on the use of immunotherapy in HCC, alone or in combinations with antiangiogenetic agents as well as local treatment. However, the majority of those trials are still ongoing and, until now, only a few combinations were approved in the clinical practice from the regulatory authorities. Additionally, decisions about the choice of the right sequence of treatments in HCC patients in the light of the "continuum of care" principles, is still hard. In fact, it requires careful consideration in a multidisciplinary context in order to ensure a tailored treatment for each patient.

Abstract: Immune checkpoint inhibitors (ICIs) represent a promising treatment for many kinds of cancers, including hepatocellular carcinoma (HCC). The rationale for using ICIs in HCC is based on the immunogenic background of hepatitis and cirrhosis and on the observation of high programmed death-ligand 1 (PD-L1) expression and tumor-infiltrating lymphocytes in this cancer. Promising data from phase I/II studies in advanced HCC, showing durable objective response rates (~20% in first- and second-line settings) and good safety profile, have led to phase III studies with ICIs as single agents or in combination therapy, both in first and second line setting. While the activity of immunotherapy agents as single agents seems to be limited to an "ill-defined" small subset of patients, the combination of the anti PD-L1 atezolizumab and anti-vascular endothelial growth factor bevacizumab revealed a benefit in the outcomes when compared to sorafenib in the first line. In addition, the activity and efficacy of the combinations between anti-PD-1/anti-PD-L1 antibody and other ICIs, tyrosine kinase inhibitors, or surgical and locoregional therapies, has also been investigated in clinical trials. In this review, we provide an overview of the role of ICIs in the management of HCC with a critical evaluation of the current status and future directions.

Keywords: hepatocellular carcinoma; immunotherapy; combination therapy; predictive markers

1. Introduction

Hepatocellular carcinoma (HCC) is the first primary liver cancer in incidence, showing 65,000 new cases/year in Europe, and the third cause of cancer related death worldwide [1]. The most important risk factors for HCC are chronic infections from B and C hepatitis virus (HBV and HCV, respectively), alcoholic cirrhosis, nonalcoholic fatty liver disease, aflatoxin B1, hemochromatosis, as well as other causes of cirrhosis [2].

Patients affected by HCC are complex and require a careful management in a multidisciplinary context involving experts in the field. In fact, they usually have concurrent diseases, such as cirrhosis or metabolic alterations, as well as history of alcohol abuse or liver interventions, which lead to a poor liver condition, eventually with portal hypertension and gastric and esophageal bleeding. Starting from this assumption, patients with HCC should be referred to dedicated centers and receive a multidisciplinary assessment at the diagnosis and during the entire treatment period. In this process, the staging evaluation is crucial in the HCC management algorithm in order to determine the outcomes and the treatment allocation. Among different staging systems, the Barcelona Clinic Liver Cancer (BCLC) is the more used, representing the accepted standard according to the study of Liver Disease (AASLD) and the European Association for the Study of the liver [3]. It combines multiple variables into an algorithm and identifies five stages for the disease: Patients with early HCC (stage 0/A) who are candidates for curative-intent radical therapies such as resection, liver transplantation and ablation; patients with multinodular tumours (stage B, intermediate) and candidate to local treatment, such as chemoembolization; those in advanced stage (stage C), eligible for systemic treatments and patients in terminal stage (stage D) for whom palliative cares are recommended. Switching to systemic therapy after locoregional treatment failure is known to have a crucial role in the decisions-making process in patients with trans-catheter arterial chemoembolization (TACE) refractory and intermediate stage HCC, since this transition has a great improvement impact on survival [4–6]. Therefore, clinicians should be careful in detecting the optimal timing for TACE refractory patients (TACE toxicity, disease progression after one or two courses of TACE, absence of a response, vascular or extrahepatic spread) to switch to systemic treatment [7].

Regarding stage C, according to international guidelines [8], sorafenib and lenvatinib represent the standard-of-care options in the first-line treatment. In patients who show a progression to first-line treatment, up to date, the multikinase inhibitors regorafenib and cabozantinib or ramucirumab, the anti-vascular-endothelial growth factor-2 (VEGF-R2), are the main choices in the second-line setting [9–11]. However, the small magnitude of survival benefit obtained with those tyrosine kinase inhibitors (TKI) and their poor tolerability have brought out the need for new therapeutic strategies.

In this context, the immune checkpoint inhibitors (ICIs) might represent the most important novelty and the future perspective also in the field of HCC. Indeed, over the last decades the understanding of the relationship between cancer and the immune system has progressed and ICIs have shown to improve the outcomes of patients in many kind of tumours, replacing the chemotherapy in some cases [12]. Regarding HCC, its peculiar immunogenic microenvironment has encouraged the use of immunotherapeutic agents, firstly revealing a potential role for pembrolizumab and nivolumab [13–15]. Recently, the regimen of atezolizumab and bevacizumab showed significantly longer overall survival (OS) and progression-free survival (PFS), as well as better patient-outcomes than sorafenib in the first-line systemic treatment [16]. These results identify not only the first therapy to improve survival beyond sorafenib over years, but also the first successful combination therapy and the first positive randomized phase III trial of ICIs in this challenging cancer. However, despite the advances in HCC treatment, only a small subset of patients respond to immunotherapy. Therefore, new tools to identify prognostic and predictive biomarkers able to select those patients who might actually benefit from ICIs -based treatment are urgently needed.

The aim of this review is to delineate an overview the biologic rationale for using immunotherapies in HCC according to BCLC stage, the current status and recent advances, alongside with the discussion of the areas for improvement and future implications.

2. Immunotherapies in HCC: The Biological Rationale of Their Use

Immunotherapy in HCC is particularly attractive for several reasons. The liver is an immunological organ that works as a biological filter against infections, which could came from the blood flow or gastrointestinal tract in which there might be a release of proteins and pro-inflammatory cytokines [17,18]. In fact, the liver is constantly exposed to many kind of antigens from food and microbiota, which can stimulate immune cells from innate and adaptive immune system. Several trials have demonstrated that the liver has developed an immune tolerability during its evolution process, due to the permanent exposure to those antigens [19,20]. This fact is supported by the evidence of the low rates of allograft rejection into the liver if compared to other organ transplants [21,22]. In addition, HCC is considered an inflammation-related cancer with a potential immunogenicity. In fact, it is know that the majority of HCC arises in liver affected by cirrhosis and hepatitis, which are considered typical inflammatory conditions [23]. Therefore, this inflammatory environment could act as "pro-neoplastic" factor, since it is involved in cancer progression through different mechanisms, such as the DNA damage and genomic aberrations.

However, even if the liver has an "immunosuppressive" basal condition, several trials showed that an immune response to tumour is possible also in HCC. In particular, in patients who developed HCC after drug-induced immunosuppression, the discontinuation of immunosuppressive treatment lead to a spontaneous tumor regression by the reactivation of cytotoxic T-cells, that are able to identify and eliminate the cancer cells [24,25]; there is an increase in programmed death 1 (PD-1) and its ligand (PD-L1) expression as well as tumour infiltrating lymphocytes (TILs) in patients with HCC [26,27], leading to the immunosuppression [28,29]. Additionally, high Cytotoxic T-Lymphocyte Antigen 4 (CTLA-4) expression on regulatory T-cells (Tregs) in peripheral blood has been recorded in HCC patients in association with a decrease in immunosuppressive cytolitic granzyme B production by CD8+ T-cells. Regarding CD4+ T-cells, CTLA-4 is essential for their activation during the priming phase of the immune response. In fact, in the physiological process, T-cell are activated after the antigen presentation, whereas CTLA-4 reduces this activity, leading to T-cell suppression by blocking the binding between CD28 and CD80- CD86. Additionally, it plays an important role in the function of Tregs; in fact, CTLA-4 expression on CD14+ dendritic cells inhibits T-cell proliferation, inducing at the same time, the apoptosis of these cells by increasing the production of IL-10 and indoleamine-2,3-dioxygenase (IDO). Based on this background, there is a strong rationale to test ICIs in HCC [30] (Figure 1).

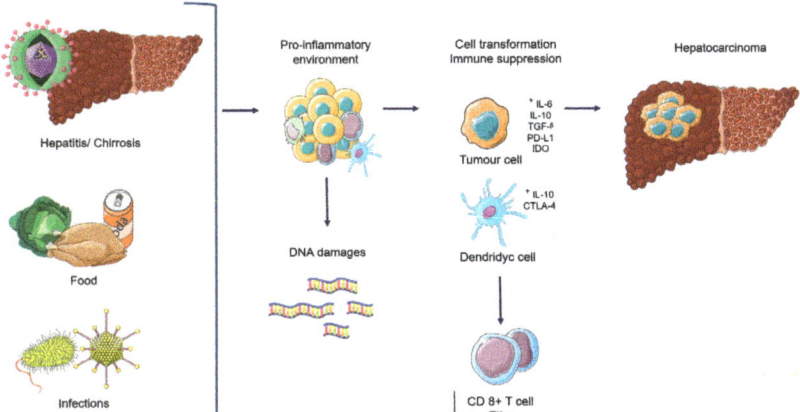

Figure 1. Biological rational to use Immune checkpoint inhibitors in Hepatocellular carcinoma. Abbreviations: IL-6: interleukin 6; IL-10: interleukin 10, TGF-β: tumor growth factor β PD-L!: programmed death ligand-1; IDO: Indoleamine Dioxygenaje, CTLA-4: Cytotoxic T-Lymphocyte Antigen 4; TILs: tumour infiltrating lymphocytes.

3. Immune Checkpoint Agents: Mechanism of Action and Their Use in HCC

Recently, the systemic management of HCC has been revolutionized by the advent of ICIs, a therapeutic class of monoclonal antibodies that blocks the immune checkpoints. These are co-inhibitory molecules physiologically expressed in different cells types, such as natural killer cells, dendritic cells, tumor-associated macrophages, monocytes and myeloid-derived suppressor cells (MDSCs)—including B and T cells, and that mantains self-tolerance [31,32]. ICIs act by applying a break that prevent the activation of these cells, limiting tissue damage. The balance between co-stimulatory signals and immune checkpoints leads to T cells activation, defining the intensity of the immune response.

ICIs have become a mainstay in the treatment of many cancers and then numerous clinical trials have been conducted and others are still ongoing with the aim to evaluate the safety and efficacy of these agents in several solid and hematological malignancies [12]. The main immune checkpoint receptors are CTLA-4, PD-1, TIM-3, BTLA, VISTA, LAG-3 and OX-40 [33].

The success of cancer immunotherapies through PD-1 and CTLA-4 mediated immunosuppression led to the developement of many clinical trials also in HCC. In this field, two classes of ICIs are currently being tested as mono or combination therapies: CTLA-4 (tremelimumab and ipilimumab) and PD-1/PD-L1 inhibitors (anti-PD-1: nivolumab, pembrolizumab, tislelizumab, camrelizumab and sintilimab; anti PD-L1: atezolizumab and durvalumab).

In the sections below we provide a short description of ICIs mechanism of actions, followed by the current state of advancement of ICIs-based therapies in HCC, according to BCLC stages.

3.1. The Checkpoint Pathways Regulated by CTLA-4 and by PD-1 Receptors

CTLA-4 is expressed on activated T cells, Tregs, and at low levels, on naïve T cells [34,35]. Its main function is to downregulate the activation of T lymphocytes by blocking the co-stimulatory signal of CD28 (for other details see Section 2).

PD-1 plays a key role in the regulation and in the maintenance of the balance between T cells activation and immune tolerance. Unlike CTLA-4, PD-1 is widely expressed, it can be detected not only on the T cell surface, especially CD8+ T cells, but also on Tregs and MDSC [36]. Additionally, whereas CTLA-4 mainly regulates the activation of T cells in lymphatic tissues, the most important action of PD-1 is to reduce the activity of T cells in peripheral tissues during the immune cell mediated or inflammatory response. Then, T cell function is influenced by the level of PD-1 activity and, when PD-1 binds to its ligands PD-L1 or PD-L2, T cell proliferation and cytokine release are inhibited.

Today, it is well-know that chronic exposure to antigens leads to the hyperexpression of PD-1 in T cells [37]. Additionally, cancer cells can turn PD-L1/PD-1 signaling to their own advantage through the expression of PD-L1 or PD-L2. This can activate PD-1 in TILs, resulting in the escaper of immune surveillance [38,39]. The hyperexpression of PD-1 reported on CD8+ T-cells in patients with HCC and the increase in PD-1+CD8+ TILs confirmed the previous theory. Additionally, the presence of those cells in HCC specimens was associated with higher rate of progression of disease after curative hepatic resection [26,27].

3.2. ICIs-Therapies in HCC Patients According to BCLC Stage

3.2.1. Early Stage HCC (BCLC stage 0 or A)

According to international guidelines [8], liver resection or ablation treatments are the standard of care for patients with BCLC stage 0 or A. Single tumors in patients with well-preserved liver function and no clinically significant portal hypertension [40] is the mainstay indication for resection, providing a survival rate of almost 60% at 5 years and no postoperative liver failure (postoperative mortality <3%). However, after liver resection, tumour recurrence can be observed in 50–70% of cases in 5 years, the majority in the first two years, and no adjuvant therapies have been shown to reduce recurrence rate in this field [5,41,42].

Ablation is preferred in patients with BCLC stage 0 or A, who are not candidates for surgery [8]. The main procedure is by percutaneous radiofrequency (RFA), which acts by causing ischemic cell damage with the release of neoantigens and promotion of immunogenic cell death [43,44].

Due to these considerations, ICIs have been thought to be beneficial in the adjuvant setting for patients with high risk of recurrence after complete resection or complete response by local ablation. Therefore, several clinical trials are ongoing in this regard. Figure 2 summarizes the treatment options for BCLC stage 0/A HCC and the ICIs ongoing clinical trials in this field.

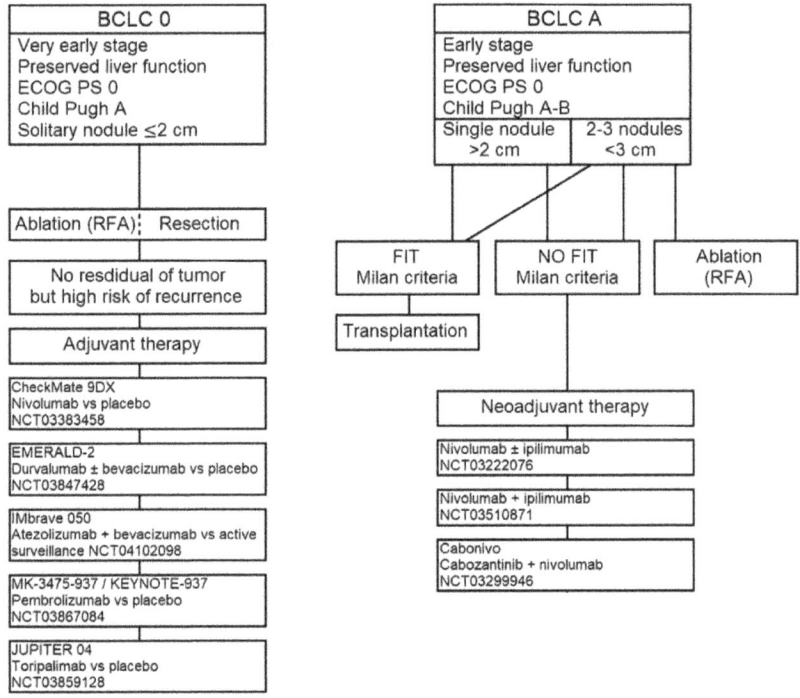

Figure 2. Treatment options and ICIs ongoing clinical trials for BCLC stage 0-A HCC.

In particular, the phase III, placebo-controlled CheckMate 9Dx trial (Clinical Trials.gov Identifier: NCT03383458) is investigating the role of nivolumab in this setting; the phase III EMERALD-2 trial (NCT03847428) is assessing the efficacy and safety of durvalumab as monotherapy or in combination with bevacizumab versus placebo; the phase III IMbrave050 (NCT04102098) is assessing the efficacy of atezolizumab plus bevacizumab versus active surveillance. Additionally, the safety and efficacy of pembrolizumab versus placebo as adjuvant therapy are being studied in the phase III trial KEYNOTE-937 (NCT03867084); the JUPITER-04 clinical trial (NCT03859128) is investigating the possible role of toripalimab (Recombinant Humanized Anti-PD-1 Monoclonal Antibody, JS001) in the improvement of relapse free survival (RFS) compared to placebo in the adjuvant treatment for patients who underwent complete liver resection, but with high risk of relapse.

It is worth remembering that liver transplantation can also be performed in patients with a limited tumor burden and fit the Milan criteria [45]. In these patients <10% recurrence rate and 70% five-year survival rate are expected [46]. However, the low availability of liver allografts is the major limitation for liver transplantation. In patients who do not fit the Milano criteria and in patients who are in the waiting list for transplant, the UNOS (United Network for Organ Sharing) allows the use of neoadjuvant treatments generally ablation or transarterial therapies; this is due to the long waiting period and the risk of tumor progression [8].

In the context of neoadjuvant setting for resectable HCC, the use of ICIs is currently being studied in several trials. In particular, the phase II randomized NCT03222076 trial is evaluating the safety and tolerability of nivolumab alone or in combination with ipilimumab; the interim analysis involving 8 patients showed a 37.5% of pathological complete response in the entire population with a good safety profile [47]. Additionally, the phase II NCT03510871 is evaluating the efficacy, in terms of tumor shrinkage, objective response rate and neoadjuvant down-stage rate, of nivolumab plus ipilimumab in this setting. Then, the phase I CaboNivo trial (NCT03299946) is assessing the feasibility and efficacy of cabozantinib plus nivolumab followed by definitive resection.

In conclusion, up to date, the treatment algorithm for patients with BCLC 0 or A HCC is unchanged and doesn't include the use of ICIs. Further evaluations, as well as the results of those ongoing trials are awaiting in order to eventually improve the treatment's choices.

3.2.2. Intermediate—Stage HCC (BCLC Stage B)

Patients with intermediate-stage tumors should be considered for TACE, according to the current guidelines indications [8]. However, all the studies that have investigated the combination of sorafenib and TACE over the last decades did not reveal to improve the OS when compared with sorafenib or TACE alone [48,49]. For example, the TACTICS study, which evaluated the efficacy of TACE plus sorafenib versus TACE in unresectable HCC, showed to only improve PFS (25.2 versus 13.5 months, $p = 0.006$) [50].

In this context, there are newly evidences supporting the use of immunotherapy in the BCLC B stage, basing on the concept that a combination of ICIs and TACE may improve the efficacy of the standard treatment. Indeed, TACE leads to tumor necrosis and cellular damage by inducing high intratumoral temperature. This mechanism of action is responsible for the higher release of neoantigens, which promote an immunogenic environment [51]. Figure 3 summarizes the treatment options for BCLC stage B HCC and the ICIs ongoing clinical trials in this field.

Preliminary results of the phase I/II PETAL clinical trial, which evaluated the safety and activity of pembrolizumab after TACE, revealed a good tolerability for the sequential treatment without cumulative side effects [52]. Tremelimumab is also being evaluated with TACE/RFA in the ongoing NCT01853618 trial. Additionally, clinical trials testing TACE plus nivolumab (NCT03143270) and durvalumab plus tremelimumab following TACE (NCT03638141) are currently running.

Then, some studies are investigating the synergistic effects of different mechanisms of action in order to improve the patients' outcome in this staging group. In fact, the ischemic cell damage related to TACE might produce an increase in the vascular endothelial growth factors (VEGFs) levels in addition to the increase in the immunogenic cell death and stimulation of a peripheral immune response. Therefore, the follow therapeutic combinations are being tested in this field: TACE plus pembrolizumab and lenvatinib (LEAP-012 trial, NCT04246177); TACE plus durvalumab and bevacizumab (EMERALD-1 study, NCT03778957).

Selective internal radiation therapy (SIRT) is another transaerterial approach used in patients with BCLC stage B tumors. It consists in the intraarterial infusion of microspheres with the radioisotope Yttrium-90. According to retrospective studies, SIRT determines objective responses similar to TACE [53]; however, no data about survival are available in this field, due to the lack of phase III comparative studies between TACE and SIRT. Regarding ICIs, several trials in combination with SIRT are currently recruiting. NCT03099564 is an open-label multi-center trial assessing the efficacy and safety of pembrolizumab with Yttrium-90; NCT03033446 trial is a phase II study with the objective is to evaluate the effect of SIRT in combination with nivolumab in Asian patients, also pre-treated with prior local therapies. Additionally, the phase I NCT02837029 trial is identifying the maximum tolerated dose of nivolumab for combination treatment of nivolumab and Yttrium-90. Then, NASIR-HCC study (NCT03380130) completed the enrollment with the aim to evaluate the safety and the antitumoral efficacy of nivolumab after SIRT for patients with unresectable HCC, who were candidates for locoregional therapies.

At least, in patients with BCLC stage B who progressed to transarterial therapies, a systemic treatment is recommended instead of multiple local therapies [8]. After progression to sorafenib, the phase II NCT03316872 study is testing the efficacy of the combination of pembrolizumab and stereotactic body radiotherapy (SBRT).

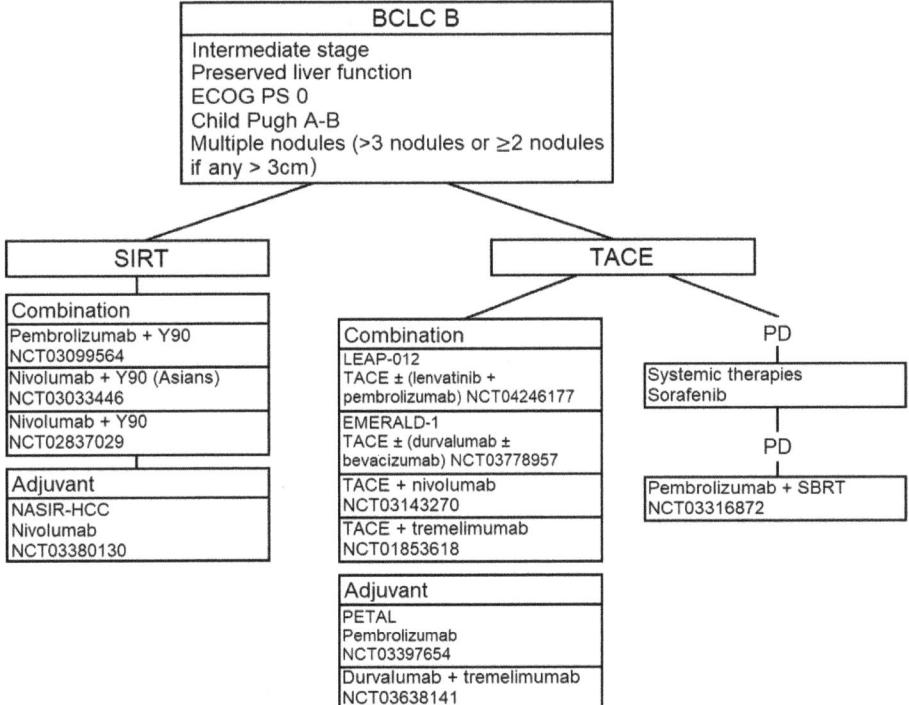

Figure 3. Treatment options and ICIs ongoing clinical trials for BCLC stage B.

3.2.3. Advanced—Stage HCC (BCLC stage C)

Systemic therapies are indicated in patients with advanced disease (BCLC stage C) or intermediate stage disease (BCLC stage B), who are not eligible for locoregional therapies or after progression to local treatment, as already mentioned. According to international guidelines [8], target therapy with TKIs is the standard of care in the first line treatment, whereas there is no indication to use chemotherapy in this setting due to the lack of efficacy. In this context, immunotherapy represents an exciting treatment alternative to explore (Figure 4).

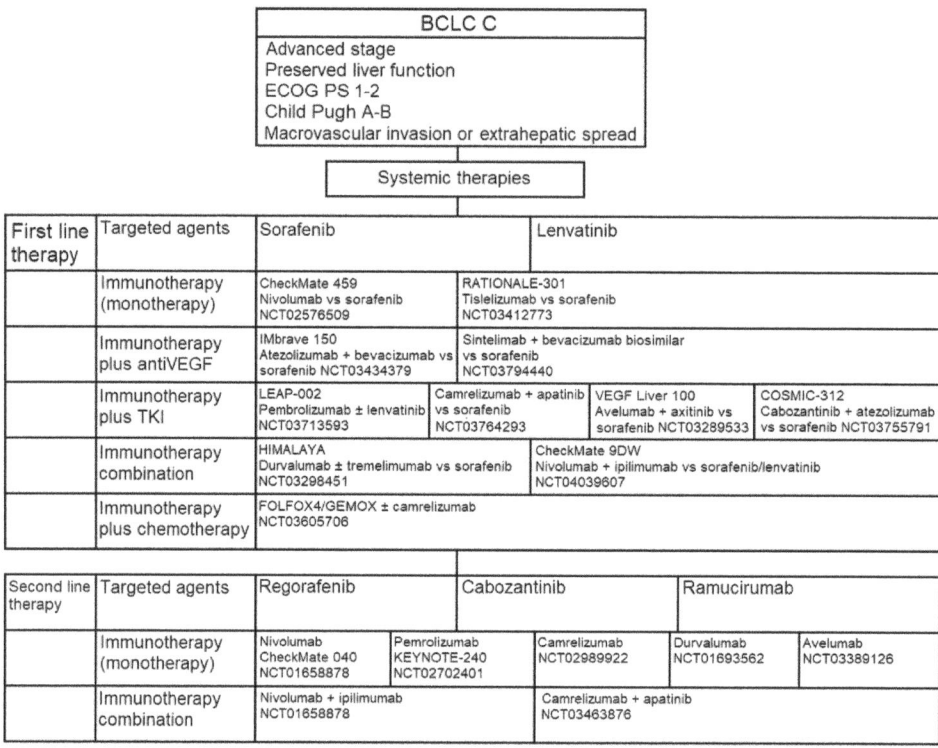

Figure 4. Treatment options and ICIs ongoing clinical trials for BCLC stage C.

First-Line Therapy: From the Standard of Care to the New Frontiers

Historically, Sorafenib was the first systemic drug approved by Food and Drug Administration (FDA) for the treatment of advanced HCC and has remained a unique and effective standard of care for frontline therapy for approximately 10 years [54]. In 2018, Lenvatinib (another TKI) received the FDA approval for advanced HCC on the basis of the phase III non-inferiority REFLECT trial, which excluded conditions with the main portal vein invasion, clear bile duct invasion and >50% of tumour total liver volume occupancy [9]. Based on this background, immunotherapy was considered a promising alternative to treatment with TKI. Therefore, the number of trials evaluating the role of ICIs in the first line therapy for advanced HCC has increased, both in monotherapy and in combination with other ICIs or targeted/antiangiogenetic agents.

More in details, the phase III CheckMate-459 trial evaluated the efficacy and safety of nivolumab (a fully human anti-PD-1 IgG4 antibody administered at 240 mg every two weeks) versus sorafenib as first line therapy in patients with unresectable HCC. The study did not reached its primary endpoint (OS: 16.4 versus 14.7 months in the experimental and standard arm, respectively; Hazard Ratio (HR): 0.85; 95% Confidence Interval (CI): 0.72–1.02; $p = 0.0752$) [55]. ORR were 15% and 7% in the nivolumab and sorafenib arm, respectively; a clinical benefit was reported in all the pre-planned subgroups, including those according to hepatitis infection status, presence of vascular invasion and/or extrahepatic spread, geographical region (Asia versus non-Asia). Of note, 140 patients (38%) in the experimental arm and 170 patients (46%) in the control arm received subsequent lines of treatment. Though the primary endpoint was not met, nivolumab showed clinically meaningful improvements in OS, ORR and complete response rate as well as favorable safety profile as first line treatment.

Tislelizumab (BGB-A317) is another humanized IgG4 antibody against PD-1 tested in the front-line treatment for advanced HCC. Based on promising results of the phase I trial, involving 61 patients with solid cancers included HCC [56], the phase III non-inferiority RATIONALE-301 trial compared tislelizumab (200 mg every three weeks) versus sorafenib [57]. The trial is currently ongoing (NCT03412773) with an expected end date on June 2021.

Despite interesting results from ICIs monotherapy studies in HCC, only a small group of patients benefits from ICIs [58]. Thus, several combination approaches have been utilized with the aim to improve the anti-tumor efficacy and survival in the whole HCC population, targeting different pathways. Well-known combinations included: anti PD-/PD-L1 antibody with non-immune-based-therapies (TKI, anti-VEGF, chemotherapies); two types of ICIs (anti PD-1/PD-L1 and anti CTLA-4 antibodies); ICIs with existing locoregional therapies (already discussed above; see Sections 3.2.1 and 3.2.2).

Atezolizumab plus bevacizumab is one of the most interesting combination tested in this field. The rationale of the combination could be found in preclinical studies, in which bevacizumab showed to enhance PD1/PD-L1 efficacy by reversing VEGF-mediated immunosuppression and by promoting tumor T-cell infiltration [59]. Indeed, during the process of carcinogenesis, the VEGF stimulates the formation of new vessels (angiogenesis), reducing simultaneously the immune response against the tumor. Therefore, the use of anti-VEGF drugs could have a double effect on cancer cells, which is antiangiogenic and immunomodulation. More in details, the VEGF would exercise its immunosuppressive role through three main ways: reducing T cell activation by inhibition of the maturation of dendritic cells through nuclear factor kB; creating aberrant tumor vessels and down-regulating the selectins and adhesion molecules (necessary for the adhesion of T cells to the vascular endothelium itself); increasing the number of inhibitory immune cells in the tumor microenvironment. In this light, bevacizumab might improve the tumour immunogenity, leading to a stronger host immune-response.

In July 2018, the FDA assigned the breakthrough therapy approval to atezolizumab in combination with bevacizumab in advanced HCC on the basis of the results of the phase Ib GO30140 Study [60]. The interim data analysis of this trial showed a response rate in 32% of patients by RECIST criteria. Responses were durable (≥6 months: 52%, ≥12 months: 26%), grade 3–4 treatment related adverse events (TRAEs) occurred in 27% of patients and hypertension was the most common (10%). Even if 2% of patients had a drug-related death, the combination was well-tolerated, having a good safety profile. Then, the phase III IMbrave150 study randomized 501 patients with unresectable or metastatic HCC, naïve to systemic therapy, to receive atezolizumab (1200 mg every three weeks) and bevacizumab (15 mg/kg every three weeks, $n = 336$ patients) or sorafenib (165 patients) [61]. The two primary endpoints were OS and independent review facility–assessed PFS per RECIST 1.1. The trial showed 42% OS improvement in the atezolizumab plus bevacizumab arm (HR= 0.58; 95% CI: 0.42–0.79, $p = 0.0006$) as well as 41% improvement in PFS (HR = 0.59; 95% CI: 0.47–0.76, $p < 0.0001$) if compared with sorafenib. It is noteworthy that, over the last decades, IMbrave150 is the first phase III positive trial, showing an improvement in both OS and PFS in this setting of disease for a new combination of drugs beyond sorafenib. Regarding the safety profile, 38% of patients had a grade 3–4 TRAEs in the combination arm; the most frequent were bleeding in the gastrointestinal tract, infections and fever [62]. The combination therapy also resulted in better quality of life outcomes (longer time to deterioration of quality of life and functioning) than treatment with sorafenib. Time to deterioration, that was the reduction of 10 points from the baseline-reported score, was 11.2 months in patients receiving atezolizumab and bevacizumab and 3.6 months in those treated with sorafenib (HR = 0.63; 95% CI: 0.46–0.85). Declines in physical and role functioning also improved in the experimental arm. Further, the physical functioning had a median delay of 13.1 versus 4.9 months for the experimental and control arm, respectively (HR = 0.53; 95% CI: 0.39–0.73), as well as the role functioning (median delay of 9.2 versus 3.6 months, respectively (HR = 0.62; 95% CI: 0.46–0.84)). Additionally, atezolizumab and bevacizumab delayed the time to deterioration. Combination treated patients reported appetite loss, fatigue, pain, and diarrhea in a lower proportion than sorafenib, experiencing less clinically meaningful deterioration in each of these symptoms. Based on those results, on January 2020 a supplemental Biologics License Application was submitted to the FDA for atezolizumab plus bevacizumab combination in the first-line treatment for advanced HCC [63]. The combination was finally approved by FDA in this setting [64].

Regarding other combinations between anti-PD-1 and anti-VEGF, the phase II/III ORIENT-32 trial (NCT03794440) is assessing the safety, tolerability and effectiveness of sintilimab in combination with IBI305 (anti-VEGF monoclonal antibody, bevacizumab biosimilar) in patients with HCC as the first-line treatment compared with sorafenib (estimated end date: December 2022).

Another intriguing combination is between ICIs and TKI. The rationale for their combination comes from the evidence that antiangiogenic mechanisms may increase tumor hypoxia, leading to the upregulation of the costimulatory molecule OX40 in T-cell-mediated immunity; OX40 promotes the survival and expansion of CD8+T cells and the recall response of CD8+memory T cells [65]. Examples of combinations of ICIs and molecular targeted therapy are: pembrolizumab plus lenvatinib; camrelizumab plus apatinib; avelumab plus axitinib; atezolizumab plus cabozantinib.

Starting from the first combination, the Keynote-524 is an open-label, phase Ib study which tested the safety of pembrolizumab and lenvatinib in patients with unresectable HCC, not amenable to locoregional treatments [66]. The trial had a safety lead-in of six patients with a subsequent expansion cohort of 24 previously untreated patients. It revealed ORR by RECIST and modified RECIST (mRECIST) of 36.7%, and 50%, respectively. Based on these promising preliminary results, the trial is actually involving 104 patients in the phase II and led to breakthrough FDA approval of the combination on July 2019. The phase III study LEAP-002 study (NCT03713593) is now evaluating lenvatinib as single agent or in association with pembrolizumab in the first line setting [67].

Regarding camrelizumab and apatinib, a phase III clinical trial is currently testing this combination versus sorafenib as first-line therapy in patients with advanced HCC (NCT03764293).

The phase Ib VEGF Liver 100 study (NCT03289533) investigated the safety of avelumab (10 mg/kg every two weeks) co-administered with the TKI axitinib (5 mg orally twice a day) as first line. The treatment was active, showing an ORR of 13.6% and 31.8% based on RECIST 1.1 or mRECIST criteria, respectively; median PFS was 5.5 and 3.8 months, based on RECIST and mRECIST, respectively. The study reported higher grade 3 TKI-TRAEs, especially hypertension (50%) and hand–foot syndrome (22.7%), in the experimental arm, but without grade ≥3 immune-related adverse events [68].

The phase III COSMIC-312 trial (NCT03755791) is comparing the association between cabozantinib and atezolizumab versus sorafenib in patients with advanced HCC naïve to systemic treatments.

The combinations between anti-PD-1 plus chemotherapy and anti-PD-L1 and anti-CTLA-4 represent the last most important combinations investigated in the field of first line treatment for advanced HCC. In the first case, the combination of camrelizumab plus FOLFOX-4 (5-fluorouracil plus oxaliplatin) or GEMOX (gemcitabine plus oxaliplatin) was tested in a phase II study, involving 34 patients [69]. The trial showed an ORR of 26.5%, a disease control rate (DCR) of 79.4% and a median PFS of 5.5 months. These data have led to investigate camrelizumab in combination with FOLFOX-4 in a phase III study (NCT03605706) in the same setting; the trial is currently running.

Regarding the combinations between anti PD-L1 and anti-CTLA-4, the rationale for their use consists in the ability of improving the immune stimulation by targeting different pathways; this strategy has been already investigated in many kinds of tumors with positive results [70]. In this subgroup the most important combinations are durvalumab plus tremelimumab (NCT03298451) and nivolumab plus ipilimumab (NCT01658878, NCT03222076, NCT03510871), both under investigation at the time of writing.

Durvalumab and tremelimumab have been investigated in a phase I/II study involving 40 patients with advanced HCC [71]. The trial used a schedule of durvalumab at the dose of 20 mg/kg and tremelimumab at the dose of 1 mg/kg every 4 weeks, followed by 20 mg/kg durvalumab as maintenance. In this study, it is important to note that 70% of the patients had received previous systemic therapies and half of the study population had no history of hepatitis. The ORR was 15% (all the responses were seen in patients without history of hepatitis); 16-weeks disease-control rate was 57%. However, there was 20% of serious TRAEs, leading to the discontinuation of treatment for toxicity in 7% of patients. Based on these results, the randomized phase III HIMALAYA study (NCT03298451) is currently running. The trial is investigating the efficacy and safety of durvalumab plus tremelimumab or durvalumab as single agent versus sorafenib as first-line treatment for patients with naïve unresectable HCC. The preliminary safety results showed that the combination was well tolerated [72]; the most common all-grade TRAEs included fatigue (27.5%), increased alanine aminotransferase (ALT; 20.0%), pruritus (22.5%), increased aspartate aminotransferase (AST; 17.5%), elevated lipase (10.0%). Twenty-five percent of patients experienced grade 3/4 TRAEs or serious AEs and no treatment-related deaths occurred. Of note, this trial represents the first phase III study that have evaluated a combination between two ICIs as first-line treatment for advanced HCC. Therefore, in January 2020, the FDA approved the combination of durvalumab and tremelimumab in this field, designing these agents as orphan drugs [73].

Last, the phase III CheckMate-9DW trial is currently investigating the efficacy of nivolumab plus ipilimumab versus standard care (sorafenib or lenvatinib) in patients with advanced HCC naïve to systemic treatment (NCT04039707); the results are awaited.

In conclusion, immunotherapy with ICI as a monotherapy or in combination seem to be promising as a first line of treatment for patients with BCLC stage C HCC. However, the majority of trials are still ongoing and only few combinations were approved in clinical practice from regulatory authorities. Additionally, the authorization was recent in the majority of cases, so we have a very few data (or no one in some cases) regarding the phase IV, as well as real life data from the every-day clinical practice. Therefore, all reported results should still be considered with caution.

Second-Line Therapy: From the Standard of Care to the New Frontiers

Second-line treatments are needed for patients with good performance status, after progression or no tolerability to first-line treatment. In recent years, new advances have been made to test new systemic treatments in the second line, even if no drugs investigated in this line was tested after progression to lenvatinib. Indeed, regorafenib, the first therapeutic agent approved by the FDA in this setting, was tested in patients progressing to treatment with sorafenib (see above).

According to international guidelines [8], regorafenib, lenvatinib and ramucirumab are the biological agents used as standard of care in this setting. Shortly, in the phase III RESORCE trial, regorafenib showed to increase OS, if compared with best supportive care, from 7.8 to 10.6 months, decreasing the risk of death by 37% [10]. The CELESTIAL trial examined cabozantinib versus placebo in patients with advanced HCC who were previously treated with sorafenib [11]. Unlike the RESORCE trial, this study included also patients intolerant to sorafenib or in progression after two lines of therapy for advanced disease. In REACH-2 trial (the first phase III biomarker-driven study), ramucirumab, a human immunoglobulin G1 monoclonal anti-VEGFR2 antibody, significantly improved median OS (from 7.3 to 8.5 months) in a subgroup of patients with serum baseline alpha fetoprotein (AFP) levels ≥400 ng/mL [12].

Moving from the standard of care, immunotherapy represents a new frontiers for HCC treatment also in the second line setting. In this context, anti-PD-1 (nivolumab, pembrolizumab and camrelizumab), anti-PD-L1 (durvalumab and avelumab) and anti-CTLA-4 (tremelimumab) are being tested.

Nivolumab was the first ICI approved for patients with advanced HCC and progressed to sorafenib, based on the results of the phase I/II CheckMate-040 study [14]. In particular, the phase I part of the trial has tested escalating doses of nivolumab in 48 patients divided into three cohorts (virus-uninfected, HBV- and HCV-infected advanced HCC). The antiviral control was mandatory only in patients with HBV infection. The most frequent TRAEs were dose-unrelated and included fatigue, rash, pruritus and an increase in liver enzyme levels. Twenty-five percent of patients had a grade 3/4 TRAEs; adrenal insufficiency, diarrhea, hepatitis, and acute kidney injury were the most important. Then, the dose expansion investigated the effect of nivolumab at 3 mg/kg in 214 subjects (HCV positive,

HBV positive and no viral hepatitis: 50/51/113, respectively; the last group was stratified in two subgroups: patients naïve to treatment or intolerant to sorafenib ($n = 56$) and patients progressed after sorafenib ($n = 57$)). ORR were 15% and 20% in the dose-escalation and expansion cohorts, respectively; the median OS was 15 months in the dose escalation group (95% CI = 9.6–20.2 months). The expression of PD-L1 in tumour cells was not related to response rate. The study revealed—for the first time—that nivolumab was effective and safe in patients with advanced HCC, like previously showed also in other types of cancer [74]. Notably, the trial showed that nivolumab can be safely used also in patients with HBV or HCV infections, reporting an impressive ORR with durable responses in the entire cohorts (uninfected, HBV-infected and HCV-infected patients).

Then, Pembrolizumab was approved through an accelerate process by FDA on 9 November 2018 for treatment of patients with HCC after a previous treatment with sorabenib. The KEYNOTE-224 trial [15] is a non-randomised, multicenter, open label phase II trial, which investigated the activity of pembrolizumab (200 mg every three weeks) in 104 patients affected by advanced HCC, who were refractory or intolerant to sorafenib (80% and 20% of the study population, respectively; all patients had a Child-Pugh A liver function score). The trial showed to improve the survival outcomes (median PFS: 4.9 months (95% CI 3.4–7.2); median OS: 12.9 months (95% CI 9.7–15.5); 1-year OS rate; 54% (95% CI 44–63)). Twenty-five percent of patients had grade 3-4 TRAEs and the most frequent was hypertransaminasemia (6%). Notably, none of the 26 HCV-positive (25% of the entire population) as well as none of the 22 HBV-positive patients (21% of the entire population) had worsening or re-activation of hepatitis. ORR was reported in 18 patients (17%), 77% of whom were long responders (>9 months). The trial evaluated also the relationship between PD-L1 expression and response to treatment, by using two indices of PD-L1 expression. The combined positive score (CPS: number ($n.$) of PD-L1-positive cells (both tumour and host immune cells)/ $n.$ of viable tumor cells $\times$ 100) and the tumor proportion score (TPS: $n.$ of PD-L1 positive tumor cells/ $n.$ of viable tumor cells $\times$ 100). CPS was positive in 22 (42%) and negative in 7 (13%) patients. ORR was 25%, with the best responses in CPS and TPS positive tumours: 32% versus 20% ($p = 0.021$) and 43% vs 22% ($p=0.088$), respectively. PFS there was significantly longer in CPS positive ($p = 0.026$) but not in TPS positive patients ($p = 0.096$). In conclusion, the trial showed that pembrolizumab leads to durable responses and favorable outcomes in patients with advanced HCC who received a previous treatment with sorafenib. Then, the phase III KEYNOTE-240 randomized 413 patients affected by advanced HCC, who were refractory or intolerant to sorafenib (all patients had a Child-Pugh A liver function score), to pembrolizumab or best supportive care [75]. The trial showed a median OS of 13.9 months (95% CI: 11.6-16.0 months) for pembrolizumab versus 10.6 months (95% CI: 8.3–13.5 months) for placebo (HR: 0.781; 95% CI: 0.611–0.998; $p = 0.0238$). Median PFS for pembrolizumab was 3.0 months (95% CI: 2.8–4.1 months) versus 2.8 months (95% CI: 1.6–3.0 months; HR: 0.718; 95% CI: 0.570–0.904; $p = 0.0022$). Therefore, pembrolizumab showed a trend of better OS and PFS in this field, even if without statistical significance. However, the results were in line with the findings of KEYNOTE-224 [15]. Additionally, it is important to note that the number of patients who received an active post-study treatment was higher in the placebo arm than in the experimental arm, probably affecting the outcomes reported in the trial.

Last, the phase III KEYNOTE-394 trial (NCT03062358) is currently testing the efficacy of Pembrolizumab versus placebo in Asian pretreated patients with advanced HCC.

Camrelizumab (also known as SHR-1210) is a human IgG4 antibody against PD-1, which showed a promising activity in 58 patients with solid cancers evaluated in a phase I trial, including HCC [76]. A phase II/III trial is ongoing in China (NCT02989922), enrolling patients with advanced HCC who had failure or intolerance to prior systemic treatment. A total of 217 patients were randomized to camrelizumab (3 mg/kg every two ($n = 109$) or three weeks ($n = 108$)). Preliminary results were promising: ORR: 13.8%, 6-month OS rate: 74.7%, median time to response: 2 months, median duration of response: not reached, DCR: 44.7%, median PFS: 2.1 months. The unique TRAE reported was reactive capillary hemangioma, even if the pathogenesis, as well as the relation to the tumor response are not clear; it was observed in 66.8% of HCC patients treated. In conclusion, camrelizumab showed interesting ORR, durable response and acceptable toxicities in this Chinese trial [77].

Durvalumab and avelumab are the most relevant anti-PD-L1 agents investigated in the field of HCC. A phase I/II trial of durvalumab monotherapy in solid tumours, including HCC ($n = 40$), showed 10% ORR and median OS of 13.2 months (NCT01693562) [78]. Avelumab is a human IgG1 antibody against PD-L1; it is currently been testing as single agent, as well as in combination for advanced HCC [79]. A phase II study of avelumab, involving 30 HCC patients after sorafenib treatment, is ongoing (NCT03389126).

The anti-CTLA-4 antibody have a role in HCC treatment by increasing the expression of tumor-associated antigens, such as interleukin (IL)-1, IL-6 and macrophage inflammatory protein-1 [80–82].

The first anti-CTLA-4 antibody investigated in the field of HCC was tremelimumab. In particular, a phase II trial (NCT01008358) assessed the activity of tremelimumab in HCC pre-treated patients with chronic HCV infection. They received the treatment at the dosage of 15 mg/kg intravenously every 90 days until tumor progression or severe toxicity [83]. The preliminary results showed a DCR in 76.4% of patients (partial response: 17.6%) and a time to progression of 6.48 months (95% CI: 3.95–9.14 months). Notably, the trial showed that the treatment was safe also in patients with Child-Pugh stage B (42.9%).

ICIs combinations are being tested also in second-line setting for advanced HCC. The combination of nivolumab and ipilimumab for patients with advanced HCC who progressed after sorafenib treatment was firstly tested in the phase I/II CheckMate-040 study [84]. In particular, the trial randomized patients with Child-Pugh A class into three arms, according to different dosages in the combination: (a) nivolumab 1 mg/kg + ipilimumab 3 mg/kg, (b) nivolumab 3 mg/kg + ipilimumab 1 mg/kg every 3 weeks (four cycles), then nivolumab 240 mg flat dose every two weeks as maintenance, (c) nivolumab 3 mg/kg + ipilimumab 1 mg/kg every 6 weeks. The treatments were continued until disease progression or toxicity. TRAEs occurred in 37% of patients and skin toxicity-related ere the most common. However, only 5% of patients discontinued the treatment due to unacceptable toxicity. The trial demonstrated that the combination improves the ORR if compared to nivolumab monotherapy (31% versus 14%, respectively), with a promising effect on outcome (median OS: 22.8 months in the combination arm). The updated results after a minimum of 28-month follow-up, showed that 33% of patients had a response to treatment in the combination arm (8% complete response and 24% partial response) [85]. There was a long duration of response (from 4.6 to 30.5 months): maintenance of responses was recorded in 88%, 56% and 31% of patients at 6, 12 and 24 months, respectively. The ORR, as assessed by blinded independent central review using RECIST criteria modified for immunotherapy, was 35% (95% CI, 22–50%); complete and partial responses were observed in 12% and 22% of patients, respectively. Overall, the DCR was 54.0% (95% CI, 39.3–68.2%). Based on these results, in November 2019 the FDA gave a positive response about the use of nivolumab in combination with ipilimumab and in March 2020 approved the combination for patients with HCC progressed after sorafenib, according to the following schedule: nivolumab 1 mg/kg followed by ipilimumab 3 mg/kg every 3 weeks for 4 doses, followed by nivolumab 240 mg every 2 weeks or 480 mg every 4 weeks [86].

Regarding combinations between ICIs and TKI, a phase I trial (NCT02942329) investigated camrelizumab (200 mg every 2 weeks) associated to apatinib (a TKI selectively acting on VEGFR2, administered at the dose of 125-500 mg once daily) in patients with advanced HCC, gastric or esophagogastric junction cancer [87]. The trial involved 18 patients with HCC, showing ORR of 50.0% and a median PFS of 5.8 months. The TRAEs were manageable and the discontinuation due to toxicities was reported in only one patient (grade 3 hyperbilirubinaemia). Then, a phase II study (NCT03463876) is exploring the efficacy and safety of the combination of apatinib (250 mg orally every day) and camrelizumab (200mg (3mg/kg for underweight patients) every 2 weeks) in this setting.

4. Future Perspectives

ICIs-based treatments welcomed new opportunities in the treatment of HCC, and not only in the advanced stage. However, despite promising results from clinical studies, only few patients benefit from ICIs [88]. Indeed, recent data showed that immunotherapies enhance survival, but their effects are limited [58]. The failure of ICI therapy might be related to the changes in the immunogenicity of cancer itself as well as of microenvironment [89–91]. Indeed, in this regard, the gut microbiome has gained significant attention since its alterations could affect the response to immunotherapies [92,93].

In addition, there is a lack of validated prognostic and predictive biomarkers able to guide the choice of the best treatment for each patient. In this context, some trials reported that high PD-L1 expression could be associated with poor outcome [58], even if its predictive role is still unclear and elusive, as proven by the responses to treatments both, in patients with high and low expression of PD-L1 [94]. Regarding tumor mutation burden (TMB), its role seems to be less important in HCC. In fact, HCC showed to be less immunogenic than other tumours, showing low TMB (median number of 5 Mut/Mb) [95–97]. Therefore, up to date, TMB is not used as potential predictive biomarker in HCC [98].

Then, other possible predictive biomarkers may be the overexpression of TIM-3 and LAG-3 in patients after receiving a previous anti-PD-1 therapy [99], whereas the epithelial-to-mesenchymal transition (EMT) could be related to resistance to immunotherapy. In fact, a study evaluating the specimens from 422 HCC patients, showed that the presence of EMT was linked to a more aggressive disease with worst outcome [89]. Wnt/CTNNB1 mutations could also be a further biomarker of ICI resistance. Thus, the identification of better predictive biomarkers, in order to improve the efficacy of ICI therapy is a hot and challenge issue.

5. Conclusions

The treatment algorithm for HCC management according to BCLC stage is evolving. In this context, ICIs represent an intriguing challenge. Therefore, several clinical trials are focusing on the use of immunotherapy in HCC, alone or in combinations with TKI/antiangiogenetic agents as well as local treatment, according to the tumour stage. However, the majority of those trials are still ongoing and, until now, only a few combinations were approved in the clinical practice from the regulatory authorities. Therefore, all the reported results should be still considered with caution.

Additionally, decisions about the choice of the right sequence of treatments in HCC patients in the light of the "continuum of care" principles, is still hard. In fact, it requires careful consideration in a multidisciplinary context in order to ensure a tailored treatment for each patient.

Author Contributions: Conceptualization, P.F.; resources, P.F., P.G.; writing—original draft preparation, P.F., A.P.; writing of particular sections: all authors; writing—review and editing, P.F., A.P., B.D.; supervision, B.D. All authors have read and agreed to the published version of the manuscript.

Funding: This research received no external funding.

Acknowledgments: We would thank K. El Bairi (Mohamed Ist University, Faculty of Medicine and Pharmacy; Oujda, Morocco) and E.F. Giunta (Medical Oncology Unit, Università degli studi della Campania "L.Vanvitelli") for the editorial assistance.

Conflicts of Interest: A.P. received personal fee from Eli-Lilly; B.D. received personal fee from Ipsen, Eisai, Eli Lilly, Astra Zeneca, Sanofi, M.S.D., Bayer, Roche, Amgen. No fees are connected with the submitted paper. The other authors declare no conflict of interest. The funders had no role in the design, writing of the manuscript, or in the decision to publish the paper.

References

1. Bray, F.; Me, J.F.; Soerjomataram, I.; Siegel, R.L.; Torre, L.A.; Jemal, A. Global cancer statistics 2018, GLOBOCAN estimates of incidence and mortality worldwide for 36 cancers in 185 countries. *CA A Cancer J. Clin.* **2018**, *68*, 394–424. [CrossRef]
2. Forner, A.; Reig, M.; Bruix, J. Hepatocellular carcinoma. *Lancet* **2018**, *391*, 1301–1314. [CrossRef]

3. Llovet, J.M.; Fuster, J.; Bruix, J. Barcelona-Clínic Liver Cancer Group. The Barcelona approach: Diagnosis, staging, and treatment of hepatocellular carcinoma. *Liver Transplant.* **2004**, *10* (Suppl. 1), S115–S120. [CrossRef] [PubMed]
4. Kudo, M.; Matsui, O.; Izumi, N.; Kadoya, M.; Okusaka, T.; MiyayamaMasashi, S.; Yamakado, K.; Tsuchiya, K.; Ueshima, K.; Hiraoka, A.; et al. Transarterial chemoembolization failure/refractoriness: JSH-LCSGJ criteria 2014 update. *Oncology* **2014**, *87* (Suppl. 1), 22–31. [CrossRef] [PubMed]
5. Raoul, J.-L.; Sangro, B.; Forner, A.; Mazzaferro, V.; Piscaglia, F.; Bolondi, L.; Lencioni, R. Evolving strategies for the management of intermediate-stage hepatocellular carcinoma: Available evidence and expert opinion on the use of transarterial chemoembolization. *Cancer Treat. Rev.* **2011**, *37*, 212–220. [CrossRef]
6. Galle, P.R.; Tovoli, F.; Foerster, F.; Wörns, M.A.; Cucchetti, A.; Bolondi, L. The treatment of intermediate stage tumours beyond TACE: From surgery to systemic therapy. *J. Hepatol.* **2017**, *67*, 173–183. [CrossRef]
7. Peck-Radosavljevic, M.; Kudo, M.; Raoul, J.-L.; Lee, H.C.; Decaens, T.; Heo, J.; Lin, S.-M.; Shan, H.; Yang, Y.; Bayh, I.; et al. Outcomes of patients (pts) with hepatocellular carcinoma (HCC) treated with transarterial chemoembolization (TACE): Global OPTIMIS final analysis. *J. Clin. Oncol.* **2018**, *36* (Suppl. 15), 4018. [CrossRef]
8. Vogel, A.; Cervantes, A.; Chau, I.; Daniele, B.; Llovet, J.; Meyer, T.; Nault, J.-C.; Neumann, U.; Ricke, J.; Sangro, B.; et al. Hepatocellular carcinoma: ESMO clinical practice guidelines for diagnosis, treatment and follow-up. *Ann. Oncol.* **2018**, *29* (Suppl. 4), iv238–iv255. [CrossRef]
9. Kudo, M.; Finn, R.S.; Qin, S.; Han, K.-H.; Ikeda, K.; Piscaglia, F.; Baron, A.; Park, J.-W.; Han, G.; Jassem, J.; et al. Lenvatinib versus sorafenib in first-line treatment of patients with unresectable hepatocellular carcinoma: A randomised phase 3 non-inferiority trial. *Lancet* **2018**, *391*, 1163–1173. [CrossRef]
10. Bruix, J.; Qin, S.; Merle, P.; Granito, A.; Huang, Y.-H.; Bodoky, G.; Pracht, M.; Yokosuka, O.; Rosmorduc, O.; Breder, V.; et al. Regorafenib for patients with hepatocellular carcinoma who progressed on sorafenib treatment (RESORCE): A randomised, double-blind, placebo-controlled, phase 3 trial. *Lancet* **2017**, *389*, 56–66. [CrossRef]
11. Abou-Alfa, G.K.; Meyer, T.; Cheng, A.-L.; El-Khoueiry, A.B.; Rimassa, L.; Ryoo, B.-Y.; Cicin, I.; Merle, P.; Chen, Y.; Park, J.-W.; et al. Cabozantinib in Patients with Advanced and Progressing Hepatocellular Carcinoma. *N. Engl. J. Med.* **2018**, *379*, 54–63. [CrossRef] [PubMed]
12. Zhu, A.X.; Khang, Y.K.; Yen, C.J.; Finn, R.S.; Galle, P.R.; Llovet, J.M.; Assenat, E.; Brandi, G.; Pracht, M.; Lim, H.Y.; et al. Ramucirumab after sorafenib in patients with advanced hepatocellular carcinoma and increased α-fetoprotein concentrations (REACH-2): A randomised, double-blind, placebo-controlled, phase 3 trial. *Lancet Oncol.* **2019**, *20*, 282–296. [CrossRef]
13. Hargadon, K.M.; Johnson, C.E.; Williams, C.J. Immune checkpoint blockade therapy for cancer: An overview of FDA-approved immune checkpoint inhibitors. *Int. Immunopharmacol.* **2018**, *62*, 29–39. [CrossRef]
14. El-Khoueiry, A.B.; Sangro, B.; Yau, T.; Crocenzi, T.S.; Kudo, M.; Hsu, C.; Kim, T.-Y.; Choo, S.-P.; Trojan, J.; Welling, T.H.; et al. Nivolumab in patients with advanced hepatocellular carcinoma (CheckMate 040): An open-label, non-comparative, phase 1/2 dose escalation and expansion trial. *Lancet* **2017**, *389*, 2492–2502. [CrossRef]
15. Zhu, A.X.; Finn, R.S.; Edeline, J.; Cattan, S.; Ogasawara, S.; Palmer, D.; Verslype, C.; Zagonel, V.; Fartoux, L.; Vogel, A.; et al. Pembrolizumab in patients with advanced hepatocellular carcinoma previously treated with sorafenib (KEYNOTE-224): A non-randomised, openlabel phase 2 trial. *Lancet Oncol.* **2018**, *19*, 940–952. [CrossRef]
16. Cheng, A.-L.; Qin, S.; Ikeda, M.; Galle, P.; Ducreux, M.; Zhu, A.; Kim, T.-Y.; Kudo, M.; Breder, V.; Merle, P.; et al. LBA3I Mbrave150, Efficacy and safety results from a ph III study evaluating atezolizumab (atezo) + bevacizumab (bev) vs sorafenib (Sor) as first treatment (tx) for patients (pts) with unresectable hepatocellular carcinoma (HCC). *Ann. Oncol.* **2019**, *30* (Suppl. 9), ix183–ix202. [CrossRef]
17. Jenne, C.N.; Kubes, P. Immune surveillance by the liver. *Nat. Immunol.* **2013**, *14*, 996–1006. [CrossRef]
18. Robinson, M.W.; Harmon, C.; O'Farrelly, C. Liver immunology and its role in inflammation and homeostasis. *Cell. Mol. Immunol.* **2016**, *13*, 267–276. [CrossRef]
19. Severi, T.; van Malenstein, H.; Verslype, C.; van Pelt, J.F. Tumor initiation and progression in hepatocellular carcinoma: Risk factors, classification, and therapeutic targets. *Acta Pharmacol. Sin.* **2010**, *31*, 1409–1420. [CrossRef]

20. Thomson, A.W.; Knolle, P.A. Antigen-presenting cell function in the tolerogenic liver environment. *Nat. Rev. Immunol.* **2010**, *10*, 753–766. [CrossRef]
21. Bowen, D.G.; Zen, M.; Holz, L.; Davis, T.; McCaughan, G.W.; Bertolino, P. The site of primary T cell activation is a determinant of the balance between intrahepatic tolerance and immunity. *J. Clin. Investig.* **2004**, *114*, 701–712. [CrossRef] [PubMed]
22. Takatsuki, M.; Uemoto, S.; Inomata, Y.; Egawa, H.; Kiuchi, T.; Fujita, S.; Hayashi, M.; Kanematsu, T.; Tanaka, K. Weaning of immunosuppression in living donor liver transplant recipients. *Transplantation* **2001**, *72*, 449–454. [CrossRef] [PubMed]
23. Starzl, T.E.; Demetris, A.J.; Trucco, M.; Murase, N.; Ricordi, C.; Ildstad, S.; Ramos, H.; Todo, S.; Tzakis, A.; Fung, J.J.; et al. Cell migration and chimerism after whole-organ transplantation: The basis of graft acceptance. *Hepatology* **1993**, *17*, 1127–1152. [CrossRef] [PubMed]
24. Kumar, A.; Le, D.T. Hepatocellular carcinoma regression after cessation of immunosuppressive therapy. *J. Clin. Oncol.* **2016**, *34*, 90–92. [CrossRef]
25. Pardoll, D.M. The blockade of immune checkpoints in cancer immunotherapy. *Nat. Rev. Cancer* **2012**, *12*, 252–264. [CrossRef] [PubMed]
26. Shi, F.; Shi, M.; Zeng, Z.; Qi, R.-Z.; Liu, Z.; Zhang, J.-Y.; Yang, Y.; Tien, P.; Wang, F.-S. PD-1 and PD-L1 upregulation promotes CD8+ Tcell apoptosis and postoperative recurrence in hepatocellular carcinoma patients. *Int. J. Cancer* **2011**, *128*, 887–896. [CrossRef]
27. Wu, K.; Kryczek, I.; Chen, L.; Zou, W.; Welling, T.H. Kupffer Cell Suppression of CD8+ T cells in Human Hepatocellular Carcinoma is Mediated by B7-H1/PD-1 Interactions. *Cancer Res.* **2009**, *69*, 8067–8075. [CrossRef]
28. Gao, Q.; Wang, X.-Y.; Qiu, S.-J.; Yamato, I.; Sho, M.; Nakajima, Y.; Zhou, J.; Li, B.-Z.; Shi, Y.-H.; Xiao, Y.-S.; et al. Overexpression of PD-L1 significantly associates with tumor aggressiveness and postoperative recurrence in human hepatocellular carcinoma. *Clin. Cancer Res.* **2009**, *15*, 971–979. [CrossRef]
29. Kalathil, S.; Lugade, A.A.; Miller, A.; Iyer, R.; Thanavala, Y. Higher frequencies of GARP+ CTLA-4+ Foxp3+ T regulatory cells and myeloid-derived suppressor cells in hepatocellular carcinoma patients are associated with impaired T-cell functionality. *Cancer Res.* **2013**, *73*, 2435–2444. [CrossRef]
30. Han, Y.; Chen, Z.; Yang, Y.; Jiang, Z.; Gu, Y.; Liu, Y.; Lin, C.; Pan, Z.; Yu, Y.; Jiang, M.; et al. Human CD14+CTLA-4+ regulatory dendritic cells suppress T-cell response by cytotoxicT-lymphocyte antigen-4dependent-IL-10andindoleamine-2, 3-dioxygenase production in hepatocellular carcinoma. *Hepatology* **2014**, *59*, 567–579. [CrossRef]
31. Iñarrairaegui, M.; Melero, I.; Sangro, B. Immunotherapy of Hepatocellular Carcinoma: Facts and Hopes. *Clin. Cancer Res.* **2018**, *24*, 1518–1524. [CrossRef] [PubMed]
32. Greten, T.F.; Sangro, B. Targets for immunotherapy of liver cancer. *J. Hepatol.* **2018**, *68*, 157–166. [CrossRef] [PubMed]
33. Marin-Acevedo, J.A.; Dholaria, B.; Soyano, A.E.; Knutson, K.L.; Chumsri, S.; Lou, Y. Next generation of immune checkpoint therapy in cancer: New developments and challenges. *J. Hematol. Oncol.* **2018**, *11*, 39. [CrossRef]
34. Manzotti, C.N.; Liu, M.K.; Burke, F.; Dussably, L.; Zheng, Y.; Sansom, D.M. Integration of CD28 and CTLA-4 function results in differential responses of T cells to CD80 and CD86. *Eur. J. Immnol.* **2006**, *36*, 1413–1422. [CrossRef] [PubMed]
35. Gavin, M.A.; Rasmussen, J.P.; Fontenot, J.D.; Vasta, V.; Manganiello, V.C.; Beavo, J.A.; Rudensky, A.Y. Foxp3-dependent programme of regulatory T-cell differentiation. *Nature* **2007**, *445*, 771–775. [CrossRef]
36. Nikolova, M.; Lelievre, J.D.; Carriere, M.; Bensussan, A.; Lévy, Y. Regulatory T cells differentially modulate the maturation and apoptosis of human CD8+ T-cell subsets. *Blood* **2009**, *113*, 4556–4565. [CrossRef]
37. Simona, S.; Labarriere, N. PD-1 expression on tumor-specific T cells: Friend or foe for immunotherapy? *OncoImmunology* **2018**, *7*, e1364828. [CrossRef]
38. Zou, W.; Chen, L. Inhibitory B7-family molecules in the tumour microenvironment. *Nat. Rev. Immunol.* **2008**, *8*, 467–477. [CrossRef]
39. Iwai, Y.; Ishida, M.; Tanaka, Y.; Okazaki, T.; Honjo, T.; Minato, N. Involvement of PD-L1 on tumor cells in the escape from host immune system and tumor immunotherapy by PD-L1 blockade. *Proc. Natl. Acad. Sci. USA* **2002**, *99*, 12293–12297. [CrossRef]

40. Roayaie, S.; Jibara, G.; Tabrizian, P.; Park, J.-W.; Yang, J.; Yan, L.; Schwartz, M.; Han, G.; Izzo, F.; Chen, M.; et al. The role of hepatic resection in the treatment of hepatocellular cancer. *Hepatology* **2015**, *62*, 440–451. [CrossRef]
41. Bruix, J.; Takayama, T.; Mazzaferro, V.; Chau, G.-Y.; Yang, J.; Kudo, M.; Cai, J.; Poon, R.T.; Han, K.-H.; Tak, W.Y.; et al. Adjuvant sorafenib for hepatocellular carcinoma after resection or ablation (STORM): A phase 3, randomised, doubleblind, placebo-controlled trial. *Lancet Oncol.* **2015**, *16*, 1344–1354. [CrossRef]
42. Galle, P.R.; Forner, A.; Llovet, J.M.; Mazzaferro, V.; Piscaglia, F.; Raoul, J.L.; Schirmacher, P.; Vilgrain, V. EASL clinical practice guidelines: Management of hepatocellular carcinoma. *J. Hepatol.* **2018**, *69*, 182–236. [CrossRef] [PubMed]
43. Mizukoshi, E.; Yamashita, T.; Arai, K.; Sunagozaka, H.; Ueda, T.; Arihara, F.; Kagaya, T.; Yamashita, T.; Fushimi, K.; Kaneko, S. Enhancement of tumorassociated antigen-specific T cell responses by radiofrequency ablation of hepatocellular carcinoma. *Hepatology* **2013**, *57*, 1448–1457. [CrossRef]
44. Sharma, P.; Allison, J.P. Immune checkpoint targeting in cancer therapy: Toward combination strategies with curative potential. *Cell* **2015**, *161*, 205–214. [CrossRef]
45. Herrero, J.I.; Sangro, B.; Pardo, F.; Quiroga, J.; Iñarrairaegui, M.; Rotellar, F.; Montiel, C.; Alegre, F.; Prieto, J. Liver transplantation in patients with hepatocellular carcinoma across Milan criteria. *Liver Transpl.* **2008**, *14*, 272–278. [CrossRef] [PubMed]
46. Mazzaferro, V.; Sposito, C.; Zhou, J.; Pinna, A.D.; De Carlis, L.; Fan, J.; Cescon, M.; Di Sandro, S.; Yi-Feng, H.; Lauterio, A.; et al. Metroticket 2. 0 model for analysis of competing risks of death after liver transplantation for hepatocellular carcinoma. *Gastroenterology* **2018**, *154*, 128–139. [CrossRef]
47. Kaseb, A.O.; Carmagnani Pestana, R.; Vence, L.M.; Blando, J.M.; Singh, S.; Ikoma, N.; Raghav, K.P.S.; Sakamuri, D.; Girard, L.; Tan, D.; et al. Randomized, open-label, perioperative phase II study evaluating nivolumab alone versus nivolumab plus ipilimumab in patients with resectable HCC. *J. Clin. Oncol.* **2019**, *37* (Suppl. 4), 185. [CrossRef]
48. Meyer, T.; Fox, R.; Ma, Y.T.; Ross, P.J.; James, M.W.; Sturgess, R.; Stubbs, C.; Stocken, D.D.; Wall, L.; Watkinson, A.; et al. Sorafenib in combination with transarterial chemoembolisation in patients with unresectable hepatocellular carcinoma (TACE 2): A randomised placebocontrolled, double-blind, phase 3 trial. *Lancet Gastroenterol. Hepatol.* **2017**, *2*, 565–575. [CrossRef]
49. Lencioni, R.; Llovet, J.M.; Han, G.; Tak, W.Y.; Gar-Yang, C.; Guglielmi, A.; Paik, S.W.; Reig, M.; Kim, D.Y.; Chau, G.-Y.; et al. Sorafenib or placebo plus TACE with doxorubicin-eluting beads for intermediate stage HCC: The SPACE trial. *J. Hepatol.* **2016**, *64*, 1090–1098. [CrossRef]
50. Kudo, M.; Ueshima, K.; Ikeda, M.; Torimura, T.; Tanabe, N.; Aikata, H.; Izumi, N.; Yamasaki, T.; Nojiri, S.; Hino, K.; et al. Randomized, open label, multicenter, phase II trial comparing transarterial chemoembolization (TACE) plus sorafenib with TACE alone in patients with hepatocellular carcinoma (HCC): TACTICS trial. *J. Clin. Oncol.* **2018**, *36* (Suppl. 4), 206. [CrossRef]
51. Greten, T.F.; Mauda-Havakuk, M.; Heinrich, B.; Korangy, F.; Wood, B.J. Combined locoregional-immunotherapy for liver cancer. *J. Hepatol.* **2019**, *70*, 999–1007. [CrossRef] [PubMed]
52. Pinato, D.J.; Cole, T.; Bengsch, B.; Tait, P. 750PA phase Ib study of pembrolizumab following trans-arterial chemoembolization (TACE) in hepatocellular carcinoma (HCC): PETAL. *Ann. Oncol.* **2019**, *30* (Suppl. 5). [CrossRef]
53. Salem, R.; Gordon, A.C.; Mouli, S.; Hickey, R.; Kallini, J.; Gabr, A.; Mulcahy, M.F.; Baker, T.; Abecassis, M.; Miller, F.H.; et al. Y90 Radioembolization Significantly Prolongs Time to Progression Compared With Chemoembolization in Patients With Hepatocellular Carcinoma. *Gastroenterology* **2016**, *151*, 1155–1163.e2. [CrossRef]
54. Llovet, J.M.; Ricci, S.; Mazzaferro, V.; Hilgard, P.; Gane, E.; Blanc, J.-F.; De Oliveira, A.C.; Santoro, A.; Raoul, J.-L.; Forner, A.; et al. Sorafenib in Advanced Hepatocellular Carcinoma. *N. Engl. J. Med.* **2008**, *359*, 378–390. [CrossRef] [PubMed]
55. Yau, T.; Park, J.; Finn, R.; Cheng, A.-L.; Mathurin, P.; Edeline, J.; Kudo, M.; Han, K.-H.; Harding, J.; Merle, P.; et al. LBA38_PR - CheckMate 459, A randomized, multi-center phase III study of nivolumab (NIVO) vs sorafenib (SOR) as first-line (1L) treatment in patients (pts) with advanced hepatocellular carcinoma (aHCC). *Ann. Oncol.* **2019**, *30*, v874–v875. [CrossRef]
56. Deva, S.; Lee, J.-S.; Lin, C.-C.; Yen, C.-J.; Millward, M.; Chao, Y.; Keam, B.; Jameson, M.; Hou, M.-M.; Kang, Y.-K.; et al. A phase Ia/Ib trial of tislelizumab, an anti-PD-1 antibody (ab), in patients (pts) with advanced solid tumors. *Ann. Oncol.* **2018**, *29* (Suppl. 10), x24–x38. [CrossRef]

57. Qin, S.; Finn, R.S.; Kudo, M.; Meyer, T.; Vogel, A.; Ducreux, M.; Mercade, T.M.; Tomasello, G.; Boisserie, F.; Hou, J.; et al. A phase 3, randomized, open-label, multicenter study to compare the efficacy and safety of tislelizumab, an anti-PD-1 antibody, versus sorafenib as first-line treatment in patients with advanced hepatocellular carcinoma. *J. Clin. Oncol.* **2018**, *36* (Suppl. 15), TPS3110. [CrossRef]
58. Xu, F.; Jin, T.; Zhu, Y.; Dai, C. Immune checkpoint therapy in liver cancer. *J. Exp. Clin. Cancer Res.* **2018**, *37*, 110. [CrossRef]
59. Wallin, J.J.; Bendell, J.C.; Funke, R.; Sznol, M.; Korski, K.; Jones, S.; Hernandez, G.; Mier, J.; He, X.; Hodi, F.S.; et al. Atezolizumab in combination with bevacizumab enhances antigen-specific T-cell migration in metastatic renal cell carcinoma. *Nat. Commun.* **2016**, *7*, 12624. [CrossRef]
60. Pishvaian, M.J.; Lee, M.S.; Ryoo, B.-Y.; Stein, S.; Lee, K.-H.; Verret, W.; Spahn, J.; Shao, H.; Liu, B.; Iizuka, K.; et al. Updated safety and clinical activity results from a phase Ib study of atezolizumab + bevacizumab in hepatocellular carcinoma (HCC). *Ann. Oncol.* **2018**, *29* (Suppl. 8), viii718–viii719. [CrossRef]
61. Finn, R.S.; Qin, S.; Ikeda, M.; Galle, P.R. Atezolizumab plus Bevacizumab in Unresectable Hepatocellular Carcinoma. *N. Engl. J. Med.* **2020**, *382*, 1894–1905. [CrossRef] [PubMed]
62. Galle, P.R.; Finn, R.S.; Qin, S.; Ikeda, M.; Zhu, A.X.; Kim, T.-Y.; Kudo, M.; Breder, V.V.; Merle, P.; Kaseb, A.O.; et al. Patient-reported outcomes from the phase III IMbrave150 trial of atezolizumab plus bevacizumab vs sorafenib as first-line treatment for patients with unresectable hepatocellular carcinoma. *J. Clin. Oncol.* **2020**, *38*. [CrossRef]
63. Roche Submits Supplemental Biologics License Application to the FDA for Tecentriq in Combination with Avastin for the Most Common form of Liver Cancer [news release]. 2020. Available online: https://bit.ly/3aP84gz (accessed on 27 January 2020).
64. FDA. FDA Approves Atezolizumab Plus Bevacizumab for Unresectable Hepatocellular Carcinoma [News Release]. 2020. Available online: https://www.fda.gov/drugs/drug-approvals-and-databases/fda-approves-atezolizumab-plus-bevacizumab-unresectable-hepatocellular-carcinoma (accessed on 16 September 2020).
65. Fu, Y.; Lin, Q.; Zhang, Z.; Zhang, L. Therapeutic strategies for the costimulatory molecule OX40 in T-cell-mediated immunity. *Acta Pharm. Sin. B* **2020**, *10*, 414–433. [CrossRef] [PubMed]
66. Ikeda, M.; Sung, M.W.; Kudo, M.; Kobayashi, M.; Baron, A.D.; Finn, R.S.; Kaneko, S.; Zhu, A.X.; Kubota, T.; Kraljevic, S.; et al. A phase 1b trial of lenvatinib (LEN) plus pembrolizumab (PEM) in patients (pts) with unresectable hepatocellular carcinoma (uHCC). *J. Clin. Oncol.* **2018**, *36* (Suppl. 15). [CrossRef]
67. Llovet, J.M.; Kudo, M.; Cheng, A.-L.; Finn, R.S.; Galle, P.R.; Kaneko, S.; Meyer, T.; Qin, S.; Dutcus, C.E.; Chen, E.; et al. Lenvatinib (len) plus pembrolizumab (pembro) for the firstline treatment of patients (pts) with advanced hepatocellular carcinoma (HCC): Phase 3 LEAP-002 study. *J. Clin. Oncol.* **2019**, *37*, TPS4152. [CrossRef]
68. Kudo, M.; Motomura, K.; Wada, Y.; Inaba, Y.; Sakamoto, Y.; Kurosaki, M.; Umeyama, Y.; Kamei, Y.; Yoshimitsu, J.; Fujii, Y.; et al. First-line avelumab + axitinib in patients with advanced hepatocellular carcinoma: Results from a phase 1b trial (VEGF Liver 100). *J. Clin. Oncol.* **2019**, *37* (Suppl. 15), 4072. [CrossRef]
69. Qin, S.; Chen, Z.; Liu, Y.; Xiong, J.; Ren, Z.; Meng, Z.; Gu, S.; Wang, L.; Zou, J. A phase II study of anti–PD-1 antibody camrelizumab plus FOLFOX4 or GEMOX systemic chemotherapy as first-line therapy for advanced hepatocellular carcinoma or biliary tract cancer. *J. Clin. Oncol.* **2019**, *37* (Suppl. 15), 4074. [CrossRef]
70. Giannini, E.G.; Aglitti, A.; Borzio, M.; Gambato, M.; Guarino, M.; Iavarone, M.; Lai, Q.; Sandri, G.B.L.; Melandro, F.; Morisco, F.; et al. Overview of immune checkpoint inhibitors therapy for hepatocellular carcinoma, and the ITA. LI. *Cancers* **2019**, *11*, 1689. [CrossRef]
71. Kelley, R.K.; Abou-Alfa, G.K.; Bendell, J.C.; Kim, T.-Y.; Borad, M.J.; Yong, W.-P.; Morse, M.; Kang, Y.-K.; Rebelatto, M.; Makowsky, M.; et al. Phase I/II study of durvalumab and tremelimumab in patients with unresectable hepatocellular carcinoma (HCC): Phase I safety and efficacy analyses. *J. Clin. Oncol.* **2017**, *35*. [CrossRef]
72. Abou-Alfa, G.K.; Chan, S.L.; Furuse, J.; Galle, P.R.; Kelley, R.K.; Qin, S.; Armstrong, J.; Darilay, A.; Vlahovic, G.; Negro, A.; et al. A randomized, multicenter phase 3 study of durvalumab (D) and tremelimumab (T) as first-line treatment in patients with unresectable hepatocellular carcinoma (HCC): HIMALAYA study. *J. Clin. Oncol.* **2018**, *36*, TPS4144. [CrossRef]

73. AstraZeneca. Imfinzi and Tremelimumab Granted Orphan Drug Designation in the US for Liver Cancer [News Release]. 2020. Available online: https://www.astrazeneca.com/media-centre/press-releases/2020/imfinzi-and-tremelimumab-granted-orphan-drug-designation-in-the-us-for-liver-cancer-20012020.html (accessed on 21 January 2020).
74. Ribas, A.; Wolchok, J.D. Cancer immunotherapy using checkpoint blockade. *Science* 2018, *359*, 1350–1355. [CrossRef] [PubMed]
75. Finn, R.S.; Ryoo, B.-Y.; Merle, P.; Kudo, M.; Bouattour, M.; Lim, H.Y.; Breder, V.; Edeline, J.; Chao, Y.; Ogasawara, S.; et al. Pembrolizumab As Second-Line Therapy in Patients With Advanced Hepatocellular Carcinoma in KEYNOTE-240, A Randomized, Double-Blind, Phase III Trial. *J. Clin. Oncol.* 2020, *38*, 193–202. [CrossRef] [PubMed]
76. Huang, J.; Mo, H.; Wu, D.; Chen, X.; Ma, L.; Lan, B.; Qu, D.; Yang, Q.; Xu, B. Phase I study of the anti-PD-1 antibody SHR-1210 in patients with advanced solid tumors. *J. Clin. Oncol.* 2017, *35*, e15572. [CrossRef]
77. Qin, S.; Ren, Z.; Meng, Z.; Chen, Z.; Chai, X.; Xiong, J.; Bai, Y.; Yang, L.; Zhu, H.; Fang, W.; et al. A randomized multicentered phase II study to evaluate SHR-1210 (PD-1 antibody) in subjects with advanced hepatocellular carcinoma (HCC) who failed or intolerable to prior systemic treatment. *Ann. Oncol.* 2018, *29*, viii719–viii720. [CrossRef]
78. Wainberg, Z.A.; Segal, N.H.; Jaeger, D.; Lee, K.-H.; Marshall, J.; Antonia, S.J.; Butler, M.; Sanborn, R.E.; Nemunaitis, J.; Carlson, C.A.; et al. Safety and clinical activity of durvalumab monotherapy in patients with hepatocellular carcinoma (HCC). *J. Clin. Oncol.* 2017, *35*, 4071. [CrossRef]
79. Busato, D.; Mossenta, M.; Baboci, L.; Di Cintio, F.; Toffoli, G.; Bo, M.D. Novel immunotherapeutic approaches for hepatocellular carcinoma treatment. *Expert Rev. Clin. Pharmacol.* 2019, *12*, 453–470. [CrossRef] [PubMed]
80. Hato, T.; Goyal, L.; Greten, T.F.; Duda, D.G.; Zhu, A.X. Immune checkpoint blockade in hepatocellular carcinoma: Current progress and future directions. *Hepatology* 2014, *60*, 1776–1782. [CrossRef]
81. Schneider, H.; Downey, J.; Smith, A.; Zinselmeyer, B.H.; Rush, C.; Brewer, J.M.; Wei, B.; Hogg, N.; Garside, P.; Rudd, C.E. Reversal of the TCR stop signal by CTLA-4. *Science* 2006, *313*, 1972–1975. [CrossRef]
82. Mizukoshi, E.; Nakamoto, Y.; Arai, K.; Yamashita, T.; Sakai, A.; Takamura, T.; Kaneko, S. Comparative analysis of various tumor-associated antigen-specific t-cell responses in patients with hepatocellular carcinoma. *Hepatology* 2011, *53*, 1206–1216. [CrossRef]
83. Sangro, B.; Gomez-Martin, C.; De La Mata, M.; Iñarrairaegui, M.; Garralda, E.; Barrera, P.; Riezu-Boj, J.I.; Larrea, E.; Alfaro, C.; Sarobe, P.; et al. A clinical trial of CTLA-4 blockade with tremelimumab in patients with hepatocellular carcinoma and chronic hepatitis C. *J. Hepatol.* 2013, *59*, 81–88. [CrossRef]
84. Yau, T.; Kang, Y.-K.; Kim, T.-Y.; El-Khoueiry, A.B.; Santoro, A.; Sangro, B.; Melero, I.; Kudo, M.; Hou, M.-M.; Matilla, A.; et al. Nivolumab (NIVO) + ipilimumab (IPI) combination therapy in patients (pts) with advanced hepatocellular carcinoma (aHCC): Results from CheckMate 040. *J. Clin. Oncol.* 2019, 37. [CrossRef]
85. Sangro, B.; He, A.R.; Yau, T.; Hsu, C.; Kang, Y.-K.; Kim, T.-Y.; Santoro, A.; Melero, I.; Kudo, M.; Ho, M.-M.; et al. Nivolumab + ipilimumab combination therapy in patients with advanced hepatocellular carcinoma: Subgroup analysis from CheckMate 040. In Proceedings of the 2020 Gastrointestinal Cancers Symposium, San Francisco, CA, USA, 13–15 February 2020.
86. Bristol Myers Squibb Company. U.S. *Food and Drug Administration Approves Opdivo (nivolumab) + Yervoy (ipilimumab) for Patients with Hepatocellular Carcinoma (HCC) Previously Treated with Sorafenib [News Release]*; Bristol Myers Squibb Company: Princeton, NJ, USA, 2020.
87. Xu, J.; Zhang, Y.; Jia, R.; Yue, C.Y.; Chang, L.; Liu, R.-R.; Zhang, G.; Zhao, C.H.; Zhang, Y.Y.; Chen, C.X.; et al. Anti-PD-1 antibody SHR-1210 combined with apatinib for advanced hepatocellular carcinoma, gastric, or esophagogastric junction cancer: An open-label, dose escalation and expansion study. *Clin. Cancer Res.* 2019, *25*, 515–523. [CrossRef]
88. Nishida, N.; Kudo, M. Immune checkpoint blockade for the treatment of human hepatocellular carcinoma. *Hepatol. Res.* 2018, *48*, 622–634. [CrossRef] [PubMed]
89. Koyama, S.; Akbay, E.A.; Li, Y.Y.; Herter-Sprie, G.S.; Buczkowski, K.A.; Richards, W.G.; Gandhi, L.; Redig, A.J.; Rodig, S.J.; Asahina, H.; et al. Adaptive resistance to therapeutic PD-1 blockade is associated with upregulation of alternative immune checkpoints. *Nat. Commun.* 2016, *7*, 10501. [CrossRef] [PubMed]
90. Peng, W.; Chen, J.Q.; Liu, C.; Malu, S.; Creasy, C.; Tetzlaff, M.; Xu, C.; McKenzie, J.; Zhang, C.; Liang, X.; et al. Loss of PTEN Promotes Resistance to T Cell-Mediated Immunotherapy. *Cancer Discov.* 2016, *6*, 202–216. [CrossRef]

91. Gopalakrishnan, V.; Spencer, C.N.; Nezi, L.; Reuben, A.; Andrews, M.C.; Karpinets, T.V.; Prieto, P.A.; Vicente, D.; Hoffman, K.; Wei, S.C.; et al. Gut microbiome modulates response to anti-PD-1 immunotherapy in melanoma patients. *Science* **2018**, *359*, 97–103. [CrossRef]
92. Ma, C.; Han, M.; Heinrich, B.; Fu, Q.; Zhang, Q.; Sandhu, M.; Agdashian, D.; Terabe, M.; Berzofsky, J.A.; Fako, V.; et al. Gut microbiome-mediated bile acid metabolism regulates liver cancer via NKT cells. *Science* **2018**, *360*, 858. [CrossRef] [PubMed]
93. Zheng, Y.; Wang, T.; Tu, X.; Huang, Y.; Zhang, H.; Tan, D.; Jiang, W.; Cai, S.; Zhao, P.; Song, R.; et al. Gut microbiome affects the response to anti-PD-1 immunotherapy in patients with hepatocellular carcinoma. *J. Immunother. Cancer* **2019**, *7*, 193. [CrossRef] [PubMed]
94. Jung, H.I.; Jeong, D.; Ji, S.; Ahn, T.S.; Bae, S.H.; Chin, S.; Chung, J.C.; Kim, H.C.; Lee, M.S.; Baek, M.-J. Overexpression of PD-L1 and PD-L2 is associated with poor prognosis in patients with hepatocellular carcinoma. *Cancer Res. Treat.* **2017**, *49*, 246–254. [CrossRef]
95. Alexandrov, L.B.; Initiative, A.P.C.G.; Nik-Zainal, S.; Wedge, D.C.; Aparicio, S.A.J.R.; Behjati, S.; Biankin, A.V.; Bignell, G.R.; Bolli, N.; Borg, A.; et al. Signatures of mutational processes in human cancer. *Nature* **2013**, *500*, 415–421. [CrossRef] [PubMed]
96. Totoki, Y.; Tatsuno, K.; Covington, K.R.; Ueda, H.; Creighton, C.J.; Kato, M.; Tsuji, S.; A Donehower, L.; Slagle, B.L.; Nakamura, H.; et al. Trans-ancestry mutational landscape of hepatocellular carcinoma genomes. *Nat. Genet.* **2014**, *46*, 1267–1273. [CrossRef] [PubMed]
97. Yarchoan, M.; Hopkins, A.; Jaffee, E.M. Tumor mutational burden and response rate to PD-1 inhibition. *N. Engl. J. Med.* **2017**, *377*, 2500–2501. [CrossRef]
98. Ma, K.; Jin, Q.; Wang, M.; Li, X.; Zhang, Y. Research progress and clinical application of predictive biomarker for immune checkpoint inhibitors. *Expert Rev. Mol. Diagn.* **2019**, *19*, 517–529. [CrossRef] [PubMed]
99. Zucman-Rossi, J.; Villanueva, A.; Nault, J.-C.; Llovet, J.M. Genetic landscape and biomarkers of hepatocellular carcinoma. *Gastroenterology* **2015**, *149*, 1226–1239.e4. [CrossRef] [PubMed]

Publisher's Note: MDPI stays neutral with regard to jurisdictional claims in published maps and institutional affiliations.

© 2020 by the authors. Licensee MDPI, Basel, Switzerland. This article is an open access article distributed under the terms and conditions of the Creative Commons Attribution (CC BY) license (http://creativecommons.org/licenses/by/4.0/).

Review

Translational Considerations to Improve Response and Overcome Therapy Resistance in Immunotherapy for Hepatocellular Carcinoma

Sophia Heinrich [1,2,†], **Darko Castven** [2,3,†], **Peter R. Galle** [4,*] **and Jens U. Marquardt** [2,3,*]

1. Laboratory of Human Carcinogenesis, Liver Carcinogenesis Section, Center for Cancer Research, National Cancer Institute, National Institutes of Health, Bethesda, MD 20892, USA; sophia.franck@nih.gov
2. Department of Medicine I, Lichtenberg Research Group for Molecular Hepatocarcinogenesis, University Medical Center, 55131 Mainz, Germany; darko.castven@uksh.de
3. Lichtenberg Research Group for Molecular Hepatocarcinogenesis, Department of Medicine I, University Medical Center Schleswig Holstein, 23538 Luebeck, Germany
4. Department of Medicine I, University Medical Center, 55131 Mainz, Germany
* Correspondence: peter.galle@unimedizin-mainz.de (P.R.G.); Jens.Marquardt@uksh.de (J.U.M.); Tel.: +49-06131-17-7275 (P.R.G.); +49-0451-500-44101 (J.U.M.); Fax: +49-06131-17-5595 (P.R.G.); +49-0451-500-44104 (J.U.M.)
† These authors contributed equally to this work.

Received: 4 August 2020; Accepted: 31 August 2020; Published: 3 September 2020

Simple Summary: Immunotherapeutic approaches became a promising treatment option and an intensive field of research in liver cancer. Despite promising results in preclinical studies, only moderate response rates have been reported in phase III clinical trials and predictive biomarkers are still missing. Therefore, translational considerations are important to overcome resistance to immunotherapy. This article reviews potential predictors for response to immunotherapy in hepatocellular carcinoma (HCC) as well as potential mechanisms for therapy resistance. Further, we will discuss translational considerations to overcome therapy resistance in HCC and improve overall response rates.

Abstract: Over the last decade, progress in systemic therapies significantly improved the outcome of primary liver cancer. More recently, precision oncological and immunotherapeutic approaches became the focus of intense scientific and clinical research. Herein, preclinical studies showed promising results with high response rates and improvement of overall survival. However, results of phase III clinical trials revealed that only a subfraction of hepatocellular carcinoma (HCC) patients respond to therapy and display only moderate objective response rates. Further, predictive molecular characteristics are largely missing. In consequence, suitable trial design has emerged as a crucial factor for the success of a novel compound. In addition, increasing knowledge from translational studies indicate the importance of targeting the tumor immune environment to overcome resistance to immunotherapy. Thus, combination of different immunotherapies with other treatment modalities including antibodies, tyrosine kinase inhibitors, or local therapies is highly promising. However, the mechanisms of failure to respond to immunotherapy in liver cancer are still not fully understood and the modulation of the immune system and cellular tumor composition is particularly relevant in this context. Altogether, it is increasingly clear that tailoring of immunotherapy and individualized approaches are required to improve efficacy and patient outcome in liver cancer. This review provides an overview of the current knowledge as well as translational considerations to overcome therapy resistance in immunotherapy of primary liver cancer.

Keywords: hepatocellular carcinoma; immunotherapy; translational approaches; combination therapies; therapy resistance

1. Introduction

Primary liver cancer, in particular hepatocellular carcinoma (HCC) ranks among the most common malignancies worldwide with a rising incidence in the Western world [1–4]. Between 80–90% of HCC cases develop in an inflammation-associated milieu [5], i.e., on the background of a pre-existing chronic liver disease and, most commonly, an advanced fibrosis or cirrhosis. Due to demographic changes in the distribution of diabetes mellitus type II and obesity, non-alcoholic fatty liver disease, or steatohepatitis (non-alcoholic fatty liver disease (NAFLD)/non-alcoholic steatohepatitis (NASH)) show a sharp increase in HCC numbers [6] and are considered as metabolic predispositions to liver cancer [7,8]. Numerous immune suppressor mechanisms that involve different immune cell types lead to immune evasion of the tumor and have been shown to contribute to HCC initiation and progression [9,10].

Despite well known risk factors, i.e., chronic viral hepatitis, alcohol consumption, and metabolic syndrome, the majority of HCC patients are diagnosed in late, non-resectable, and non-curative stages of the disease, when a considerable phenotypic and molecular heterogeneity renders HCC highly resistant to conventional chemotherapy and/or irradiation [11]. Until 2016, only limited systemic treatment options were available in advanced stages of HCC, namely sorafenib and regorafenib, tyrosine kinase inhibitors (TKI) [12–14]. Since then, only Lenvatinib (first-line), regorafenib, cabozantinib, all TKIs, and ramucirumab (second-line), a monoclonal antibody against VEGFR, have shown efficacy in phase III clinical trials [13,15–17]. Despite the approval of new and targeted therapy, patients' prognosis remained limited to 12–13 months in first-line and 9–11 months in second-line therapy, and besides alpha-fetoprotein (AFP), there is no biomarker available for patient stratification [18].

Given the inflammatory background of HCC, the hepatic tumor microenvironment (TME) plays a pivotal role in tumor initiation, modulation of tumor invasiveness, metastatic spread as well as tumor suppression and immune surveillance of cancer cells [19]. Therefore, modern therapeutic approaches that focus on modulation of the TME are particularly promising.

The liver is an immune tolerant organ due to its prominent role in protection against inappropriate immune responses. The inflammatory stimuli emerge as a consequence to exposition with major inflammatory processes mediated by a large antigenic load from the gastrointestinal tract trough blood from the portal vein [20]. In addition, the setting of a chronic liver inflammation or cirrhosis further reinforces the hepatic immune tolerance [21]. On a single cell level, it has been demonstrated that HCCs show a higher abundance of regulatory T cells (T_{regs}) as well as their local clonal expansion within the tumor. Furthermore, a higher abundance of exhausted CD8 T cells is present in the tumor tissue [22]. This has a significant influence on tumor surveillance. Decreased number of tumor attacking immune cells such as T effector cells and more tumor supporting cells, e.g., MDSCs and T_{regs} lead to a disruption of the cellular composition during chronic liver diseases and is associated with patient outcome [23–28]. During hepatocarcinogenesis, several immunosuppressive effects have been detected that are associated with patient survival. Immune cell composition leading to anti-tumor immunity or tolerance is crucial for tumor growth or cell death. T_{regs} as well as myeloid derived suppressor cells (MDSC) accumulate in the liver and suppress antitumor immunity in HCC [9,29]. Macrophages, in the liver called Kupffer cells, suppress early HCC development; however, undergo a switch from M1 to M2 during tumor progression, which leads to a suppression of the adaptive immune system and support of the tumor [10,30–33]. Tumor associated macrophages (TAM) represent the predominant component of the innate immune system and promote tumor proliferation, angiogenesis and invasion [34,35] Furthermore, parenchymal cells such as endothelial cells, hepatic stellate cells (HSC), and hepatocytes influence effector functions of infiltrating lymphocytes [21]. This leads to an intratumoral loss of cytotoxic T cells, which is associated with tumor progression [21,35,36]. Natural killer (NK) cell, important players of innate immunity in the liver, show an impaired function in HCCs [29,37]. This dysfunctional and imbalanced immune system is a hallmark of cancer progression in HCC and is associated with patient prognosis. [38,39]

After the approval of immune checkpoint inhibitors (ICI) in melanoma and non-small cell lung cancer (NSCLC), immunotherapies have raised significant interest in other solid tumors including HCC. In 2017 and 2018, the FDA granted accelerated approval for the first immunotherapy agents, nivolumab and pembrolizumab or the combination of nivolumab and ipilimumab, for patients with advanced HCC after progression under sorafenib after promising results from phase II clinical trials [40–42]. Other checkpoint inhibitors are currently being investigated in clinical trials as single agents as well as in combination therapies [42–45]. A detailed list of currently approved immunotherapeutic agents can be found in Table 1. Nevertheless, immunotherapy in liver cancer has been challenging. Objective response rates are still low. Given the fact that only some patients respond to therapy, the various degrees of side effects such as autoimmune reactions need to be taken into account [40,46,47]. Thus, predictive biomarkers are urgently needed. Furthermore, there are no long-term data for those patients responding to therapy and even though there are some studies addressing a neoadjuvant treatment option, we do not have any strong data in curative settings yet. However, first results from combination therapies show a significant improvement in all clinical endpoints including overall survival and quality of life, which raises optimism for the future of this approach in primary liver cancer [48]. Even scenarios in adjuvant or neoadjuvant use are now under current discussion [49,50], but our overall understanding of the treatment response remains limited.

Table 1. Currently approved immunotherapy in hepatocellular carcinoma (HCC)

Target Molecule	Drug Name	Company
PD-1	Nivolumab	Bristol Meyer Squibb
PD-1	Pembrolizumab	Merck
PD-L1	Atezolizumab (in combination with bevacizumab)	Roche
CTLA-4	Ipilimumab	Bristol Meyer Squibb/Medarex

Abbreviations: PD-1 (programmed cell death protein 1), PD-L1 (programmed cell death ligand 1), CTLA-4 (cytotoxic T-lymphcyte-associatet protein 4).

Given the success of immunotherapy in several tumor entities, we here review the potential predictors for response to immunotherapy in HCC. In addition, we are addressing potential mechanisms for therapy resistance. Finally, we discuss translational considerations to overcome therapy resistance in HCC.

2. General Strategies for Immunomodulatory Treatments in Primary Liver Cancer

There are different strategies to induce antitumor immune response that are currently under investigation in primary liver cancer involving both innate and adaptive immune systems. Specifically, targeting of checkpoint molecules as well as the interaction of T cells and antigen-presenting cells (APCs) have been of interest in recent years [51]. Neoantigens expressed on the tumor itself can also be used as targets for immunotherapy [52]. Local therapies and oncolytic viruses can promote neoantigen release even more, thereby further enhancing the antitumor immune response [53,54]. In addition, detailed information on tumor neoantigens can be explored to develop anti-tumor vaccines and autologous T cells can be manipulated and/or stimulated ex vivo before retransfer, e.g., chimeric antigen receptor (CAR) T cells or cytokine-induced killer cells (Figure 1) [55,56].

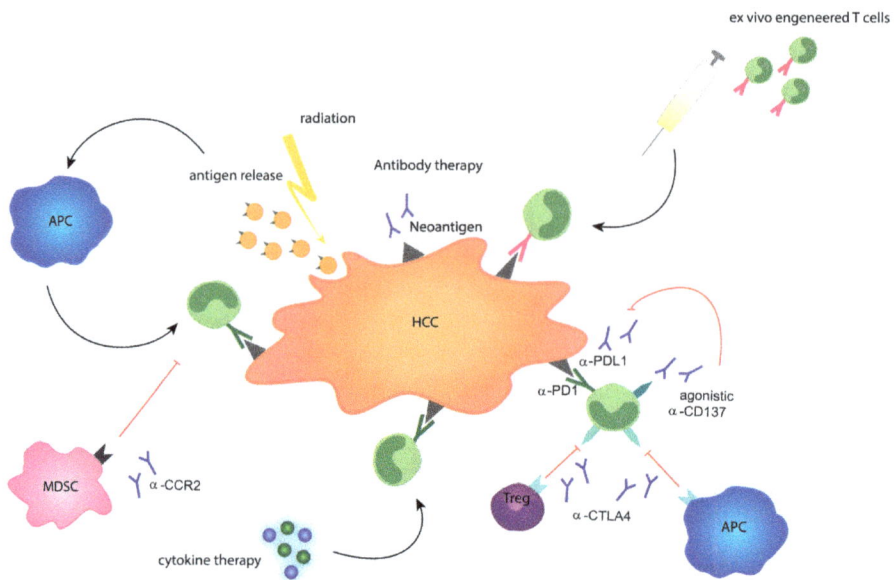

Figure 1. Overview of therapeutic approaches in immunotherapy. Targeted antibody therapy can block inhibitory signals, e.g., CTLA-4 or PD-1 and "unleash" anti-tumor immunity or block immunosuppressive mechanisms of the adaptive as well as the innate immune system. Besides a direct anti tumoral effect, irradiation leads to an antigen release that promotes antigen presentation by APCs and enhances anti-tumoral T cell response. Cytokine therapy is an option to enhance a general T cell response in the tumor. Ex-vivo engineered T cells or antibodies against tumor specific neoantigens induce a targeted anti-tumor response. Abbreviations: Myeloid derived suppressor cell (MDSC), antigen presenting cell (APC), regulatory T cell (Treg), cytotoxic T lymphocyte antigen 4 (CTLA4), programmed death protein (PD1), programmed death ligand 1 (PDL1).

However, it is well known that immune escape and evasion of immune-mediated cytotoxicity are among the hallmarks of cancers and are often mediated by induction of an immunosuppressive microenvironment [57,58]. To overcome escape from immunosurveillance by cancer cells, therapeutic approaches focus on boosting antitumor response either by activation of cytotoxic immune cells or elimination of immune-suppressing cells. Furthermore, tumors also evade from the immune system by upregulation of programmed cell death ligand 1 (PD-L1) on cancer cells. Tumor immune cell interactions are based on two phases of T cell activation: an early priming phase in the lymph node and an effector phase in the tumor tissue. Involved in this process are APCs, that bind cancer antigens, migrate to the lymph node, and activate immature T cells. Activation of T cells in the priming phase can be blocked by upregulation of the checkpoint molecule cytotoxic T lymphocyte antigen 4 (CTLA-4) on T cells. CTLA-4 is also highly expressed on T_{regs} that inhibit antigen presentation on dendritic cells (DC). This is a cycle, that leads to less cytotoxic, more exhausted T cells and, thus, impaired anti-tumor response. Activation of T cells in the effector phase can be blocked by programmed death protein 1 (PD-1)/ programmed death ligand 1(PD-L1) that is expressed in tumor cell interaction. Both "breaks" can be effectively released by anti-PD-1, anti PD-L1, or anti CTLA-4 therapy and enhance anti-tumor immune response (Figure 2) [40,43,59,60].

In HCC, immunotherapy is an intensively studied field encompassing all the above mentioned antibody-based, cell-based, and vaccine-based treatment options [61]. In addition, the combination of different therapy regimes may provide a significant benefit (Figure 1) [43].

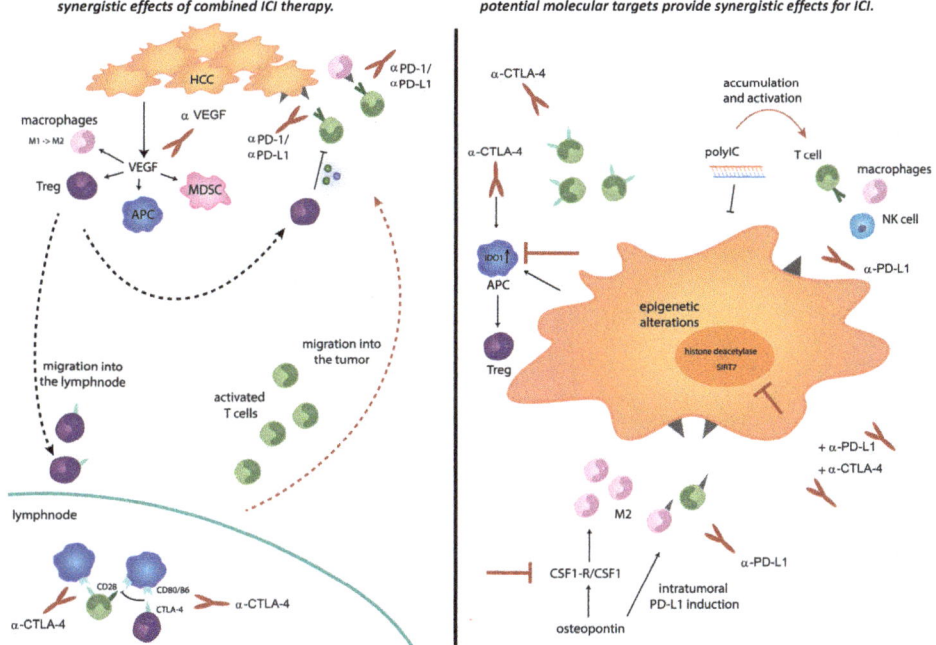

Figure 2. Translational consideration to overcome therapy resistance. Therapeutic approaches for sensitization to immunotherapy. **Left panel**: Anti-CTLA-4 or anti-angiogenic therapy increases recruitment and migration of activated T cells into the tumor. Anti-PD-1/anti PD-L1 therapy enhances cytolytic activity of T cells. **Right panel**: Anti-CTLA-4 treatment induced IDO1 expression in dendritic cells (DC). Indoleamine 2,3-dioxygenase 1 (IDO1) leads to activation of T_{regs} and causes resistance to anti-CTLA-4 therapy, while blocking of IDO could interrupt this mechanism of resistance. PolyIC inhibits tumor growth and leads to an accumulation and activation of immune cell subsets, whereas anti-PD-L1 therapy could provide synergistic effects. Osteopontin induces M2 migration into the tumor as well as PD-L1 induction. Targeted therapy of CSF1 in combination with anti-PD-L1 therapy might provide synergistic effects. Epigenetic regulations as synergistic effect for ICI therapy.

3. Predictors for Response or Resistance to Current Immune-Modulatory Therapies

Immunotherapy as a modern approach for cancer treatment has become a key topic in translational research over the last decade. After approval of the first PD-1/PD-L1 and CTLA-4 blocking antibodies for melanoma, checkpoint inhibitors are under intense investigation in many tumor entities. Unleashing of the immune system to attack the tumor seems to be an effective anti-tumor treatment. Many immunotherapies have been shown to be effective as monotherapies but also in combination with other immune-based and targeted approaches in preclinical and clinical studies [62–66]. However, despite good clinical efficacy in other tumor entities, response rates in HCC as well as cholangiocarcinoma are surprisingly low [40,67–70]. A common observation in HCC is the missing significance or lack of surrogate markers of response utilized in other entities. Thus, improved strategies to estimate therapy response would enable to stratify patients according to their clinical benefit and prevent unnecessary side effects caused by the therapy [40,70–73].

Mechanisms of resistance to immunotherapy are still not fully understood. Especially in the context of a possible pseudoprogression or even hyperprogression under immunotherapy, predictive biomarkers are urgently needed [74].

3.1. Tumor Characteristics and Tumor Infiltrating Lymphocytes as Predictors for Response

Multiple studies revealed potential molecular characteristics that are associated with immunotherapy response. However, up to now, no biomarker for HCC has been prospectively validated in authentic human patients. The most prominent biomarkers are PD-1 and PD-L1 expression on tumor tissue as well as on infiltrating immune cells.

Expression of PD-1/PD-L1 in HCC have been described in 17% (PD-L1) and 27% (PD-1) on immune and 10–20% (PD-L1) on tumor cells, using immunohistochemistry [40,41,75–77]. High PD-L1 expression in tumors itself is associated with more aggressive HCCs independent of immunotherapy [76,78].

Several translational studies investigated numbers of immune cells and respective activation of checkpoint molecules as possible biomarkers for immunotherapy response in HCC. In other entities such as NSCLC, PD-1^{high} T cells showed a higher capacity for tumor recognition, recruit other immune cells, and are predictive for response and overall survival under PD-1 therapy, which demonstrates that a distinct T cell subtype is needed for response to PD-1/PD-L1 therapy [79]. In HCC patients, high PD-1 expression in tumor tissue is connected to an exhausted immune cell phenotype with impaired effector function of tumor infiltrating lymphocytes (TIL), which contributes to immune evasion [75,80–82]. A recent study further demonstrated that PD-1, LAG3 (lymphocyte activation gene 3), TIM3 (T cell membrane protein 3), and CTLA-4 positive TILs are exhausted and functionally compromised, thus, induce lower levels of effector cytokines. Conversely, this phenotype could be reversed back to an effector phenotype with ICI [82].

Using sequencing and TCR analysis, another study investigated the distribution of mutational and neoantigen burden in different tumor regions as a possible driver for immune cell heterogeneity. Analysis of peptide binding affinity of these neoantigens revealed a correlation of the higher ones with TIL heterogeneity. However, the region with the highest TIL heterogeneity showed the lowest putatively immunogenic neoepitopes, suggesting that the adaptive immune response has edited the tumor to be less immunogenic [83].

Another study stratified HCC patients into CD8$^+$PD-1^{high} and CD8$^+$PD-1^{low}. A gene signature that effectively predicted anti-PD-1 therapy response in several tumor entities was significantly enriched in corresponding PD-1^{high} expressers [75]. Furthermore, high frequencies of CD14$^+$CD16$^-$HLA-DR^{high} monocytes was shown to predict therapy response in melanoma patients [84]. This phenotype was also elevated in PD-1^{high} expressers [75]. Both findings might provide an indirect surrogate of therapy response in PD-1^{high} HCC patients. Consistently, the PD-1^{high} HCCs also expressed markers such as LAG3 and TIM3 confirming the exhausted phenotype of the cells and delineating the rational of targeting these markers in liver cancer. In vitro experiments could further show that blocking of PD-1 increased IFN production and effectively enhanced the immune response. However, this effect was only present in PD-1^{high} HCCs [75].

Recently, single cell sequencing approaches became affordable and promising tools for translational science. These investigations are ideal to dissecting immune cell populations in the context of the diseased hepatic microenvironment as well as immunotherapies. A recent single cell sequencing analysis demonstrated a complex composition of highly diverse T cell subpopulations in HCC tumors [22]. A subgroup expressing high levels of exhaustion markers such as CTLA4 and PDCD1 was identified that stratified patients according to the clinical outcome [22]. Furthermore, complex composition of immune cells could be revealed and shown to be spatially different between intratumoral regions, extra-tumoral regions, ascites, and the peripheral blood [85]. While modulation of this immune cell contexture could be highly promising in a therapeutic setting, the clinical use of cellular compositions as predictors for therapy response needs to be evaluated.

Only one single cell study focused on the malignant cells in HCC so far. Analysis of the tumor and the TME identified VEGFA^{high} tumors that drive the TME reprogramming [86]. Consequently, further single cell analysis of T cells revealed different transcriptomic profiles in VEGFA^{high} tumors. These observations imply that a combination of vascular endothelial growth factor (VEGF) therapy and immunotherapy might help to overcome some non-response mechanisms.

Overall, results of these preclinical studies suggest that it is probably not enough to screen for widely expressed markers in the tissue and underline the importance of detailed characterization of the cellular compositions to shed light into cellular interactions to reveal context-dependent response mechanisms to immunotherapy.

For objective comparison of PD-L1 expression in clinical trials, mainly two different scoring systems have been established [87,88]. The tumor proportion score (TPS) calculates the percentage of PD-L1 tumor cells of all viable tumor cells, whereas the combined positive score (CPS) calculates the percentage of all PD-L1 positive cells (tumor cells, macrophages, lymphocytes) divided by all viable tumor cells [87,88]. PD-1/PD-L1 expression in tissue is associated with therapy response in melanoma, NSCLC, renal cancer and gastric cancer in large clinical trials [59,87,89–91].

Despite the promising results from the above-mentioned translational studies, explorative investigations performed on patients in clinical trials have failed to identify robust predictive markers that clearly identify patients likely to benefit from immunotherapy in HCC up to now.

Clinical trials for HCC using ICI included both of the mentioned scores to predict response. The CHECKMATE-40 trial, investigating the anti-PD-1 antibody Nivolumab as a second line therapy in HCC reported response rates regardless of PD-L1 expression rates. PD-L1 expression was calculated using the TPS score (overall response rate (ORR) 26% in patients with PD-L1 expression >1% and ORR 19% of patients with PD-L1 expression <1%,). However, PD-L1 expression >1% could only be detected in 20% of the patient population. The lack of robust association indicates that PD-L1 expression on tumor cells cannot be used as a single binary marker for therapy decisions [40]. The phase II clinical trial KEYNOTE-224 used the anti-PD-1 antibody pembrolizumab after progression under sorafenib. Response to therapy was assessed using TPS as well as CPS score [41]. Only CPS score showed significant association with response to therapy. The proportion of CPS score positive patients in the KEYNOTE cohort has been reported as 42% [41]. Although the follow-up phase III study KEYNOTE-240 did not reach its clinical endpoint of OS, knowledge of PD-L1 expression and CPS score can be highly instrumental for future studies and are urgently awaited [70]. Noteworthily, different cutoffs and definitions about PD-L1 positivity have been used in clinical trials, which might have limited the comparability of the findings [92].

High tumor mutational burden (TMB), generally defined as over 10 mutations/mb, or microsatellite instability (MSI) are hypothesized to be intrinsically immunogenic [93]. Hence, TMB or MSI status were predictive for response to therapy with PD-1 checkpoint-inhibitors in several tumor entities [66,94–96]. However, compared to other tumor entities, HCC mainly has a low TMB of <10 mutations/Mb^{low} and MSI rates below 1% [75,96–99]. Given the low prevalence and only limited predictive ability of TMB, it emphasizes the need for more comprehensive molecular biomarkers [97].

Circulating immune cells and corresponding expression of checkpoint molecules have been intensively evaluated as predictive biomarker. Isolation and subsequent characterization would enable a closer and non-invasive therapy monitoring, which is not possible using tissue samples. However, only one study could identify an association of circulating immune cells and response to therapy so far. A higher expression of CD4$^+$PD1$^+$ cells in circulating peripheral blood mononuclear cells (PBMC) at baseline may predict a better response to tremelimumab treatment in HCC patients [100]. However, more recently, results from several clinical trials suggest that induction of a CD8 T cell response after CTLA-4 priming might enhance the anti-tumor efficacy of PD-1 inhibition [46]. This interesting observation should be pursued in future studies.

Furthermore, high soluble PD-L1 levels are associated with a poor prognosis in HCC patients [101]. However, soluble PD-L1 could not be shown to be predictive under immunotherapy in HCC in contrast to other tumor entities [102,103].

Finally, studies have shown that the microbiome influences the immune system. Mice with liver tumors showed a better immune response and lower tumor burden when treated with antibiotics that reduced the overall bacterial burden in the gut but favor *Clostridium scindens*. Reduction of bacteria through antibiotics alters the composition of bile acids, which subsequently resulted in increased infiltration of NKT cells with anti-tumor function into the liver. On the other hand, gut microbiota has been shown to promote obesity-associated liver cancer by driving prostaglandin E2 (PGE2) production through higher expression of COX2. PGE2 eventually suppressed antitumor immunity and resulted in higher tumor burden of obesity-driven HCC [104]. Several studies have shown that the microbiome influences not only immune cells but also the efficiency of immunotherapy. Anti-PD1 therapy could be significantly improved by combing it with oral administration of *Bifidobacterium*, which resulted in reduced tumor growth of B16.SIY melanoma tumors [105]. Another study found the fecal transplantation of *Akkermansia muciniphia* can restore efficacy of anti-PD-1 immunotherapy, which was mediated by increasing the recruitment of $CCR9^+$ $CXCR3^+$ $CD4^+$ T lymphocytes [105]. In human melanoma, anti-CTLA-4 therapy was associated with outgrowth of *Bacteroides fragilis*. Oral feeding of *Bacteroides fragilis* in germ-free mice resulted in restored therapeutic response to anti-CTLA-4 treatment [106]. Notably, a recent study focused on fecal samples from patients under immunotherapy as a predictive parameter and revealed a higher species richness in responding patients than in non-responders [107]. Furthermore, other studies suggest an association between commensal microbial composition and therapy response to immune therapy treatment in melanoma as well as HCC, whereas patient numbers were very limited ($N = 8$) [107,108]. Thus, data on the microbiome should be assessed as adjuvant information in future studies to identify its potential as a biomarker [109]. Data is mixed but it is clear that the composition of bacteria in the gut has influence and might predict response to immunotherapy and cannot be neglected. Sample acquisition in a hospitalized setting seems easy so that specifically response assessment and subsequent alteration of the treatment strategy based on the microbiome status seems to be reasonable.

Overall, while not yet conclusive in HCC, these findings provide the first mechanistic explanations of tumor cell biodiversity and why some patients may respond to therapy and others do not [86].

3.2. Molecular Subtyping of HCC

In the past, exome sequencing enabled a precise description of the mutational landscape in HCC including the identification of the most relevant oncogenic drivers (TERT, TP53, CTNNB1, AXIN1, ARID1A and ARID2) [18,110,111]. In 28% of all HCCs, potential targetable mutations were identified [112]. However, despite strong efforts, none of these potential biomarkers showed a significant survival benefit and could be implemented in clinical trials [18].

Analysis of the immune composition as well as the transcriptomic profile in HCC lead to the classification of inflamed "hot" tumors and non-inflamed "cold" tumors based on the presence of T cells, macrophages, B cells, PD1 signaling, and cytotoxic cytokines. Interestingly, "cold" tumors co-occur with WNT/CTNNB1 as well as chromosomal alterations of the tumor [18,111,113].

A retrospective analysis of genomic alterations of HCC patients undergoing immunotherapy revealed WNT1/CTNNB1 mutations to be associated with lower disease control rates (0% vs. 53%), shorter median progression free survival (PFS) (2.0 vs. 7.4 months), and shorter median OS [18,114]. This possible CTNNB1 immune excluded class could recently be confirmed in a translational mouse model [114,115]. Upregulation of β-catenin leads to an immune exclusion of the tumor and also resistance to anti-PD-1 therapy. These results conclusively illustrate, that other therapy modalities might be more suitable for cold or immune excluded and, potentially, other subclasses of HCC, and challenges the design of recent clinical investigations. In this context, molecular stratification of patients will become increasingly important and should be mandatory for future clinical trials.

4. Combination Treatments to Improve Therapy Response in HCC

4.1. Combination Therapies of Checkpoint Inhibitors

Given that the response to immunotherapy is restricted to 15–30% of the patients, the majority of the patients are not objectively responding or show a primary resistance to ICI. After initial studies on effectiveness of immunomodulatory drugs, new studies are focusing on mechanisms to increase therapy response [42]. The rationale behind combinations therapies is based on synergistic effects by CTLA4 induction followed by PD-1/PD-L1 blockade (Figure 2). Combination of different ICI blocks immune cell activation at different steps in their activation process. CTLA4 increases CD8 T cell activation in the priming phase in the lymph node as well as CD8 cell infiltration into the tumor. This enhances the effect of PD-1/PD-L1 blockade in the tumor microenvironment. The number of pretreatment or treatment induced intratumoral T cell infiltration correlates with clinical response to therapy, which emphasizes that the crucial factor for response to immunotherapy lies in releasing tumor-specific T cells [116].

The combination of checkpoint inhibitors antiPD-1/anti-PD-L1 plus anti-CTLA-4 antibodies have shown promising response rates of 40–60% in melanoma, NSCLC, and renal cancer [132–134]. Based on this, combination therapies of CTLA-4 and PD-1/PD-L1 blockade are currently under investigation [42].

In HCC, these combinations are also being actively pursed in clinical trials [45,135]. ORR rates for advanced non-resectable HCC in a phase II clinical trial (durvalumab (anti-PD-L1) and tremelimumab (anti-CTLA-4)) have been reported recently as 22% with 35% of the patients showing adverse events [136]. The phase III clinical trial (HIMALAYA) is currently underway [45]. However, the CHECKMATE-040 trial investigating nivolumab and ipilimumab could show overall response rates of 32% [42]. Further studies are required focusing on effectiveness versus increased adverse events. For a detailed list see Table 2.

4.2. Combination Therapies of Checkpoint Inhibitors and Anti-Angiogenesis

Another approach to enhance response to therapy explores additive effect of MKIs and ICI. It is well known that high VEGF levels in the TME modulate immunosuppressive T_{regs}, macrophages and MDSCs, whereby promoting tumor growth [86]. Anti-angiogenic effects of MKIs mediated by VEGF inhibition can synergistically enhance the anti-tumor effects of ICI. Furthermore, Sorafenib effectively inhibits macrophage migration, macrophage induces epithelial-mesenchymal transition as well as macrophage-NK cell crosstalk in the liver [34,137]. In line with this, combination of pembrolizumab (anti PD-1) and Lenvatinib (MKI) reduced the secretion of immunosuppressive cytokines such as TGF-β and IL-10 and inhibited expression of PD-1 and Tim3, which enhanced antitumor immune response in a mouse model of hepatocarcinogenesis [138].

Further, the IMbrave150 phase III clinical trial confirmed the promising effects for the combination of atezolizumab plus bevacizumab, a direct VEGF inhibitor, in a first line treatment in HCC patients [48]. The experimental arm showed an ORR 33% versus 13% for sorafenib arm and median OS at 12 months was 67% versus 55%. These results have led to an FDA approval for the combination of bevacizumab and atezolizumab for advanced HCC and will likely become the new standard of care in advanced HCCs. Many other combination studies are currently underway (Table 2). Similar to the findings from the IMbrave150 study, ORR for pembrolizumab plus Lenvatinib have been reported 36% in a phase Ib clinical trial. Notably, 36% had serious treatment related adverse events [122]. Nevertheless, combination of ICI and MKIs show promising anti-tumor response rates. Many other studies are currently underway. For a detailed list see Table 2.

Table 2. Summary of clinical trials for immunotherapy (mono- and combination therapies/completed and ongoing) in HCC.

Author (Year)	Phase/Trial Name	Target	Therapy Regimen	ORR	PFS (Months)	pts	DCR	Additional Information	Status
Monotherapy									
Sangro (2013) [44]	II	CTLA-4	Tremilimumab	17.6% (3 PR)	6.48	21	76.4%		completed
El-Khoueiry (2017) [40]	II CHECKMATE-40 Second line	PD-1	Nivolumab	20% (3 CR, 39 PR)	4.0	214	64% (37% over 6 months)	PD-1^{high} ORR 26%, PD-1^{low} ORR 19%, 9 months OS 74%, KM median not reached yet	completed
Zhu (2018) [41]	II KEYNOTE-224 Second line	PD-1	Pembrolizumab	17% (1 CR, 17 PR)	NR	104	64%	Positive correlation of ORR and TPS score	completed
Finn (2020) [117]	III KEYNOTE-240 Second line	PD-1	Pembrolizumab	18.3% (6 CR, 45 PR)	3.0	413	62.2%, (31% over 6 months)	OS 13.9 months	negative trial
Yau (2019) [73]	III CHECKMATE459 First line	PD-1	Nivolumab vs. Sorafenib	15% (14 CR, 43 PR)	3.7	743	-	PD-L1^{high} ORR 28%, PD-L1^{low} ORR 12%	negative trial
Qin (2019) [118]	III Rationale 301 First line	PD-L1/PD-L2	Tislelizumab vs. Sorafenib	-	-	-	-		ongoing
Exposito (2018) [119]	III CHECKMATE-9DX Adjuvant	PD-1	Nivolumab	-	-	530	-		ongoing
Combination of immunotherapies									
Yau (2019) [42]	I/II CHECKMATE-40	PD-1 + CTLA-4	Nivolumab + Ipilimumab	32% (4 CR, 12 PR)	-	148	54%	12 months OS 61% PD-1^{high} and PD-1^{low}: no difference	ongoing
Kelley (2020) [46]	I/II	PD-L1, CTLA-4	Durvalumab + Tremelimumab	22%	-	75	-	Median OS 18.7 months	ongoing
Abou-Alfa (2018) [45]	III HIMALAYA	PD-L1, CTLA-4	Durvalumab + Tremelimumab vs. Durvalumab vs. Sorafenib	-	-	-	-		ongoing
Kaseb (2019) [120]	II Neoadjuvant + adjuvant	PD-1, CTLA-4	Nivolumab + Ipilimumab	37.5% (3 CR)	-	8	-		ongoing

Table 2. Cont.

Author (Year)	Phase/Trial Name	Target	Therapy Regimen	ORR	PFS (Months)	pts	DCR	Additional Information	Status
Combination with MKI									
Bang (2019) [121]	Ib	PD-L1 + VEGF	Durvalumab + Ramucirumab	11% (3 CR+PR)	4.4	28	61%	PD-L1^{high} ORR 18%, DCR 73%	ongoing
Zhu (2020) [122]	Ib KEYNOTE 524 First line	PD-1 + MKI	Pembrolizumab + Lenvatinib	36% (1 CR, 35 PR)	8.6	30	60%		ongoing
Llovet (2019) [123]	III LEAP002 First line	PD-1 + MKI	Lenvatinib + Pembrolizumab vs. Lenvatinib	-	-	750	-		ongoing
Xu (2019) [124]	I Second line	PD-1 + MKI	Camrelizumab + Apatinib	50% (8 PR)	5.8	39 (16 HCC)	93.8%	OS NR	ongoing
	II IMMUNIB First line	PD-1 + MKI	Nivolumab + Lenvatinib	-	-	est. 50	-		ongoing
Pishvaian (2018) [125]	Ib	PD-L1 + VEGF	Atezolizumab + Bevacizumab	34% (1 CR, 22 PR)	14.9	68	78% (50% over 6 months)		ongoing
Finn (2018) [48]	III IMbrave150 First line	PD-L1 + VEGF	Atezolizumab + Bevacizumab vs. Sorafenib	33% (33 CR, 75 PR)	6.8	325	72.3%		ongoing
Yau (2020) [47]	II CHECKMATE 40	PD-1 + CTLA-4 + MKI	Nivolumab + Cabozantinib vs. Nivolumab + Ipilimumab + Cabozantinib	26% (9 PR)	6.8	71	83%	71% grade III-IV AEs, discontinuation in 20%	ongoing
Kudo (2019) [126]	Ib VEGF Liver 100 First line	PD-L1 + MKI	Avelumab + Axitinib	13.6%	5.5	22	68.2%	OS 12.7 months	ongoing
Kelley (2019) [127]	III COSMIC-312 First line	PD-L1 + MKI	Atezolizumab + Cabozantinib vs. Sorafenib	-	-	640	-		ongoing
Knox (2019) [128]	III EMERALD 2 Adjuvant	PD-L1 + VEGF	Durvalumab + Bevacizumab	-	-	-	-		ongoing

Table 2. *Cont.*

Author (Year)	Phase/Trial Name	Target	Therapy Regimen	ORR	PFS (Months)	pts	DCR	Additional Information	Status
Combination with locoregional therapy									
Duffy (2017) [43]	I/II	CTLA-4 + locoregional	Tremilimumab + TACE/RFA	26.3%	7.4	32	-	OS 12.3 months	completed
Sangro (2020) [129]	III EMERALD 1 adjuvant	PD-L1 + VEGF + locoregional	Durvalumab + Bevacizumab + TACE	-	-	600	-		ongoing
Charalampos (2019) [130]	II adjuvant	PD-L1 + CTLA-4 + locoregional	Durvalumab + Tremilimumab + TACE/RFA/cryoablation	20% (2 PR)	7.8	22 (10 HCC)	60%	OS 15.9 months	ongoing
	II IMMULAB	PD-1 + locoregional	Pembrolizumab + RFA/MWA	-	-	-	-		ongoing
	II PLTHCC	PD-1 + MKI + locoregional	Immunotherapy + Lenvatinib + TCAE	-	-	-	-		ongoing
Popovic (2019) [131]	Ib CaboNivo Neoadjuvant in locally advanced HCC	PD-1 + MKI + resection	Nivolumab + Cabozantinib	-	-	15	-		ongoing

4.3. Combination of Immunotherapy and Locoregional Therapy

A different approach to improve the response is to modulate the immunogenicity of tumors or to boost the immune system by combination of locoregional and/or radiotherapy with immunotherapy. This approach is based on releasing tumor antigens through cell death induced by locoregional therapy, which subsequently improves immunotherapy due to better antigen presentation. Thus, this combination is also discussed for neoadjuvant settings, when tumor burden is still high. In particular, antigen release and immunological response after irradiation has been extensively studied [139–141].

In 2004, de Broke et al. [142] could already show that RFA plus blocking CTLA-4 with tremelimumab causes a strong and durable antitumor response in a mouse model of B16 OVA melanoma cells. The same group showed that cryoablation and radiofrequency enables antigen loading of dendritic cells, which induced antitumor immunity [143], indicating that locoregional therapies could have more effects than just the local tumor elimination. The immunomodulatory effects caused by local therapies are of particular interest in the era of immunotherapies [144]. Different types of cell death can cause an immunogenic or non-immunogenic influence on the environment, whereas immunogenic cell death includes the release of calreticulin and other proteins of the endoplasmatic reticulum, which leads to activation of dendritic cells and improved tumor-antigen presentation for cytotoxic T cells [144]. A classical immunogenic cell death inducing chemotherapeutic is doxorubicin, which is most commonly used in TACE procedures in HCC patients [145]. MDSCs, which are increased in HCC patients, stimulate T_{regs} and correlate with HCC progression, [146,147] are decreased after RFA. However, patients with increased frequencies were more likely to recur after treatment. The effect of TACE or RFA on T cells seems to be stronger than surgery alone. After locoregional therapy, patients had a significant increase in GPC3 specific CTLs compared to patients undergoing surgery [140]. Radioembolization (Y90) on the other hand seems to have a sustained local as well as systemic immune response, that could be shown by an increase in TNFα in CD4, CD8 T cells, and APCs. The group could further demonstrate a prediction model based on peripheral blood samples before Y90 therapy [148].

However, response immunological response rates after locoregional therapy alone was not durable enough to prevent recurrence, underlining the potential of combination with immunotherapy [54]. The first combination therapy of tremelimumab and TACE, RFA, or cryoablation showed a good tolerability and an increase in intratumoral accumulation of CD8 T cells with good clinical response [43]. Remarkably, only lesions that were not directly treated were counted as tumor response, i.e., "abscopal effect" [43]. One combined clinical trial for HCC and CCA is investigating a combined immune checkpoint inhibition with ablative therapies (Durvalumab, Tremelimumab, TACE, RFA OR Cryoablation) (NCT02821754) [130]. For a detailed list of current clinical trials see Table 2.

While preclinical and early clinical data provides a clear rational for combination therapies, several open questions remain. In the context of combination therapy, the timing and sequence of corresponding therapies and identification of the best locoregional therapy in combination with the best immunotherapy are of particular interest. Further translational studies are also needed to improve the understanding of the exact molecular mechanisms involved in the response or failure of these combinations.

5. Translational Studies to Overcome Resistance to Immunotherapies

To detect molecular and cellular predictors of positive response to immunotherapy, animal models are widely employed in preclinical investigations, particularly syngeneic, genetically engineered, and humanized mice [149]. All of them harbor certain advantages and disadvantages, which should be carefully considered to accurately address the respective questions concerning immunotherapy.

5.1. Checkpoint Inhibitors

Investigation of immune checkpoint inhibitors using suitable models represents an important aspect of translational cancer research and is required for transitioning of crucial findings from bench to bedside. Detailed investigations on factors that are assisting immune evasion and contributing to the failure of classic chemotherapy are crucial [150]. Importance of CTLA-4 and PD-1/PD-L1 was thoroughly investigated in pre-clinical and early clinical models. Results revealed interesting and useful data for further translational implications and supported currently-used strategies in clinical trials [151,152].

Study of Brown et al. [153] tried to address mechanism of adaptive resistance to immunotherapy in the context of CTLA-4 checkpoint blockade. Results of this important study suggest that induction of Indoleamine 2,3-dioxygenase 1 (IDO1) typically appears in HCCs that are resistant to anti-CTLA-4 treatment, and that it is regulated in an IFN-γ dependent manner. These observations emphasized the importance of IDO1 as a regulator of adaptive resistance against anti-CTLA-4 treatment. Thus, combined therapy of IDO1 inhibitor and anti-CTLA-4 treatment emerges as a rational approach to improve the checkpoint-based treatments for the resistant types of HCC (Figure 2) [153]. In addition to increasing numbers of studies related to CTLA-4 therapy resistance, many new investigations aimed to delineate the fundamental mechanisms of PD-1/PD-L1-dependant immune tolerance in HCC [71,154]. In a chemically-induced HCC mouse model, exhaustion of tumor-antigen-specific CD8$^+$ T cells, accumulation of PD-1 CD8$^+$ T cells as well as T_{regs} was reported at the time of late tumor progression [71]. These findings encouraged authors to investigate a combination therapy of sunitinib and anti-PD-1 antibodies. This approach not only repressed adverse tumor features like immune evasion, but also directly reduced tumor burden and activated antitumor immunity [71].

To overcome immune tolerance, it is further important to explore more precise approaches to identify molecular components involved in immune evasion in HCC [155–157]. Polyinosinic-polycytidylic acid (polyIC), a double-stranded RNA, was firstly introduced as a molecule with potent liver tumor-inhibitory role only at the pre-cancer stage [155]. However, the potency of polyIC to treat advanced HCC was identified in a later study when it was combined with anti-PD-L1 antibody [156]. The mechanism of therapy response based on the ability of polyIC to enhance accumulation and activation of innate immune cells in the liver, particularly natural killer (NK) cells and macrophages, as well as to modulate adaptive immune functions by upregulation of PD-L1 in liver sinusoidal endothelial cells. These conditions sensitized the hepatic response to PD-L1 blockade and induced accumulation of active CD8$^+$ T cells (Figure 2) [156]. These studies clearly imply that modulation of specific pathways can lead to sensitization of the tumors to PD-L1 blockade and improve the response in HCC mouse models. These interesting findings should be pursued in future pre-clinical and clinical trials.

Further efficacy improvements of checkpoint inhibitors could be achieved through disruption of pathways involved in epigenetic regulation. For example, combination of histone deacetylase inhibitor belinostat with anti-CTLA-4, or combination of anti-CTLA-4 plus anti-PD-1 antibodies could lead to complete tumor rejection in a mouse HCC model [158]. Moreover, another study suggests that PD-L1 blockade and SIRT7 inhibition could be a more efficient clinical option to target HCC (Figure 2) [159]. Overall, these results provide a rationale for testing epigenetic modulators in combination with checkpoint inhibitors to enhance their therapeutic activity in patients with HCC.

All together, these animal studies clearly demonstrate the importance of the cellular composition and balance of pro- and anti-tumor immune cells for effectiveness of immunotherapy. Results clearly delineate capacity of epigenetic regulators to improve the immunotherapy response.

5.2. Application of Neoantigens and Oncolytic Viruses in Immunotherapy

One of the strategies to induce a positive immune response against cancer is the activation of CD8$^+$ T cells, either by antigen-presenting or by tumor cells. In this context, particularly interesting are the neoantigens that arise as a result of tumor-specific mutations, which could be effectively used for development of novel therapeutic approaches [52]. An effective way to increase neoantigen presentation to CD8$^+$ T cells in the tumor-/microenvironment is induction of cellular death by using various approaches, such as local ablation therapy or oncolytic viruses (OV) [53,160].

In a recent study, release of neoantigens was induced in an orthotopic mouse HCC model by applying image-guided stereotactic radiation. The treatment generated insufficient CD8$^+$ T cell mediated immune response due to feedback inhibition of T cells by increased PD-L1 expression on macrophages. Interestingly, antitumor effect was enhanced when combining stereotactic radiation with anti-PD-1 treatment. This approach promoted adaptive immunity and infiltration of CD8$^+$ cytotoxic T cells in the tumor, but only in a transient manner [72].

Great potential of OVs for the cancer treatment has been recognized in preclinical animal models as well as in human cancer patients [161,162]. Particularly interesting is application of oncolytic viruses in immunotherapies, which are specifically designed to selectively lyse cancer cells and to induce specific anti-tumor immunity. However, despite of a number of OVs that were examined in the preclinical studies, a low number entered into the clinical trials [161,162]. The most advanced of them is JX-594 (Pexa-Vec), which has entered a phase 3 randomized clinical study (PHOCUS). In this trial, the main objective is to determine if treatment with JX-594 and sorafenib increases survival in patients with advanced HCC who did not previously receive systemic therapy (NCT02562755). Therefore, development of new preclinical models to evaluate the effects of oncolytic viruses in HCC will pave the road for advanced clinical trials and speed up development of new cancer treatments. In line with that, new generations of OVs have been developed with greater potential to specifically target tumor cells and stimulate the immune response [163,164]. Recently, Nakatake et al. [163] examined the antitumor activities and immune response of third-generation HSV T-01 in HCC cell lines and mouse xenograft models. Application of the virus successfully led to increased expression of MHC class I molecules on tumor cells, which further stimulated CD8$^+$ T cell-mediated immune response. Importantly, viral treatment induced only antitumor effects without affecting normal cells, demonstrating great potential and specificity of this approach [163]. The capability of HSV-1 was further examined in a study where a novel HSV-1 vector, Ld0-GFP, was developed. Administration of the vector clearly showed increased tumor selectivity and oncolytic capacity against HCC by enhancing cell apoptosis in different mouse models. Overall, viral-induced oncolysis provoked strong immunogenic cell death by activating the immunogenic cell death pathway [165]. Despite the above mentioned OVs, several other viruses have also been explored in the context of HCC.

Overall, both exploration of neoantigens and direct tumor lysis by OV, show great translational value, as some of the investigated models and are currently investigated in clinical trials.

5.3. Targeting HCC Biomarkers–Vaccines, Antibodies, and Cytokines

Targeting a specific marker or a component of immune defense in HCC could be an effective way to overcome resistance commonly observed with classic chemotherapies [111]. New opportunities are emerging as specialized anti-cancer vaccines are developed and tested in animal models [166]. Most compelling are the vaccines that specifically target HCC-associated markers such as AFP and GPC3 (approach known as "antigen-defined") [167–169]. Many studies exploited the potential of AFP for designing an effective HCC vaccine [170–173]. In order to induce immune response, different approaches such as application of AFP plasmid DNA, dendritic cell (DC) transduction with viral vectors, or a combination of AFP with heat shock proteins have been evaluated [170–173]. However, the most promising results of AFP cancer immunization were achieved through production of epitope-optimized AFP, which effectively activated CD8$^+$ T cells and generated potent antitumor effects in HCC mouse model [174]. Several studies tried to target the activation of GPC3, a glycoprotein overexpressed in many HCC tissues, in order to design an effective vaccine [169]. Preclinical evidence suggests that intravenous injection of the GPC3-coupled lymphocytes can induce a strong anti-HCC effect by regulating systemic and local immune responses [169].

In addition to the above-mentioned vaccine-based approaches, a growing number of antibodies are produced to eradicate or neutralize specific molecular or cellular targets [61]. Several antibodies were also successfully targeted including GPC3, a member of the TNF receptor family CD137, transmembrane four L6 family member 5 protein, and fibrinogen-like protein 1 [175–177]. These investigations also demonstrated various degrees of anti-tumor and immune-modulatory capacity. In addition, immune modulation directed against liver cancer can be initiated by a release of cytokines involved in cellular antitumor response [178,179]. For instance, IL-33 release in murine HCC showed to markedly inhibit tumor growth via activated CD4$^+$ and CD8$^+$ T cells, in IL-33-expressing tumor-bearing mice, while IL-18/IL-12 cytokine therapy was effective in tumor regression prompted by induction of NK cells [178–180].

Taken together, development of different strategies to target specific HCC biomarkers and to modulate cytokine release shows big potential in immunotherapy of HCC.

5.4. Adoptive Cell Transfer and CAR T Cells

The basic principle of adoptive cell transfer is to disrupt the immune tolerance of tumors and, consequently, to suppress the growth and survival of tumor cells. This is achieved when lymphocytes are extracted from the patients, with the purpose of modification and amplification in vitro, and, subsequently, transferred back into the patient. This method enhances the overall specific antitumor effect [181]. Most of the recent studies on adoptive cell transfer were focused on targeting GPC3 [182]. In a seminal study, GPC3-specific CD8$^+$ T cells were engineered and subsequent antitumor capabilities in HCC xenograft mice were tested. This approach showed only partial response, as CD8$^+$ T cells were only able to slow down tumor growth in whole-body irradiated mouse model. Further, immunodeficient model displayed higher suppression of tumor growth. In this model, failure of significant tumor response was consequence of a lack of CD8$^+$ T cell infiltration into the tumor and by mosaic-pattern of GPC3 expression which could be enhanced in future studies [182].

However, more recently, CAR T cell-based therapy gained increasing attention as a potentially more efficient method to target tumor cells [183–186]. Earlier studies have already proven the potential of CAR T cells to effectively target GPC3$^+$ HCC cells in vivo. Anti-GPC3 CAR T cells successfully suppressed tumorigenesis in subcutaneous tumors and significantly affected tumor growth in subcutaneous and orthotopic xenografts [183]. Similar observation was noted in a patient-derived xenograft model. CAR T cells directed against GPC3 eradicated tumors from patient derived xenografts that showed less aggressive phenotype and lacked PD-L1 expression, while on the contrary, GPC3 CAR T cells were less potent in aggressive tumors with high PD-L1 expression. This all emphasized the potential of combination therapy with immune checkpoint inhibitors [185]. Except of combining GPC3-CAR T cells with checkpoint inhibitors, Wu et al. investigated potential application of sorafenib to induce additive effects. The authors reported that sorafenib enhanced the antitumor effects of CAR T cells, partially by promoting IL12 secretion by TAMs as well as promotion of apoptosis in immunocompetent and immunodeficient mouse models of HCC [186]. It is also important to mention that NK cells were investigated in the context of chimeric antigen receptor with promising results. This makes NK cell-based therapy as a novel treatment option for patients with GPC3$^+$ HCC [184].

Major studies on CAR T cells in HCC have been conducted with the main focus on GPC3. They shed more light on this complex topic and provided evidence for further investigations to define new targets for CAR T treatments. However, heterogeneous intra- and inter-tumoral expression of surface antigens as targets for CAR T-based approaches including GPC3 severely complicate this approach in human HCC.

5.5. Targeting Cross-Communication between MDSCs and the TME

The chronically altered tumor microenvironment in HCC, particularly liver fibrosis, significantly shapes and modulates the course of HCC development specifically by reprogramming an immunosuppressive mechanism [187]. Accumulation of monocytic MDSCs (M-MDSC) in fibrotic tumor microenvironment in orthotopic mouse model can significantly reduce the number of TILs and increase tumorigenicity [187]. Recent investigations have revealed that contribution to immune tolerance and higher tumorigenicity was closely connected to the interaction between HSC from the fibrotic microenvironment and M-MDSC [187]. Namely, HSC could induce M-MDSC accumulation and immunosuppression through p38 MAPK-mediated enhancer reprogramming. Treatment with BET bromodomain inhibitor significantly reduced the level of M-MDSC and increased the level of tumor-infiltrating $CD8^+$ T cell. When BET bromodomain inhibitor treatment was combined with anti-PD-L1 therapy, synergistic effects of the treatments led to tumor eradication and prolonged survival in this fibrotic-HCC mouse model. Therefore, targeting cross-communication (HSC-M-MDSC) in fibrotic liver could be a novel therapeutic strategy that could improve the efficacy of anti-PD-L1 therapy [187]. More evidence on how the response to PD-1/PD-L1 therapy could be further improved is presented by indirect modulation of IL-6 signaling, a major immune-modulatory cytokine in the liver [188]. Inhibition of *Ccrk* and CCRK/EZH2/NF-κB/IL-6 signaling cascade can bypass MDSC-mediated IFN-γ^+ TNF-α^+CD8$^+$ T cell exhaustion and cause reduction in tumorigenicity. More importantly, inactivation of this signaling cascade paralleled with administration of anti-PD-L1 therapy could improve efficacy of checkpoint inhibitors in orthotopic HCC model and prevent immune evasion [188].

5.6. Targeting MDSC, TAMs, and Innate Immunity Interaction for HCC Prevention

Given the fact that macrophages promote HCC progression, therapeutic manipulation of this interaction is of major interest. This includes the inhibition of monocyte recruitment into the liver, polarization from M1 to M2 macrophages, inhibition of TAM associated cytokines, or direct inhibition of macrophages present in the tumor [189–192]. Blocking of CCL2-CCR2, which inhibits monocyte recruitment, was revealed to be effective in HCC mouse models. Namely, this approach increased tumor infiltrating macrophage numbers, promoted polarization into a M2 phenotype as well as enhanced a T cell antitumor response [193,194]. Moreover, treatment with Mi-RNA-26a effectively suppressed tumor growth by downregulating colony stimulating factor-1 (CSF1 or M CSF), which further inhibited macrophage recruitment [195]. Blocking of CSF1 and CSF1 receptor (CSF1R) has also been demonstrated to enhance the effectiveness of immune checkpoint inhibitors [157]. A recent study has reported that Osteopontin facilitates chemotactic migration and M2-like polarization of macrophages and promotes the expression of PD-L1 in HCC. These events are mediated via activation of CSF1-CSF1R pathway in macrophages, which leads to increase of immunosuppressive cytokine levels. Therefore, blocking the CSF1/CSF1R pathway could effectively prevent macrophage recruitment and M2 phenotype polarization, activate $CD8^+$ T cells, and sensitize HCC to anti-PD-L1 immune checkpoint blockade (Figure 2) [157]. PLX3397 also inhibits CSF-1R and could prevent tumor growth in a murine HCC model by macrophage reprogramming [192]. Another agent, baicalin (a flanonoid), repolarized macrophages into M1-like macrophages in an orthotopic mouse model of liver cancer [196]. All these translational findings suggest potential combination therapies to reprogram the immunological TME.

From the perspective of innate immunity, NK cells are considered to be one of the key players in the prevention of HCC [29,197]. They exert a critical role in the antitumor immunity by modulating both, innate immunity as well as activation of adaptive immunity, by cross-talking with DCs and promoting a T helper cell (Th)1-mediated immunity [29]. However, positive role of NK cells in fight against cancer has often been impaired in HCC [198,199]. It was already shown that MDSCs in patients with HCC suppress the innate immune system by diminishing autologous NK cell cytotoxicity and cytokine secretion. These events activate immune suppressor network and allow the tumors to evade the host immune response [200]. Earlier studies in mice determined that inhibition of NK cell cytotoxicity is contact-dependent, where MDSCs inhibit IL-2-mediated NK cell activation, by dysregulating Stat5 signaling [201]. More evidence on the dysregulation of NK cells by MDSCs was obtained in the liver cancer-bearing mouse model. Results showed that increased levels of MDSC directly influenced NK cell function by inhibition of their cytotoxicity and IFN-γ production. The main mediator of NK cell suppression was membrane-bound TGF-β1 on MDSC [29]. Taken together, disruption of interaction between MDSC and components of innate immunity, particularly NK cells, represents an attractive approach to confront development of HCC.

6. Conclusions and Future Direction

Primary liver cancer develops in a fine-tuned and very complex microenvironment. Immune cell composition and interactions with tumor as well as stromal cells play a crucial role in development and progression of liver cancer. Modern immune-oncological approaches in HCC significantly expanded the landscape of active compounds in HCC over the recent years. However, efficacy of targeting individual aspects of immune response, including checkpoint molecules, remain decisively low. Furthermore, predictive biomarkers for therapy response are still largely missing. Thus, implementation of results and different approaches from preclinical, translational studies might be of utmost importance to identify novel cellular or molecular targets that synergistically could improve currently used strategies. Herein, an improved understanding of the landscape of immune-oncological alterations and rationale for subsequent molecularly-guided combination therapies are urgently needed. Up till now, our current understanding remains incomplete and precise dissection of intra- and inter-tumoral heterogeneity using single cell sequencing approaches still is in its infancy for HCC. In addition, detailed knowledge on the immune-cell contexture will add additional layers of complexity that requires detailed preclinical models that closely resemble authentic human HCC. However, a better understanding of molecular interaction and pathways on a cellular level is imperative to develop new treatment regimens or combination of regimes. As the knowledge on molecular and immune-modulatory pathways evolve, the corresponding context of application and genetic background will need to be tightly controlled to ultimately implement the translational finding, overcome therapy resistance, and increase clinical response rates. Nevertheless, recent findings form clinical trials on different combination therapies are highly promising and will likely further shape the therapeutic landscape and enter the clinical practice of HCC treatment.

Author Contributions: S.H. and D.C. contributed equally to the literature research and writing of the manuscript. Figures were generated by S.H., and P.R.G. and J.U.M. designed and supervised the research process and the manuscript preparation. All authors have read and agreed to the published version of the manuscript.

Funding: J.U.M. is supported by grants from the German Research Foundation (MA 4443/2-2; SFB1292), the Volkswagen Foundation (Lichtenberg program), and by a grant from the Wilhelm-Sander Foundation (2017.007.1).

Conflicts of Interest: The authors declare no conflict of interest.

Abbreviations

Ab	Antibody
AFP	alpha-fetoprotein
APC	antigen presenting cells
CCRK	cell cycle-related kinase
CPS	combined positive score
CSF1	colony stimulating factor 1
CSF1R	colony stimulating factor receptor 1
CTLA-4	cytotoxic T lymphocyte antigen 4
CXCR-4	CXC receptor type 4
DC	dendritic cell
GPC3	glypican-3
HCC	hepatocellular carcinoma
HSC	hepatic stellate cell
ICI	immune checkpoint inhibitor
IDO1	indoleamine 2,3-dioxygenase 1
LAG3	lymphocyte activation gene 3
M-MDSC	monocytic MDSC
MDSC	myeloid-derived suppressor cells
NAFLD	non-alcoholic fatty liver disease
NASH	non-alcoholic steato hepatitis

NF-κB	nuclear factor-κB
NK cells	natural killer cells
NSCLC	non-small cell lung cancer
OR	overall response rate
OS	overall survival
PBMC	peripheral blood mononuclear cells
PD-1	programmed death protein 1
PD-L1	programmed death ligand 1
PFS	progression free survival
PSC	primary sclerosing cholangitis
TAM	tumor-associated macrophages
Th	T helper
TIL	tumor infiltrating lymphocytes
TIM3	T cell membrane protein 3
TKI	tyrosine kinase inhibitor
TME	tumor microenvironment
TPS	tumor proportion score
T_{reg}	regulatory T cells
VEGF	vascular endothelial growth factor

References

1. Khan, S.A.; Taylor-Robinson, S.D.; Toledano, M.B.; Beck, A.; Elliott, P.; Thomas, H.C. Changing international trends in mortality rates for liver, biliary and pancreatic tumours. *J. Hepatol.* **2002**, *37*, 806–813. [CrossRef]
2. Saha, S.K.; Zhu, A.X.; Fuchs, C.S.; Brooks, G.A. Forty-Year Trends in Cholangiocarcinoma Incidence in the U.S.: Intrahepatic Disease on the Rise. *Oncologist* **2016**, *21*, 594–599. [CrossRef]
3. Ryerson, A.B.; Eheman, C.R.; Altekruse, S.F.; Ward, J.W.; Jemal, A.; Sherman, R.L.; Henley, S.J.; Holtzman, D.; Lake, A.; Noone, A.M.; et al. Annual Report to the Nation on the Status of Cancer, 1975-2012, featuring the increasing incidence of liver cancer. *Cancer* **2016**, *122*, 1312–1337. [CrossRef]
4. Bray, F.; Ferlay, J.; Soerjomataram, I.; Siegel, R.L.; Torre, L.A.; Jemal, A. Global cancer statistics 2018: GLOBOCAN estimates of incidence and mortality worldwide for 36 cancers in 185 countries. *CA Cancer J. Clin.* **2018**, *68*, 394–424. [CrossRef]
5. Budhu, A.; Wang, X.W. The role of cytokines in hepatocellular carcinoma. *J. Leukoc. Biol.* **2006**, *80*, 1197–1213. [CrossRef]
6. Liu, Z.; Jiang, Y.; Yuan, H.; Fang, Q.; Cai, N.; Suo, C.; Jin, L.; Zhang, T.; Chen, X. The trends in incidence of primary liver cancer caused by specific etiologies: Results from the Global Burden of Disease Study 2016 and implications for liver cancer prevention. *J. Hepatol.* **2019**, *70*, 674–683. [CrossRef]
7. Sun, B.; Karin, M. Obesity, inflammation, and liver cancer. *J. Hepatol.* **2012**, *56*, 704–713. [CrossRef]
8. Michelotti, G.A.; Machado, M.V.; Diehl, A.M. NAFLD, NASH and liver cancer. *Nat. Rev. Gastroenterol. Hepatol.* **2013**, *10*, 656–665. [CrossRef]
9. Greten, T.F.; Wang, X.W.; Korangy, F. Current concepts of immune based treatments for patients with HCC: From basic science to novel treatment approaches. *Gut* **2015**, *64*, 842–848. [CrossRef]
10. Ambade, A.; Satishchandran, A.; Saha, B.; Gyongyosi, B.; Lowe, P.; Kodys, K.; Catalano, D.; Szabo, G. Hepatocellular carcinoma is accelerated by NASH involving M2 macrophage polarization mediated by hif-1αinduced IL-10. *Oncoimmunology* **2016**, *5*, e1221557. [CrossRef]
11. Galle, P.R.; Tovoli, F.; Foerster, F.; Worns, M.A.; Cucchetti, A.; Bolondi, L. The treatment of intermediate stage tumours beyond TACE: From surgery to systemic therapy. *J. Hepatol.* **2017**. [CrossRef] [PubMed]
12. Duffy, A.G.; Greten, T.F. Liver cancer: Regorafenib as second-line therapy in hepatocellular carcinoma. *Nat. Rev. Gastroenterol. Hepatol.* **2017**, *14*, 141–142. [CrossRef] [PubMed]
13. Bruix, J.; Qin, S.; Merle, P.; Granito, A.; Huang, Y.H.; Bodoky, G.; Pracht, M.; Yokosuka, O.; Rosmorduc, O.; Breder, V.; et al. Regorafenib for patients with hepatocellular carcinoma who progressed on sorafenib treatment (RESORCE): A randomised, double-blind, placebo-controlled, phase 3 trial. *Lancet* **2017**, *389*, 56–66. [CrossRef]

14. Llovet, J.M.; Ricci, S.; Mazzaferro, V.; Hilgard, P.; Gane, E.; Blanc, J.F.; de Oliveira, A.C.; Santoro, A.; Raoul, J.L.; Forner, A.; et al. Sorafenib in advanced hepatocellular carcinoma. *N. Engl. J. Med.* **2008**, *359*, 378–390. [CrossRef] [PubMed]
15. Kudo, M.; Finn, R.S.; Qin, S.; Han, K.H.; Ikeda, K.; Piscaglia, F.; Baron, A.; Park, J.W.; Han, G.; Jassem, J.; et al. Lenvatinib versus sorafenib in first-line treatment of patients with unresectable hepatocellular carcinoma: A randomised phase 3 non-inferiority trial. *Lancet* **2018**, *391*, 1163–1173. [CrossRef]
16. Abou-Alfa, G.K.; Meyer, T.; Cheng, A.L.; El-Khoueiry, A.B.; Rimassa, L.; Ryoo, B.Y.; Cicin, I.; Merle, P.; Chen, Y.; Park, J.W.; et al. Cabozantinib in Patients with Advanced and Progressing Hepatocellular Carcinoma. *N. Engl. J. Med.* **2018**, *379*, 54–63. [CrossRef]
17. Zhu, A.X.; Kang, Y.K.; Yen, C.J.; Finn, R.S.; Galle, P.R.; Llovet, J.M.; Assenat, E.; Brandi, G.; Pracht, M.; Lim, H.Y.; et al. Ramucirumab after sorafenib in patients with advanced hepatocellular carcinoma and increased α-fetoprotein concentrations (REACH-2): A randomised, double-blind, placebo-controlled, phase 3 trial. *Lancet. Oncol.* **2019**, *20*, 282–296. [CrossRef]
18. Pinyol, R.; Sia, D.; Llovet, J.M. Immune Exclusion-Wnt/CTNNB1 Class Predicts Resistance to Immunotherapies in HCC. *Clin. Cancer Res. Off. J. Am. Assoc. Cancer Res.* **2019**, *25*, 2021–2023. [CrossRef]
19. Yin, C.; Evason, K.J.; Asahina, K.; Stainier, D.Y. Hepatic stellate cells in liver development, regeneration, and cancer. *J. Clin. Investig.* **2013**, *123*, 1902–1910. [CrossRef]
20. Jenne, C.N.; Kubes, P. Immune surveillance by the liver. *Nat. Immunol.* **2013**, *14*, 996–1006. [CrossRef]
21. Makarova-Rusher, O.V.; Medina-Echeverz, J.; Duffy, A.G.; Greten, T.F. The yin and yang of evasion and immune activation in HCC. *J. Hepatol.* **2015**, *62*, 1420–1429. [CrossRef] [PubMed]
22. Zheng, C.; Zheng, L.; Yoo, J.K.; Guo, H.; Zhang, Y.; Guo, X.; Kang, B.; Hu, R.; Huang, J.Y.; Zhang, Q.; et al. Landscape of Infiltrating T Cells in Liver Cancer Revealed by Single-Cell Sequencing. *Cell* **2017**, *169*, 1342–1356.e16. [CrossRef] [PubMed]
23. Kapanadze, T.; Gamrekelashvili, J.; Ma, C.; Chan, C.; Zhao, F.; Hewitt, S.; Zender, L.; Kapoor, V.; Felsher, D.W.; Manns, M.P.; et al. Regulation of accumulation and function of myeloid derived suppressor cells in different murine models of hepatocellular carcinoma. *J. Hepatol.* **2013**, *59*, 1007–1013. [CrossRef] [PubMed]
24. Mizukoshi, E.; Yamashita, T.; Arai, K.; Terashima, T.; Kitahara, M.; Nakagawa, H.; Iida, N.; Fushimi, K.; Kaneko, S. Myeloid-derived suppressor cells correlate with patient outcomes in hepatic arterial infusion chemotherapy for hepatocellular carcinoma. *Cancer Immunol. Immunother.* **2016**, *65*, 715–725. [CrossRef] [PubMed]
25. Yang, X.H.; Yamagiwa, S.; Ichida, T.; Matsuda, Y.; Sugahara, S.; Watanabe, H.; Sato, Y.; Abo, T.; Horwitz, D.A.; Aoyagi, Y. Increase of CD4+ CD25+ regulatory T-cells in the liver of patients with hepatocellular carcinoma. *J. Hepatol.* **2006**, *45*, 254–262. [CrossRef]
26. Iwata, T.; Kondo, Y.; Kimura, O.; Morosawa, T.; Fujisaka, Y.; Umetsu, T.; Kogure, T.; Inoue, J.; Nakagome, Y.; Shimosegawa, T. PD-L1(+)MDSCs are increased in HCC patients and induced by soluble factor in the tumor microenvironment. *Sci. Rep.* **2016**, *6*, 39296. [CrossRef]
27. Foerster, F.; Hess, M.; Gerhold-Ay, A.; Marquardt, J.U.; Becker, D.; Galle, P.R.; Schuppan, D.; Binder, H.; Bockamp, E. The immune contexture of hepatocellular carcinoma predicts clinical outcome. *Sci. Rep.* **2018**, *8*, 5351. [CrossRef]
28. Brunner, S.M.; Rubner, C.; Kesselring, R.; Martin, M.; Griesshammer, E.; Ruemmele, P.; Stempfl, T.; Teufel, A.; Schlitt, H.J.; Fichtner-Feigl, S. Tumor-infiltrating, interleukin-33-producing effector-memory CD8(+) T cells in resected hepatocellular carcinoma prolong patient survival. *Hepatology* **2015**, *61*, 1957–1967. [CrossRef]
29. Li, H.; Han, Y.; Guo, Q.; Zhang, M.; Cao, X. Cancer-expanded myeloid-derived suppressor cells induce anergy of NK cells through membrane-bound TGF-beta 1. *J. Immunol.* **2009**, *182*, 240–249. [CrossRef] [PubMed]
30. Prieto, J.; Melero, I.; Sangro, B. Immunological landscape and immunotherapy of hepatocellular carcinoma. *Nat. Rev. Gastroenterol. Hepatol.* **2015**, *12*, 681–700. [CrossRef]
31. Schreiber, R.D.; Old, L.J.; Smyth, M.J. Cancer immunoediting: Integrating immunity's roles in cancer suppression and promotion. *Science* **2011**, *331*, 1565–1570. [CrossRef] [PubMed]
32. Yu, L.X.; Ling, Y.; Wang, H.Y. Role of nonresolving inflammation in hepatocellular carcinoma development and progression. *NPJ Precis. Oncol.* **2018**, *2*, 6. [CrossRef] [PubMed]
33. Eggert, T.; Wolter, K.; Ji, J.; Ma, C.; Yevsa, T.; Klotz, S.; Medina-Echeverz, J.; Longerich, T.; Forgues, M.; Reisinger, F.; et al. Distinct Functions of Senescence-Associated Immune Responses in Liver Tumor Surveillance and Tumor Progression. *Cancer Cell* **2016**, *30*, 533–547. [CrossRef] [PubMed]

34. Sprinzl, M.F.; Reisinger, F.; Puschnik, A.; Ringelhan, M.; Ackermann, K.; Hartmann, D.; Schiemann, M.; Weinmann, A.; Galle, P.R.; Schuchmann, M.; et al. Sorafenib perpetuates cellular anticancer effector functions by modulating the crosstalk between macrophages and natural killer cells. *Hepatology* **2013**, *57*, 2358–2368. [CrossRef]
35. Ma, C.; Kesarwala, A.H.; Eggert, T.; Medina-Echeverz, J.; Kleiner, D.E.; Jin, P.; Stroncek, D.F.; Terabe, M.; Kapoor, V.; ElGindi, M.; et al. NAFLD causes selective CD4(+) T lymphocyte loss and promotes hepatocarcinogenesis. *Nature* **2016**, *531*, 253–257. [CrossRef]
36. Wu, Y.; Kuang, D.M.; Pan, W.D.; Wan, Y.L.; Lao, X.M.; Wang, D.; Li, X.F.; Zheng, L. Monocyte/macrophage-elicited natural killer cell dysfunction in hepatocellular carcinoma is mediated by CD48/2B4 interactions. *Hepatology* **2013**, *57*, 1107–1116. [CrossRef]
37. Sui, Q.; Zhang, J.; Sun, X.; Zhang, C.; Han, Q.; Tian, Z. NK cells are the crucial antitumor mediators when STAT3-mediated immunosuppression is blocked in hepatocellular carcinoma. *J. Immunol.* **2014**, *193*, 2016–2023. [CrossRef]
38. Zhou, J.; Ding, T.; Pan, W.; Zhu, L.Y.; Li, L.; Zheng, L. Increased intratumoral regulatory T cells are related to intratumoral macrophages and poor prognosis in hepatocellular carcinoma patients. *Int. J. Cancer* **2009**, *125*, 1640–1648. [CrossRef]
39. Dong, P.; Ma, L.; Liu, L.; Zhao, G.; Zhang, S.; Dong, L.; Xue, R.; Chen, S. CD86$^+$/CD206$^+$, Diametrically Polarized Tumor-Associated Macrophages, Predict Hepatocellular Carcinoma Patient Prognosis. *Int. J. Mol. Sci.* **2016**, *17*, 320. [CrossRef] [PubMed]
40. El-Khoueiry, A.B.; Sangro, B.; Yau, T.; Crocenzi, T.S.; Kudo, M.; Hsu, C.; Kim, T.Y.; Choo, S.P.; Trojan, J.; Welling, T.H.R.; et al. Nivolumab in patients with advanced hepatocellular carcinoma (CheckMate 040): An open-label, non-comparative, phase 1/2 dose escalation and expansion trial. *Lancet* **2017**, *389*, 2492–2502. [CrossRef]
41. Zhu, A.X.; Finn, R.S.; Edeline, J.; Cattan, S.; Ogasawara, S.; Palmer, D.; Verslype, C.; Zagonel, V.; Fartoux, L.; Vogel, A.; et al. Pembrolizumab in patients with advanced hepatocellular carcinoma previously treated with sorafenib (KEYNOTE-224): A non-randomised, open-label phase 2 trial. *Lancet. Oncol.* **2018**, *19*, 940–952. [CrossRef]
42. Yau, T.; Kang, Y.-K.; Kim, T.-Y.; El-Khoueiry, A.B.; Santoro, A.; Sangro, B.; Melero, I.; Kudo, M.; Hou, M.-M.; Matilla, A.; et al. Nivolumab (NIVO) + ipilimumab (IPI) combination therapy in patients (pts) with advanced hepatocellular carcinoma (aHCC): Results from CheckMate 040. *J. Clin. Oncol.* **2019**, *37*, 4012. [CrossRef]
43. Duffy, A.G.; Ulahannan, S.V.; Makorova-Rusher, O.; Rahma, O.; Wedemeyer, H.; Pratt, D.; Davis, J.L.; Hughes, M.S.; Heller, T.; ElGindi, M.; et al. Tremelimumab in combination with ablation in patients with advanced hepatocellular carcinoma. *J. Hepatol.* **2017**, *66*, 545–551. [CrossRef] [PubMed]
44. Sangro, B.; Gomez-Martin, C.; de la Mata, M.; Inarrairaegui, M.; Garralda, E.; Barrera, P.; Riezu-Boj, J.I.; Larrea, E.; Alfaro, C.; Sarobe, P.; et al. A clinical trial of CTLA-4 blockade with tremelimumab in patients with hepatocellular carcinoma and chronic hepatitis C. *J. Hepatol.* **2013**, *59*, 81–88. [CrossRef]
45. Abou-Alfa, G.K.; Chan, S.L.; Furuse, J.; Galle, P.R.; Kelley, R.K.; Qin, S.; Armstrong, J.; Darilay, A.; Vlahovic, G.; Negro, A.; et al. A randomized, multicenter phase 3 study of durvalumab (D) and tremelimumab (T) as first-line treatment in patients with unresectable hepatocellular carcinoma (HCC): HIMALAYA study. *J. Clin. Oncol.* **2018**, *36*, TPS4144. [CrossRef]
46. Kelley, R.K.; Sangro, B.; Harris, W.P.; Ikeda, M.; Okusaka, T.; Kang, Y.-K.; Qin, S.; Tai, W.M.D.; Lim, H.Y.; Yau, T.; et al. Efficacy, tolerability, and biologic activity of a novel regimen of tremelimumab (T) in combination with durvalumab (D) for patients (pts) with advanced hepatocellular carcinoma (aHCC). *J. Clin. Oncol.* **2020**, *38*, 4508. [CrossRef]
47. Yau, T.; Zagonel, V.; Santoro, A.; Acosta-Rivera, M.; Choo, S.P.; Matilla, A.; He, A.R.; Gracián, A.C.; El-Khoueiry, A.B.; Sangro, B.; et al. Nivolumab (NIVO) + ipilimumab (IPI) + cabozantinib (CABO) combination therapy in patients (pts) with advanced hepatocellular carcinoma (aHCC): Results from CheckMate 040. *J. Clin. Oncol.* **2020**, *38*, 478. [CrossRef]
48. Finn, R.S.; Qin, S.; Ikeda, M.; Galle, P.R.; Ducreux, M.; Kim, T.Y.; Kudo, M.; Breder, V.; Merle, P.; Kaseb, A.O.; et al. Atezolizumab plus Bevacizumab in Unresectable Hepatocellular Carcinoma. *N. Engl. J. Med.* **2020**, *382*, 1894–1905. [CrossRef]
49. Brown, Z.J.; Greten, T.F.; Heinrich, B. Adjuvant Treatment of Hepatocellular Carcinoma: Prospect of Immunotherapy. *Hepatology* **2019**, *70*, 1437–1442. [CrossRef]

50. Adcock, C.S.; Puneky, L.V.; Campbell, G.S. Favorable Response of Metastatic Hepatocellular Carcinoma to Treatment with Trans-arterial Radioembolization Followed by Sorafenib and Nivolumab. *Cureus* **2019**, *11*, e4083. [CrossRef]
51. Crispe, I.N. The liver as a lymphoid organ. *Annu. Rev. Immunol.* **2009**, *27*, 147–163. [CrossRef] [PubMed]
52. Schumacher, T.N.; Schreiber, R.D. Neoantigens in cancer immunotherapy. *Science* **2015**, *348*, 69–74. [CrossRef] [PubMed]
53. Bernstein, M.B.; Krishnan, S.; Hodge, J.W.; Chang, J.Y. Immunotherapy and stereotactic ablative radiotherapy (ISABR): A curative approach? *Nat. Rev. Clin. Oncol.* **2016**, *13*, 516–524. [CrossRef] [PubMed]
54. Mizukoshi, E.; Yamashita, T.; Arai, K.; Sunagozaka, H.; Ueda, T.; Arihara, F.; Kagaya, T.; Yamashita, T.; Fushimi, K.; Kaneko, S. Enhancement of tumor-associated antigen-specific T cell responses by radiofrequency ablation of hepatocellular carcinoma. *Hepatology* **2013**, *57*, 1448–1457. [CrossRef]
55. Zhang, Q.; Zhang, Z.; Peng, M.; Fu, S.; Xue, Z.; Zhang, R. CAR-T cell therapy in gastrointestinal tumors and hepatic carcinoma: From bench to bedside. *Oncoimmunology* **2016**, *5*, e1251539. [CrossRef]
56. Lee, J.H.; Lee, J.H.; Lim, Y.S.; Yeon, J.E.; Song, T.J.; Yu, S.J.; Gwak, G.Y.; Kim, K.M.; Kim, Y.J.; Lee, J.W.; et al. Adjuvant immunotherapy with autologous cytokine-induced killer cells for hepatocellular carcinoma. *Gastroenterology* **2015**, *148*, 1383–1391.e16. [CrossRef]
57. Zou, W. Immunosuppressive networks in the tumour environment and their therapeutic relevance. *Nat. Rev. Cancer* **2005**, *5*, 263–274. [CrossRef]
58. Hogdall, D.; Lewinska, M.; Andersen, J.B. Desmoplastic Tumor Microenvironment and Immunotherapy in Cholangiocarcinoma. *Trends Cancer* **2018**, *4*, 239–255. [CrossRef]
59. Topalian, S.L.; Hodi, F.S.; Brahmer, J.R.; Gettinger, S.N.; Smith, D.C.; McDermott, D.F.; Powderly, J.D.; Carvajal, R.D.; Sosman, J.A.; Atkins, M.B.; et al. Safety, activity, and immune correlates of anti-PD-1 antibody in cancer. *N. Engl. J. Med.* **2012**, *366*, 2443–2454. [CrossRef]
60. Tumeh, P.C.; Harview, C.L.; Yearley, J.H.; Shintaku, I.P.; Taylor, E.J.; Robert, L.; Chmielowski, B.; Spasic, M.; Henry, G.; Ciobanu, V.; et al. PD-1 blockade induces responses by inhibiting adaptive immune resistance. *Nature* **2014**, *515*, 568–571. [CrossRef]
61. Brown, Z.J.; Heinrich, B.; Greten, T.F. Mouse models of hepatocellular carcinoma: An overview and highlights for immunotherapy research. *Nat. Rev. Gastroenterol. Hepatol.* **2018**, *15*, 536–554. [CrossRef] [PubMed]
62. Schachter, J.; Ribas, A.; Long, G.V.; Arance, A.; Grob, J.J.; Mortier, L.; Daud, A.; Carlino, M.S.; McNeil, C.; Lotem, M.; et al. Pembrolizumab versus ipilimumab for advanced melanoma: Final overall survival results of a multicentre, randomised, open-label phase 3 study (KEYNOTE-006). *Lancet* **2017**, *390*, 1853–1862. [CrossRef]
63. Gandhi, L.; Rodríguez-Abreu, D.; Gadgeel, S.; Esteban, E.; Felip, E.; De Angelis, F.; Domine, M.; Clingan, P.; Hochmair, M.J.; Powell, S.F.; et al. Pembrolizumab plus Chemotherapy in Metastatic Non-Small-Cell Lung Cancer. *N. Engl. J. Med.* **2018**, *378*, 2078–2092. [CrossRef] [PubMed]
64. Brahmer, J.; Reckamp, K.L.; Baas, P.; Crinò, L.; Eberhardt, W.E.; Poddubskaya, E.; Antonia, S.; Pluzanski, A.; Vokes, E.E.; Holgado, E.; et al. Nivolumab versus Docetaxel in Advanced Squamous-Cell Non-Small-Cell Lung Cancer. *N. Engl. J. Med.* **2015**, *373*, 123–135. [CrossRef] [PubMed]
65. Larkin, J.; Chiarion-Sileni, V.; Gonzalez, R.; Grob, J.J.; Cowey, C.L.; Lao, C.D.; Schadendorf, D.; Dummer, R.; Smylie, M.; Rutkowski, P.; et al. Combined Nivolumab and Ipilimumab or Monotherapy in Untreated Melanoma. *N. Engl. J. Med.* **2015**, *373*, 23–34. [CrossRef] [PubMed]
66. Le, D.T.; Durham, J.N.; Smith, K.N.; Wang, H.; Bartlett, B.R.; Aulakh, L.K.; Lu, S.; Kemberling, H.; Wilt, C.; Luber, B.S.; et al. Mismatch repair deficiency predicts response of solid tumors to PD-1 blockade. *Science* **2017**, *357*, 409–413. [CrossRef] [PubMed]
67. Xie, C.; Duffy, A.G.; Mabry-Hrones, D.; Wood, B.; Levy, E.; Krishnasamy, V.; Khan, J.; Wei, J.S.; Agdashian, D.; Tyagi, M.; et al. Tremelimumab in Combination With Microwave Ablation in Patients With Refractory Biliary Tract Cancer. *Hepatology* **2019**, *69*, 2048–2060. [CrossRef]
68. Gou, M.; Zhang, Y.; Si, H.; Dai, G. Efficacy and safety of nivolumab for metastatic biliary tract cancer. *Onco Targets Ther.* **2019**, *12*, 861–867. [CrossRef]
69. Kim, R.D.; Kim, D.W.; Alese, O.B.; Li, D.; Shah, N.; Schell, M.J.; Zhou, J.M.; Chung, V. A phase II study of nivolumab in patients with advanced refractory biliary tract cancers (BTC). *J. Clin. Oncol.* **2019**, *37*, 4097. [CrossRef]

70. Finn, R.S.; Ryoo, B.-Y.; Merle, P.; Kudo, M.; Bouattour, M.; Lim, H.-Y.; Breder, V.V.; Edeline, J.; Chao, Y.; Ogasawara, S.; et al. Results of KEYNOTE-240: Phase 3 study of pembrolizumab (Pembro) vs best supportive care (BSC) for second line therapy in advanced hepatocellular carcinoma (HCC). *Am. Soc. Clin. Oncol.* **2019**, *37*, 4004. [CrossRef]
71. Li, G.; Liu, D.; Cooper, T.K.; Kimchi, E.T.; Qi, X.; Avella, D.M.; Li, N.; Yang, Q.X.; Kester, M.; Rountree, C.B.; et al. Successful chemoimmunotherapy against hepatocellular cancer in a novel murine model. *J. Hepatol.* **2017**, *66*, 75–85. [CrossRef] [PubMed]
72. Friedman, D.; Baird, J.R.; Young, K.H.; Cottam, B.; Crittenden, M.R.; Friedman, S.; Gough, M.J.; Newell, P. Programmed cell death-1 blockade enhances response to stereotactic radiation in an orthotopic murine model of hepatocellular carcinoma. *Hepatol. Res. Off. J. Jpn. Soc. Hepatol.* **2017**, *47*, 702–714. [CrossRef] [PubMed]
73. Yau, T.; Park, J.W.; Finn, R.S.; Cheng, A.-L.; Mathurin, P.; Edeline, J.; Kudo, M.; Han, K.-H.; Harding, J.J.; Merle, P.; et al. LBA38_PRCheckMate 459: A randomized, multi-center phase III study of nivolumab (NIVO) vs sorafenib (SOR) as first-line (1L) treatment in patients (pts) with advanced hepatocellular carcinoma (aHCC). *Ann. Oncol.* **2019**, *30*. [CrossRef]
74. Champiat, S.; Ferrara, R.; Massard, C.; Besse, B.; Marabelle, A.; Soria, J.C.; Ferte, C. Hyperprogressive disease: Recognizing a novel pattern to improve patient management. *Nat. Rev. Clin. Oncol.* **2018**, *15*, 748–762. [CrossRef]
75. Kim, H.D.; Song, G.W.; Park, S.; Jung, M.K.; Kim, M.H.; Kang, H.J.; Yoo, C.; Yi, K.; Kim, K.H.; Eo, S.; et al. Association Between Expression Level of PD1 by Tumor-Infiltrating CD8(+) T Cells and Features of Hepatocellular Carcinoma. *Gastroenterology* **2018**, *155*, 1936–1950.e17. [CrossRef]
76. Calderaro, J.; Rousseau, B.; Amaddeo, G.; Mercey, M.; Charpy, C.; Costentin, C.; Luciani, A.; Zafrani, E.S.; Laurent, A.; Azoulay, D.; et al. Programmed death ligand 1 expression in hepatocellular carcinoma: Relationship With clinical and pathological features. *Hepatology* **2016**, *64*, 2038–2046. [CrossRef]
77. Zhao, P.; Li, L.; Jiang, X.; Li, Q. Mismatch repair deficiency/microsatellite instability-high as a predictor for anti-PD-1/PD-L1 immunotherapy efficacy. *J. Hematol. Oncol.* **2019**, *12*, 54. [CrossRef]
78. Gao, Q.; Wang, X.Y.; Qiu, S.J.; Yamato, I.; Sho, M.; Nakajima, Y.; Zhou, J.; Li, B.Z.; Shi, Y.H.; Xiao, Y.S.; et al. Overexpression of PD-L1 significantly associates with tumor aggressiveness and postoperative recurrence in human hepatocellular carcinoma. *Clin. Cancer Res. Off. J. Am. Assoc. Cancer Res.* **2009**, *15*, 971–979. [CrossRef]
79. Thommen, D.S.; Koelzer, V.H.; Herzig, P.; Roller, A.; Trefny, M.; Dimeloe, S.; Kiialainen, A.; Hanhart, J.; Schill, C.; Hess, C.; et al. A transcriptionally and functionally distinct PD-1(+) CD8(+) T cell pool with predictive potential in non-small-cell lung cancer treated with PD-1 blockade. *Nat. Med.* **2018**, *24*, 994–1004. [CrossRef]
80. Wang, B.J.; Bao, J.J.; Wang, J.Z.; Wang, Y.; Jiang, M.; Xing, M.Y.; Zhang, W.G.; Qi, J.Y.; Roggendorf, M.; Lu, M.J.; et al. Immunostaining of PD-1/PD-Ls in liver tissues of patients with hepatitis and hepatocellular carcinoma. *World J. Gastroenterol.* **2011**, *17*, 3322–3329. [CrossRef]
81. Shi, F.; Shi, M.; Zeng, Z.; Qi, R.Z.; Liu, Z.W.; Zhang, J.Y.; Yang, Y.P.; Tien, P.; Wang, F.S. PD-1 and PD-L1 upregulation promotes CD8(+) T-cell apoptosis and postoperative recurrence in hepatocellular carcinoma patients. *Int. J. Cancer* **2011**, *128*, 887–896. [CrossRef] [PubMed]
82. Zhou, G.; Sprengers, D.; Boor, P.P.C.; Doukas, M.; Schutz, H.; Mancham, S.; Pedroza-Gonzalez, A.; Polak, W.G.; de Jonge, J.; Gaspersz, M.; et al. Antibodies Against Immune Checkpoint Molecules Restore Functions of Tumor-Infiltrating T Cells in Hepatocellular Carcinomas. *Gastroenterology* **2017**, *153*, 1107–1119.e10. [CrossRef] [PubMed]
83. Losic, B.; Craig, A.J.; Villacorta-Martin, C.; Martins-Filho, S.N.; Akers, N.; Chen, X.; Ahsen, M.E.; von Felden, J.; Labgaa, I.; D'Avola, D.; et al. Intratumoral heterogeneity and clonal evolution in liver cancer. *Nat. Commun.* **2020**, *11*, 291. [CrossRef] [PubMed]
84. Krieg, C.; Nowicka, M.; Guglietta, S.; Schindler, S.; Hartmann, F.J.; Weber, L.M.; Dummer, R.; Robinson, M.D.; Levesque, M.P.; Becher, B. High-dimensional single-cell analysis predicts response to anti-PD-1 immunotherapy. *Nat. Med.* **2018**, *24*, 144–153. [CrossRef]
85. Zhang, Q.; He, Y.; Luo, N.; Patel, S.J.; Han, Y.; Gao, R.; Modak, M.; Carotta, S.; Haslinger, C.; Kind, D.; et al. Landscape and Dynamics of Single Immune Cells in Hepatocellular Carcinoma. *Cell* **2019**, *179*, 829–845.e20. [CrossRef]

86. Ma, L.; Hernandez, M.O.; Zhao, Y.; Mehta, M.; Tran, B.; Kelly, M.; Rae, Z.; Hernandez, J.M.; Davis, J.L.; Martin, S.P.; et al. Tumor Cell Biodiversity Drives Microenvironmental Reprogramming in Liver Cancer. *Cancer Cell* **2019**, *36*, 418–430.e6. [CrossRef]
87. Kulangara, K.; Zhang, N.; Corigliano, E.; Guerrero, L.; Waldroup, S.; Jaiswal, D.; Ms, M.J.; Shah, S.; Hanks, D.; Wang, J.; et al. Clinical Utility of the Combined Positive Score for Programmed Death Ligand-1 Expression and the Approval of Pembrolizumab for Treatment of Gastric Cancer. *Arch. Pathol. Lab. Med.* **2019**, *143*, 330–337. [CrossRef]
88. Garon, E.B.; Rizvi, N.A.; Hui, R.; Leighl, N.; Balmanoukian, A.S.; Eder, J.P.; Patnaik, A.; Aggarwal, C.; Gubens, M.; Horn, L.; et al. Pembrolizumab for the treatment of non-small-cell lung cancer. *N. Engl. J. Med.* **2015**, *372*, 2018–2028. [CrossRef]
89. Taube, J.M.; Klein, A.; Brahmer, J.R.; Xu, H.; Pan, X.; Kim, J.H.; Chen, L.; Pardoll, D.M.; Topalian, S.L.; Anders, R.A. Association of PD-1, PD-1 ligands, and other features of the tumor immune microenvironment with response to anti-PD-1 therapy. *Clin. Cancer Res. Off. J. Am. Assoc. Cancer Res.* **2014**, *20*, 5064–5074. [CrossRef]
90. Herbst, R.S.; Soria, J.C.; Kowanetz, M.; Fine, G.D.; Hamid, O.; Gordon, M.S.; Sosman, J.A.; McDermott, D.F.; Powderly, J.D.; Gettinger, S.N.; et al. Predictive correlates of response to the anti-PD-L1 antibody MPDL3280A in cancer patients. *Nature* **2014**, *515*, 563–567. [CrossRef]
91. Daud, A.I.; Wolchok, J.D.; Robert, C.; Hwu, W.J.; Weber, J.S.; Ribas, A.; Hodi, F.S.; Joshua, A.M.; Kefford, R.; Hersey, P.; et al. Programmed Death-Ligand 1 Expression and Response to the Anti-Programmed Death 1 Antibody Pembrolizumab in Melanoma. *J. Clin. Oncol. Off. J. Am. Soc. Clin. Oncol.* **2016**, *34*, 4102–4109. [CrossRef] [PubMed]
92. Udall, M.; Rizzo, M.; Kenny, J.; Doherty, J.; Dahm, S.; Robbins, P.; Faulkner, E. PD-L1 diagnostic tests: A systematic literature review of scoring algorithms and test-validation metrics. *Diagn. Pathol.* **2018**, *13*, 12. [CrossRef] [PubMed]
93. Pinato, D.J.; Guerra, N.; Fessas, P.; Murphy, R.; Mineo, T.; Mauri, F.A.; Mukherjee, S.K.; Thursz, M.; Wong, C.N.; Sharma, R.; et al. Immune-based therapies for hepatocellular carcinoma. *Oncogene* **2020**, *39*, 3620–3637. [CrossRef] [PubMed]
94. Cristescu, R.; Mogg, R.; Ayers, M.; Albright, A.; Murphy, E.; Yearley, J.; Sher, X.; Liu, X.Q.; Lu, H.; Nebozhyn, M.; et al. Pan-tumor genomic biomarkers for PD-1 checkpoint blockade-based immunotherapy. *Science* **2018**, *362*. [CrossRef] [PubMed]
95. Le, D.T.; Uram, J.N.; Wang, H.; Bartlett, B.R.; Kemberling, H.; Eyring, A.D.; Skora, A.D.; Luber, B.S.; Azad, N.S.; Laheru, D.; et al. PD-1 Blockade in Tumors with Mismatch-Repair Deficiency. *N. Engl. J. Med.* **2015**, *372*, 2509–2520. [CrossRef]
96. Yarchoan, M.; Hopkins, A.; Jaffee, E.M. Tumor Mutational Burden and Response Rate to PD-1 Inhibition. *N. Engl. J. Med.* **2017**, *377*, 2500–2501. [CrossRef]
97. Ang, C.; Klempner, S.J.; Ali, S.M.; Madison, R.; Ross, J.S.; Severson, E.A.; Fabrizio, D.; Goodman, A.; Kurzrock, R.; Suh, J.; et al. Prevalence of established and emerging biomarkers of immune checkpoint inhibitor response in advanced hepatocellular carcinoma. *Oncotarget* **2019**, *10*, 4018–4025. [CrossRef]
98. Goumard, C.; Desbois-Mouthon, C.; Wendum, D.; Calmel, C.; Merabtene, F.; Scatton, O.; Praz, F. Low Levels of Microsatellite Instability at Simple Repeated Sequences Commonly Occur in Human Hepatocellular Carcinoma. *Cancer Genom. Proteom.* **2017**, *14*, 329–339. [CrossRef]
99. Totoki, Y.; Tatsuno, K.; Covington, K.R.; Ueda, H.; Creighton, C.J.; Kato, M.; Tsuji, S.; Donehower, L.A.; Slagle, B.L.; Nakamura, H.; et al. Trans-ancestry mutational landscape of hepatocellular carcinoma genomes. *Nat. Genet.* **2014**, *46*, 1267–1273. [CrossRef]
100. Agdashian, D.; ElGindi, M.; Xie, C.; Sandhu, M.; Pratt, D.; Kleiner, D.E.; Figg, W.D.; Rytlewski, J.A.; Sanders, C.; Yusko, E.C.; et al. The effect of anti-CTLA4 treatment on peripheral and intra-tumoral T cells in patients with hepatocellular carcinoma. *Cancer Immunol. Immunother.* **2019**, *68*, 599–608. [CrossRef]
101. Finkelmeier, F.; Canli, Ö.; Tal, A.; Pleli, T.; Trojan, J.; Schmidt, M.; Kronenberger, B.; Zeuzem, S.; Piiper, A.; Greten, F.R.; et al. High levels of the soluble programmed death-ligand (sPD-L1) identify hepatocellular carcinoma patients with a poor prognosis. *Eur. J. Cancer* **2016**, *59*, 152–159. [CrossRef] [PubMed]

102. Feun, L.G.; Li, Y.Y.; Wu, C.; Wangpaichitr, M.; Jones, P.D.; Richman, S.P.; Madrazo, B.; Kwon, D.; Garcia-Buitrago, M.; Martin, P.; et al. Phase 2 study of pembrolizumab and circulating biomarkers to predict anticancer response in advanced, unresectable hepatocellular carcinoma. *Cancer* **2019**, *125*, 3603–3614. [CrossRef] [PubMed]
103. Zhou, J.; Mahoney, K.M.; Giobbie-Hurder, A.; Zhao, F.; Lee, S.; Liao, X.; Rodig, S.; Li, J.; Wu, X.; Butterfield, L.H.; et al. Soluble PD-L1 as a Biomarker in Malignant Melanoma Treated with Checkpoint Blockade. *Cancer Immunol. Res.* **2017**, *5*, 480–492. [CrossRef] [PubMed]
104. Loo, T.M.; Kamachi, F.; Watanabe, Y.; Yoshimoto, S.; Kanda, H.; Arai, Y.; Nakajima-Takagi, Y.; Iwama, A.; Koga, T.; Sugimoto, Y.; et al. Gut Microbiota Promotes Obesity-Associated Liver Cancer through PGE(2)-Mediated Suppression of Antitumor Immunity. *Cancer Discov.* **2017**, *7*, 522–538. [CrossRef] [PubMed]
105. Sivan, A.; Corrales, L.; Hubert, N.; Williams, J.B.; Aquino-Michaels, K.; Earley, Z.M.; Benyamin, F.W.; Lei, Y.M.; Jabri, B.; Alegre, M.L.; et al. Commensal Bifidobacterium promotes antitumor immunity and facilitates anti-PD-L1 efficacy. *Science* **2015**, *350*, 1084–1089. [CrossRef] [PubMed]
106. Vétizou, M.; Pitt, J.M.; Daillère, R.; Lepage, P.; Waldschmitt, N.; Flament, C.; Rusakiewicz, S.; Routy, B.; Roberti, M.P.; Duong, C.P.; et al. Anticancer immunotherapy by CTLA-4 blockade relies on the gut microbiota. *Science* **2015**, *350*, 1079–1084. [CrossRef] [PubMed]
107. Zheng, Y.; Wang, T.; Tu, X.; Huang, Y.; Zhang, H.; Tan, D.; Jiang, W.; Cai, S.; Zhao, P.; Song, R.; et al. Gut microbiome affects the response to anti-PD-1 immunotherapy in patients with hepatocellular carcinoma. *J. Immunother Cancer* **2019**, *7*, 193. [CrossRef]
108. Matson, V.; Fessler, J.; Bao, R.; Chongsuwat, T.; Zha, Y.; Alegre, M.L.; Luke, J.J.; Gajewski, T.F. The commensal microbiome is associated with anti-PD-1 efficacy in metastatic melanoma patients. *Science* **2018**, *359*, 104–108. [CrossRef]
109. Ma, C.; Han, M.; Heinrich, B.; Fu, Q.; Zhang, Q.; Sandhu, M.; Agdashian, D.; Terabe, M.; Berzofsky, J.A.; Fako, V.; et al. Gut microbiome-mediated bile acid metabolism regulates liver cancer via NKT cells. *Science* **2018**, *360*. [CrossRef]
110. Ally, A.; Balasundaram, M.; Carlsen, R.; Chuah, E.; Clarke, A.; Dhalla, N.; Holt, R.A.; Jones, S.J.; Lee, D.; Ma, Y.; et al. Comprehensive and Integrative Genomic Characterization of Hepatocellular Carcinoma. *Cell* **2017**, *169*, 1327–1341.e23. [CrossRef]
111. Llovet, J.M.; Montal, R.; Sia, D.; Finn, R.S. Molecular therapies and precision medicine for hepatocellular carcinoma. *Nat. Rev. Clin. Oncol.* **2018**. [CrossRef] [PubMed]
112. Schulze, K.; Imbeaud, S.; Letouzé, E.; Alexandrov, L.B.; Calderaro, J.; Rebouissou, S.; Couchy, G.; Meiller, C.; Shinde, J.; Soysouvanh, F.; et al. Exome sequencing of hepatocellular carcinomas identifies new mutational signatures and potential therapeutic targets. *Nat. Genet.* **2015**, *47*, 505–511. [CrossRef] [PubMed]
113. Sia, D.; Jiao, Y.; Martinez-Quetglas, I.; Kuchuk, O.; Villacorta-Martin, C.; Castro de Moura, M.; Putra, J.; Camprecios, G.; Bassaganyas, L.; Akers, N.; et al. Identification of an Immune-specific Class of Hepatocellular Carcinoma, Based on Molecular Features. *Gastroenterology* **2017**, *153*, 812–826. [CrossRef] [PubMed]
114. Harding, J.J.; Nandakumar, S.; Armenia, J.; Khalil, D.N.; Albano, M.; Ly, M.; Shia, J.; Hechtman, J.F.; Kundra, R.; El Dika, I.; et al. Prospective Genotyping of Hepatocellular Carcinoma: Clinical Implications of Next-Generation Sequencing for Matching Patients to Targeted and Immune Therapies. *Clin. Cancer Res. Off. J. Am. Assoc. Cancer Res.* **2019**, *25*, 2116–2126. [CrossRef]
115. Ruiz de Galarreta, M.; Bresnahan, E.; Molina-Sánchez, P.; Lindblad, K.E.; Maier, B.; Sia, D.; Puigvehi, M.; Miguela, V.; Casanova-Acebes, M.; Dhainaut, M.; et al. β-Catenin Activation Promotes Immune Escape and Resistance to Anti-PD-1 Therapy in Hepatocellular Carcinoma. *Cancer Discov.* **2019**, *9*, 1124–1141. [CrossRef]
116. Hugo, W.; Zaretsky, J.M.; Sun, L.; Song, C.; Moreno, B.H.; Hu-Lieskovan, S.; Berent-Maoz, B.; Pang, J.; Chmielowski, B.; Cherry, G.; et al. Genomic and Transcriptomic Features of Response to Anti-PD-1 Therapy in Metastatic Melanoma. *Cell* **2016**, *165*, 35–44. [CrossRef]
117. Finn, R.S.; Ryoo, B.-Y.; Merle, P.; Kudo, M.; Bouattour, M.; Lim, H.Y.; Breder, V.; Edeline, J.; Chao, Y.; Ogasawara, S.; et al. Pembrolizumab As Second-Line Therapy in Patients With Advanced Hepatocellular Carcinoma in KEYNOTE-240: A Randomized, Double-Blind, Phase III Trial. *J. Clin. Oncol.* **2020**, *38*, 193–202. [CrossRef]

118. Qin, S.; Finn, R.S.; Kudo, M.; Meyer, T.; Vogel, A.; Ducreux, M.; Macarulla, T.M.; Tomasello, G.; Boisserie, F.; Hou, J.; et al. RATIONALE 301 study: Tislelizumab versus sorafenib as first-line treatment for unresectable hepatocellular carcinoma. *Future Oncol.* **2019**, *15*, 1811–1822. [CrossRef]
119. Jimenez Exposito, M.J.; Akce, M.; Alvarez, J.L.M.; Assenat, E.; Balart, L.A.; Baron, A.D.; Decaens, T.; Heurgue-Berlot, A.; Martin, A.O.; Paik, S.W.; et al. CA209-9DX: Phase III, randomized, double-blind study of adjuvant nivolumab vs placebo for patients with hepatocellular carcinoma (HCC) at high risk of recurrence after curative resection or ablation. *Ann. Oncol.* **2018**, *29*, ix65. [CrossRef]
120. Kaseb, A.O.; Pestana, R.C.; Vence, L.M.; Blando, J.M.; Singh, S.; Ikoma, N.; Vauthey, J.-N.; Allison, J.P.; Sharma, P. Randomized, open-label, perioperative phase II study evaluating nivolumab alone versus nivolumab plus ipilimumab in patients with resectable HCC. *J. Clin. Oncol.* **2019**, *37*, 185. [CrossRef]
121. Bang, Y.-J.; Golan, T.; Lin, C.-C.; Dahan, L.; Fu, S.; Moreno, V.; Geva, R.; Reck, M.; Wasserstrom, H.A.; Mi, G.; et al. Ramucirumab (Ram) and durvalumab (Durva) treatment of metastatic non-small cell lung cancer (NSCLC), gastric/gastroesophageal junction (G/GEJ) adenocarcinoma, and hepatocellular carcinoma (HCC) following progression on systemic treatment(s). *J. Clin. Oncol.* **2019**, *37*, 2528. [CrossRef]
122. Zhu, A.X.; Finn, R.S.; Ikeda, M.; Sung, M.W.; Baron, A.D.; Kudo, M.; Okusaka, T.; Kobayashi, M.; Kumada, H.; Kaneko, S.; et al. A phase Ib study of lenvatinib (LEN) plus pembrolizumab (PEMBRO) in unresectable hepatocellular carcinoma (uHCC). *J. Clin. Oncol.* **2020**, *38*, 4519. [CrossRef]
123. Llovet, J.M.; Kudo, M.; Cheng, A.-L.; Finn, R.S.; Galle, P.R.; Kaneko, S.; Meyer, T.; Qin, S.; Dutcus, C.E.; Chen, E.; et al. Lenvatinib (len) plus pembrolizumab (pembro) for the first-line treatment of patients (pts) with advanced hepatocellular carcinoma (HCC): Phase 3 LEAP-002 study. *J. Clin. Oncol.* **2019**, *37*, TPS4152. [CrossRef]
124. Xu, J.; Zhang, Y.; Jia, R.; Yue, C.; Chang, L.; Liu, R.; Zhang, G.; Zhao, C.; Zhang, Y.; Chen, C.; et al. Anti-PD-1 Antibody SHR-1210 Combined with Apatinib for Advanced Hepatocellular Carcinoma, Gastric, or Esophagogastric Junction Cancer: An Open-label, Dose Escalation and Expansion Study. *Clin. Cancer Res. Off. J. Am. Assoc. Cancer Res.* **2019**, *25*, 515–523. [CrossRef]
125. Pishvaian, M.J.; Lee, M.S.; Ryoo, B.Y.; Stein, S.; Lee, K.H.; Verret, W.; Spahn, J.; Shao, H.; Liu, B.; Iizuka, K.; et al. Updated safety and clinical activity results from a phase Ib study of atezolizumab + bevacizumab in hepatocellular carcinoma (HCC). *Ann. Oncol.* **2018**, *29*, viii718–viii719. [CrossRef]
126. Kudo, M.; Motomura, K.; Wada, Y.; Inaba, Y.; Sakamoto, Y.; Kurosaki, M.; Umeyama, Y.; Kamei, Y.; Yoshimitsu, J.; Fujii, Y.; et al. First-line avelumab + axitinib in patients with advanced hepatocellular carcinoma: Results from a phase 1b trial (VEGF Liver 100). *J. Clin. Oncol.* **2019**, *37*, 4072. [CrossRef]
127. Kelley, R.K.; Cheng, A.-L.; Braiteh, F.S.; Park, J.-W.; Benzaghou, F.; Milwee, S.; Borgman, A.; El-Khoueiry, A.B.; Kayali, Z.K.; Zhu, A.X.; et al. Phase 3 (COSMIC-312) study of cabozantinib (C) in combination with atezolizumab (A) versus sorafenib (S) in patients (pts) with advanced hepatocellular carcinoma (aHCC) who have not received previous systemic anticancer therapy. *J. Clin. Oncol.* **2019**, *37*, TPS4157. [CrossRef]
128. Knox, J.; Cheng, A.; Cleary, S.; Galle, P.; Kokudo, N.; Lencioni, R.; Park, J.; Zhou, J.; Mann, H.; Morgan, S.; et al. A phase 3 study of durvalumab with or without bevacizumab as adjuvant therapy in patients with hepatocellular carcinoma (HCC) who are at high risk of recurrence after curative hepatic resection. *Ann. Oncol.* **2019**, *30*. [CrossRef]
129. Sangro, B.; Kudo, M.; Qin, S.; Ren, Z.; Chan, S.; Erinjeri, J.; Arai, Y.; He, P.; Morgan, S.; Cohen, G.; et al. P-347 A phase 3, randomized, double-blind, placebo-controlled study of transarterial chemoembolization combined with durvalumab or durvalumab plus bevacizumab therapy in patients with locoregional hepatocellular carcinoma: EMERALD-1. *Ann. Oncol.* **2020**, *31*, S202–S203. [CrossRef]
130. Floudas, C.S.; Xie, C.; Brar, G.; Morelli, M.P.; Fioravanti, S.; Walker, M.; Mabry-Hrones, D.; Wood, B.J.; Levy, E.B.; Krishnasamy, V.P.; et al. Combined immune checkpoint inhibition (ICI) with tremelimumab and durvalumab in patients with advanced hepatocellular carcinoma (HCC) or biliary tract carcinomas (BTC). *J. Clin. Oncol.* **2019**, *37*, 336. [CrossRef]
131. Popovic, A.; Sugar, E.; Ferguson, A.; Wilt, B.; Durham, J.N.; Kamel, I.R.; Kim, A.; Philosophe, B.; Anders, R.A.; Jaffee, E.M.; et al. Abstract CT207: Feasibility of neoadjuvant cabozantinib plus nivolumab followed by definitive resection for patients with locally advanced hepatocellular carcinoma: A Phase Ib trial (NCT03299946). *Cancer Res.* **2019**, *79*, CT207. [CrossRef]

132. Motzer, R.J.; Tannir, N.M.; McDermott, D.F.; Aren Frontera, O.; Melichar, B.; Choueiri, T.K.; Plimack, E.R.; Barthelemy, P.; Porta, C.; George, S.; et al. Nivolumab plus Ipilimumab versus Sunitinib in Advanced Renal-Cell Carcinoma. *N. Engl. J. Med.* **2018**, *378*, 1277–1290. [CrossRef] [PubMed]
133. Wolchok, J.D.; Chiarion-Sileni, V.; Gonzalez, R.; Rutkowski, P.; Grob, J.J.; Cowey, C.L.; Lao, C.D.; Wagstaff, J.; Schadendorf, D.; Ferrucci, P.F.; et al. Overall Survival with Combined Nivolumab and Ipilimumab in Advanced Melanoma. *N. Engl. J. Med.* **2017**, *377*, 1345–1356. [CrossRef] [PubMed]
134. Hellmann, M.D.; Paz-Ares, L.; Bernabe Caro, R.; Zurawski, B.; Kim, S.W.; Carcereny Costa, E.; Park, K.; Alexandru, A.; Lupinacci, L.; de la Mora Jimenez, E.; et al. Nivolumab plus Ipilimumab in Advanced Non-Small-Cell Lung Cancer. *N. Engl. J. Med.* **2019**, *381*, 2020–2031. [CrossRef]
135. Kaseb, A.; Vence, L.; Blando, J.; Yadav, S.; Ikoma, N.; Pestana, R.; Vauthey, J.; Cao, H.; Chun, Y.; Sakamura, D.; et al. Randomized, open-label, perioperative phase II study evaluating nivolumab alone versus nivolumab plus ipilimumab in patients with resectable HCC. *Ann. Oncol. Off. J. Eur. Soc. Med. Oncol.* **2019**, *30* (Suppl. 4), iv112. [CrossRef]
136. Kelley, R.K.; Abou-Alfa, G.K.; Bendell, J.C.; Kim, T.-Y.; Borad, M.J.; Yong, W.-P.; Morse, M.; Kang, Y.-K.; Rebelatto, M.; Makowsky, M.; et al. Phase I/II study of durvalumab and tremelimumab in patients with unresectable hepatocellular carcinoma (HCC): Phase I safety and efficacy analyses. *J. Clin. Oncol.* **2017**, *35*, 4073. [CrossRef]
137. Deng, Y.R.; Liu, W.B.; Lian, Z.X.; Li, X.; Hou, X. Sorafenib inhibits macrophage-mediated epithelial-mesenchymal transition in hepatocellular carcinoma. *Oncotarget* **2016**, *7*, 38292–38305. [CrossRef]
138. Kato, Y.; Tabata, K.; Kimura, T.; Yachie-Kinoshita, A.; Ozawa, Y.; Yamada, K.; Ito, J.; Tachino, S.; Hori, Y.; Matsuki, M.; et al. Lenvatinib plus anti-PD-1 antibody combination treatment activates CD8+ T cells through reduction of tumor-associated macrophage and activation of the interferon pathway. *PLoS ONE* **2019**, *14*, e0212513. [CrossRef]
139. Zerbini, A.; Pilli, M.; Laccabue, D.; Pelosi, G.; Molinari, A.; Negri, E.; Cerioni, S.; Fagnoni, F.; Soliani, P.; Ferrari, C.; et al. Radiofrequency thermal ablation for hepatocellular carcinoma stimulates autologous NK-cell response. *Gastroenterology* **2010**, *138*, 1931–1942. [CrossRef]
140. Nobuoka, D.; Motomura, Y.; Shirakawa, H.; Yoshikawa, T.; Kuronuma, T.; Takahashi, M.; Nakachi, K.; Ishii, H.; Furuse, J.; Gotohda, N.; et al. Radiofrequency ablation for hepatocellular carcinoma induces glypican-3 peptide-specific cytotoxic T lymphocytes. *Int. J. Oncol.* **2012**, *40*, 63–70. [CrossRef]
141. Hansler, J.; Wissniowski, T.T.; Schuppan, D.; Witte, A.; Bernatik, T.; Hahn, E.G.; Strobel, D. Activation and dramatically increased cytolytic activity of tumor specific T lymphocytes after radio-frequency ablation in patients with hepatocellular carcinoma and colorectal liver metastases. *World J. Gastroenterol.* **2006**, *12*, 3716–3721. [CrossRef] [PubMed]
142. den Brok, M.H.; Sutmuller, R.P.; van der Voort, R.; Bennink, E.J.; Figdor, C.G.; Ruers, T.J.; Adema, G.J. In situ tumor ablation creates an antigen source for the generation of antitumor immunity. *Cancer Res.* **2004**, *64*, 4024–4029. [CrossRef] [PubMed]
143. Den Brok, M.H.; Sutmuller, R.P.; Nierkens, S.; Bennink, E.J.; Frielink, C.; Toonen, L.W.; Boerman, O.C.; Figdor, C.G.; Ruers, T.J.; Adema, G.J. Efficient loading of dendritic cells following cryo and radiofrequency ablation in combination with immune modulation induces anti-tumour immunity. *Br. J. Cancer* **2006**, *95*, 896–905. [CrossRef] [PubMed]
144. Greten, T.F.; Mauda-Havakuk, M.; Heinrich, B.; Korangy, F.; Wood, B.J. Combined locoregional-immunotherapy for liver cancer. *J. Hepatol.* **2019**, *70*, 999–1007. [CrossRef] [PubMed]
145. Apetoh, L.; Mignot, G.; Panaretakis, T.; Kroemer, G.; Zitvogel, L. Immunogenicity of anthracyclines: Moving towards more personalized medicine. *Trends Mol. Med.* **2008**, *14*, 141–151. [CrossRef]
146. Hoechst, B.; Ormandy, L.A.; Ballmaier, M.; Lehner, F.; Kruger, C.; Manns, M.P.; Greten, T.F.; Korangy, F. A new population of myeloid-derived suppressor cells in hepatocellular carcinoma patients induces CD4(+)CD25(+)Foxp3(+) T cells. *Gastroenterology* **2008**, *135*, 234–243. [CrossRef]
147. Arihara, F.; Mizukoshi, E.; Kitahara, M.; Takata, Y.; Arai, K.; Yamashita, T.; Nakamoto, Y.; Kaneko, S. Increase in CD14+HLA-DR -/low myeloid-derived suppressor cells in hepatocellular carcinoma patients and its impact on prognosis. *Cancer Immunol. Immunother.* **2013**, *62*, 1421–1430. [CrossRef]
148. Chew, V.; Lee, Y.H.; Pan, L.; Nasir, N.J.M.; Lim, C.J.; Chua, C.; Lai, L.; Hazirah, S.N.; Lim, T.K.H.; Goh, B.K.P.; et al. Immune activation underlies a sustained clinical response to Yttrium-90 radioembolisation in hepatocellular carcinoma. *Gut* **2019**, *68*, 335–346. [CrossRef]

149. Bresnahan, E.; Lindblad, K.E.; Ruiz de Galarreta, M.; Lujambio, A. Mouse models of oncoimmunology in hepatocellular carcinoma. *Clin. Cancer Res. Off. J. Am. Assoc. Cancer Res.* **2020**. [CrossRef]
150. Olson, B.; Li, Y.; Lin, Y.; Liu, E.T.; Patnaik, A. Mouse Models for Cancer Immunotherapy Research. *Cancer Discov.* **2018**, *8*, 1358–1365. [CrossRef]
151. Wolchok, J.D. PD-1 Blockers. *Cell* **2015**, *162*, 937. [CrossRef] [PubMed]
152. El Dika, I.; Khalil, D.N.; Abou-Alfa, G.K. Immune checkpoint inhibitors for hepatocellular carcinoma. *Cancer* **2019**, *125*, 3312–3319. [CrossRef]
153. Brown, Z.J.; Yu, S.J.; Heinrich, B.; Ma, C.; Fu, Q.; Sandhu, M.; Agdashian, D.; Zhang, Q.; Korangy, F.; Greten, T.F. Indoleamine 2,3-dioxygenase provides adaptive resistance to immune checkpoint inhibitors in hepatocellular carcinoma. *Cancer Immunol. Immunother.* **2018**, *67*, 1305–1315. [CrossRef] [PubMed]
154. Kimura, T.; Kato, Y.; Ozawa, Y.; Kodama, K.; Ito, J.; Ichikawa, K.; Yamada, K.; Hori, Y.; Tabata, K.; Takase, K.; et al. Immunomodulatory activity of lenvatinib contributes to antitumor activity in the Hepa1-6 hepatocellular carcinoma model. *Cancer Sci.* **2018**, *109*, 3993–4002. [CrossRef] [PubMed]
155. Lee, J.; Liao, R.; Wang, G.; Yang, B.H.; Luo, X.; Varki, N.M.; Qiu, S.J.; Ren, B.; Fu, W.; Feng, G.S. Preventive Inhibition of Liver Tumorigenesis by Systemic Activation of Innate Immune Functions. *Cell Rep.* **2017**, *21*, 1870–1882. [CrossRef] [PubMed]
156. Wen, L.; Xin, B.; Wu, P.; Lin, C.H.; Peng, C.; Wang, G.; Lee, J.; Lu, L.F.; Feng, G.S. An Efficient Combination Immunotherapy for Primary Liver Cancer by Harmonized Activation of Innate and Adaptive Immunity in Mice. *Hepatology* **2019**, *69*, 2518–2532. [CrossRef] [PubMed]
157. Zhu, Y.; Yang, J.; Xu, D.; Gao, X.M.; Zhang, Z.; Hsu, J.L.; Li, C.W.; Lim, S.O.; Sheng, Y.Y.; Zhang, Y.; et al. Disruption of tumour-associated macrophage trafficking by the osteopontin-induced colony-stimulating factor-1 signalling sensitises hepatocellular carcinoma to anti-PD-L1 blockade. *Gut* **2019**, *68*, 1653–1666. [CrossRef]
158. Llopiz, D.; Ruiz, M.; Villanueva, L.; Iglesias, T.; Silva, L.; Egea, J.; Lasarte, J.J.; Pivette, P.; Trochon-Joseph, V.; Vasseur, B.; et al. Enhanced anti-tumor efficacy of checkpoint inhibitors in combination with the histone deacetylase inhibitor Belinostat in a murine hepatocellular carcinoma model. *Cancer Immunol. Immunother.* **2019**, *68*, 379–393. [CrossRef]
159. Xiang, J.; Zhang, N.; Sun, H.; Su, L.; Zhang, C.; Xu, H.; Feng, J.; Wang, M.; Chen, J.; Liu, L.; et al. Disruption of SIRT7 Increases the Efficacy of Checkpoint Inhibitor via MEF2D Regulation of Programmed Cell Death 1 Ligand 1 in Hepatocellular Carcinoma Cells. *Gastroenterology* **2020**, *158*, 664–678.e24. [CrossRef]
160. Kaufman, H.L.; Kohlhapp, F.J.; Zloza, A. Oncolytic viruses: A new class of immunotherapy drugs. *Nat. Rev. Drug Discov.* **2015**, *14*, 642–662. [CrossRef]
161. Yoo, S.Y.; Badrinath, N.; Woo, H.Y.; Heo, J. Oncolytic Virus-Based Immunotherapies for Hepatocellular Carcinoma. *Mediat. Inflamm.* **2017**, *2017*. [CrossRef] [PubMed]
162. Heo, J.; Reid, T.; Ruo, L.; Breitbach, C.J.; Rose, S.; Bloomston, M.; Cho, M.; Lim, H.Y.; Chung, H.C.; Kim, C.W.; et al. Randomized dose-finding clinical trial of oncolytic immunotherapeutic vaccinia JX-594 in liver cancer. *Nat. Med.* **2013**, *19*, 329–336. [CrossRef] [PubMed]
163. Nakatake, R.; Kaibori, M.; Nakamura, Y.; Tanaka, Y.; Matushima, H.; Okumura, T.; Murakami, T.; Ino, Y.; Todo, T.; Kon, M. Third-generation oncolytic herpes simplex virus inhibits the growth of liver tumors in mice. *Cancer Sci.* **2018**, *109*, 600–610. [CrossRef] [PubMed]
164. Fukuhara, H.; Ino, Y.; Todo, T. Oncolytic virus therapy: A new era of cancer treatment at dawn. *Cancer Sci.* **2016**, *107*, 1373–1379. [CrossRef]
165. Luo, Y.; Lin, C.; Ren, W.; Ju, F.; Xu, Z.; Liu, H.; Yu, Z.; Chen, J.; Zhang, J.; Liu, P.; et al. Intravenous Injections of a Rationally Selected Oncolytic Herpes Virus as a Potent Virotherapy for Hepatocellular Carcinoma. *Mol. Ther. Oncolytics* **2019**, *15*, 153–165. [CrossRef]
166. Maeng, H.; Terabe, M.; Berzofsky, J.A. Cancer vaccines: Translation from mice to human clinical trials. *Curr. Opin. Immunol.* **2018**, *51*, 111–122. [CrossRef]
167. Si, C.; Xu, M.; Lu, M.; Yu, Y.; Yang, M.; Yan, M.; Zhou, L.; Yang, X. In vivo antitumor activity evaluation of cancer vaccines prepared by various antigen forms in a murine hepatocellular carcinoma model. *Oncol. Lett.* **2017**, *14*, 7391–7397. [CrossRef]
168. Huang, F.; Chen, J.; Lan, R.; Wang, Z.; Chen, R.; Lin, J.; Zhang, L.; Fu, L. delta-Catenin peptide vaccines repress hepatocellular carcinoma growth via CD8(+) T cell activation. *Oncoimmunology* **2018**, *7*, e1450713. [CrossRef]

169. Wu, Q.; Pi, L.; Le Trinh, T.; Zuo, C.; Xia, M.; Jiao, Y.; Hou, Z.; Jo, S.; Puszyk, W.; Pham, K.; et al. A Novel Vaccine Targeting Glypican-3 as a Treatment for Hepatocellular Carcinoma. *Mol. Ther. J. Am. Soc. Gene Ther.* **2017**, *25*, 2299–2308. [CrossRef]
170. Hanke, P.; Serwe, M.; Dombrowski, F.; Sauerbruch, T.; Caselmann, W.H. DNA vaccination with AFP-encoding plasmid DNA prevents growth of subcutaneous AFP-expressing tumors and does not interfere with liver regeneration in mice. *Cancer Gene Ther.* **2002**, *9*, 346–355. [CrossRef]
171. Tian, G.; Yi, J.L.; Xiong, P. Specific cellular immunity and antitumor responses in C57BL/6 mice induced by DNA vaccine encoding murine AFP. *Hepatobiliary Pancreat. Dis. Int.* **2004**, *3*, 440–443. [PubMed]
172. Tan, X.H.; Zhu, Q.; Liu, C.; Liu, X.L.; Shao, X.T.; Wei, B. Immunization with dendritic cells infected with human AFP adenovirus vector effectively elicits immunity against mouse hepatocellular carcinomas. *Zhonghua Zhong Liu Za Zhi* **2006**, *28*, 13–16. [PubMed]
173. Lan, Y.H.; Li, Y.G.; Liang, Z.W.; Chen, M.; Peng, M.L.; Tang, L.; Hu, H.D.; Ren, H. A DNA vaccine against chimeric AFP enhanced by HSP70 suppresses growth of hepatocellular carcinoma. *Cancer Immunol. Immunother.* **2007**, *56*, 1009–1016. [CrossRef] [PubMed]
174. Hong, Y.; Peng, Y.; Guo, Z.S.; Guevara-Patino, J.; Pang, J.; Butterfield, L.H.; Mivechi, N.F.; Munn, D.H.; Bartlett, D.L.; He, Y. Epitope-optimized alpha-fetoprotein genetic vaccines prevent carcinogen-induced murine autochthonous hepatocellular carcinoma. *Hepatology* **2014**, *59*, 1448–1458. [CrossRef]
175. Ahn, H.M.; Ryu, J.; Song, J.M.; Lee, Y.; Kim, H.J.; Ko, D.; Choi, I.; Kim, S.J.; Lee, J.W.; Kim, S. Anti-cancer Activity of Novel TM4SF5-Targeting Antibodies through TM4SF5 Neutralization and Immune Cell-Mediated Cytotoxicity. *Theranostics* **2017**, *7*, 594–613. [CrossRef]
176. Piotrowska, D.; Baczynska, K. Incidence of positive Waaler-Rose test in women exposed to urogenital infections. *Reumatologia* **1978**, *16*, 21–24.
177. Feng, M.; Gao, W.; Wang, R.; Chen, W.; Man, Y.G.; Figg, W.D.; Wang, X.W.; Dimitrov, D.S.; Ho, M. Therapeutically targeting glypican-3 via a conformation-specific single-domain antibody in hepatocellular carcinoma. *Proc. Natl. Acad. Sci. USA* **2013**, *110*, E1083–E1091. [CrossRef]
178. Jin, Z.; Lei, L.; Lin, D.; Liu, Y.; Song, Y.; Gong, H.; Zhu, Y.; Mei, Y.; Hu, B.; Wu, Y.; et al. IL-33 Released in the Liver Inhibits Tumor Growth via Promotion of CD4(+) and CD8(+) T Cell Responses in Hepatocellular Carcinoma. *J. Immunol.* **2018**, *201*, 3770–3779. [CrossRef]
179. Zhuang, L.; Fulton, R.J.; Rettman, P.; Sayan, A.E.; Coad, J.; Al-Shamkhani, A.; Khakoo, S.I. Activity of IL-12/15/18 primed natural killer cells against hepatocellular carcinoma. *Hepatol. Int.* **2019**, *13*, 75–83. [CrossRef]
180. Subleski, J.J.; Hall, V.L.; Back, T.C.; Ortaldo, J.R.; Wiltrout, R.H. Enhanced antitumor response by divergent modulation of natural killer and natural killer T cells in the liver. *Cancer Res.* **2006**, *66*, 11005–11012. [CrossRef]
181. Zhang, R.; Zhang, Z.; Liu, Z.; Wei, D.; Wu, X.; Bian, H.; Chen, Z. Adoptive cell transfer therapy for hepatocellular carcinoma. *Front. Med.* **2019**, *13*, 3–11. [CrossRef] [PubMed]
182. Dargel, C.; Bassani-Sternberg, M.; Hasreiter, J.; Zani, F.; Bockmann, J.H.; Thiele, F.; Bohne, F.; Wisskirchen, K.; Wilde, S.; Sprinzl, M.F.; et al. T Cells Engineered to Express a T-Cell Receptor Specific for Glypican-3 to Recognize and Kill Hepatoma Cells In Vitro and in Mice. *Gastroenterology* **2015**, *149*, 1042–1052. [CrossRef] [PubMed]
183. Gao, H.; Li, K.; Tu, H.; Pan, X.; Jiang, H.; Shi, B.; Kong, J.; Wang, H.; Yang, S.; Gu, J.; et al. Development of T cells redirected to glypican-3 for the treatment of hepatocellular carcinoma. *Clin. Cancer Res. Off. J. Am. Assoc. Cancer Res.* **2014**, *20*, 6418–6428. [CrossRef] [PubMed]
184. Yu, M.; Luo, H.; Fan, M.; Wu, X.; Shi, B.; Di, S.; Liu, Y.; Pan, Z.; Jiang, H.; Li, Z. Development of GPC3-Specific Chimeric Antigen Receptor-Engineered Natural Killer Cells for the Treatment of Hepatocellular Carcinoma. *Mol. Ther.* **2018**, *26*, 366–378. [CrossRef]
185. Jiang, Z.; Jiang, X.; Chen, S.; Lai, Y.; Wei, X.; Li, B.; Lin, S.; Wang, S.; Wu, Q.; Liang, Q.; et al. Anti-GPC3-CAR T Cells Suppress the Growth of Tumor Cells in Patient-Derived Xenografts of Hepatocellular Carcinoma. *Front. Immunol.* **2016**, *7*, 690. [CrossRef]
186. Wu, X.; Luo, H.; Shi, B.; Di, S.; Sun, R.; Su, J.; Liu, Y.; Li, H.; Jiang, H.; Li, Z. Combined Antitumor Effects of Sorafenib and GPC3-CAR T Cells in Mouse Models of Hepatocellular Carcinoma. *Mol. Ther.* **2019**, *27*, 1483–1494. [CrossRef]

187. Liu, M.; Zhou, J.; Liu, X.; Feng, Y.; Yang, W.; Wu, F.; Cheung, O.K.; Sun, H.; Zeng, X.; Tang, W.; et al. Targeting monocyte-intrinsic enhancer reprogramming improves immunotherapy efficacy in hepatocellular carcinoma. *Gut* **2020**, *69*, 365–379. [CrossRef]
188. Zhou, J.; Liu, M.; Sun, H.; Feng, Y.; Xu, L.; Chan, A.W.H.; Tong, J.H.; Wong, J.; Chong, C.C.N.; Lai, P.B.S.; et al. Hepatoma-intrinsic CCRK inhibition diminishes myeloid-derived suppressor cell immunosuppression and enhances immune-checkpoint blockade efficacy. *Gut* **2018**, *67*, 931–944. [CrossRef]
189. Mantovani, A.; Marchesi, F.; Malesci, A.; Laghi, L.; Allavena, P. Tumour-associated macrophages as treatment targets in oncology. *Nat. Reviews. Clin. Oncol.* **2017**, *14*, 399–416. [CrossRef]
190. Degroote, H.; Van Dierendonck, A.; Geerts, A.; Van Vlierberghe, H.; Devisscher, L. Preclinical and Clinical Therapeutic Strategies Affecting Tumor-Associated Macrophages in Hepatocellular Carcinoma. *J. Immunol. Res.* **2018**, *2018*. [CrossRef]
191. Zheng, X.; Turkowski, K.; Mora, J.; Brüne, B.; Seeger, W.; Weigert, A.; Savai, R. Redirecting tumor-associated macrophages to become tumoricidal effectors as a novel strategy for cancer therapy. *Oncotarget* **2017**, *8*, 48436–48452. [CrossRef] [PubMed]
192. Ao, J.Y.; Zhu, X.D.; Chai, Z.T.; Cai, H.; Zhang, Y.Y.; Zhang, K.Z.; Kong, L.Q.; Zhang, N.; Ye, B.G.; Ma, D.N.; et al. Colony-Stimulating Factor 1 Receptor Blockade Inhibits Tumor Growth by Altering the Polarization of Tumor-Associated Macrophages in Hepatocellular Carcinoma. *Mol. Cancer Ther.* **2017**, *16*, 1544–1554. [CrossRef] [PubMed]
193. Li, X.; Yao, W.; Yuan, Y.; Chen, P.; Li, B.; Li, J.; Chu, R.; Song, H.; Xie, D.; Jiang, X.; et al. Targeting of tumour-infiltrating macrophages via CCL2/CCR2 signalling as a therapeutic strategy against hepatocellular carcinoma. *Gut* **2017**, *66*, 157–167. [CrossRef] [PubMed]
194. Yao, W.; Ba, Q.; Li, X.; Li, H.; Zhang, S.; Yuan, Y.; Wang, F.; Duan, X.; Li, J.; Zhang, W.; et al. A Natural CCR2 Antagonist Relieves Tumor-associated Macrophage-mediated Immunosuppression to Produce a Therapeutic Effect for Liver Cancer. *EBioMedicine* **2017**, *22*, 58–67. [CrossRef] [PubMed]
195. Chai, Z.T.; Zhu, X.D.; Ao, J.Y.; Wang, W.Q.; Gao, D.M.; Kong, J.; Zhang, N.; Zhang, Y.Y.; Ye, B.G.; Ma, D.N.; et al. microRNA-26a suppresses recruitment of macrophages by down-regulating macrophage colony-stimulating factor expression through the PI3K/Akt pathway in hepatocellular carcinoma. *J. Hematol. Oncol.* **2015**, *8*, 56. [CrossRef]
196. Tan, H.Y.; Wang, N.; Man, K.; Tsao, S.W.; Che, C.M.; Feng, Y. Autophagy-induced RelB/p52 activation mediates tumour-associated macrophage repolarisation and suppression of hepatocellular carcinoma by natural compound baicalin. *Cell Death Dis.* **2015**, *6*, e1942. [CrossRef]
197. Liu, P.; Chen, L.; Zhang, H. Natural Killer Cells in Liver Disease and Hepatocellular Carcinoma and the NK Cell-Based Immunotherapy. *J. Immunol. Res.* **2018**, *2018*, 1206737. [CrossRef]
198. Mikulak, J.; Bruni, E.; Oriolo, F.; Di Vito, C.; Mavilio, D. Hepatic Natural Killer Cells: Organ-Specific Sentinels of Liver Immune Homeostasis and Physiopathology. *Front. Immunol.* **2019**, *10*, 946. [CrossRef]
199. Juengpanich, S.; Shi, L.; Iranmanesh, Y.; Chen, J.; Cheng, Z.; Khoo, A.K.; Pan, L.; Wang, Y.; Cai, X. The role of natural killer cells in hepatocellular carcinoma development and treatment: A narrative review. *Transl. Oncol.* **2019**, *12*, 1092–1107. [CrossRef]
200. Hoechst, B.; Voigtlaender, T.; Ormandy, L.; Gamrekelashvili, J.; Zhao, F.; Wedemeyer, H.; Lehner, F.; Manns, M.P.; Greten, T.F.; Korangy, F. Myeloid derived suppressor cells inhibit natural killer cells in patients with hepatocellular carcinoma via the NKp30 receptor. *Hepatology* **2009**, *50*, 799–807. [CrossRef]
201. Liu, C.; Yu, S.; Kappes, J.; Wang, J.; Grizzle, W.E.; Zinn, K.R.; Zhang, H.G. Expansion of spleen myeloid suppressor cells represses NK cell cytotoxicity in tumor-bearing host. *Blood* **2007**, *109*, 4336–4342. [CrossRef] [PubMed]

© 2020 by the authors. Licensee MDPI, Basel, Switzerland. This article is an open access article distributed under the terms and conditions of the Creative Commons Attribution (CC BY) license (http://creativecommons.org/licenses/by/4.0/).

Article

Programmed Death 1 Ligand Expression in the Monocytes of Patients with Hepatocellular Carcinoma Depends on Tumor Progression

Akira Asai *, Hidetaka Yasuoka, Masahiro Matsui, Yusuke Tsuchimoto, Shinya Fukunishi and Kazuhide Higuchi

The Second Department of Internal Medicine, Osaka Medical College, Takatsuki 5698686, Japan; yh0403.4351@gmail.com (H.Y.); masa1987_11_18@yahoo.co.jp (M.M.); in2141@osaka-med.ac.jp (Y.T.); in2104@osaka-med.ac.jp (S.F.); higuchi@osaka-med.ac.jp (K.H.)
* Correspondence: in2108@osaka-med.ac.jp; Tel.: +81-(726)-83-1221

Received: 28 June 2020; Accepted: 8 August 2020; Published: 14 August 2020

Abstract: Monocytes ($CD14^+$ cells) from advanced-stage hepatocellular carcinoma (HCC) patients express programmed death 1 ligand (PD-L)/PD-1 and suppress the host antitumor immune response. However, it is unclear whether cancer progression is associated with $CD14^+$ cells. We compared $CD14^+$ cell properties before and after cancer progression in the same HCC patients and examined their role in antitumor immunity. $CD14^+$ cells were isolated from 15 naïve early-stage HCC patients before treatment initiation and after cancer progression to advanced stages. Although $CD14^+$ cells from patients at early HCC stages exhibited antitumor activity in humanized murine chimera, $CD14^+$ cells from the same patients after progression to advanced stages lacked this activity. Moreover, $CD14^+$ cells from early HCC stages scantly expressed PD-L1 and PD-L2 and produced few cytokines, while $CD14^+$ cells from advanced stages showed increased PD-L expression and produced IL-10 and CCL1. $CD14^+$ cells were also isolated from five naïve advanced-stage HCC patients before treatment as well as after treatment-induced tumor regression. The $CD14^+$ cells from patients with advanced-stage HCC expressed PD-L expressions, produced IL-10 and CCL1, and exhibited minimal tumoricidal activity. After treatment-induced tumor regression, $CD14^+$ cells from the same patients did not express PD-Ls, failed to produce cytokines, and recovered tumoricidal activity. These results indicate that PD-L expression as well as $CD14^+$ cell phenotype depend on the tumor stage in HCC patients. PD-L expressions of monocytes may be used as a new marker in the classification of cancer progression in HCC.

Keywords: $CD14^+$ cells; hepatocellular carcinoma; programmed death 1 ligands

1. Introduction

Hepatocellular carcinoma (HCC) is the fourth most common cause of cancer-related death worldwide. The World Health Organization estimates that more than 1 million individuals will die from HCC in 2030 [1]. The majority of cases occur in patients with liver disease, mostly as a result of hepatitis B or C virus (HBV or HCV) infection, alcohol abuse, or nonalcoholic steatohepatitis. The five-year survival rate of HCC patients is only 18%, making it the second most lethal tumor [2]. Therapeutic options are primarily selected on the basis of tumor stage and the extent of liver dysfunction. In the past, HCC was treated by surgical resection or radiofrequency ablation. However, frequent synchronous or metachronous recurrence in the form of new tumors or intrahepatic metastases led to high mortality rates in HCC patients [3].

Recently, immune checkpoint inhibitors have become available as a new treatment option for advanced-stage HCC [4]. These compounds target the programmed death 1 (PD-1)/programmed death ligand (PD-L) axis, which is involved in cancer immune escape or evasion [5]. PD-L1, which is expressed in HCC tumor cells, interacts with PD-1 receptors on activated T cells, leading to their inactivation [6–8] and ultimately suppressing the antitumor immune response of effector cells [9]. However, it has been reported that some PD-L1-positive tumors do not respond to the anti-PD-1 antibody, while a proportion of PD-L1-negative tumors do [10,11]. These discrepancies are not fully understood, although several mechanisms have been proposed [12,13].

The tumor microenvironment plays an important role in the establishment and progression of tumors. Among the immune cells in the HCC tissue, tumor-associated macrophages (Mϕ, TAMs) sustain tumor progression and are recruited from circulating $CD14^+$ cells (monocytes) to the tumor microenvironment through tumor-derived signals [14]. In response to microbial stimuli and IFN-γ, M1 polarization occurs, thus leading to tumoricidal activity by producing high amounts of toxic intermediates. However, once tumors are progressed, the Mϕ that infiltrate the tumor tissue differentiate into M2Mϕ, which promote growth, invasion, and metastasis of tumor cells, thus inducing angiogenesis and suppressing antitumor immunity. TAMs often play a central role in tumor progression, and many TAMs have the property of M2Mϕ. Three different M2Mϕ phenotypes (M2aMϕ, M2bMϕ, and M2cMϕ) have been described [15] that are distinguished from each other based on their gene expression profiles, chemokine production, and surface marker expression [16,17]. Specifically, IL-10- and CCL17-producing Mϕ with mannose receptor gene expression are identified as M2aMϕ, IL-10- and CCL1-producing Mϕ are classified as M2bMϕ, and IL-10- and CXCL13-producing Mϕ with mannose receptor gene expression are recognized as M2cMϕ.

In a previous study, $CD14^+$ cells detected in peripheral blood of patients with advanced stages of HCC were characterized as the M2b phenotype and were found to be a significant contributor to tumor growth promotion [18]. Meanwhile, in our recent report, we showed that the monocytes of patients with advanced-stage HCC expressed PD-L1 and PD-L2 and suppressed the antitumor immune response of other effector cells. Notably, these patients had a poor prognosis [19]. However, the impact of these cells on the host immune response against HCC is unknown. Moreover, it is also unclear whether these cells are induced by cancer progression or whether they actively contribute to this process. In this study, we investigated the relationship between disease progression and PD-L expression in monocytes of HCC patients. Finally, the tumoricidal activity of these monocytes against HCC was examined.

2. Results

2.1. Growth of HepG2 Cell-Derived Tumors in Chimeric Mice with $CD14^+$ Cells from HCC Patients

To compare the antitumor activity of $CD14^+$ cells from HCC patients at different tumor stages, xNSG mice (NSG mice exposed to whole-body X-ray radiation) were used to generate humanized murine chimeras. Specifically, the mice were injected with HepG2 cells and then with $CD14^+$ cells from the same patients at early tumor stages (at the time of initial treatment) or at advanced tumor stages (Figure 1). The growth of solid tumors was measured weekly for six weeks. At the end of this period, the size of tumors in xNSG mice not transplanted with $CD14^+$ cells was 253 ± 93.9 mm^3. Notably, tumor growth was not detected in xNSG mice transplanted with $CD14^+$ cells from early-stage HCC patients. However, solid tumors were observed in xNSG mice that had been transplanted with $CD14^+$ cells collected at advanced tumor stages (229 ± 88.9 mm^3). These results indicate that the antitumor activity of $CD14^+$ cells in the same patients changed with tumor progression.

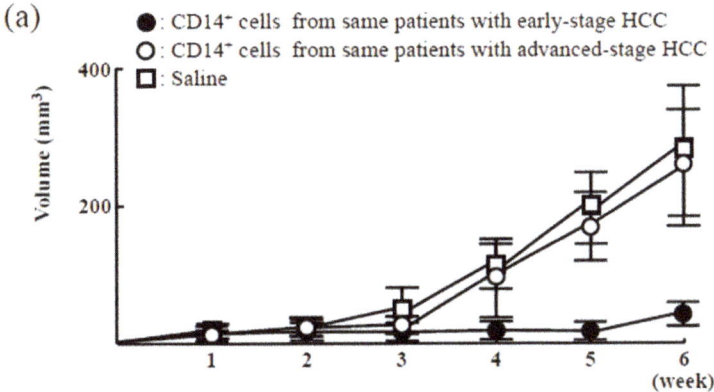

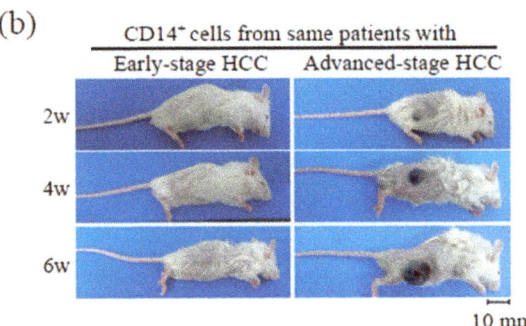

Figure 1. Growth of human HepG2 cells in chimeric mice transplanted with CD14$^+$ cells from hepatocellular carcinoma (HCC) patients. (**a**,**b**) Chimeric mice were created in xNSG mice by transplantation of CD14$^+$ cells (1 × 10^6 cells/chimera) from patients with early-stage HCC (closed circles; n = 5). CD14$^+$ cells were also isolated from those patients who were later diagnosed with advanced-stage HCC, and new chimeric mice were created from xNSG mice by the same method (open circles; n = 5). Both groups of chimeric mice were subcutaneously inoculated with HepG2 cells (2 × 10^6 cells/mice). xNSG control mice received saline along with tumor cell inoculation (open squares; n = 3). Tumor size was measured with a microcaliper, and tumor volume is expressed in mm^3.

2.2. Differences in the Properties of CD14$^+$ Cells in the Same Patients with Early and Advanced HCC

Tumor stage-related changes in the properties of CD14$^+$ cells were examined. CD14$^+$ cells were stained for PD-L1 and PD-L2 (Figure 2a,b). CD14$^+$ cells isolated from patients with early-stage HCC expressed PD-L1 (41.2 ± 9.7%) and PD-L2 (24.5 ± 4.5%). However, the expression of these PD-Ls was higher in CD14$^+$ cells obtained from the same patients at advanced HCC stages (PD-L1: 50.0 ± 7.8%, PD-L2: 36.2 ± 3.9%). Further, CD14$^+$ cells (1 × 10^6 cells/mL) from the same patients with early stages and advanced stages were cultured for 48 h, and the culture media was assayed by ELISA for production of IL-12, IL-10, CCL17, CCL1, and CXCL13 (Figure 2c). IL-12 was not produced by either population of CD14$^+$ cells. However, IL-10 production was higher in CD14$^+$ cells from advanced-stage HCC patients (212.9 ± 146.9 pg/mL) than in those from early-stage patients (64.5 ± 49.6 pg/mL). The production levels of CCL17 and CXCL13 did not differ between the two groups of CD14$^+$ cells. CCL1 production was significantly higher at advanced HCC stages (2.84 ± 3.49 ng/mL) compared to early tumor stages (0.03 ± 0.08 ng/mL). In conclusion, CD14$^+$ cells from patients with early-stage HCC exhibited poor PD-L, IL-12, and IL-10 expression and were negative for CCL17, CCL1, and CXCL13 (considered a quiescent Mϕ phenotype). Meanwhile, CD14+ cells from the same patients at advanced HCC stages acquired PD-L1 and PD-L2 expression and exhibited a IL-12⁻IL-10+CCL17⁻CCL1+CXCL13⁻ phenotype (considered a M2bMϕ phenotype).

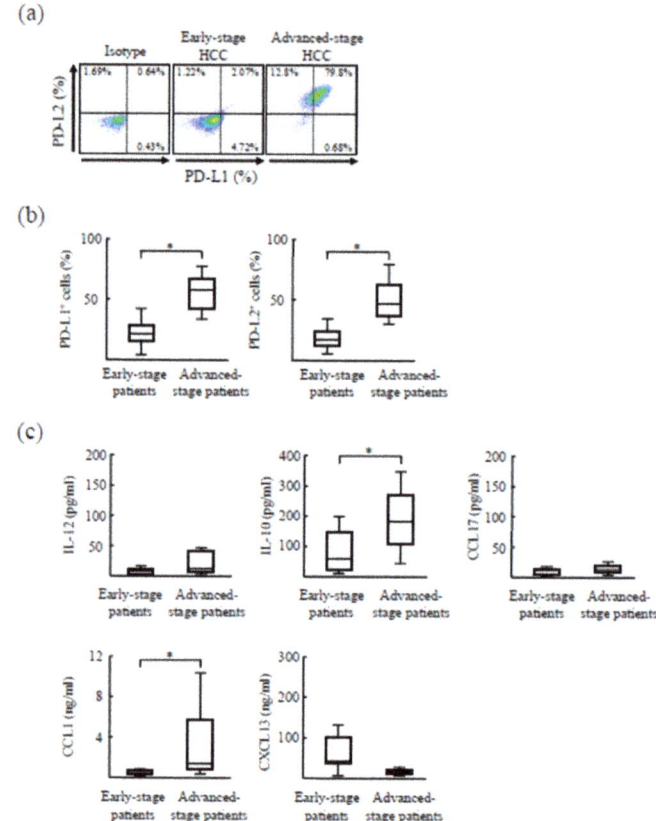

Figure 2. Differences in the properties of CD14+ cells in the same patients at early and advanced HCC stages. (**a,b**) CD14+ cells were isolated from HCC patients ($n = 12$) at early tumor stages before initial treatment and at advanced stages after treatment initiation. These cells were stained with anti-PD-L1 and anti-PD-L2 antibodies and assayed by flow cytometry. (**c**) CD14+ cells were cultured for 24 h, and the culture media was assayed by ELISA for IL-12, IL-10, CCL17, CCL1, and CXCL13 production. * $p < 0.05$. PD-L, programmed death ligand.

2.3. Tumor Regression Is Associated with Restoration of CD14+ Cell Properties

In a small subset of patients with advanced-stage HCC, the treatments induced tumor regression. CD14+ cells isolated before treatment initiation from these responsive patients exhibited high expression of both PD-L1 (84.1 ± 15.7%) and PD-L2 (81.7 ± 13.9%). Notably, in CD14+ cells collected from the same patients with early stages of HCC after initial treatment, expression levels of both PD-Ls were found to be decreased (PD-L1, 18.7 ± 4.5%; PD-L2, 50.3 ± 35.7%) (Figure 3a). Moreover, although IL-10 (189 ± 78 pg/mL) and CCL1 (3.2 ± 1.9 ng/mL) were detected in the culture media of CD14+ cells from patients with advanced disease, the production of these cytokines was abolished after treatment and tumor regression (Figure 3b).

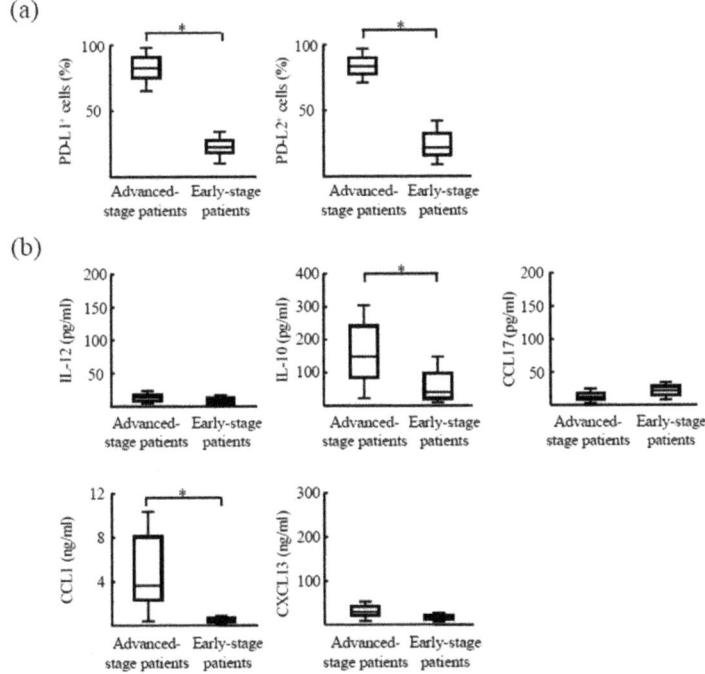

Figure 3. Changes in CD14$^+$ cell properties related to treatment-induced tumor regression in patients with advanced-stage HCC. (**a**) CD14$^+$ cells were isolated from patients with advanced HCC before treatment initiation and after tumor regression (n = 5). CD14$^+$ cells were stained with anti-PD-L1 and anti-PD-L2 antibodies and assayed by flow cytometry. (**b**) CD14$^+$ cells were cultured for 24 h, and their culture media was analyzed by ELISA for IL-12, IL-10, CCL17, CCL1, and CXCL13 detection. * p < 0.05.

We also investigated PD-L expression in CD14$^+$ cells from the peripheral blood of an additional 89 HCC patients with various stages of HCC. CD14$^+$ cells from patients with early stages of HCC had weak expression of PD-L1 or PD-L2. Meanwhile, CD14$^+$ cells from advanced stages strongly expressed both PD-Ls (Figure 4). Subsequently, the cytotoxic effect of CD14$^+$ cells against HepG2 cells was examined. CD14$^+$ cells exhibited weaker tumoricidal activity in patients with advanced HCC compared to those with early-stage HCC. Moreover, treatment-induced tumor regression was associated with the restoration of antitumor activity in CD14$^+$ cells. These results indicate that the properties of CD14$^+$ cells depend on the state of HCC progression (Figure 5).

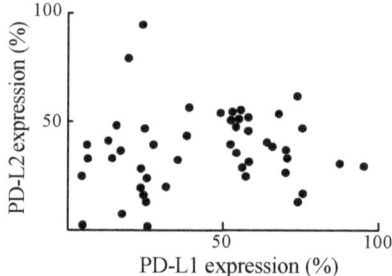

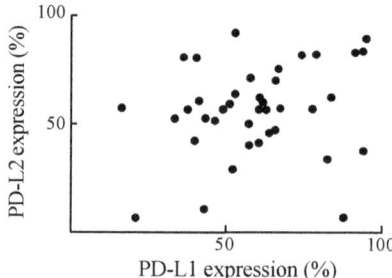

Figure 4. PD-L expressions in CD14$^+$ cells from patients with several stages of HCC. CD14$^+$ cells were isolated from peripheral blood of patients with early stages of HCC (n = 48) (**a**) and advanced stages (n = 41) (**b**). These cells were stained for PD-L1 and PD-L2 and assayed for flow cytometry.

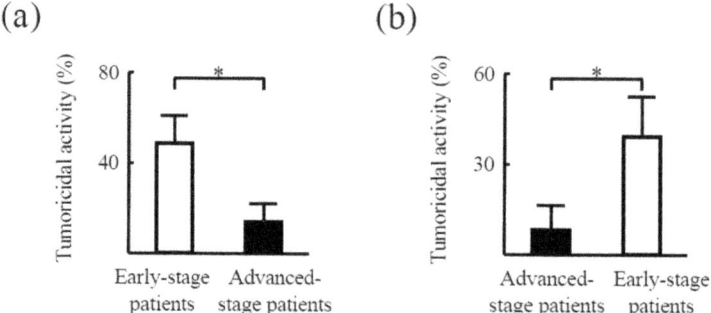

Figure 5. Tumoricidal activity of CD14$^+$ cells depends on cancer progression. CD14$^+$ cells isolated from patients before and after tumor regression were examined for cytotoxic activity against HepG2 cells by lactate dehydrogenase (LDH) release assay. (**a**) Changes related to tumor progression from early stages to advanced stages (n = 5). (**b**) Changes related to tumor progression from advanced stages to early stages (n = 5). * $p < 0.05$.

3. Discussion

In this study, the antitumor activity of monocytes was investigated in relation to cancer progression. Monocytes are innate immune cells that serve as important regulators of cancer development and progression. These cells appear to play a dichotomous role depending on the cancer type/tissue of origin as well as the tumor microenvironment and stage of tumor growth [20]. CCL2, produced by tumor cells and associated stromal cells, is one of the best characterized tumor-derived factors that induces chemotaxis in monocytes, causing circulating monocytes to be recruited from the peripheral blood into the tumor sites [14,21]. During early stages of tumor growth, recruited monocytes directly kill tumor cells via cytokine-mediated induction of cell death and phagocytosis (M1Mϕ). Specifically, the recruited monocytes exposed to IFN-γ produce tumor necrosis factor-related apoptosis-inducing ligand (TRAIL), which induces cell death in TRAIL-sensitive tumor cells and stimulates secretion of CCL2 and IL-8 from tumor cells. Meanwhile, tumors that manage to escape immune surveillance mechanisms will progress, and M1Mϕ cells will become reprogrammed within the tumor environment to limit their cytotoxicity and differentiate into TAMs (M2Mϕ) [22]. These cells then produce IL-10 and TGF-β, which function to suppress the activities of other antitumor effector immune cells [15,23]. In this study, we investigated the phenotype of $CD14^+$ cells (monocytes) from the peripheral blood before they were recruited to the tissues.

Results showed that $CD14^+$ cells from patients with early stages of HCC were PD-L1$^-$PD-L2$^-$IL-12$^-$IL-10$^-$CCL17$^-$CCL1$^-$CXCL13$^-$ (considered as a quiescent phenotype), while $CD14^+$ cells from the same patients after tumor progression were PD-L1$^+$PD-L2$^+$IL-12$^-$IL-10$^+$CCL17$^-$CCL1$^+$CXCL13$^-$ (considered to be the M2b phenotype). These results suggest that circulating monocytes from patients with advanced stages of HCC had already become skewed toward the M2b phenotype in the peripheral blood by the influence of tumor-associating factors. When these monocytes then become recruited to the tumor tissues by CCL2, they facilitate tumorigenesis by promoting immune suppression, extracellular matrix remodeling, angiogenesis, and tumor cell intravasation into the vasculature. Similarly, in other cancers, it has been reported that properties of peripheral blood monocytes are associated with better survival [24–26]. These results suggest that discrimination of monocyte phenotypes may provide a diagnostic or prognostic marker for HCC. Furthermore, the development of treatments that target monocyte differentiation may prove effective.

The generation of these M2b monocytes/Mϕ during cancer progression could be affected by immune complex formation as various factors capable of inducing M2bMϕ have been described, including immune complexes [27,28]. Moreover, immune complexes targeting cancer antigens have been detected in the serum of patients with various cancers [29,30]. Therefore, it is reasonable to postulate that as cancer progresses, serum immune complexes increase, promoting the production of M2b monocytes from quiescent monocytes. It is further speculated that the M2b monocytes then suppress antitumor effector cells by producing cytokines, thereby promoting cancer progression.

Another mechanism of tumor progression includes the expression of PD-Ls on tumors, which help facilitate their escape from antitumor immunity by binding to PD-1 on various antitumor effector cells, including $CD8^+$ cells. In this study, PD-Ls were found to be expressed in $CD14^+$ cells in HCC patients over time as the cancer progressed. We have previously reported the influence of PD-L1$^+$PD-L2$^+$CD14$^+$ cells from patients with advanced stages of HCC on the antitumor activity of $CD8^+$ cells. Specifically, PD-L1$^+$PD-L2$^+$CD14$^+$ cells were cocultured with $CD8^+$ cells isolated from syngeneic patients, resulting in the antitumor activity of $CD8^+$ cells being suppressed against HepG2 cells and Huh7 cells. Meanwhile, their antitumor activity was restored following treatment with a PD-1 antibody, that is, $CD14^+$ cells suppressed the antitumor activity of $CD8^+$ cells via the PD-L/PD-1 pathway, which can be restored by PD-1 antibodies [19]. Similarly, it has been reported that PD-L1 on dendritic cells mediates $CD8^+$ T-cell antitumor activity [31]. In many tumor types, PD-L1 expression is reportedly correlated with PD-1/PD-L1 inhibition. However, patients with very low or absent PD-L1 expression in tumor cells may still derive some benefit from treatment with PD-1 antibody [32]. Hence, the expression of PD-Ls may be responsible for the effect of $CD14^+$ cells, which warrants further investigation into the precise associated mechanism.

The expression of PD-L1 is controlled by different mechanisms. PD-L1 constitutive expression in cancer cells may be due to several oncogenic pathways, including chromosome 9 amplification [33,34], PTEN deletions, PI3K/AKT [35], and EGFR mutations [36], MYC overexpression [37], CDK5 disruption [38], and increased PD-L1 transcription [39,40]. The expression of PD-L1 and PD-L2 in tissue Mϕ has been detected in HCC and other cancers [41–43]. However, the mechanism of their expression in peripheral blood monocytes remains unclear. Recently, it was reported that the intracellular transfer of cell surface proteins from Reed–Sternberg cells to monocytes, a process known as "trogocytosis", is induced by direct contact between these cells. Trogocytosis mediates the transfer of PD-L1/L2 from lymphoma cells to monocytes within 1 h [44,45]. Therefore, trogocytosis from cancer cells to monocytes may also occur in HCC patients. However, additional experiments are needed to clarify this issue.

There are certain limitations to note in this study. First, the sample size was relatively small. Notably, this was a longitudinal study, which means repeated blood sampling from the same patients with HCC over a long period of time. Therefore, enrollment was a time-consuming process. Second, patients who had undergone various treatments against HCC were included in the study, and an influence of these treatments on monocyte function cannot be excluded. Therefore, long-term studies with a larger number of patients and more homogeneous cohorts are needed. Third, PD-L1/L2 expression in other antitumor effector cells, such as CD4 T cells, CD8 T cells, B cells, and mast cells, were not examined. However, we have previously reported that the expression of both PD-L1 and PD-L2 in monocytes is associated with poor prognosis [19]. We conclude that monocytes expressing both PD-L1 and PD-L2 may play a key role in antitumor immunity.

4. Materials and Methods

4.1. Ethics Statement

The study was approved by the Institutional Review Board of the Osaka Medical College. All subjects gave their informed consent for inclusion before they participated in the study. The study was conducted in accordance with the Declaration of Helsinki, and the protocol was approved by the Ethics Committee of Osaka Medical College (approval number: 2125). All animal experiments were carried out in compliance with Japanese regulations. The local institutional animal ethics board of Osaka Medical College approved all mouse experiments (approval number: 26022).

4.2. Patients and Specimens

Patients with HCC were classified as "early stage" based on diagnosis of very early- and early-stage HCC and as "advanced stage" based on diagnosis of intermediate- and advanced-stage HCC, according to the Barcelona Clinic Liver Cancer (BCLC) staging system [46]. A total of 168 naïve patients, pathologically confirmed as HCC, were hospitalized in the Osaka Medical College Hospital from April 2010 to January 2018. Fourteen patients with primary or secondary immunodeficiencies (e.g., other cancers, autoimmune diseases, hematologic diseases, infections, chronic heart failures, chronic renal failures, and multiple organ failures) or receiving multikinase inhibitors or immunosuppressive agents were excluded. Among the included patients, 15 patients with early-stage HCC received the initial treatment (radiofrequency ablation) for the purpose of radical cure and cancer progressed to advanced stages after the initial treatment. Blood samples were drawn twice from each patient: at admission for the purpose of initial HCC treatment and after diagnosis of advanced-stage HCC. Clinical characteristics of these patients are shown in Table 1.

Table 1. Changes in the clinical characteristics of the same patients with early- and advanced-stage HCC.

(n = 15)	Early Stages	Advanced Stages	p-Value
Age (year, range)	72.4 (62–82)		
Gender (Male/Female)	14/1		
Etiology (%)			
HBV	2 (6.7)		
HCV	5 (33.3)		
Others	8 (53.3)		
Child-Pugh class			
A/B/C	15/0/0	15/0/0	
TNM stage			<0.001
I/II/III/IV	3/12/0/0	0/0/14/1	
WBC ($\times 10^6$/mL, mean ± SD)	4.69 ± 1.23	4.64 ± 2.01	0.948
Neutrophils ($\times 10^6$/mL, mean ± SD)	2.47 ± 0.87	2.69 ± 1.40	0.794
Lymphocytes ($\times 10^6$/mL, mean ± SD)	1.45 ± 0.85	1.35 ± 0.73	0.788
Monocytes ($\times 10^5$/mL, mean ± SD)	3.58 ± 1.02	3.73 ± 1.40	0.796
Platelets ($\times 10^4$/mL, mean ± SD)	12.8 ± 6.0	10.5 ± 4.5	0.483
AST (IU/L, mean ± SD)	56.5 ± 28.2	54.5 ± 27.2	0.873
ALT (IU/L, mean ± SD)	47.8 ± 28.7	38.5 ± 21.3	0.417
Albumin (g/dL, mean ± SD)	3.55 ± 0.47	3.53 ± 0.53	0.928
Total bilirubin (mg/dL, mean ± SD)	0.98 ± 0.56	1.0 ± 0.87	0.935
Prothrombin time (%, mean ± SD)	87.0 ± 13.5	91.2 ± 18.7	0.360
CRP (mg/dL, mean ± SD)	0.35 ± 0.53	0.17 ± 0.15	0.367
AFP (ng/mL, range)	50.8 (3.7–237.0)	63.1 (4.5–185.0)	0.713
DCP (mAU/mL, range)	252.4 (8.0–1750)	378.8 (14.1–2910)	0.691

HBV, hepatitis B virus; HCV, hepatitis C virus; WBC, white blood cells; SD, standard deviation; AST, aspartate aminotransferase; ALT, alanine aminotransferase; CRP, C-reactive protein; AFP, alpha fetoprotein; DCP, des-gamma-carboxyl prothrombin.

Conversely, the study also included five patients with advanced-stage HCC who received the initial treatment that induced tumor regression. Similarly, blood samples were drawn twice from each of these five patients: at admission for the purpose of initial HCC treatment and after diagnosis of early-stage HCC. Clinical characteristics of these patients are shown in Table 2.

Table 2. Changes in the clinical characteristics of same patients following tumor regression.

(n = 5)	Advanced Stages	Early Stages	p-Value
Age (year, range)	74.4 (61–80)		
Gender (Male/Female)	4/1		
Etiology (%)			
HBV	0 (0.0)		
HCV	4 (80.0)		
Others	1 (20.0)		
Child-Pugh class			
A/B/C	5/0/0	5/0/0	
TNM stage			0.128
I/II/III/IV	0/1/4/0	2/3/0/0	
WBC ($\times 10^6$/mL, mean ± SD)	4.87 ± 1.10	4.63 ± 1.09	0.764
Neutrophils ($\times 10^6$/mL, mean ± SD)	3.01 ± 0.69	2.79 ± 1.20	0.762
Lymphocytes ($\times 10^6$/mL, mean ± SD)	1.28 ± 0.40	1.25 ± 0.54	0.939
Monocytes ($\times 10^5$/mL, mean ± SD)	3.01 ± 0.51	2.89 ± 1.24	0.861
Platelets ($\times 10^4$/mL, mean ± SD)	15.5 ± 5.5	14.3 ± 4.6	0.483
AST (IU/L, mean ± SD)	28.4 ± 11.9	29.2 ± 14.1	0.873
ALT (IU/L, mean ± SD)	25.2 ± 16.5	23.6 ± 14.0	0.417
Albumin (g/dL, mean ± SD)	4.26 ± 0.29	4.0 ± 0.34	0.350
Total bilirubin (mg/dL, mean ± SD)	0.72 ± 0.31	0.76 ± 0.34	0.935
Prothrombin time (%, mean ± SD)	86.2 ± 8.7	72.8 ± 10.3	0.360

Table 2. *Cont.*

(n = 5)	Advanced Stages	Early Stages	p-Value
CRP (mg/dL, mean ± SD)	0.12 ± 0.12	0.19 ± 0.14	0.41
AFP (ng/mL, range)	50.8 (2.6–171.3)	8.3 (2.3–21.5)	0.933
DCP (mAU/mL, range)	203.7 (52.0–809)	65.9 (21.7–115.9)	0.57

HBV, hepatitis B virus; HCV, hepatitis C virus; WBC, white blood cells; SD, standard deviation; AST, aspartate aminotransferase; ALT, alanine aminotransferase; CRP, C-reactive protein; AFP, alpha fetoprotein; DCP, des-gamma-carboxyl prothrombin.

4.3. Reagents, Media, and Cells

Anti-CD14 magnetic particles-DM and IMag buffer were purchased from BD Biosciences (San Jose, CA, USA). Phycoerythrin-conjugated anti-human PD-L1 monoclonal antibodies (mAbs), allophycocyanin-conjugated anti-human PD-L2 mAbs, IL-12 ELISA MAX kits, and IL-10 ELISA MAX kits were purchased from Biolegend (San Diego, CA, USA). Human rCCL1, rCCL17, and rCXCL13 were purchased from Peprotech (Rocky Hill, NJ, USA). Anti-CCL17 mAbs, anti-CCL1 mAbs, and anti-CXCL13 mAbs were purchased from R&D Systems (Minneapolis, MN, USA). The kits for assessment of cytotoxicity (LDH releasing assay) were from Roche Diagnostics (Mannheim, Germany). HepG2 cells (human hepatoblastoma cells), from DS Pharma Biomedical (Osaka, Japan), were cultured at 37 °C in HepG2 human hepatocellular carcinoma expansion medium (Cellular Engineering Technologies Inc., Coralville, IA, USA). RPMI-1640 medium supplemented with 10% fetal bovine serum was used for $CD14^+$ cells.

4.4. $CD14^+$ Cell Characterization

Ten milliliters of whole blood were drawn into a vacutainer tube containing a small amount of sodium heparin at admission. Peripheral blood mononuclear cells (PBMC) were isolated from heparinized whole blood by Lymphocyte Separation Medium 1077 density gradient centrifugation. PBMC (5×10^6 cells/mL) in IMag buffer were incubated with magnetic beads coated with anti-CD14 mAb (40 min at 4 °C); then, $CD14^+$ cells were magnetically harvested. $CD14^+$ cells obtained by this procedure were >97% pure, as assessed by flow cytometry [47]. $CD14^+$ cells were incubated in fluorescence-activated cell sorting buffer with PE-conjugated anti-human PD-L1, APC-conjugated anti-human PD-L2, or isotype control mAb for 15 min at 4 °C. After washing, PD-L1 and PD-L2 expression was measured using a FACSAria flow cytometer and analyzed by FlowJo software version 10.6.0. In some experiments, $CD14^+$ cells (1×10^6 cells/mL) were cultured for 24 h. The culture media were assayed by ELISA for IL-12 (M1Mϕ biomarker), IL-10 (M2Mϕ biomarker), CCL17 (M2aMϕ biomarker), CCL1 (M2bMϕ biomarker), and CXCL13 (M2cMϕ biomarker) [48].

4.5. Tumoricidal Activity of $CD14^+$ Cells against HepG2 Cell In Vitro

Next, $CD14^+$ cells (5×10^5 cells/mL) were stimulated with HepG2 homogenates for 24 h. HepG2 homogenates were made by crushing HepG2 cells (2×10^6 cells/mL) in phosphate-buffered saline with an ultrasonic crusher for 15 min. After washing, $CD14^+$ cells were cocultured with HepG2 cells (1×10^5 cells/mL) for 24 h. The tumoricidal activity of $CD14^+$ cells against HepG2 cells was measured by LDH release assay [18].

4.6. Tumoricidal Activity of $CD14^+$ Cells against HepG2 Cells in Humanized Murine Chimeras

Pathogen-free, male NOD.Cg-PrkcscidIl12rgtm1Wjl/SzJ (NSG) mice aged 7–10 weeks were purchased from Jackson Laboratory (Bar harbor, ME, USA). NSG mice lack functional T, B, and NK cells [49–51], and their macrophages exhibit defective phagocytosis, digestion, and antigen presentation [52]. The NSG mice were exposed to whole-body X-irradiation (4 Gy) to deplete neutrophils [53] and were defined as xNSG mice in this study. Neutrophils in these animals did not recover for 4 weeks after irradiation. xNSG mice were utilized for the creation of humanized murine chimeras. Specifically, they were inoculated with HepG2 cells in the right groin (2×10^6 cells/mouse). Then, the mice were intravenously inoculated every two weeks with CD14$^+$ cells (1×10^6 cells/mouse) isolated from patients with early or advanced-stage HCC. Before inoculation, CD14$^+$ cells from early-stage patients were analyzed by flow cytometry for the expression of IL-12, while CD14$^+$ cells from advanced-stage patients were analyzed for IL-10 and CCL1 expression. In the chimeras, the inoculated cells spread throughout the body within 2 days of inoculation and were functional for at least 6 weeks. The tumor volume was measured with a microcaliper once a week for 6 weeks and expressed in mm^3.

4.7. Statistical Analyses

Statistical analyses were performed using JMP Pro software version 14 (Tokyo, Japan). Quantitative values are expressed as means. Differences in quantitative values between the two groups were analyzed by Mann–Whitney U test. Differences in the ratio of some parameters were analyzed by Fisher's exact test. Differences with p value < 0.05 were considered statistically significant.

5. Conclusions

CD14$^+$ monocytes from patients with early-stage HCC expressed low levels of PD-L and exhibited antitumor activity. However, CD14$^+$ cells from the same patients whose HCC progressed to advanced stages expressed a higher level of PD-L and lacked tumoricidal effects. These findings indicate that the properties of CD14$^+$ cells are strongly related to the state of tumor progression in patients with HCC.

Author Contributions: Conception and design of the study, A.A.; acquisition of data, H.Y., M.M., S.F., and Y.T.; statistical analysis, A.A.; interpretation of data and drafting manuscript, A.A.; critical revision and study supervision, K.H. All authors have read and agreed to the published version of the manuscript.

Funding: This research was funded by the JSPS KAKENHI, grant numbers JP 18K08018, 18KK0456, and 19K08942.

Conflicts of Interest: The authors declare no conflict of interest.

References

1. Yang, J.D.; Hainaut, P.; Gores, G.J.; Amadou, A.; Plymoth, A.; Roberts, L.R. A global view of hepatocellular carcinoma: Trends, risk, prevention and management. *Nat. Rev. Gastroenterol. Hepatol.* **2019**, *16*, 589–604. [CrossRef]
2. Villanueva, A. Hepatocellular Carcinoma. *N. Engl. J. Med.* **2019**, *380*, 1450–1462. [CrossRef]
3. Lin, S.; Hoffmann, K.; Schemmer, P. Treatment of hepatocellular carcinoma: A systematic review. *Liver Cancer* **2012**, *1*, 144–158. [CrossRef] [PubMed]
4. Pinato, D.J.; Guerra, N.; Fessas, P.; Murphy, R.; Mineo, T.; Mauri, F.A.; Mukherjee, S.K.; Thursz, M.; Wong, C.N.; Sharma, R.; et al. Immune-based therapies for hepatocellular carcinoma. *Oncogene* **2020**, *39*, 3620–3637. [CrossRef] [PubMed]
5. Martins, F.; Sofiya, L.; Sykiotis, G.P.; Lamine, F.; Maillard, M.; Fraga, M.; Shabafrouz, K.; Ribi, C.; Cairoli, A.; Guex-Crosier, Y.; et al. Adverse effects of immune-checkpoint inhibitors: Epidemiology, management and surveillance. *Nat. Rev. Clin. Oncol.* **2019**, *16*, 563–580. [CrossRef] [PubMed]
6. Liu, G.M.; Li, X.G.; Zhang, Y.M. Prognostic role of PD-L1 for HCC patients after potentially curative resection: A meta-analysis. *Cancer Cell Int.* **2019**, *19*, 22. [CrossRef] [PubMed]
7. Im, S.J.; Hashimoto, M.; Gerner, M.Y.; Lee, J.; Kissick, H.T.; Burger, M.C.; Shan, Q.; Hale, J.S.; Lee, J.; Nasti, T.H.; et al. Defining CD8+ T cells that provide the proliferative burst after PD-1 therapy. *Nature* **2016**, *537*, 417–421. [CrossRef]
8. Chikuma, S.; Terawaki, S.; Hayashi, T.; Nabeshima, R.; Yoshida, T.; Shibayama, S.; Okazaki, T.; Honjo, T. PD-1-mediated suppression of IL-2 production induces CD8+ T cell anergy in vivo. *J. Immunol. (Baltimore Md. 1950)* **2009**, *182*, 6682–6689. [CrossRef]
9. Iwai, Y.; Ishida, M.; Tanaka, Y.; Okazaki, T.; Honjo, T.; Minato, N. Involvement of PD-L1 on tumor cells in the escape from host immune system and tumor immunotherapy by PD-L1 blockade. *Proc. Natl. Acad. Sci. USA* **2002**, *99*, 12293–12297. [CrossRef]

10. El-Khoueiry, A.B.; Sangro, B.; Yau, T.; Crocenzi, T.S.; Kudo, M.; Hsu, C.; Kim, T.Y.; Choo, S.P.; Trojan, J.; Welling, T.H.R.; et al. Nivolumab in patients with advanced hepatocellular carcinoma (CheckMate 040): An open-label, non-comparative, phase 1/2 dose escalation and expansion trial. *Lancet* **2017**, *389*, 2492–2502. [CrossRef]
11. Kang, Y.K.; Boku, N.; Satoh, T.; Ryu, M.H.; Chao, Y.; Kato, K.; Chung, H.C.; Chen, J.S.; Muro, K.; Kang, W.K.; et al. Nivolumab in patients with advanced gastric or gastro-oesophageal junction cancer refractory to, or intolerant of, at least two previous chemotherapy regimens (ONO-4538-12, ATTRACTION-2): A randomised, double-blind, placebo-controlled, phase 3 trial. *Lancet* **2017**, *390*, 2461–2471. [CrossRef]
12. Sun, C.; Mezzadra, R.; Schumacher, T.N. Regulation and Function of the PD-L1 Checkpoint. *Immunity* **2018**, *48*, 434–452. [CrossRef] [PubMed]
13. Shen, X.; Zhao, B. Efficacy of PD-1 or PD-L1 inhibitors and PD-L1 expression status in cancer: Meta-analysis. *BMJ* **2018**, *362*, k3529. [CrossRef] [PubMed]
14. Shi, C.; Pamer, E.G. Monocyte recruitment during infection and inflammation. *Nat. Rev. Immunol.* **2011**, *11*, 762–774. [CrossRef] [PubMed]
15. Mantovani, A.; Sica, A.; Sozzani, S.; Allavena, P.; Vecchi, A.; Locati, M. The chemokine system in diverse forms of macrophage activation and polarization. *Trends Immunol.* **2004**, *25*, 677–686. [CrossRef]
16. Filardy, A.A.; Pires, D.R.; Nunes, M.P.; Takiya, C.M.; Freire-de-Lima, C.G.; Ribeiro-Gomes, F.L.; DosReis, G.A. Proinflammatory clearance of apoptotic neutrophils induces an IL-12(low)IL-10(high) regulatory phenotype in macrophages. *J. Immunol. (Baltimore Md. 1950)* **2010**, *185*, 2044–2050. [CrossRef]
17. Edwards, J.P.; Zhang, X.; Frauwirth, K.A.; Mosser, D.M. Biochemical and functional characterization of three activated macrophage populations. *J. Leukoc Biol.* **2006**, *80*, 1298–1307. [CrossRef]
18. Asai, A.; Tsuchimoto, Y.; Ohama, H.; Fukunishi, S.; Tsuda, Y.; Kobayashi, M.; Higuchi, K.; Suzuki, F. Host antitumor resistance improved by the macrophage polarization in a chimera model of patients with HCC. *Oncoimmunology* **2017**, *6*, e1299301. [CrossRef]
19. Yasuoka, H.; Asai, A.; Ohama, H.; Tsuchimoto, Y.; Fukunishi, S.; Higuchi, K. Increased both PD-L1 and PD-L2 expressions on monocytes of patients with hepatocellular carcinoma was associated with a poor prognosis. *Sci. Rep.* **2020**, *10*, 10377. [CrossRef]
20. Olingy, C.E.; Dinh, H.Q.; Hedrick, C.C. Monocyte heterogeneity and functions in cancer. *J. Leukoc Biol.* **2019**, *106*, 309–322. [CrossRef]
21. Serbina, N.V.; Pamer, E.G. Monocyte emigration from bone marrow during bacterial infection requires signals mediated by chemokine receptor CCR2. *Nat. Immunol.* **2006**, *7*, 311–317. [CrossRef] [PubMed]
22. Richards, D.M.; Hettinger, J.; Feuerer, M. Monocytes and macrophages in cancer: Development and functions. *Cancer Microenviron.* **2013**, *6*, 179–191. [CrossRef]
23. Solinas, G.; Germano, G.; Mantovani, A.; Allavena, P. Tumor-associated macrophages (TAM) as major players of the cancer-related inflammation. *J. Leukoc. Biol.* **2009**, *86*, 1065–1073. [CrossRef]
24. Tadmor, T.; Bari, A.; Marcheselli, L.; Sacchi, S.; Aviv, A.; Baldini, L.; Gobbi, P.G.; Pozzi, S.; Ferri, P.; Cox, M.C.; et al. Absolute Monocyte Count and Lymphocyte-Monocyte Ratio Predict Outcome in Nodular Sclerosis Hodgkin Lymphoma: Evaluation Based on Data From 1450 Patients. *Mayo Clin. Proc.* **2015**, *90*, 756–764. [CrossRef] [PubMed]
25. Sanford, D.E.; Belt, B.A.; Panni, R.Z.; Mayer, A.; Deshpande, A.D.; Carpenter, D.; Mitchem, J.B.; Plambeck-Suess, S.M.; Worley, L.A.; Goetz, B.D.; et al. Inflammatory monocyte mobilization decreases patient survival in pancreatic cancer: A role for targeting the CCL2/CCR2 axis. *Clin. Cancer Res. Off. J. Am. Assoc. Cancer Res.* **2013**, *19*, 3404–3415. [CrossRef] [PubMed]
26. Macek Jilkova, Z.; Aspord, C.; Decaens, T. Predictive Factors for Response to PD-1/PD-L1 Checkpoint Inhibition in the Field of Hepatocellular Carcinoma: Current Status and Challenges. *Cancers* **2019**, *11*, 1554. [CrossRef] [PubMed]
27. Wang, L.X.; Zhang, S.X.; Wu, H.J.; Rong, X.L.; Guo, J. M2b macrophage polarization and its roles in diseases. *J. Leukoc Biol.* **2019**, *106*, 345–358. [CrossRef]
28. Nakamura, K.; Ito, I.; Kobayashi, M.; Herndon, D.N.; Suzuki, F. Orosomucoid 1 drives opportunistic infections through the polarization of monocytes to the M2b phenotype. *Cytokine* **2015**, *73*, 8–15. [CrossRef]

29. Ohyama, K.; Yoshimi, H.; Aibara, N.; Nakamura, Y.; Miyata, Y.; Sakai, H.; Fujita, F.; Imaizumi, Y.; Chauhan, A.K.; Kishikawa, N.; et al. Immune complexome analysis reveals the specific and frequent presence of immune complex antigens in lung cancer patients: A pilot study. *Int. J. Cancer. J. Int. Du Cancer* **2017**, *140*, 370–380. [CrossRef]
30. Guo, L.; Wei, R.; Lin, Y.; Kwok, H.F. Clinical and Recent Patents Applications of PD-1/PD-L1 Targeting Immunotherapy in Cancer Treatment-Current Progress, Strategy, and Future Perspective. *Front. Immunol.* **2020**, *11*, 1508. [CrossRef]
31. Oh, S.A.; Wu, D.-C.; Cheung, J.; Navarro, A.; Xiong, H.; Cubas, R.; Totpal, K.; Chiu, H.; Wu, Y.; Comps-Agrar, L.; et al. PD-L1 expression by dendritic cells is a key regulator of T-cell immunity in cancer. *Nat. Cancer* **2020**, *1*, 681–691. [CrossRef]
32. Patel, S.P.; Kurzrock, R. PD-L1 Expression as a Predictive Biomarker in Cancer Immunotherapy. *Mol Cancer Ther.* **2015**, *14*, 847–856. [CrossRef]
33. Rooney, M.S.; Shukla, S.A.; Wu, C.J.; Getz, G.; Hacohen, N. Molecular and genetic properties of tumors associated with local immune cytolytic activity. *Cell* **2015**, *160*, 48–61. [CrossRef] [PubMed]
34. Ansell, S.M.; Lesokhin, A.M.; Borrello, I.; Halwani, A.; Scott, E.C.; Gutierrez, M.; Schuster, S.J.; Millenson, M.M.; Cattry, D.; Freeman, G.J.; et al. PD-1 blockade with nivolumab in relapsed or refractory Hodgkin's lymphoma. *N. Engl. J. Med.* **2015**, *372*, 311–319. [CrossRef] [PubMed]
35. Lastwika, K.J.; Wilson, W., 3rd; Li, Q.K.; Norris, J.; Xu, H.; Ghazarian, S.R.; Kitagawa, H.; Kawabata, S.; Taube, J.M.; Yao, S.; et al. Control of PD-L1 Expression by Oncogenic Activation of the AKT-mTOR Pathway in Non-Small Cell Lung Cancer. *Cancer Res.* **2016**, *76*, 227–238. [CrossRef] [PubMed]
36. Akbay, E.A.; Koyama, S.; Carretero, J.; Altabef, A.; Tchaicha, J.H.; Christensen, C.L.; Mikse, O.R.; Cherniack, A.D.; Beauchamp, E.M.; Pugh, T.J.; et al. Activation of the PD-1 pathway contributes to immune escape in EGFR-driven lung tumors. *Cancer Discov.* **2013**, *3*, 1355–1363. [CrossRef]
37. Casey, S.C.; Tong, L.; Li, Y.; Do, R.; Walz, S.; Fitzgerald, K.N.; Gouw, A.M.; Baylot, V.; Gütgemann, I.; Eilers, M.; et al. MYC regulates the antitumor immune response through CD47 and PD-L1. *Science* **2016**, *352*, 227–231. [CrossRef]
38. Dorand, R.D.; Nthale, J.; Myers, J.T.; Barkauskas, D.S.; Avril, S.; Chirieleison, S.M.; Pareek, T.K.; Abbott, D.W.; Stearns, D.S.; Letterio, J.J.; et al. Cdk5 disruption attenuates tumor PD-L1 expression and promotes antitumor immunity. *Science* **2016**, *353*, 399–403. [CrossRef]
39. Kataoka, K.; Shiraishi, Y.; Takeda, Y.; Sakata, S.; Matsumoto, M.; Nagano, S.; Maeda, T.; Nagata, Y.; Kitanaka, A.; Mizuno, S.; et al. Aberrant PD-L1 expression through 3′-UTR disruption in multiple cancers. *Nature* **2016**, *534*, 402–406. [CrossRef]
40. Ribas, A.; Hu-Lieskovan, S. What does PD-L1 positive or negative mean? *J. Exp. Med.* **2016**, *213*, 2835–2840. [CrossRef]
41. Zong, Z.; Zou, J.; Mao, R.; Ma, C.; Li, N.; Wang, J.; Wang, X.; Zhou, H.; Zhang, L.; Shi, Y. M1 Macrophages Induce PD-L1 Expression in Hepatocellular Carcinoma Cells Through IL-1β Signaling. *Front. Immunol.* **2019**, *10*, 1643. [CrossRef] [PubMed]
42. Horlad, H.; Ma, C.; Yano, H.; Pan, C.; Ohnishi, K.; Fujiwara, Y.; Endo, S.; Kikukawa, Y.; Okuno, Y.; Matsuoka, M.; et al. An IL-27/Stat3 axis induces expression of programmed cell death 1 ligands (PD-L1/2) on infiltrating macrophages in lymphoma. *Cancer Sci.* **2016**, *107*, 1696–1704. [CrossRef] [PubMed]
43. Schultheis, A.M.; Scheel, A.H.; Ozretić, L.; George, J.; Thomas, R.K.; Hagemann, T.; Zander, T.; Wolf, J.; Buettner, R. PD-L1 expression in small cell neuroendocrine carcinomas. *Eur. J. Cancer.* **2015**, *51*, 421–426. [CrossRef] [PubMed]
44. Kawashima, M.; Carreras, J.; Higuchi, H.; Kotaki, R.; Hoshina, T.; Okuyama, K.; Suzuki, N.; Kakizaki, M.; Miyatake, Y.; Ando, K.; et al. PD-L1/L2 protein levels rapidly increase on monocytes via trogocytosis from tumor cells in classical Hodgkin lymphoma. *Leukemia* **2020**. [CrossRef]
45. Davis, D.M. Intercellular transfer of cell-surface proteins is common and can affect many stages of an immune response. *Nat. Rev. Immunol.* **2007**, *7*, 238–243. [CrossRef]
46. Bruix, J.; Sherman, M. Management of hepatocellular carcinoma: An update. *Hepatology (Baltimore Md.)* **2011**, *53*, 1020–1022. [CrossRef]
47. Tsuchimoto, Y.; Asai, A.; Tsuda, Y.; Ito, I.; Nishiguchi, T.; Garcia, M.C.; Suzuki, S.; Kobayashi, M.; Higuchi, K.; Suzuki, F. M2b Monocytes Provoke Bacterial Pneumonia and Gut Bacteria-Associated Sepsis in Alcoholics. *J. Immunol. (Baltimore Md. 1950)* **2015**, *195*, 5169–5177. [CrossRef]

48. Asai, A.; Nakamura, K.; Kobayashi, M.; Herndon, D.N.; Suzuki, F. CCL1 released from M2b macrophages is essentially required for the maintenance of their properties. *J. Leukoc. Biol.* **2012**, *92*, 859–867. [CrossRef]
49. Chen, K.; Ahmed, S.; Adeyi, O.; Dick, J.E.; Ghanekar, A. Human solid tumor xenografts in immunodeficient mice are vulnerable to lymphomagenesis associated with Epstein-Barr virus. *PLoS ONE* **2012**, *7*, e39294. [CrossRef]
50. Shultz, L.D.; Lyons, B.L.; Burzenski, L.M.; Gott, B.; Chen, X.; Chaleff, S.; Kotb, M.; Gillies, S.D.; King, M.; Mangada, J.; et al. Human lymphoid and myeloid cell development in NOD/LtSz-scid IL2R gamma null mice engrafted with mobilized human hemopoietic stem cells. *J. Immunol. (Baltimore Md. 1950)* **2005**, *174*, 6477–6489. [CrossRef]
51. Agliano, A.; Martin-Padura, I.; Mancuso, P.; Marighetti, P.; Rabascio, C.; Pruneri, G.; Shultz, L.D.; Bertolini, F. Human acute leukemia cells injected in NOD/LtSz-scid/IL-2Rgamma null mice generate a faster and more efficient disease compared to other NOD/scid-related strains. *Int. J. Cancer* **2008**, *123*, 2222–2227. [CrossRef] [PubMed]
52. Hu, Z.; Van Rooijen, N.; Yang, Y.G. Macrophages prevent human red blood cell reconstitution in immunodeficient mice. *Blood* **2011**, *118*, 5938–5946. [CrossRef] [PubMed]
53. Ohama, H.; Asai, A.; Ito, I.; Suzuki, S.; Kobayashi, M.; Higuchi, K.; Suzuki, F. M2b macrophage elimination and improved resistance of mice with chronic alcohol consumption to opportunistic infections. *Am. J. Pathol.* **2015**, *185*, 420–431. [CrossRef] [PubMed]

 © 2020 by the authors. Licensee MDPI, Basel, Switzerland. This article is an open access article distributed under the terms and conditions of the Creative Commons Attribution (CC BY) license (http://creativecommons.org/licenses/by/4.0/).

Article

Early Change in the Plasma Levels of Circulating Soluble Immune Checkpoint Proteins in Patients with Unresectable Hepatocellular Carcinoma Treated by Lenvatinib or Transcatheter Arterial Chemoembolization

Naoshi Odagiri, Hoang Hai, Le Thi Thanh Thuy, Minh Phuong Dong, Maito Suoh, Kohei Kotani, Atsushi Hagihara, Sawako Uchida-Kobayashi, Akihiro Tamori, Masaru Enomoto * and Norifumi Kawada

Department of Hepatology, Graduate School of Medicine, Osaka City University, Osaka 545-8585, Japan; m2055463@med.osaka-cu.ac.jp (N.O.); hhai@med.osaka-cu.ac.jp (H.H.); thuylt@med.osaka-cu.ac.jp (L.T.T.T.); dongminhphuong15@gmail.com (M.P.D.); maito55jp@gmail.com (M.S.); kouhei-k@med.osaka-cu.ac.jp (K.K.); hagy@med.osaka-cu.ac.jp (A.H.); sawako@med.osaka-cu.ac.jp (S.U.-K.); atamori@med.osaka-cu.ac.jp (A.T.); kawadanori@med.osaka-cu.ac.jp (N.K.)
* Correspondence: enomoto-m@med.osaka-cu.ac.jp; Tel.: +81-6-6645-3897

Received: 25 June 2020; Accepted: 21 July 2020; Published: 24 July 2020

Abstract: Immune checkpoint inhibitors, combined with anti-angiogenic agents or locoregional treatments (e.g., transarterial chemoembolization (TACE)), are expected to become standard-of-care for unresectable hepatocellular carcinoma (HCC). We measured the plasma levels of 16 soluble checkpoint proteins using multiplexed fluorescent bead-based immunoassays in patients with HCC who underwent lenvatinib ($n = 24$) or TACE ($n = 22$) treatment. In lenvatinib-treated patients, plasma levels of sCD27 (soluble cluster of differentiation 27) decreased ($p = 0.040$) and levels of sCD40 ($p = 0.014$) and sTIM-3 ($p < 0.001$) were increased at Week 1, while levels of sCD27 ($p < 0.001$) were increased significantly at Weeks 2 through 4. At Week 1 of TACE, in addition to sCD27 ($p = 0.028$), sCD40 ($p < 0.001$), and sTIM-3 (soluble T-cell immunoglobulin and mucin domain–3) ($p < 0.001$), levels of sHVEM (soluble herpesvirus entry mediator) ($p = 0.003$), sTLR-2 (soluble Toll-like receptor 2) ($p = 0.009$), sCD80 ($p = 0.036$), sCTLA-4 (soluble cytotoxic T-lymphocyte antigen 4) ($p = 0.005$), sGITR (soluble glucocorticoid-induced tumor necrosis factor receptor) ($p = 0.030$), sGITRL (soluble glucocorticoid-induced TNFR-related ligand) ($p = 0.090$), and sPD-L1 (soluble programmed death-ligand 1) ($p = 0.070$) also increased. The fold-changes in soluble checkpoint receptors and their ligands, including sCTLA-4 with sCD80/sCD86 and sPD-1 (soluble programmed cell death domain–1) with sPD-L1 were positively correlated in both the lenvatinib and TACE treatment groups. Our results suggest that there are some limited differences in immunomodulatory effects between anti-angiogenic agents and TACE. Further studies from multicenters may help to identify an effective combination therapy.

Keywords: HCC; liver cancer; molecular-targeted agent; TACE; tyrosine kinase inhibitor

1. Introduction

Primary liver cancer, of which hepatocellular carcinoma (HCC) is the most common type, is the third most common cause of cancer-related deaths worldwide [1,2]. Patients with early-stage HCC can potentially receive curative treatments, such as surgical resection, transplantation, or ablation, and patients at the intermediate stage can undergo transarterial chemoembolization (TACE), but those at an advanced stage of disease are only likely to benefit from systemic therapies [3–5]. Molecular-targeted therapies, the two anti-angiogenic tyrosine kinase inhibitors sorafenib [6] and lenvatinib [7] are used as first-line treatments for patients with advanced-stage HCC. Meanwhile, other anti-angiogenic agents like regorafenib [8], cabozantinib [9], and ramucirumab (if α-fetoprotein > 400 ng/mL) [10] have been licensed as second-line treatments. Immune checkpoint inhibitors, such as anti–PD-1 (programmed cell death domain–1) pembrolizumab [11], nivolumab [12] or nivolumab combined with anti–CTLA-4 (cytotoxic T-lymphocyte antigen 4) (ipilimumab) therapy [13], may also be indicated for sorafenib-refractory patients [14–16].

However, therapeutic responses to immunotherapy alone are obtained in a minority of patients with HCC. Current clinical trials are therefore focusing on whether combinations of different types of treatments, including anti-angiogenic agents and TACE, may be promising for the enhancement of the antitumor effects of immune checkpoint inhibitors [17]. For example, the combinations of lenvatinib plus pembrolizumab [18] and bevacizumab plus anti–PD-L1 atezolizumab therapy [19] are currently being tested in clinical investigations. In addition, tremelimumab, an anti–CTLA-4 therapy, was more effective when paired with TACE than as monotherapy [20].

In our previous study [21], sorafenib treatment in patients with unresectable HCC provoked dynamic changes in soluble immune checkpoint protein levels as revealed using multiplexed fluorescent bead-based immunoassays. To date, circulating soluble checkpoint proteins, which are part of a family of full-length receptors produced by messenger RNA expression or by the cleavage of membrane-bound proteins, have been studied extensively in various cancers, but not in HCC [22]. Further, changes in the plasma levels of soluble proteins during the early days of treatment with other anti-angiogenic agents or locoregional treatments (e.g., TACE) have yet to be determined.

We hypothesized that lenvatinib and TACE therapies as well as sorafenib would affect plasma levels of soluble immune checkpoint proteins. An understanding of the effects of these therapies on soluble immune checkpoint proteins may provide insight into their immunomodulative characteristics, which would help to develop more effective combination immunotherapies for HCC. In this study, we measured the concentrations of 16 soluble immune checkpoint proteins (Table A1) in plasma samples obtained over four weeks of treatment from patients with unresectable HCC who underwent lenvatinib or TACE treatment.

2. Results

2.1. Patient Characteristics

The baseline characteristics of patients with HCC in the lenvatinib ($n = 24$) and TACE ($n = 22$) groups are described in Table 1. In brief, the median age of patients was about 75 years (75 vs. 76 years; $p = 0.530$) and males accounted for the majority (75.0% vs. 68.2%; $p = 0.746$) of the population in both treatment groups, with no significant difference apparent between the two groups. Most patients in the TACE group had Barcelona Clinic Liver Cancer (BCLC) [23] stage A (54.5%) cancer, while all patients had BCLC stage B (54.2%) or stage C (45.8%) cancer in the lenvatinib group ($p < 0.001$). Some patients in the TACE group underwent the therapy as their initial HCC treatment, but none in the lenvatinib group (36.4% vs. 0%; $p = 0.001$). No significant differences were found between the lenvatinib and TACE groups with respect to the etiology of liver disease, Eastern Cooperative Oncology Group performance status score, aspartate and alanine aminotransferase (AST and ALT) levels, gamma-glutamyl transferase level, hepatic reserve (Child–Pugh class or albumin–bilirubin score [24,25]) and α-fetoprotein and des-γ-carboxy prothrombin levels as tumor markers.

Table 1. Characteristics of hepatocellular carcinoma (HCC) patients at baseline.

Characteristics		Lenvatinib ($n = 24$)	TACE ($n = 22$)	p Value
Age	–	75 (69, 78)	76 (69, 80)	0.530
Sex	Male	18 (75.0)	15 (68.2)	0.746
	Female	6 (25.0)	7 (31.8)	
Etiology	Alcohol	5 (20.8)	3 (13.6)	0.536
	HBV	3 (12.5)	1 (4.5)	
	HCV	10 (41.7)	13 (59.1)	
	HBV + HCV	1 (4.2)	0 (0.0)	
	NASH	1 (4.2)	3 (13.6)	
	Unknown	4 (16.7)	2 (9.1)	
ECOG Perfomance Status	0 or 1	23 (95.8)	19 (86.4)	0.336
	2	1 (4.2)	3 (13.6)	
Aspartate aminotransferase		39 (27, 59)	39 (30, 50)	0.826
Alanine aminotransferase		27 (20, 52)	23 (16, 46)	0.567
Gamma-glutamyl transferase		62 (30, 122)	61 (32, 120)	0.930
Child-Pugh class	A	21 (87.5)	18 (81.8)	0.694
	B	3 (12.5)	4 (18.2)	
ALBI grade	1	8 (33.3)	8 (36.4)	0.999
	2	16 (66.7)	14 (63.6)	
α-Fetoprotein		59.4 (11.4, 1123.7)	21.2 (5.3, 55.9)	0.126
Des-γ-carboxy prothrombin		105 (72, 1312)	137 (56, 282)	0.605
BCLC stage	A	0 (0.0)	12 (54.5)	<0.001
	B	13 (54.2)	7 (31.8)	
	C	11 (45.8)	3 (13.6)	
Previous therapies	None	0 (0.0)	8 (36.4)	0.001
	Resection	5 (20.8)	2 (9.1)	0.418
	RFA/PEI	11 (45.8)	7 (31.8)	0.378
	TACE	23 (95.8)	9 (40.9)	<0.001
	HAIC	3 (12.5)	3 (13.6)	0.999
	Radiation	2 (8.3)	0 (0.0)	0.490
	Chemotherapy	0 (0.0)	2 (9.1)	0.223

Data are shown as median [interquartile range] or number (%). Abbreviations: HBV, hepatitis B virus; HCV, hepatitis C virus; NASH, non-alcoholic steatohepatitis; ECOG, Eastern Cooperative Oncology Group; ALBI, albumin–bilirubin; BCLC, Barcelona Clinic Liver Cancer; RFA, radiofrequency ablation; PEI, percutaneous ethanol injection; TACE, transarterial chemoembolization; HAIC, hepatic arterial infusion chemotherapy.

2.2. Changes in Plasma Soluble Checkpoint Protein Levels at Week 1 after the Initiation of Lenvatinib

First, we investigated whether lenvatinib affects the plasma levels of immune checkpoint proteins in the early phase of treatment. As previously reported [21], we analyzed soluble checkpoint protein levels in patients with HCC at Week 1 of lenvatinib treatment. Ultimately, a significant decrease was observed in the level of soluble cluster of differentiation 27 (sCD27) ($p = 0.040$) and significant increases were found in the levels of sCD40 ($p = 0.014$) and soluble T-cell immunoglobulin and mucin domain–3 (sTIM-3) ($p < 0.001$) when compared with at baseline. Meanwhile, no significant changes were found in soluble herpesvirus entry mediator (sHVEM) and the other 12 immune checkpoint proteins (Figures 1 and A1).

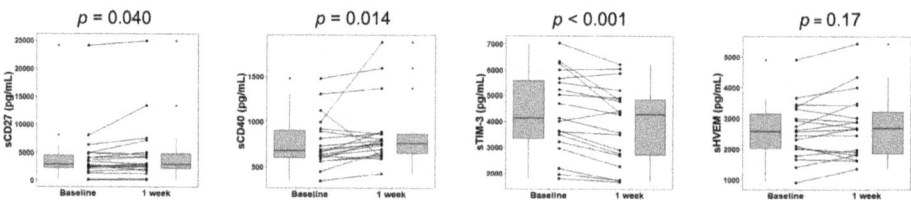

Figure 1. Changes in soluble immune checkpoint protein levels in plasma after 1 week of lenvatinib treatment. Box plots show the sCD27, sCD40, sTIM-3, and sHVEM levels in HCC patients at baseline and Week 1 of treatment. The vertical lengths of the boxes indicate the interquartile ranges and the lines in the boxes suggest the median values, while the error bars show the minimum and maximum values (in a range). The Wilcoxon signed-rank test was used. HCC, hepatocellular carcinoma; sTIM-3, soluble T-cell immunoglobulin and mucin domain–3; sHVEM, soluble herpesvirus entry mediator; sCD, soluble cluster of differentiation.

2.3. Changes in Plasma Soluble Checkpoint Protein Levels at Weeks 2 through 4 after the Initiation of Lenvatinib

Next, we sought to reveal changes in immune checkpoint protein levels at the later stage of lenvatinib therapy. We analyzed soluble checkpoint protein levels in the plasma of patients with HCC at Weeks 2 through 4 of lenvatinib treatment. The increase in sCD27 level was significant ($p < 0.001$). A trend toward increasing sCD40 and sHVEM levels was also observed, but no change reached statistical significance ($p = 0.070$ and 0.090). Also, the change in sTIM-3 was no longer significant at this time point (Figure 2), while the other 12 immune checkpoint proteins showed similar outcomes (Appendix A Figure A2).

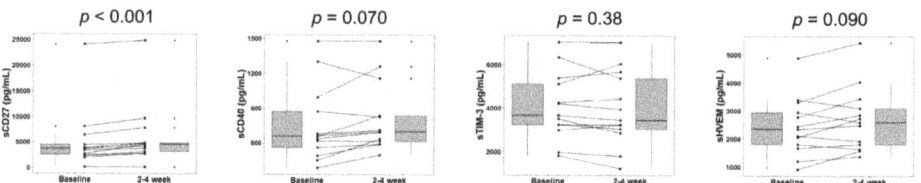

Figure 2. Changes in soluble immune checkpoint protein levels in plasma after 2 to 4 weeks of lenvatinib treatment. Box plots demonstrate the sCD27, sCD40, sTIM-3, and sHVEM levels in HCC patients at baseline and Weeks 2 through 4 of treatment. The vertical lengths of the boxes indicate the interquartile ranges and the lines in the boxes display the median values, while the error bars show the minimum and maximum values (in a range). The Wilcoxon signed-rank test was used.

2.4. Changes in Plasma Soluble Checkpoint Protein Levels at Week 1 after TACE

Lenvatinib can suppress tumor blood flow by its pharmacological anti-angiogenic effects. To establish a contrast with those who received lenvatinib, we also investigated patients who underwent conventional TACE, which disrupts tumor blood flow via the artificial embolization of hepatic arteries. We analyzed soluble immune checkpoint protein levels in the plasma of patients with HCC at 1 week after TACE. In addition to sCD27 ($p = 0.028$), sCD40 ($p < 0.001$), sTIM-3 ($p < 0.001$), and sHVEM ($p = 0.003$) levels, which exhibited significant (or marginal) changes at Week 1 or Weeks 2 through 4 of lenvatinib treatment, the levels of another six proteins—namely, soluble Toll-like receptor 2 (sTLR-2) ($p = 0.009$), sCD80 ($p = 0.036$), sCTLA-4 (soluble cytotoxic T-lymphocyte antigen 4) ($p = 0.005$), soluble glucocorticoid-induced TNFR-related protein (sGITR) ($p = 0.030$), soluble glucocorticoid-induced TNFR-related ligand (sGITRL) ($p = 0.090$), and sPD-L1 (soluble programmed death-ligand 1) ($p = 0.070$)—were also increased (Figure 3). However, the levels of the remaining six immune checkpoint proteins showed no significant changes (Appendix A Figure A3).

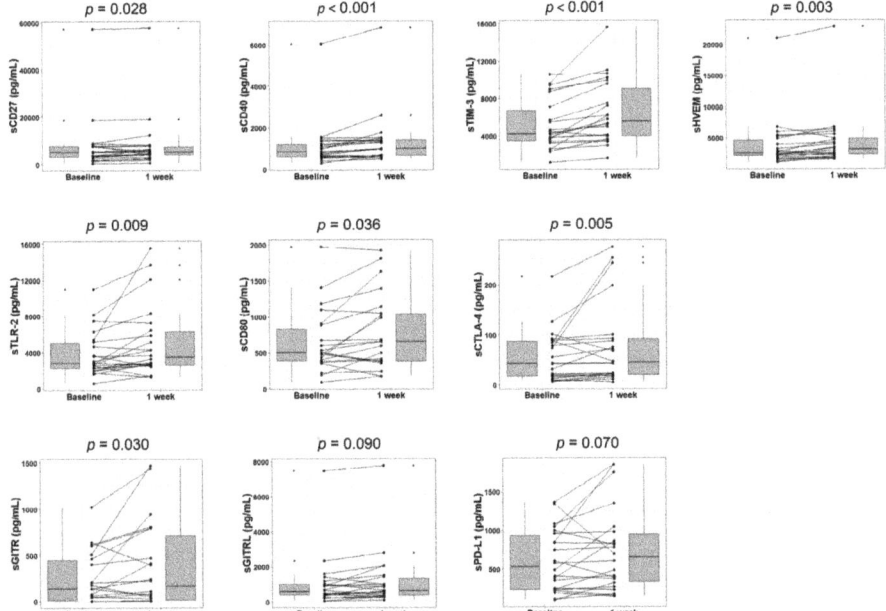

Figure 3. Changes in soluble immune checkpoint protein levels in plasma after 1 week of TACE treatment. Box plots display the sCD27, sCD40, sTIM-3, sHVEM, sTLR-2, sCD80, sCTLA-4, sGITR, sGITRL, and sPD-L1 levels in HCC patients at baseline and Week 1 of treatment. The vertical lengths of the boxes indicate the interquartile ranges and the lines in the boxes suggest the median values, while the error bars show the minimum and maximum values (range). The Wilcoxon signed-rank test was used. TACE, transarterial chemoembolization; sTLR-2, soluble Toll-like receptor 2; sCTLA-4, soluble cytotoxic T-lymphocyte antigen 4; sGITR, soluble glucocorticoid-induced tumor necrosis factor receptor; sGITRL, soluble glucocorticoid-induced TNFR-related ligand; sPD-L1, soluble programmed death-ligand 1.

2.5. Relationships between Fold-Changes in Plasma Soluble Immune Checkpoint Protein Levels

The correlations between the fold-changes in the soluble forms of immune checkpoint proteins in plasma at Week 1 of lenvatinib treatment are shown in Figure 4. The fold-changes in soluble checkpoint receptors and their ligands, including sCTLA-4 with sCD80 ($p < 0.001$; $r = 0.82$)/sCD86 ($p < 0.001$; $r = 0.78$), and sPD-1 with sPD-L1 ($p < 0.001$; $r = 0.91$), were positively correlated (Figure 4a). Among the three soluble checkpoint proteins with significant change at Week 1, sCD40 was positively correlated with some proteins in fold-changes; sCD86 ($p = 0.005$; $r = 0.62$), sPD-1 ($p = 0.010$; $r = 0.58$), sPD-L1 ($p = 0.046$; $r = 0.48$) (Figure 4b), sCD28 ($p = 0.012$; $r = 0.57$), sTLR-2 ($p = 0.048$; $r = 0.46$), and sHVEM ($p = 0.042$; $r = 0.47$).

At Weeks 2 to 4 of lenvatinib treatment, the fold-changes in soluble immune checkpoint receptors and their ligands in plasma, including sCTLA-4 with sCD80 ($p = 0.055$; $r = 0.55$)/sCD86 ($p = 0.009$; $r = 0.71$) and sPD-1 with sPD-L1 ($p < 0.001$; $r = 0.86$), were also positively correlated (Appendix A Figure A4a). Separately, the fold-change in sCD27 was positively correlated with those of sCD86 ($p = 0.030$; $r = 0.61$), sPD-1 ($p = 0.010$; $r = 0.70$), and sPD-L1 ($p = 0.050$; $r = 0.56$), respectively (Appendix A Figure A4b).

The correlations between the fold-changes in soluble forms of immune checkpoint proteins in plasma at Week 1 of TACE are shown in Figure 5. Again, the fold-changes in soluble checkpoint receptors and their ligands, including sCTLA-4 with sCD80 ($p = 0.018$; $r = 0.50$)/sCD86 ($p < 0.001$; $r = 0.72$) and sPD-1 with sPD-L1 ($p < 0.001$; $r = 0.89$), were positively correlated (Figure 5a). A strong correlation was noted between the fold-changes in the sCD80 and sTLR-2 levels ($p < 0.001$; $r = 0.89$), sCD40 and sHVEM levels ($p < 0.001$; $r = 0.78$), and sHVEM and sTIM-3 levels ($p < 0.001$; $r = 0.72$), respectively (Figure 5b).

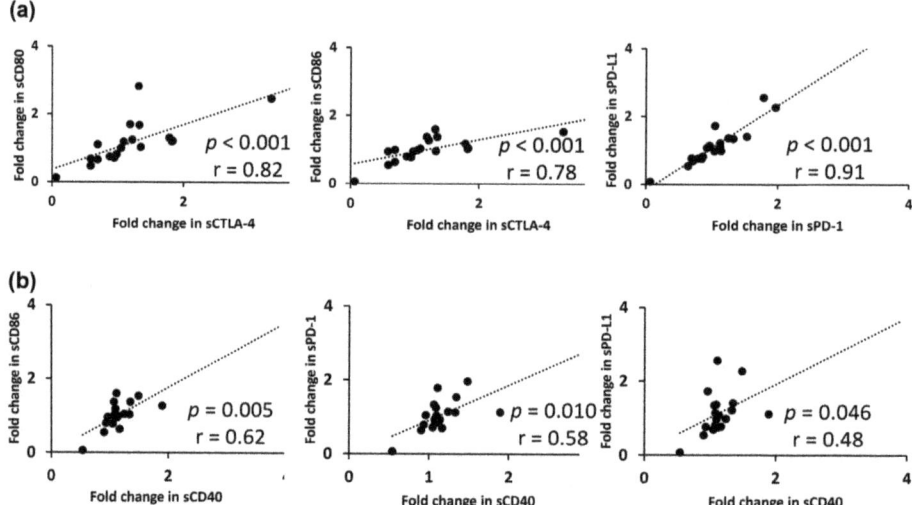

Figure 4. The relationships between fold-changes in soluble immune checkpoint protein levels in plasma at 1 week of lenvatinib treatment are presented. (**a**) A positive correlation was found between fold-changes in soluble checkpoint receptors and their ligands, including sCTLA-4 with sCD80/sCD86 and sPD-1 with sPD-L1, respectively. (**b**) sCD40 was positively correlated with some proteins in fold-changes, including sCD86, sPD-1, and sPD-L1. Spearman's rank correlation test was used. A p-value of less than 0.05 was considered to be statistically significant.

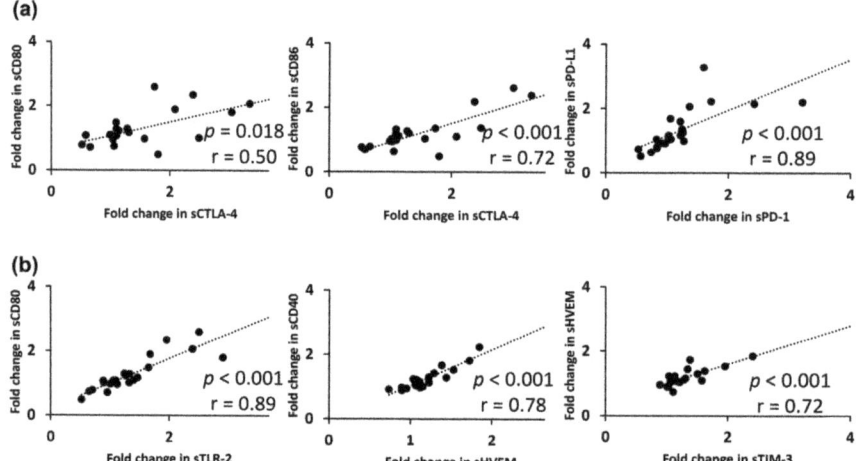

Figure 5. The relationships between fold-changes in soluble immune checkpoint protein levels in plasma at 1 week of receiving TACE are shown. (**a**) A positive correlation was found between the fold-changes of soluble checkpoint receptors and their ligands, including sCTLA-4 with sCD80/sCD86 and sPD-1 with sPD-L1. (**b**) Separately, a strong correlation was observed between the fold changes in sCD80 and sTLR-2 levels, sCD40 and sHVEM levels, and sHVEM and sTIM-3 levels. Spearman's rank correlation test was used. A p-value of less than 0.05 was considered to be statistically significant.

3. Discussion

This study attempted the simultaneous quantification of 16 soluble immune checkpoint proteins in the plasma of patients with HCC who were observed during the early phase of treatment with lenvatinib or TACE. These immune checkpoint proteins include soluble forms of stimulatory or inhibitory factors, which modulate T-cell activation/proliferation and compose the cancer-immunity cycle [26]. In our previous report [21], we examined changes to these immune checkpoint proteins in sorafenib-treated HCC. The current results could offer additional data to compare changes in soluble checkpoint protein levels among tyrosine kinase inhibitors and TACE, which may clarify the immunomodulatory aspects of these treatments.

In this study, the plasma level of sCD27 was decreased and those of sCD40 and sTIM-3 were increased significantly at Week 1, while the level of sCD27 was increased significantly at Weeks 2 through 4 of lenvatinib treatment (Figures 1 and 2). Although CD27 and CD40 are stimulatory factors and TIM-3 is an inhibitory factor, respectively, in the cancer-immunity cycle, soluble forms of stimulatory/inhibitory factors do not necessarily have set positive/negative immune effects; currently, the functions of these soluble proteins are yet to be fully defined. In short, (1) CD27 supports the antigen-specific expansion of naïve T-cells and is vital for the generation of T-cell memory [27]. A previous study found that circulating sCD27 constitutes a functional protein directly involved in $CD8^+$ T-cell activation [28]. A persistent increase in the sCD27 level during Weeks 2 through 4 of lenvatinib treatment may reflect the $CD8^+$ T-cell–related immunomodulatory effect of lenvatinib, which is in line with a previous study [29] revealing that the antitumor activity of lenvatinib was significantly diminished by $CD8^+$ T-cell depletion. Further, (2) CD40 plays an important role mainly in the signaling pathways for the functioning of B-cells, monocytes, and dendritic cells [30]. In recent research [31], plasma sCD40 levels were upregulated in correlation with disease severity in patients with alcoholic hepatitis, indicating the existence of dysregulation of the immune system in chronic liver disease. Also, (3) TIM-3 promotes the exhaustion of T-cells in various types of cancer [32]. One study suggests circulating sTIM-3 might competitively bind to galectin-9 (a ligand of TIM-3), preventing a TIM-3/galectin-9–mediated immune response [33]. The increase in plasma sTIM-3 during lenvatinib treatment may restore immune exhaustion.

In vitro and in vivo preclinical studies have provoked the thoughts that there may be different immunomodulatory effects among similar tyrosine kinase inhibitors [34]. Lenvatinib may decrease tumor-associated macrophages, facilitate polarization from an M2-like phenotype toward an M1-like phenotype, and enhance $CD4^+$ and $CD8^+$ T-cell tumor infiltration, while sorafenib may have the opposite effects [35]. In our previous study of sorafenib [21], 11 of 16 soluble immune checkpoint proteins—most of which were inhibitory factors—experienced significant increases in plasma at two weeks of treatment. In this study, sCD27 displayed a significant change at two weeks. Both sorafenib and lenvatinib basically inhibit vascular endothelial growth factor receptors (VEGFRs) 1 through 3, fibroblast growth factor receptors (FGFRs) 1 through 4, platelet-derived growth factor receptor (PDGFR)-α, RET, and KIT. One potential explanation for the discrepancy is the existence of variable inhibitory profiles among the drugs against VEGFRs and FGFRs; specifically, lenvatinib is known to show more potent inhibitory activities than sorafenib against these receptor tyrosine kinases [36]. VEGFRs are particularly important because they activate various key components such as regulatory T-cells, tumor-associated macrophages, and myeloid-derived suppressor cells. Cytokines released by these cells inhibit natural killer cell activation and $CD8^+$ T-cell proliferation, driving the emergence of an immunosuppressive microenvironment [37].

The tumor microenvironment actively participates in drug-resistance acquisition [38] in both primary and metastatic lesions [39] of HCC and other solid tumors [40]. Previous studies reported that an inflammatory microenvironment, circulating immune cells, and cytokines etc. play a significant role in the prognosis of HCC [41,42]. For example, the B-type Raf kinase (BRAF) mutation, one of the prognostic factors, could play a role in the response to tyrosine kinase inhibitors [43]. However, we found no significant difference in changes in soluble immune checkpoint protein levels between groups classified according to the response to sorafenib [21] or lenvatinib treatment, possibly due to the small sample size. Further investigation is required in a large group of patients to establish a determinant in driving clinical decision-making, which today are an unmet clinical need and a challenge for immunotherapy.

As part of our research, we also analyzed the levels of soluble immune checkpoint proteins in TACE-treated patients. Hepatic arterial embolization induces tumor necrosis and focal inflammation by blocking the tumor blood supply, which can impact cancer immunity by creating a source of tumor-associated antigens and enhancing the tumor-specific T-cell response [44]. Tampaki et al. [45] reported that TACE provokes a significant increase in the sTIM-3 level in plasma within the first week posttreatment, suggesting a reactive expansion of TIM-3 expression by T-cells as a negative feedback mechanism in response to intense immune stimulation following tumor necrosis. In our research encompassing a comprehensive measurement of 16 immune checkpoints, not only did the level of sTIM-3 but also those of many more soluble proteins changed significantly in the first week after TACE (Figure 3). However, some confounding factors may accompany TACE. When compared with systemic therapies, TACE presents a greater likelihood of inducing more sudden hypoxia in treated lesions and produces more hypoxia-related factors, which can influence components of cancer-immunity in the short-term. Therefore, the interpretation of results among TACE-treated patients is difficult and further research is necessary to better understand the treatment's immunomodulatory effect.

Correlation analyses involving two soluble immune checkpoint proteins in plasma revealed that the fold-changes in soluble receptors and their ligands were positively correlated in both lenvatinib-treated and TACE-treated patients (Figures 4a and 5a), suggesting that multiplexed fluorescent immunoassays are capable of accurate and simultaneous quantification of small amounts of the proteins. More interestingly, sCD40 at 1 week and sCD27 at two to four weeks, respectively, were positively correlated with sCTLA-4, sCD86, sPD-1, and sPD-L1 (Figure 4b and Appendix A Figure A4b). While the mechanisms regulating these proteins remain unknown, sCD40 and sCD27 may be upregulated in cooperation with major checkpoint molecules such as sCTLA-4, sCD86, sPD-1, and sPD-L1. At Week 1 of TACE treatment, many more checkpoint pairs showed correlations unexpectedly (Figure 5b). For example, CD80 and TLR-2 have both been generally used as specific M1 macrophage surface markers in previous research [46]. Elsewhere, CD40 and HVEM messenger RNAs were coexpressed in the bioinformatic analysis of bladder cancer in the Cancer Genome Atlas database [47]. HVEM is a ligand of B- and T-lymphocyte attenuator (BTLA) and soluble BTLA in combination with sTIM-3 may be able to predict the rates of disease recurrence and survival among renal cell cancer patients [48].

Our study has several limitations. First, this is not a prospective study but instead a retrospective observational study. Moreover, randomization was not performed. In current guidelines [3–5], TACE is essentially indicated as a treatment option for patients with intermediate-stage HCC, while systemic therapy is suggested for those with advanced-stage HCC. Along these lines, in this study, the baseline characteristics of patients in the lenvatinib and TACE groups were not similar to one another (Table 1). It is therefore difficult to accurately compare the changes in soluble immune checkpoint levels in plasma between these treatment groups. Second, we could not collect plasma samples at Week 2 through 4 after TACE because of the less frequent hospital visits made by TACE-treated patients. We observed dynamic changes in soluble checkpoint levels in plasma at Week 1 of TACE but could not determine how long they persisted. Third, the sample size was small. Sample size affects the power or ability of all statistical tests to detect a relationship between two variables when it truly exists. The limitations of statistical analyses, such as the Wilcoxon signed-rank test and Spearman's rank correlation test, are equally worth noting [49]. Lastly, we could not investigate associations between changes in soluble checkpoints and the outcomes of HCC patients because a majority of the included patients are still alive. Further studies are needed to clarify the clinical significance of circulating soluble checkpoint proteins in patients with unresectable HCC.

4. Materials and Methods

4.1. Patients

Between May 2018 and February 2020, we initiated lenvatinib treatment in 55 patients (46 males and nine females; median age: 73 (range: 42–85) years) with unresectable HCC at our institute. In this retrospective cohort study, 24 patients who received lenvatinib as a first-line systemic therapy (18 males and six females; median age: 75 (range: 55–89) years) were included. Lenvatinib therapy was principally indicated for patients with a good performance status and compensated liver disease for whom locoregional therapies were not indicated either due to the presence of vascular invasion, distant metastasis, or a TACE-refractory state, in accordance with available guidelines [3–5]. Lenvatinib (Lenvima®; Eisai Co. Ltd., Tokyo, Japan) was orally given to patients with unresectable HCC. The dose of lenvatinib was determined based on body weight, with initial dosages of 12 mg/day and 8 mg/day given to those over 60 kg and those under 60 kg, respectively. Plasma samples were collected at baseline ($n = 24$), during the first week (days 3–7 after initiating therapy; $n = 19$), and during the second to fourth weeks (days 8–28 after initiating therapy; $n = 13$).

As part of an effort to compare with lenvatinib treatment, we also studied 22 patients (15 males and seven females; median age: 76 (range: 44–86) years) who underwent conventional TACE during the same period. TACE was principally indicated for those with three or less HCC masses measuring greater than 3 cm or for four or more HCC masses in patients with Child–Pugh A/B hepatic reserve without extrahepatic metastasis or vascular invasion, in accordance with available guidelines [3–5]. Conventional TACE consisted of intra-arterial injection of lipiodol plus epirubicin followed by the injection of an embolic agent, Gelpart (Nippon Kayaku Co. Ltd., Tokyo, Japan), to interrupt blood flow. Plasma samples were collected at baseline and during the first week (days 3–7 after initiating therapy).

4.2. Soluble Immune Checkpoint Protein Assays

The plasma levels of 16 soluble immune checkpoint proteins were measured using multiplexed fluorescent bead-based immunoassays with the Milliplex Map Kit (EMD Millipore Corporation, Burlington, MA, USA) and the Luminex Bio-Plex-200 system (Bio-Rad Laboratories, Hercules, CA, USA): namely, sBTLA, sCD27, sCD28, sCD40, sCD80, sCD86, sCTLA-4, sGITR, sGITRL, sHVEM, sPD-1, sPD-L1, sTIM-3, sTLR-2, soluble lymphocyte-activation gene 3 (sLAG-3), and soluble inducible T-cell costimulator (sICOS). This kit enables simultaneous measurement of the concentration of these 16 checkpoint proteins in the plasma sample. In brief, the capture antibody-coupled beads were first incubated with antigen standards or samples for a specific time. The plate was washed to remove unbound materials, followed by incubation with biotinylated detection antibodies. After washing away the unbound biotinylated antibodies, the beads were incubated with a reporter streptavidin–phycoerythrin conjugate (SA–PE). Following the removal of the excess SA–PE, the beads were passed through the array reader, which measures the fluorescence of the bound SA–PE [50]. According to the manufacturer's instructions, 12.5 µL of plasma was used for each measurement and all samples were assayed in duplicate; mean values were then adopted for further analysis. For values that were lower than the limit of detection, we used 10% of the lowest recorded value as a substitute [51].

4.3. Ethical Considerations

All patients supplied informed consent and the present study was conducted in accordance with the Declaration of Helsinki and was approved by the ethical committee of Osaka City University (#3719, approved on 12 July 2017).

4.4. Statistical Analysis

Analysis was conducted in Easy R (EZR) [52] (Saitama Medical Center, Jichi Medical University, Saitama, Japan) or R [53] (The R Foundation for Statistical Computing, Vienna, Austria) and figures were produced using the package ggplot2 [54]. Categorical variables were compared using Fisher's exact test or the chi-squared test, when appropriate. Continuous variables were tested using the Mann–Whitney U test. Wilcoxon signed-rank tests were chosen to compare changes in soluble immune checkpoint concentrations during the early treatment period. Correlations of fold-changes in the levels of two proteins were determined with Spearman's rank correlation test. A p-value of less than 0.05 was considered to be statistically significant.

5. Conclusions

The present study investigated changes in 16 soluble immune checkpoint proteins in the plasma of patients with HCC treated by lenvatinib or TACE. In the 24 lenvatinib-treated patients, the plasma level of sCD27 decreased but those of sCD40 and sTIM-3 increased significantly at Week 1, while the level of sCD27 was increased significantly at Weeks 2 to 4. These changes in soluble checkpoint protein levels during lenvatinib treatment were different from those seen during sorafenib treatment in our previous study [21], suggesting the two drugs present different inhibitory profiles against the receptor tyrosine kinases. Meanwhile, in the 22 TACE-treated patients, alongside the levels of sCD27, sCD40, and sTIM-3, those of sHVEM, sTLR-2, sCD80, sCTLA-4, sGITR, sGITRL, and sPD-L1 also increased. However, interpretation of the results among TACE-treated patients is difficult because TACE may be accompanied by some confounding factors, as discussed above. Further study is needed to better understand the immunomodulatory effect of the treatments, which may help future investigators to establish an effective combination immunotherapy.

Author Contributions: Conceptualization, N.O., L.T.T.T., and M.E.; investigation, N.O., H.H., L.T.T.T., M.P.D., M.S., K.K., A.H., S.U.-K., A.T., and M.E.; formal analysis, N.O., M.P.D., and M.E.; writing-original draft preparation, N.O. and M.E.; supervision, N.K. All authors have read and agreed to the published version of the manuscript.

Funding: This research was funded in a part by the Japan Society for the Promotion of Science (grant No. J192640023) and a Grant-in-Aid for Scientific Research from the Japan Society for the Promotion of Science (grant No. J192640002).

Acknowledgments: The authors are grateful to Kazuhiro Matsumoto for his help in the statistical analyses, and to Sanae Deguchi, Rie Yasuda, Ayano Fujikawa, and the Osaka City University Graduate School of Medicine research support platform for their technical assistance.

Conflicts of Interest: Norifumi Kawada has received lecture fees from Eisai Co. Ltd.

Appendix A

Table A1. The 16 immune checkpoint proteins whose soluble forms were analyzed in this study.

Immune Checkpoint Proteins		Function
BTLA	B- and T-lymphocyte attenuator	A ligand of HVEM
CD27	Cluster of differentiation 27	Interaction with CD70 activates antigen-presenting cells
CD28	Cluster of differentiation 28	Interaction with CD80 or CD86 provides stimulatory signals required for T-cell activation and survival
CD40	Cluster of differentiation 40	Interaction with CD40L activates antigen-presenting cells
CD80	Cluster of differentiation 80	A ligand of CD28 and CTLA-4
CD86	Cluster of differentiation 86	A ligand of CD28 and CTLA-4
CTLA-4	Cytotoxic T-lymphocyte antigen 4	Interaction with either CD80 or CD86 transmits an inhibitory signal to T-cells
GITR	Glucocorticoid-induced tumor necrosis factor receptor	Interaction with GITRL inhibits the activity of Tregs and extends the survival of T-effector cells
GITRL	Glucocorticoid-induced tumor necrosis factor receptor ligand	A ligand of GITR
HVEM	Herpes virus entry mediator	Interaction with LIGHT stimulates the proliferation of T-cells, while interaction with BTLA negatively regulates T-cell responses

Table A1. Cont.

Immune Checkpoint Proteins		Function
ICOS	Inducible T-cell costimulator	Interaction with ICOS-L provides a costimulatory signal for T-cell proliferation and survival
LAG-3	Lymphocyte-activation gene 3	Interaction with MHC class II inhibits the activation of T-cells
PD-1	Programmed cell death domain–1	Interaction with PD-L1 leads to the inhibition of cytotoxic T-cells
PD-L1	Programmed death-ligand 1	A ligand of PD-1
TIM-3	T-cell immunoglobulin mucin–3	Interaction with galectin-9 negatively regulates Th1 function by triggering cell death
TLR-2	Toll-like receptor 2	TLR-2 recognizes many bacterial, fungal, and other endogenous substances and plays a role in innate immune responses

Abbreviations: Tregs, regulatory T-cells; LIGHT, homologous to lymphotoxin, exhibits inducible expression and competes with HSV glycoprotein D for binding to herpesvirus entry mediator, a receptor expressed on T-lymphocytes; MHC, major histocompatibility complex.

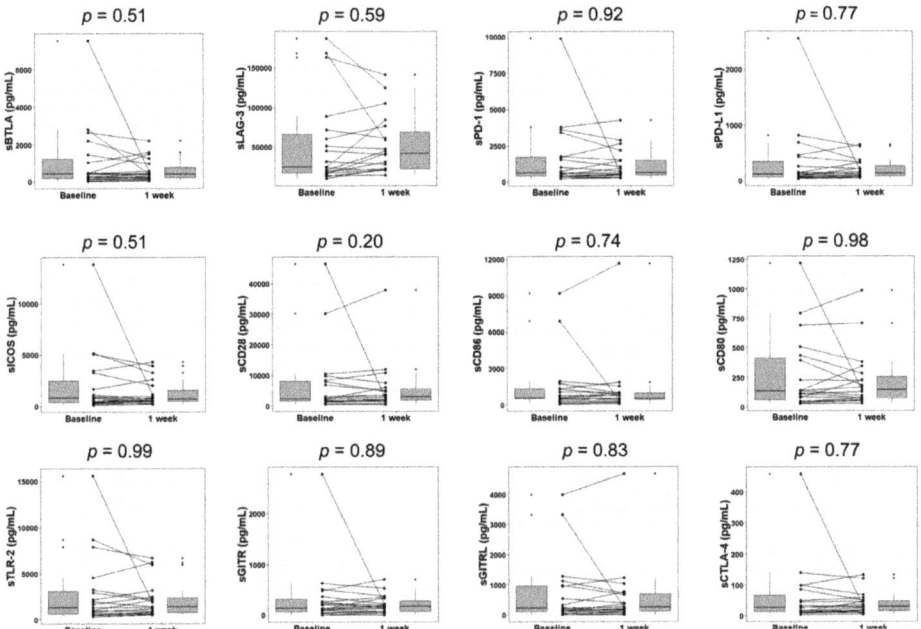

Figure A1. Changes in soluble immune checkpoint protein levels after 1 week of lenvatinib treatment. Box plots display the sBTLA, sLAG-3, sPD-1, sPD-L1, sICOS, sCD28, sCD86, sCD80, sTLR-2, sGITR, sGITRL, and sCTLA-4 levels in HCC patients at baseline and Week 1 of treatment. The vertical lengths of the boxes indicate the interquartile ranges and the lines in the boxes suggest the median values, while the error bars show the minimum and maximum values (range). The Wilcoxon signed-rank test was used.

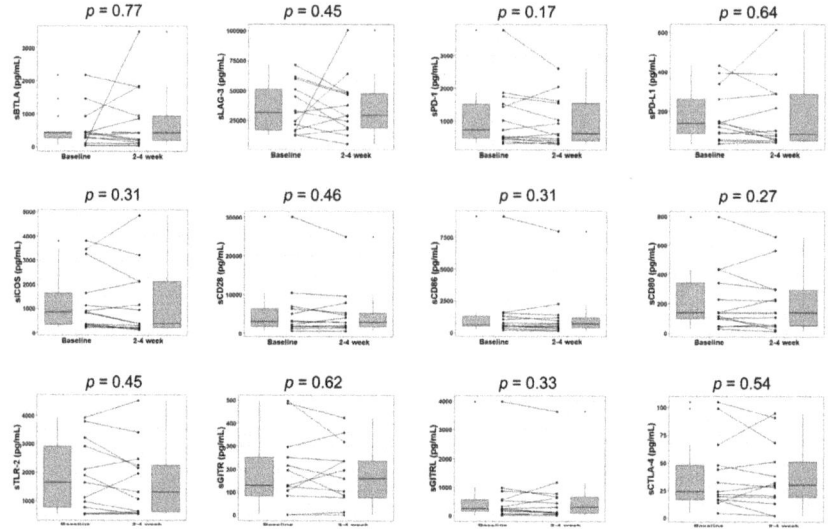

Figure A2. Changes in soluble immune checkpoint protein levels after two to four weeks of lenvatinib treatment. Box plots display the sBTLA, sLAG-3, sPD-1, sPD-L1, sICOS, sCD28, sCD86, sCD80, sTLR-2, sGITR, sGITRL, and sCTLA-4 levels in HCC patients at baseline and Week 2 through 4 of treatment. The vertical lengths of the boxes indicate the interquartile ranges and the lines in the boxes show the median values, while the error bars reveal the minimum and maximum values (range). The Wilcoxon signed-rank test was used.

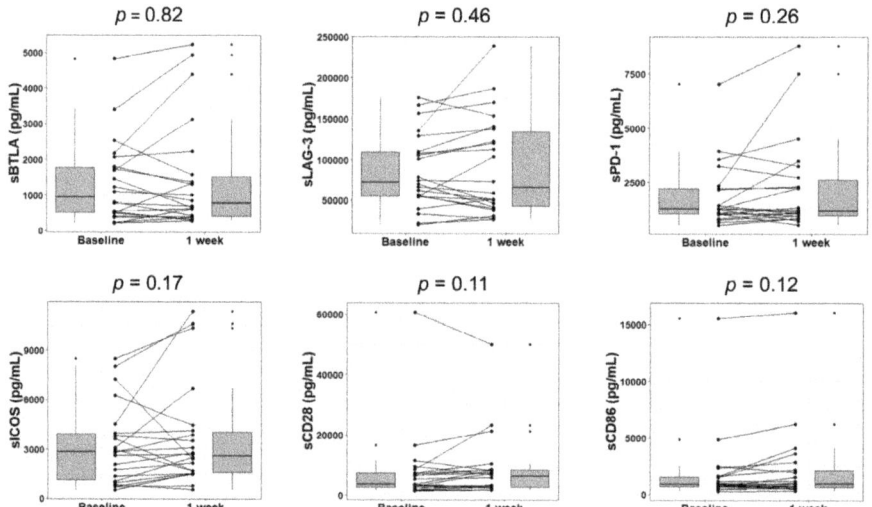

Figure A3. Changes in soluble immune checkpoint protein levels after 1 week of TACE treatment. Box plots display the sBTLA, sLAG-3, sPD-1, sICOS, sCD28, and sCD86 levels in HCC patients at baseline and Week 1 of treatment. The vertical lengths of the boxes reveal the interquartile ranges and the lines in the boxes show the median values, while the error bars suggest the minimum and maximum values (range). The Wilcoxon signed-rank test was used.

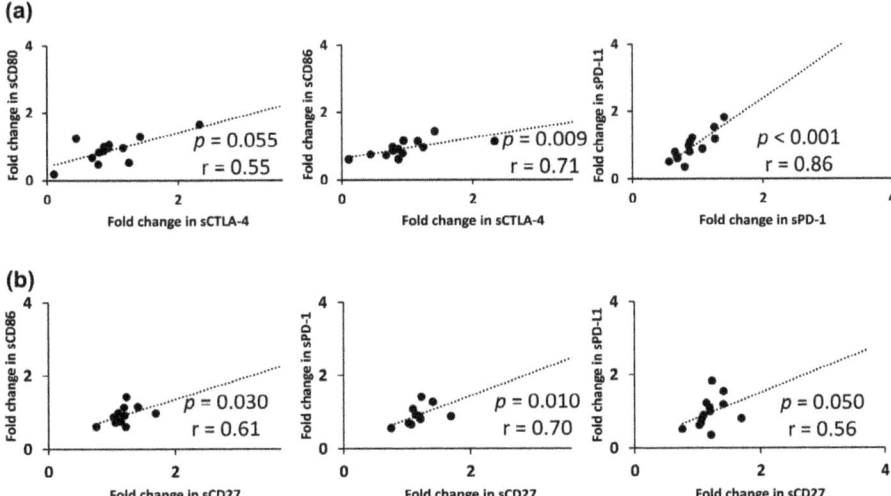

Figure A4. The relationships between fold-changes in soluble immune checkpoint protein levels in plasma at two to four weeks of lenvatinib treatment are seen. (**a**) A positive correlation was found between the fold-changes of soluble checkpoint receptors and their ligands, including sCTLA-4 with sCD80/sCD86 and sPD-1 with sPD-L1, respectively. (**b**) The fold-change of sCD27 was positively correlated with those of sCD86, sPD-1, and sPD-L1. Spearman's rank correlation test was used. A p-value of less than 0.05 was considered to be statistically significant.

References

1. Global Burden of Disease Cancer Collaboration; Fitzmaurice, C.; Abate, D.; Abbasi, N.; Abbastabar, H.; Abd-Allah, F.; Abdel-Rahman, O.; Abdelalim, A.; Abdoli, A.; Abdollahpour, I.; et al. Global, regional, and national cancer incidence, mortality, years of life lost, years lived with disability, and disability-adjusted life-years for 29 cancer groups, 1990 to 2017: A systematic analysis for the global burden of disease study. *JAMA Oncol.* **2019**. [CrossRef] [PubMed]
2. Siegel, R.L.; Miller, K.D.; Jemal, A. Cancer statistics, 2020. *CA Cancer J. Clin.* **2020**, *70*, 7–30. [CrossRef] [PubMed]
3. Marrero, J.A.; Kulik, L.M.; Sirlin, C.B.; Zhu, A.X.; Finn, R.S.; Abecassis, M.M.; Roberts, L.R.; Heimbach, J.K. Diagnosis, staging, and management of hepatocellular carcinoma: 2018 practice guidance by the american association for the study of liver diseases. *Hepatology* **2018**, *68*, 723–750. [CrossRef] [PubMed]
4. European Association for the Study of the Liver. EASL clinical practice guidelines: Management of hepatocellular carcinoma. *J. Hepatol.* **2018**, *69*, 182–236. [CrossRef]
5. Kokudo, N.; Takemura, N.; Hasegawa, K.; Takayama, T.; Kubo, S.; Shimada, M.; Nagano, H.; Hatano, E.; Izumi, N.; Kaneko, S.; et al. Clinical practice guidelines for hepatocellular carcinoma: The Japan Society of Hepatology 2017 (4th JSH-HCC guidelines) 2019 update. *Hepatol. Res.* **2019**, *49*, 1109–1113. [CrossRef]
6. Llovet, J.M.; Ricci, S.; Mazzaferro, V.; Hilgard, P.; Gane, E.; Blanc, J.F.; de Oliveira, A.C.; Santoro, A.; Raoul, J.L.; Forner, A.; et al. Sorafenib in advanced hepatocellular carcinoma. *N. Engl. J. Med.* **2008**, *359*, 378–390. [CrossRef]
7. Kudo, M.; Finn, R.S.; Qin, S.; Han, K.H.; Ikeda, K.; Piscaglia, F.; Baron, A.; Park, J.W.; Han, G.; Jassem, J.; et al. Lenvatinib versus sorafenib in first-line treatment of patients with unresectable hepatocellular carcinoma: A randomised phase 3 non-inferiority trial. *Lancet* **2018**, *391*, 1163–1173. [CrossRef]
8. Bruix, J.; Qin, S.; Merle, P.; Granito, A.; Huang, Y.H.; Bodoky, G.; Pracht, M.; Yokosuka, O.; Rosmorduc, O.; Breder, V.; et al. Regorafenib for patients with hepatocellular carcinoma who progressed on sorafenib treatment (RESORCE): A randomised, double-blind, placebo-controlled, phase 3 trial. *Lancet* **2017**, *389*, 56–66. [CrossRef]

9. Abou-Alfa, G.K.; Meyer, T.; Cheng, A.L.; El-Khoueiry, A.B.; Rimassa, L.; Ryoo, B.Y.; Cicin, I.; Merle, P.; Chen, Y.; Park, J.W.; et al. Cabozantinib in Patients with Advanced and Progressing Hepatocellular Carcinoma. *N. Engl. J. Med.* **2018**, *379*, 54–63. [CrossRef]
10. Zhu, A.X.; Kang, Y.K.; Yen, C.J.; Finn, R.S.; Galle, P.R.; Llovet, J.M.; Assenat, E.; Brandi, G.; Pracht, M.; Lim, H.Y.; et al. Ramucirumab after sorafenib in patients with advanced hepatocellular carcinoma and increased alpha-fetoprotein concentrations (REACH-2): A randomised, double-blind, placebo-controlled, phase 3 trial. *Lancet Oncol.* **2019**, *20*, 282–296. [CrossRef]
11. Zhu, A.X.; Finn, R.S.; Edeline, J.; Cattan, S.; Ogasawara, S.; Palmer, D.; Verslype, C.; Zagonel, V.; Fartoux, L.; Vogel, A.; et al. Pembrolizumab in patients with advanced hepatocellular carcinoma previously treated with sorafenib (KEYNOTE-224): A non-randomised, open-label phase 2 trial. *Lancet Oncol.* **2018**, *19*, 940–952. [CrossRef]
12. Yau, T.; Hsu, C.; Kim, T.Y.; Choo, S.P.; Kang, Y.K.; Hou, M.M.; Numata, K.; Yeo, W.; Chopra, A.; Ikeda, M.; et al. Nivolumab in advanced hepatocellular carcinoma: Sorafenib-experienced Asian cohort analysis. *J. Hepatol.* **2019**, *71*, 543–552. [CrossRef] [PubMed]
13. He, A.R.; Yau, T.; Hsu, C.; Kang, Y.-K.; Kim, T.-Y.; Santoro, A.; Sangro, B.; Melero, I.; Kudo, M.; Hou, M.-M.; et al. Nivolumab (NIVO) + ipilimumab (IPI) combination therapy in patients (pts) with advanced hepatocellular carcinoma (aHCC): Subgroup analyses from CheckMate 040. *J. Clin. Oncol.* **2020**, *38*, 512. [CrossRef]
14. Kudo, M. Systemic Therapy for Hepatocellular Carcinoma: Latest Advances. *Cancers* **2018**, *10*, 412. [CrossRef]
15. Cheng, H.; Sun, G.; Chen, H.; Li, Y.; Han, Z.; Li, Y.; Zhang, P.; Yang, L.; Li, Y. Trends in the treatment of advanced hepatocellular carcinoma: Immune checkpoint blockade immunotherapy and related combination therapies. *Am. J. Cancer Res.* **2019**, *9*, 1536–1545.
16. Liu, Z.; Lin, Y.; Zhang, J.; Zhang, Y.; Li, Y.; Liu, Z.; Li, Q.; Luo, M.; Liang, R.; Ye, J. Molecular targeted and immune checkpoint therapy for advanced hepatocellular carcinoma. *J. Exp. Clin. Cancer Res.* **2019**, *38*, 447. [CrossRef]
17. Nishida, N.; Kudo, M. Immune checkpoint blockade for the treatment of human hepatocellular carcinoma. *Hepatol. Res.* **2018**, *48*, 622–634. [CrossRef]
18. Ikeda, M.; Sung, M.W.; Kudo, M.; Kobayashi, M.; Baron, A.D.; Finn, R.S.; Kaneko, S.; Zhu, A.X.; Kubota, T.; Kraljevic, S.; et al. A phase 1b trial of lenvatinib (LEN) plus pembrolizumab (PEM) in patients (pts) with unresectable hepatocellular carcinoma (uHCC). *J. Clin. Oncol.* **2018**, *36*, 4076. [CrossRef]
19. Finn, R.S.; Qin, S.; Ikeda, M.; Galle, P.R.; Ducreux, M.; Kim, T.Y.; Kudo, M.; Breder, V.; Merle, P.; Kaseb, A.O.; et al. Atezolizumab plus Bevacizumab in Unresectable Hepatocellular Carcinoma. *N. Engl. J. Med.* **2020**, *382*, 1894–1905. [CrossRef]
20. Duffy, A.G.; Ulahannan, S.V.; Makorova-Rusher, O.; Rahma, O.; Wedemeyer, H.; Pratt, D.; Davis, J.L.; Hughes, M.S.; Heller, T.; ElGindi, M.; et al. Tremelimumab in combination with ablation in patients with advanced hepatocellular carcinoma. *J. Hepatol.* **2017**, *66*, 545–551. [CrossRef]
21. Dong, M.P.; Enomoto, M.; Thuy, L.T.T.; Hai, H.; Hieu, V.N.; Hoang, D.V.; Iida-Ueno, A.; Odagiri, N.; Amano-Teranishi, Y.; Hagihara, A.; et al. Clinical significance of circulating soluble immune checkpoint proteins in sorafenib-treated patients with advanced hepatocellular carcinoma. *Sci. Rep.* **2020**, *10*, 3392. [CrossRef] [PubMed]
22. Gu, D.; Ao, X.; Yang, Y.; Chen, Z.; Xu, X. Soluble immune checkpoints in cancer: Production, function and biological significance. *J. Immunother. Cancer* **2018**, *6*, 132. [CrossRef] [PubMed]
23. Llovet, J.M.; Bru, C.; Bruix, J. Prognosis of hepatocellular carcinoma: The BCLC staging classification. *Semin. Liver Dis.* **1999**, *19*, 329–338. [CrossRef] [PubMed]
24. Johnson, P.J.; Berhane, S.; Kagebayashi, C.; Satomura, S.; Teng, M.; Reeves, H.L.; O'Beirne, J.; Fox, R.; Skowronska, A.; Palmer, D.; et al. Assessment of liver function in patients with hepatocellular carcinoma: A new evidence-based approach-the ALBI grade. *J. Clin. Oncol.* **2015**, *33*, 550–558. [CrossRef] [PubMed]
25. Ueshima, K.; Nishida, N.; Hagiwara, S.; Aoki, T.; Minami, T.; Chishina, H.; Takita, M.; Minami, Y.; Ida, H.; Takenaka, M.; et al. Impact of Baseline ALBI Grade on the Outcomes of Hepatocellular Carcinoma Patients Treated with Lenvatinib: A Multicenter Study. *Cancers* **2019**, *11*, 952. [CrossRef]
26. Chen, D.S.; Mellman, I. Oncology meets immunology: The cancer-immunity cycle. *Immunity* **2013**, *39*, 1–10. [CrossRef]
27. Hendriks, J.; Gravestein, L.A.; Tesselaar, K.; van Lier, R.A.; Schumacher, T.N.; Borst, J. CD27 is required for generation and long-term maintenance of T cell immunity. *Nat. Immunol.* **2000**, *1*, 433–440. [CrossRef]

28. Huang, J.; Jochems, C.; Anderson, A.M.; Talaie, T.; Jales, A.; Madan, R.A.; Hodge, J.W.; Tsang, K.Y.; Liewehr, D.J.; Steinberg, S.M.; et al. Soluble CD27-pool in humans may contribute to T cell activation and tumor immunity. *J. Immunol.* **2013**, *190*, 6250–6258. [CrossRef]
29. Zhang, Q.; Liu, H.; Wang, H.; Lu, M.; Miao, Y.; Ding, J.; Li, H.; Gao, X.; Sun, S.; Zheng, J. Lenvatinib promotes antitumor immunity by enhancing the tumor infiltration and activation of NK cells. *Am. J. Cancer Res.* **2019**, *9*, 1382–1395.
30. O'Sullivan, B.; Thomas, R. CD40 and dendritic cell function. *Crit. Rev. Immunol.* **2003**, *23*, 83–107. [CrossRef]
31. Li, W.; Xia, Y.; Yang, J.; Guo, H.; Sun, G.; Sanyal, A.J.; Shah, V.H.; Lou, Y.; Zheng, X.; Chalasani, N.; et al. Immune Checkpoint Axes Are Dysregulated in Patients with Alcoholic Hepatitis. *Hepatol. Commun.* **2020**, *4*, 588–605. [CrossRef] [PubMed]
32. He, Y.; Cao, J.; Zhao, C.; Li, X.; Zhou, C.; Hirsch, F.R. TIM-3, a promising target for cancer immunotherapy. *Oncol. Targets Ther.* **2018**, *11*, 7005–7009. [CrossRef] [PubMed]
33. Wu, M.; Zhu, Y.; Zhao, J.; Ai, H.; Gong, Q.; Zhang, J.; Zhao, J.; Wang, Q.; La, X.; Ding, J. Soluble costimulatory molecule sTim3 regulates the differentiation of Th1 and Th2 in patients with unexplained recurrent spontaneous abortion. *Int. J. Clin. Exp. Med.* **2015**, *8*, 8812–8819. [PubMed]
34. Sasaki, R.; Kanda, T.; Fujisawa, M.; Matsumoto, N.; Masuzaki, R.; Ogawa, M.; Matsuoka, S.; Kuroda, K.; Moriyama, M. Different mechanisms of action of regorafenib and lenvatinib on toll-like receptor-signaling pathways in human hepatoma cell lines. *Int. J. Mol. Sci.* **2020**, *21*, 349. [CrossRef]
35. Lin, Y.Y.; Tan, C.T.; Chen, C.W.; Ou, D.L.; Cheng, A.L.; Hsu, C. Immunomodulatory effects of current targeted therapies on hepatocellular carcinoma: Implication for the future of immunotherapy. *Semin. Liver Dis.* **2018**, *38*, 379–388. [CrossRef]
36. Tohyama, O.; Matsui, J.; Kodama, K.; Hata-Sugi, N.; Kimura, T.; Okamoto, K.; Minoshima, Y.; Iwata, M.; Funahashi, Y. Antitumor activity of lenvatinib (e7080): An angiogenesis inhibitor that targets multiple receptor tyrosine kinases in preclinical human thyroid cancer models. *J. Thyroid. Res.* **2014**, *2014*, 638747. [CrossRef]
37. Voron, T.; Marcheteau, E.; Pernot, S.; Colussi, O.; Tartour, E.; Taieb, J.; Terme, M. Control of the immune response by pro-angiogenic factors. *Front. Oncol.* **2014**, *4*, 70. [CrossRef]
38. Longo, V.; Brunetti, O.; Gnoni, A.; Licchetta, A.; Delcuratolo, S.; Memeo, R.; Solimando, A.G.; Argentiero, A. Emerging role of Immune Checkpoint Inhibitors in Hepatocellular Carcinoma. *Medicina* **2019**, *55*, 698. [CrossRef]
39. Gnoni, A.; Brunetti, O.; Longo, V.; Calabrese, A.; Argentiero, A.L.; Calbi, R.; Solimando Antonio, G.; Licchetta, A. Immune system and bone microenvironment: Rationale for targeted cancer therapies. *Oncotarget* **2020**, *11*, 480–487. [CrossRef]
40. Argentiero, A.; De Summa, S.; Di Fonte, R.; Iacobazzi, R.M.; Porcelli, L.; Da Via, M.; Brunetti, O.; Azzariti, A.; Silvestris, N.; Solimando, A.G. Gene Expression Comparison between the Lymph Node-Positive and -Negative Reveals a Peculiar Immune Microenvironment Signature and a Theranostic Role for WNT Targeting in Pancreatic Ductal Adenocarcinoma: A Pilot Study. *Cancers* **2019**, *11*, 942. [CrossRef]
41. Brunetti, O.; Gnoni, A.; Licchetta, A.; Longo, V.; Calabrese, A.; Argentiero, A.; Delcuratolo, S.; Solimando, A.G.; Casadei-Gardini, A.; Silvestris, N. Predictive and Prognostic Factors in HCC Patients Treated with Sorafenib. *Medicina* **2019**, *55*, 707. [CrossRef] [PubMed]
42. Kawamura, Y.; Kobayashi, M.; Shindoh, J.; Kobayashi, Y.; Kasuya, K.; Sano, T.; Fujiyama, S.; Hosaka, T.; Saitoh, S.; Sezaki, H.; et al. (18)F-Fluorodeoxyglucose Uptake in Hepatocellular Carcinoma as a Useful Predictor of an Extremely Rapid Response to Lenvatinib. *Liver Cancer* **2020**, *9*, 84–92. [CrossRef] [PubMed]
43. Gnoni, A.; Licchetta, A.; Memeo, R.; Argentiero, A.; Solimando, A.G.; Longo, V.; Delcuratolo, S.; Brunetti, O. Role of BRAF in Hepatocellular Carcinoma: A Rationale for Future Targeted Cancer Therapies. *Medicina* **2019**, *55*, 754. [CrossRef] [PubMed]
44. Flecken, T.; Schmidt, N.; Hild, S.; Gostick, E.; Drognitz, O.; Zeiser, R.; Schemmer, P.; Bruns, H.; Eiermann, T.; Price, D.A.; et al. Immunodominance and functional alterations of tumor-associated antigen-specific CD8+ T-cell responses in hepatocellular carcinoma. *Hepatology* **2014**, *59*, 1415–1426. [CrossRef] [PubMed]
45. Tampaki, M.; Ionas, E.; Hadziyannis, E.; Deutsch, M.; Malagari, K.; Koskinas, J. Association of TIM-3 with BCLC Stage, Serum PD-L1 Detection, and Response to Transarterial Chemoembolization in Patients with Hepatocellular Carcinoma. *Cancers* **2020**, *12*, 212. [CrossRef] [PubMed]

46. Trombetta, A.C.; Soldano, S.; Contini, P.; Tomatis, V.; Ruaro, B.; Paolino, S.; Brizzolara, R.; Montagna, P.; Sulli, A.; Pizzorni, C.; et al. A circulating cell population showing both M1 and M2 monocyte/macrophage surface markers characterizes systemic sclerosis patients with lung involvement. *Respir. Res.* **2018**, *19*, 186. [CrossRef]
47. Dobosz, P.; Stempor, P.A.; Roszik, J.; Herman, A.; Layani, A.; Berger, R.; Avni, D.; Sidi, Y.; Leibowitz-Amit, R. Checkpoint genes at the cancer side of the immunological synapse in bladder cancer. *Transl. Oncol.* **2020**, *13*, 193–200. [CrossRef]
48. Wang, Q.; Zhang, J.; Tu, H.; Liang, D.; Chang, D.W.; Ye, Y.; Wu, X. Soluble immune checkpoint-related proteins as predictors of tumor recurrence, survival, and T cell phenotypes in clear cell renal cell carcinoma patients. *J. Immunother. Cancer* **2019**, *7*, 334. [CrossRef]
49. Twomey, P.J.; Viljoen, A. Limitations of the Wilcoxon matched pairs signed ranks test for comparison studies. *J. Clin. Pathol.* **2004**, *57*, 783.
50. Houser, B. Bio-Rad's Bio-Plex(R) suspension array system, xMAP technology overview. *Arch. Physiol. Biochem.* **2012**, *118*, 192–196. [CrossRef]
51. Iida-Ueno, A.; Enomoto, M.; Uchida-Kobayashi, S.; Hagihara, A.; Teranishi, Y.; Fujii, H.; Morikawa, H.; Murakami, Y.; Tamori, A.; Thuy, L.T.T.; et al. Changes in plasma interleukin-8 and tumor necrosis factor-alpha levels during the early treatment period as a predictor of the response to sorafenib in patients with unresectable hepatocellular carcinoma. *Cancer Chemother. Pharmacol.* **2018**, *82*, 857–864. [CrossRef] [PubMed]
52. Kanda, Y. Investigation of the freely available easy-to-use software 'EZR' for medical statistics. *Bone Marrow Transplant.* **2013**, *48*, 452–458. [CrossRef] [PubMed]
53. R Core Team. *R: A Language and Environment for Statistical Computing*; R Foundation for Statistical Computing: Vienna, Austria, 2019.
54. Wickham, H. *Ggplot2: Elegant Graphics for Data Analysis*; Springer: New York, NY, USA, 2016.

© 2020 by the authors. Licensee MDPI, Basel, Switzerland. This article is an open access article distributed under the terms and conditions of the Creative Commons Attribution (CC BY) license (http://creativecommons.org/licenses/by/4.0/).

Article

Effectiveness and Safety of Nivolumab in Child–Pugh B Patients with Hepatocellular Carcinoma: A Real-World Cohort Study

Won-Mook Choi [1], Danbi Lee [1], Ju Hyun Shim [1], Kang Mo Kim [1], Young-Suk Lim [1], Han Chu Lee [1], Changhoon Yoo [2], Sook Ryun Park [2], Min-Hee Ryu [2], Baek-Yeol Ryoo [2] and Jonggi Choi [1],*

[1] Department of Gastroenterology, Liver Center, Asan Medical Center, University of Ulsan College of Medicine, Seoul 05505, Korea; j.choi@amc.seoul.kr (W.-M.C.); leighdb@hanmail.net (D.L.); s5854@amc.seoul.kr (J.H.S.); kimkm70@amc.seoul.kr (K.M.K.); limys@amc.seoul.kr (Y.-S.L.); hch@amc.seoul.kr (H.C.L.)
[2] Department of Oncology, Asan Medical Center, University of Ulsan College of Medicine, Seoul 05505, Korea; yooc@amc.seoul.kr (C.Y.); srpark@amc.seoul.kr (S.R.P.); miniryu@amc.seoul.kr (M.-H.R.); ryooby@amc.seoul.kr (B.-Y.R.)
* Correspondence: jkchoi0803@gmail.com; Tel.: +82-02-3010-1328

Received: 31 May 2020; Accepted: 18 July 2020; Published: 20 July 2020

Abstract: Nivolumab has shown durable response and safety in patients with hepatocellular carcinoma (HCC) in previous trials. However, real-world data of nivolumab in HCC patients, especially those with Child–Pugh class B, are limited. To investigate the effectiveness and safety of nivolumab in a real-world cohort of patients with advanced HCC, we retrospectively evaluated 203 patients with HCC who were treated with nivolumab between July 2017 and February 2019. Of 203 patients, 132 patients were classified as Child–Pugh class A and 71 patients were Child–Pugh class B. Objective response rate was lower in patients with Child–Pugh class B than A (2.8% vs. 15.9%; $p = 0.010$). Child–Pugh class B was an independent negative predictor for objective response. Median overall survival was shorter in Child–Pugh B patients (11.3 vs. 42.9 weeks; adjusted hazard ratio [AHR], 2.10; $p < 0.001$). In Child–Pugh B patients, overall survival of patients with Child–Pugh score of 8 or 9 was worse than patients with Child–Pugh score of 7 (7.4 vs. 15.3 weeks; AHR, 1.93; $p < 0.020$). In conclusion, considering the unsatisfactory response in Child–Pugh B patients, nivolumab may not be used in unselected Child–Pugh B patients. Further studies are needed in this patient population.

Keywords: liver cancer; immune checkpoint inhibitor; effectiveness; safety

1. Introduction

Hepatocellular carcinoma (HCC) is the sixth most prevalent cancer, and the second most common cause of cancer deaths in Korea and worldwide, leading to nearly 745,000 deaths globally each year [1,2]. Many patients are newly diagnosed with advanced HCC despite regular surveillance of patients at risk, and disease recurrence or progression after initial treatment, which requires systemic therapy, is common [3,4]. Sorafenib, an oral multikinase inhibitor which improved overall survival (OS) compared to placebo in the SHARP trial [5], has been the only viable treatment for HCC over the last decade, but recent successful phase 2/3 trials of first- or second-line therapies have expanded the treatment landscape for patients with advanced HCC [6–9]. However, the majority of systemic therapies have been studied in Child–Pugh A populations. Because most trials of systemic therapies for HCC excluded patients with poor liver function (Child–Pugh B or greater hepatic dysfunction). Therefore, limited data are currently available for systemic therapies in patients with advanced liver cirrhosis. In real-world practice, liver function of patients with advanced HCC who require systemic therapy is often poor due to the tumor itself, or it has been deteriorated by previous treatments for HCC. Thus, real-world data regarding the safety and clinical outcomes of systemic therapy in HCC patients with poor liver function are of importance to guide the use of systemic therapy in this population.

Nivolumab, an immune checkpoint inhibitor that blocks programmed cell death protein-1, showed durable responses and prolonged long-term survival in the CheckMate 040 trial [6]. Although patients with Child–Pugh class B disease were included in the CheckMate 040 study, the efficacy and safety of nivolumab have not been established in HCC patients with advanced cirrhosis. Eligibility was restricted to patients with Child–Pugh scores of 7 or 8 and patients with ascites requiring paracentesis were excluded from the study. A few retrospective cohort studies and case series have reported on the safety and effectiveness of immune checkpoint inhibitors for advanced HCC patients with poor liver function [10–12]; however, those studies were limited by small numbers of patients. Here we report real-world data on the clinical outcomes and safety of nivolumab using a large retrospective cohort of patients with advanced HCC, including a large number of Child–Pugh B patients.

2. Results

2.1. Baseline Characteristics of the Study Cohort

A total of 203 patients were included in the study. Information on patient demographics, liver function characteristics, and cancer staging is presented in Table 1. Of the included patients, 132 patients had Child–Pugh class A disease and 71 patients had Child–Pugh class B disease. Most of the baseline characteristics between Child–Pugh A and B patients were significantly different. The Eastern Cooperative Oncology Group (ECOG) performance status was higher in Child–Pugh B patients than in Child–Pugh A patients. Moreover, Child–Pugh B patients had more aggressive tumor characteristics at baseline than Child–Pugh A patients, including higher levels of α-fetoprotein and protein induced by vitamin K absence or antagonist-II, and Child–Pugh B patients had more patients with portal vein invasion (Table 1).

2.2. Treatment Outcomes of Patients Receiving Nivolumab

Over a maximum follow-up period of 37.0 months with a median follow-up duration of 5.6 months (interquartile range [IQR], 2.3–11.4), 146 patients died, and 150 patients experienced disease progression after nivolumab treatment. The treatment overview of the study population is summarized in Table 2. The median duration of nivolumab treatment was 1.6 (IQR, 0.9–5.0) and 0.9 (IQR, 0.5–1.9) months with a median of four (ranged 1–57) and three (ranged 1–34) cycles; 17 (12.9%) and eight (11.3%) patients remained on treatment at the time of the last follow-up in Child–Pugh A and B groups, respectively. One hundred and fifteen (87.1%) and 63 (88.7%) patients discontinued treatment in Child–Pugh A and B groups, respectively; the reasons for treatment discontinuation were disease progression in 103 (78.0%) and 46 (64.8%) patients, death in seven (5.3%) and 16 (22.5%) patients, and adverse events (AEs) in five (3.8%) and one (1.4%) patients in Child–Pugh A and B groups, respectively. During treatment, one patient (0.5%) achieved a complete response, while 22 (10.8%) patients achieved a partial response, and 49 (24.1%) patients had stable disease, with an 11.3% objective response rate (ORR) in the total study population.

Table 1. Baseline characteristics of the study population.

Characteristics	Child–Pugh A (n = 132)	Child–Pugh B (n = 71)	p Value
Demographics			
Age, mean ± SD, y	56.9 ± 11.2	56.0 ± 9.4	0.576
Male sex, n (%)	115 (87.1)	56 (78.9)	0.182
Etiology, HBV/HCV/Other, n (%)	111/4/17 (84.1/3.0 /12.9)	57/4/10 (80.3/5.6/14.1)	0.630
ECOG performance status, n (%)			0.004
0/1/2	64/57/11 (48.5/43.2/8.3)	18/41/12 (25.4/57.7/16.9)	
Tumor characteristics			
BCLC stage, n (%)			0.822
Intermediate/Advanced	6/126 (4.5/95.5)	2/69 (2.8/97.2)	
Portal vein invasion, n (%)	46 (34.8)	42 (59.2)	0.001
Extrahepatic metastasis, n (%)	120 (90.9)	64 (90.1)	>0.999
Involved disease sites, n (%)			
Liver	104 (78.8)	66 (93.0)	0.016
Lung	79 (59.8)	48 (67.6)	0.349
Number of involved disease sites, n (%)			0.959
1–2/≥3	80/52 (60.6/69.4)	42/29 (59.2/40.8)	
α-Fetoprotein, median (IQR), ng/mL	311 (10, 3392)	2698 (44, 53727)	0.001
PIVKA-II, median (IQR), mAU/mL	1439 (150, 9129)	6846 (771, 57522)	<0.001
Immunotherapy as systemic, n (%)			0.218
First-line	2 (1.5)	1 (1.4)	
Second-line	89 (67.4)	56 (78.9)	
Third-line or more	41 (31.1)	14 (19.7)	
Liver function			
Child–Pugh score, n (%)			–
5/6	67/65 (50.8/40.2)	–	
7/8/9	–	41/15/15 (57.7/21.2/21.2)	
Platelet count, n (%) ≥150,000/μL <150,000/μL	65 (40.2) 67 (50.8)	26 (36.6) 45 (63.4)	0.115
Ascites, present, n (%)	6 (4.5)	48 (67.6)	<0.001
Albumin, median (IQR), g/dL	3.6 (3.2, 3.9)	2.9 (2.6, 3.2)	<0.001
>3.5/2.8–3.5/<2.8, n (%)	81/51/0 (61.4/38.6 /0.0)	8/33/30 (11.3/46.5/42.3)	
Total bilirubin, median (IQR), mg/dL	0.7 (0.5, 0.9)	1.3 (0.9, 2.0)	<0.001
<2/2–3 />3, n (%)	130/2/0 (98.5/1.5/0.0)	52/12/7 (73.2/16.9/9.9)	
ALBI grade, mean ± SD	−2.33 ± 0.37	−1.54 ± 0.37	<0.001
1/2/3, n (%)	35/97/0 (26.5/73.5/0.0)	0/49/22 (0.0/69.0/31.0)	

Abbreviations: ALBI, albumin-bilirubin; BCLC, Barcelona Clinic Liver Cancer; ECOG, Eastern Cooperative Oncology Group; HBV, hepatitis B virus; HCV, hepatitis C virus; IQR, interquartile range; PIVKA, protein induced by vitamin K absence or antagonist-II; SD, standard deviation.

Table 2. Treatment summary of the study population.

Disposition Characteristics	Child–Pugh A (n = 132)	Child–Pugh B (n = 71)
Treatment duration, months (IQR)	1.6 (0.9, 5.0)	0.9 (0.5, 1.9)
Treatment duration, cycles, median (range)	4 (1–57)	3 (1–34)
mean ± SD	8.5 ± 9.9	4.3 ± 5.3
Continuing treatment, n (%)	17 (12.9)	8 (11.3)
Discontinued treatment, n (%)	115 (87.1)	63 (88.7)
Disease progression, n (%)	103 (78.0)	46 (64.8)
Death, n (%)	7 (5.3)	16 (22.5)
Adverse events, n (%)	5 (3.8)	1 (1.4)

Abbreviations: IQR, interquartile range; SD, standard deviation.

2.3. Treatment Outcomes Stratified by Child–Pugh Class

When stratified by Child–Pugh class, ORR was significantly lower in the Child–Pugh B group than in the Child–Pugh A group (2.8% vs. 15.9%; $p = 0.010$) (Table 3). Two Child–Pugh B patients (2.8%) achieved partial response, with response ongoing for over six months at the time of last follow-up. Disease control rate (DCR) in the total study population was 35.5%. DCR was lower in the Child–Pugh B group than in the Child–Pugh A group (22.5% vs. 42.4%; $p = 0.008$) (Table 3).

Table 3. Tumor responses in the study population according to modified Response Evaluation Criteria in Solid Tumors (mRECIST) criteria.

Tumor Responses	Entire Cohort		p Value [a]
	Child–Pugh A (n = 132)	Child–Pugh B (n = 71)	
Best overall response, n (%)			
Complete response	1 (0.8)	0 (0.0)	
Partial response	20 (15.2)	2 (2.8)	
Stable disease	35 (26.5)	14 (19.7)	
Progressive disease	69 (52.3)	40 (56.3)	
Not evaluable [b]	7 (5.3)	15 (21.1)	
Objective response [c], n (%)	21 (15.9)	2 (2.8)	0.010
Disease control rate [d], n (%)	56 (42.4)	16 (22.5)	0.008

[a] By χ^2 test or Fisher exact test, as appropriate, for radiologic response. [b] Due to death without radiologic disease progression or early drug discontinuation due to a severe adverse drug reaction. [c] Objective response rate, defined as the proportion of patients who had complete response or partial response. [d] Disease control rate, defined as the proportion of patients who had complete response, partial response, or stable disease.

OS was longer in the Child–Pugh A group than in the Child–Pugh B group (42.9 vs. 11.3 weeks; hazard ratio [HR], 3.02; 95% confidence interval [CI], 2.15–4.24; $p < 0.001$; Figure 1A); consistent results were also seen in the multivariable analyses (adjusted hazard ratio [AHR], 2.10; 95% CI, 1.38–3.19; $p < 0.001$) (Table 4). In addition to Child–Pugh class, ECOG performance status, albumin-bilirubin grade of 3, and α-fetoprotein were other independent prognostic factors for OS of the study population in the multivariable analysis (Table S1). Median progression-free survival (PFS) was longer in the Child–Pugh A group than in the Child–Pugh B group in the univariate analysis (7.4 vs. 6.0 weeks; HR, 1.67; 95% CI, 1.22–2.29; $p = 0.014$; Figure 1B); however, this difference was not statistically significant after multivariable adjustment (AHR, 1.17; 95% CI, 0.79–1.72; $p = 0.430$) (Table 4). ECOG performance status and liver involvement of HCC were poor prognostic factors for PFS in the multivariable analysis (Table S2). Median time to progression (TTP) was 7.9 weeks (95% CI, 7.1–11.6) for the Child–Pugh A group and 6.9 weeks (95% CI, 6.0–10.1) for the Child–Pugh B group. There was no difference in TTP between the two groups in the univariate (HR, 1.35; 95% CI, 0.95–1.92; $p = 0.093$) and multivariable (AHR, 1.00; 95% CI, 0.66–1.50; $p = 0.992$) analyses (Table 4).

(**A**) Overall survival

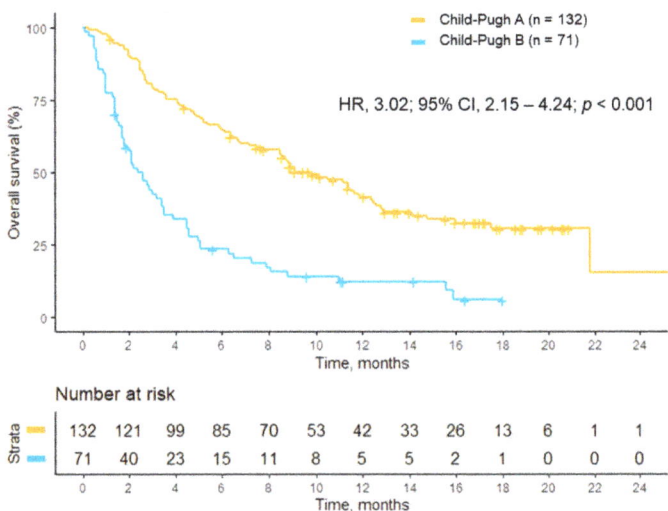

(**B**) Progression-free survival

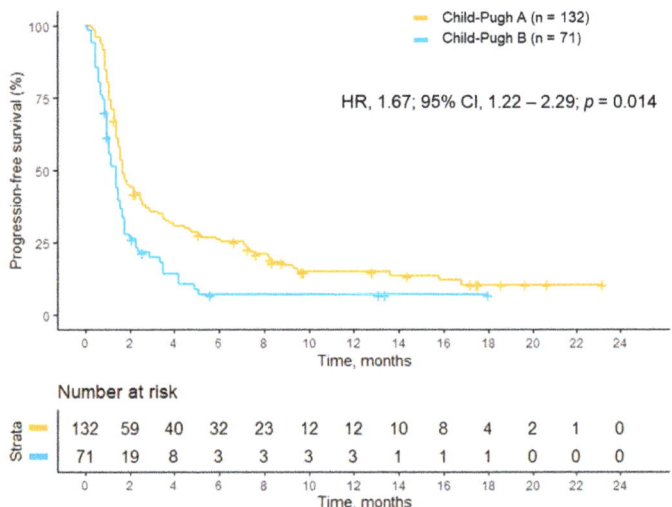

Figure 1. Kaplan–Meier analyses of survival outcomes between Child–Pugh A and B patients. (**A**) overall survival and (**B**) progression-free survival.

Table 4. Survival outcomes of the study population.

Outcome	Entire Cohort					
	Median Time, Week (95% CI)		Univariate Analysis		Multivariable Analysis	
	Child–Pugh A (n = 132)	Child–Pugh B (n = 71)	HR (95% CI) [a]	p Value	AHR (95% CI) [a]	p Value
Overall survival	42.9 (34.1–54.3)	11.3 (7.7–15.4)	3.02 (2.15–4.24)	<0.001	2.10 (1.38–3.19)	<0.001
Progression-free survival	7.4 (7.0–11.0)	6.0 (4.7–7.6)	1.67 (1.22–2.29)	0.014	1.17 (0.79–1.72)	0.430
Time to progression	7.9 (7.1–11.6)	6.9 (6.0–10.1)	1.35 (0.95–1.92)	0.093	1.04 (0.72–1.51)	0.834

[a] Cox proportional hazard regression model for the Child–Pugh B group with the Child–Pugh A group as a reference. Abbreviations: AHR, adjusted hazard ratio; CI, confidence interval; HR, hazard ratio.

2.4. Treatment Outcomes of Child–Pugh B Patients Receiving Nivolumab

Of 71 patients with Child–Pugh class B disease, 41 patients had a Child–Pugh score of 7 and the remaining 30 patients had Child–Pugh scores of 8 or 9. Marginally longer OS was observed in patients with a Child–Pugh score of 7 compared to patients with Child–Pugh scores of 8 or 9 in the univariate analysis (15.3 vs. 7.4 weeks; HR, 1.64; 95% CI, 0.98–2.72; $p = 0.058$; Figure 2A), and this difference became statistically significant after multivariable adjustment (AHR, 1.93; 95% CI, 1.11–3.35; $p = 0.020$) (Table 5). There were no significant differences in PFS (6.3 vs. 4.8 weeks; HR, 1.23; 95% CI, 0.74–2.04; $p = 0.416$; Figure 2B and AHR, 1.53; 95% CI, 0.86–2.58; $p = 0.153$) and TTP (6.9 vs. 6.1 weeks; HR, 1.04; 95% CI, 0.57–1.88; $p = 0.895$ and AHR, 1.30; 95% CI, 0.70–2.40; $p = 0.408$) between the two groups both in the univariate and multivariable analyses (Table 5). ECOG performance status and lung involvement of HCC were independent risk factors for poor OS and PFS in the multivariate analysis (Tables S3 and S4).

Table 5. Survival outcomes of Child–Pugh B patients.

Outcome	Entire Cohort					
	Median Time, Week (95% CI)		Univariate Analysis		Multivariable Analysis	
	Child–Pugh B7 (n = 41)	Child–Pugh B8/9 (n = 30)	HR (95% CI) [a]	p Value	AHR (95% CI) [a]	p Value
Overall survival	15.3 (9.3–22.3)	7.4 (6.4–14.9)	1.64 (0.98–2.72)	0.058	1.93 (1.11–3.35)	0.020
Progression-free survival	6.3 (5.0–8.0)	4.8 (3.7–7.6)	1.23 (0.74–2.04)	0.416	1.53 (0.86–2.58)	0.153
Time to progression	6.9 (6.0–12.6)	6.1 (4.6–NA)	1.04 (0.57–1.88)	0.895	1.30 (0.70–2.40)	0.408

[a] Cox proportional hazard regression model for the Child–Pugh B8/9 group with the Child–Pugh B7 group as a reference. Abbreviations: AHR, adjusted hazard ratio; CI, confidence interval; HR, hazard ratio; NA, not applicable.

2.5. Predictive Factors Associated with Treatment Response

Regarding predictive factors associated with treatment response (i.e., complete response and partial response) in patients receiving nivolumab, patients with advanced liver disease (Child–Pugh class B vs. Child–Pugh class A), high levels of tumor markers and liver involvement of HCC were poorly responsive to treatment by univariate analysis. After inclusion of predictive factors with a p value < 0.05 from the univariate analysis in the multivariable-adjusted model, Child–Pugh class (B vs. A; adjusted odds ratio [AOR], 0.21; 95% CI, 0.05–0.93; $p = 0.040$) and liver involvement of HCC (AOR, 0.34; 95% CI, 0.13–0.92; $p = 0.034$) remained as significant independent negative predictors for treatment response (Table 6).

(**A**) Overall survival

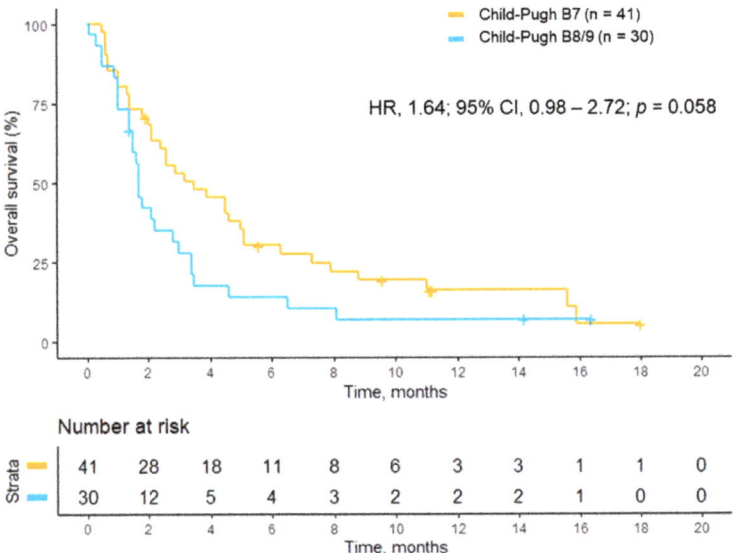

(**B**) Progression-free survival

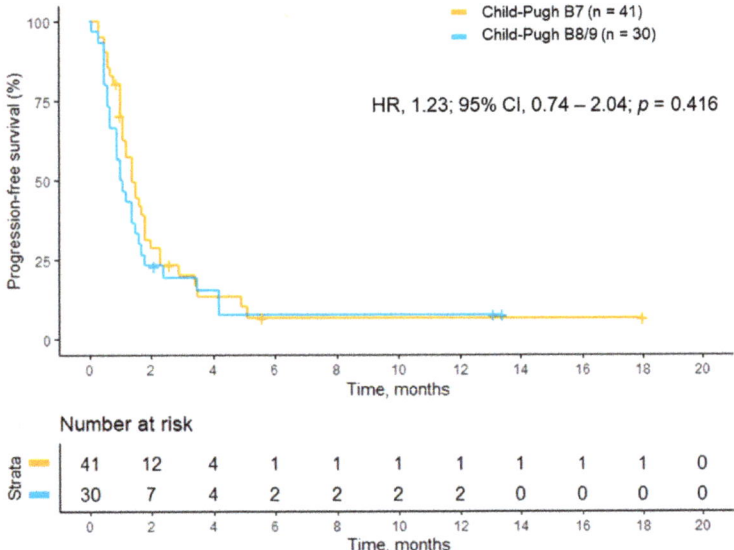

Figure 2. Kaplan–Meier analyses of survival outcomes between Child–Pugh B7 and B8/9 patients. (**A**) overall survival and (**B**) progression-free survival.

Table 6. Predictive factors for treatment response.

Characteristics	Univariate Analysis		Multivariable Analysis	
	OR (95% CI)	p Value	AOR (95% CI)	p Value
Child–Pugh class				0.040
A	1 (reference)	0.013 *	1 (reference)	
B	0.15 (0.03–0.67)		0.21 (0.05–0.93)	
Age	1.02 (0.98–1.07)	0.305	-	-
Sex				
Female	1 (reference)	0.333	-	-
Male	2.10 (0.47–9.43)			
Ascites, present	0.23 (0.05–1.04)	0.056	-	-
α-Fetoprotein, ng/mL				
<400	1 (reference)	0.349	-	-
≥400	0.66 (0.27–1.58)			
PIVKA-II, mAU/mL				
<2000	1 (reference)	0.035	1 (reference)	0.234
≥2000	0.37 (0.14–0.93)		0.55 (0.21–1.47)	
Albumin (per 1 g/dL increase)	1.96 (0.83–4.62)	0.125	-	-
Total bilirubin (per 1 mg/dL increase)	0.82 (0.48–1.40)	0.469	-	-
ALBI grade				
1	1 (reference)		-	-
2	0.79 (0.27–2.31)	0.668		
3	0.29 (0.03–2.63)	0.268		
Etiology				
HBV	1 (reference)	0.082	-	-
Non-HBV etiology	2.38 (0.90–6.30)			
Portal vein invasion, present	0.53 (0.21–1.36)	0.190	-	-
Extrahepatic metastasis, present	2.44 (0.31–19.22)	0.396	-	-
Involved disease sites, present				
Liver	0.24 (0.09–0.61)	0.003	0.34 (0.13–0.92)	0.034
Lung	1.14 (0.46–2.83)	0.780	-	-
Number of involved disease sites per patient				
1–2	1 (reference)	0.067	-	-
≥3	0.38 (0.14–1.07)			

Abbreviations: AOR; adjusted odds ratio; ALBI, albumin-bilirubin; CI, confidence interval; HBV, hepatitis B virus; OR, odds ratio; PIVKA, protein induced by vitamin K absence or antagonist-II. * $p < 0.05$.

Considering the low response rate in Child–Pugh class B patients, predictive factors associated with disease control (i.e., complete response, partial response, and stable disease) were assessed instead of treatment response. Characteristics associated with higher tumor burden including the presence of extrahepatic metastasis, lung involvement of HCC, and ≥ 3 numbers of involved disease sites were significant negative predictors for disease control in the univariate analysis. Among them, lung involvement of HCC (AOR, 0.14; 95% CI, 0.03–0.64; $p = 0.011$) remained a significant independent negative predictor for disease control in the multivariable analysis (Table S5).

2.6. Safety of Nivolumab

During the treatment, five (3.8%) patients in the Child–Pugh A group and one (1.4%) patient in the Child–Pugh B group had grade 3 or higher toxicities that were probably attributable to nivolumab, leading to drug discontinuation. In the Child–Pugh A group, two patients developed immune-mediated hepatitis, three patients developed immune-mediated pneumonitis. One patient in the Child–Pugh B group suffered from severe anorexia (Table 7). Eleven (8.3%) and 11 (15.5%) patients in the Child–Pugh A and Child–Pugh B groups, respectively, required dose delay due to AEs (Table 7).

Table 7. Adverse events requiring discontinuation or dose delay.

Adverse Events	Child–Pugh A (n = 132)		Child–Pugh B (n = 71)	
	Any Grade	Grade ≥ 3	Any Grade	Grade ≥ 3
Hepatitis	3 (2.3)	2 (1.5)	3 (4.2)	
Pneumonitis	3 (2.3)	3 (2.3)		
Anorexia	3 (2.3)		6 (8.5)	1 (1.4)
Nausea	1 (0.8)		1 (1.4)	
Pain	1 (0.8)		3 (4.2)	
Anemia	3 (2.3)		3 (4.2)	
Fatigue	1 (0.8)		5 (7.0)	
Rash	2 (1.5)		1 (1.4)	
Insomnia			1 (1.4)	

3. Discussion

We evaluated the effectiveness and safety of nivolumab in a large real-world cohort of advanced HCC patients including Child–Pugh B patients. ORR and DCR were lower in Child–Pugh B patients than in Child–Pugh A patients and Child–Pugh class B was an independent negative predictor for objective response in our patients. OS was shorter in Child–Pugh B patients. However, TTP and PFS were comparable between Child–Pugh A and B patients by multivariable-adjusted analysis. In the subgroup analysis of Child–Pugh B patients, patients with a Child–Pugh score of 7 survived longer than patients with Child–Pugh scores of 8 or 9; however, there were no differences in PFS and TTP between the two groups. Regarding predictors for nivolumab response in Child–Pugh B patients, lung involvement of HCC, which might represent tumor spread and burden based on the fact that most of the Child–Pugh B patients in our study had liver involvement of HCC, was the only significant negative predictor for disease control. No significant differences were observed in the safety measures of nivolumab between the two groups. Rather, immune-mediated serious AEs due to nivolumab treatment were found to occur less frequently in Child–Pugh B patients than in Child–Pugh A patients.

A lower ORR and DCR in Child–Pugh B patients compared with Child–Pugh A patients observed in our study can be interpreted in two ways. First, Child–Pugh B patients received fewer cycles and had shorter durations of nivolumab treatment than Child–Pugh A patients. Indeed, 22.5% of Child–Pugh B patients discontinued treatment due to death mostly resulting from liver function deterioration, whereas only 5.3% of Child–Pugh A patients ceased the treatment due to death. These facts imply that some patients with poor liver function may not have had enough time to maintain nivolumab treatment because of progressive liver dysfunction. Moreover, these may adversely affect the overall poorer outcomes in Child–Pugh B patients compared with Child–Pugh A patients.

Second, it is well-established that cirrhosis is associated with innate and adaptive immune dysfunction. Moreover, the immune function becomes more impaired as underlying liver cirrhosis progresses [13]. A previous study showed that the cyclooxygenase-derived prostaglandin E2 drives cirrhosis-associated immunosuppression [14]. In addition, patients with decompensated cirrhosis are more vulnerable to endotoxemia or bacteremia, resulting in the up-regulation of prostaglandin E2, and comorbidity with hypoalbuminemia in these patients also provokes increased levels of free prostaglandin E2, causing pathological immune impairment [15]. Cirrhosis alters the number and function of monocytes, NK cells, and T lymphocytes, which play a key role in killing tumor cells [13,16,17]. Thus, proper tumor-killing may not be possible even if T cell reinvigoration is induced by nivolumab, which may explain the poorer ORR in Child–Pugh B patients than in Child–Pugh A patients. A lower incidence of immune-mediated AEs in Child–Pugh B patients also supports this hypothesis.

Recently, results of clinical trials evaluating the efficacy and safety of immune checkpoint inhibitors alone or combination therapies for HCC have been published. In the KEYNOTE-240 trial, although pembrolizumab in a second-line setting after prior sorafenib therapy improved OS (HR, 0.78; 95% CI, 0.61–1.00; one-sided p = 0.024) and PFS (HR, 0.72; 95% CI, 0.57–0.90; one-sided p = 0.002) compared to placebo, the outcomes did not reach statistical significance per specified criteria [18]. In the CheckMate-459 trial, nivolumab showed an improved OS compared to sorafenib; however, this difference also was not statistically significant [19]. As a combination therapy, atezolimumab with bevacizumab led to better OS (HR, 0.58; 95% CI, 0.42–0.79) than sorafenib in patients with unresectable HCC [20]. However, none of those clinical trials included Child–Pugh class B patients due to competing risk of death from underlying cirrhosis. Moreover, most of the ongoing clinical trials of immune checkpoint inhibitors target Child–Pugh class A disease.

There is insufficient evidence for the use of systemic therapy in Child–Pugh class B patients. The most widely reported systemic therapy in this population is sorafenib. A meta-analysis of thirty studies demonstrated that Child–Pugh B liver function is associated with worse OS compared to Child–Pugh A liver function despite similar response rate, safety, and tolerability [21]. Several previous studies have evaluated the efficacy and safety of nivolumab in Child–Pugh class B patients. In the Child–Pugh B cohort of CheckMate 040 trial, outcomes were much better than was seen in our patients: median OS was 7.6 months, ORR was 10.2%, and the DCR was 55.1% [22]. However, it is important to note that the CheckMate 040 Child–Pugh B cohort excluded patients with Child–Pugh scores of 9 points and patients with ECOG performance status 2 or recent history of paracentesis for ascites or hepatic encephalopathy. In contrast, the current study cohort included patients with more advanced disease with a Child–Pugh score of 9 reported in approximately 21.1% of patients and ECOG performance status 2 reported in approximately 16.9% of patients, thus providing a real-world data of the effectiveness and safety of nivolumab in a wider range of patients. Several retrospective case series or cohort studies of Child–Pugh B patients reported better ORR, from 11.8% to 20%, and longer median OS, from 5.9 to 8.6 months, compared to our Child–Pugh B cohort [10–12,23]. However, it is questionable how the ORR in those retrospective studies was better than in the CheckMate 040 Child–Pugh B cohort, notwithstanding the fact that patients with more advanced liver disease were included and the tumor response was evaluated by Response Evaluation Criteria in Solid Tumors (RECIST) 1.1 criteria instead of modified RECIST (mRECIST) in those retrospective studies. Selection bias might also be an issue because the previous retrospective studies had very small patient numbers. Besides, the patients included in our study had more aggressive tumor features with higher proportions of macroscopic vascular invasion and extrahepatic spread compared to the patients included in the previous studies.

Adverse effects of immune checkpoint inhibitors such as nivolumab are different from those of systemic chemotherapy. Clinicians should be aware of immune-mediated AEs such as immune-mediated hepatitis and pneumonitis when using nivolumab. As observed in the clinical trials and previous studies, the AEs of nivolumab in our cohort were manageable and nivolumab appeared to be safe even in Child–Pugh B patients overall. Grade 1 or 2 AEs occurred more frequently in Child–Pugh B patients but were attributed to comorbid liver disease rather than nivolumab treatment. Interestingly, we observed a lower incidence of immune-mediated AEs in Child–Pugh B patients compared to Child–Pugh A patients albeit the incidence of immune-mediated AEs was very low.

This study had several limitations. First, as a retrospective study, it has inherent limitations including bias and confounding. Considering that most clinical trials only include patients with Child–Pugh class A to avoid competing risks of death from liver cirrhosis on the overall outcome, this retrospective cohort study may provide valuable information for evaluating the effectiveness and safety of nivolumab in a real-world setting where the patients tend to be more heterogeneous than patients in clinical trials. Second, as a single-center study, this study may have limited generalizability. Third, most of the HCC cases were caused by hepatitis B virus infection, which may be associated with poorer prognosis [24]. However, there was no evidence that underlying HCC etiology affects the efficacy of nivolumab treatment [25]. Finally, since the data were collected retrospectively from electronic medical records, only adverse events resulting in drug discontinuation or dose delay could be identified in detail. However, considering that nivolumab was well-tolerated, with the exception of rarely occurring severe immune-mediated adverse events seen in previous studies [6,10–12,18,21], we believe that the information on adverse events of our study contains clinically meaningful information despite the lack of detailed adverse event information.

4. Materials and Methods

4.1. Study Population

From July 2017 to February 2019, 221 consecutive patients received nivolumab treatment for unresectable HCC at Asan Medical Center and were retrospectively enrolled in this study. HCC diagnosis was based on multiphase computed tomography and/or magnetic resonance imaging or pathological confirmation in selected cases according to the current international guidelines of HCC [26]. Patients were excluded if they had Child–Pugh class C liver function ($n = 5$), had ECOG performance status > 2 ($n = 2$), had received liver transplantation ($n = 5$), or had been followed-up for less than one cycle of nivolumab ($n = 6$). After excluding 18 patients, 203 patients were included in the final analyses. Nivolumab was administered at 3 mg/kg body weight every 2 weeks intravenously until disease progression, severe adverse events, or death occurred. Dosage delays were permitted according to individual patient tolerability. Patient information, including demographic characteristics, laboratory results, safety assessment and grading, and clinical outcomes were collected from electronic medical records. The response evaluation was carried out every 6-8 weeks during nivolumab treatment, and additional image examinations were allowed when clinically indicated.

This study was approved by the Institutional Review Board of Asan Medical Center (IRB No. 2019-0605) and the informed consent of enrolled patients was waived owing to the retrospective nature of the study.

4.2. Outcome Assessment

Clinical tumor response was assessed by the mRECIST criteria [27]. ORR, defined as the proportion of patients with complete or partial response, and DCR, defined as the proportion of patients with complete response, partial response, or stable disease, were evaluated. Other oncological outcomes included TTP, defined as the time from nivolumab treatment to radiological or clinical progression; PFS, defined as the time from nivolumab treatment to progression or death due to any cause; and OS, defined as the time from nivolumab treatment to death due to any cause. Safety assessment and grading were recorded in patients' electronic medical records when treatment-related adverse events led to dose reduction or discontinuation of nivolumab.

4.3. Statistical Analysis

Categorical variables were analyzed as frequency and percentages and were compared using Fisher's exact test or the chi-square test, as appropriate. Continuous variables were expressed as median and IQR or mean and standard deviation and were compared using unpaired two-tailed t tests. Survival outcomes were estimated by the Kaplan–Meier method and compared with the log-rank test. In addition, univariate and multivariable Cox proportional hazard models were used to calculate HRs for survival outcomes and their 95% CIs. To identify the predictive factors associated with treatment response of nivolumab, univariate and multivariable logistic regression models were applied. Variables with p values less than 0.05 in the univariate analysis were used in the multivariable analysis.

All statistical analyses were performed using R statistical software, version 3.6.1 (R Foundation Inc; http://cran.r-project.org/). For all analyses, p values < 0.05 were considered statistically significant.

5. Conclusions

In the present study, the ORR and DCR were lower and the OS was shorter in Child–Pugh class B patients than those in Child–Pugh class A patients. Moreover, Child–Pugh class B was an independent negative predictor for nivolumab response. In particular, Child–Pugh B patients who had high tumor burden with lung involvement or Child–Pugh scores of 8 or 9 may not benefit from nivolumab treatment. Considering the unsatisfactory treatment response and poor prognosis in Child–Pugh B patients, nivolumab may not be beneficial in unselected patients of this patient population. Further investigation in this patient population is needed to confirm our findings.

Supplementary Materials: The following are available online at http://www.mdpi.com/2072-6694/12/7/1968/s1, Table S1: Univariate and multivariable analyses for overall survival of the study population, Table S2: Univariate and multivariable analyses for progression-free survival of the study population, Table S3: Univariate and multivariable analyses for overall survival of Child-Pugh B patients, Table S4: Univariate and multivariable analyses for progression-free survival of Child-Pugh B patients, Table S5: Predictive factors for disease control in Child-Pugh B patients.

Author Contributions: Conceptualization, W.-M.C. and J.C.; data curation, W.-M.C., J.C., D.L., J.H.S., K.M.K., Y.-S.L., H.C.L., C.Y., S.R.P., M.-H.R., and B.-Y.R.; formal analysis, W.-M.C. and J.C.; writing—original draft preparation, W.-M.C. and J.C.; writing—review and editing, J.C.; supervision, J.C. All authors have read and agreed to the published version of the manuscript.

Conflicts of Interest: The authors declare no conflict of interest.

Abbreviations

AE	adverse event
AHR	adjusted hazard ratio
AOR	adjusted odds ratio
CI	confidence interval
DCR	disease control rate
ECOG	Eastern Cooperative Oncology Group
HCC	hepatocellular carcinoma
HR	hazard ratio
IQR	interquartile range
mRECIST	modified Response Evaluation Criteria in Solid Tumors
OS	overall survival
ORR	objective response rate
PFS	progression-free survival
RECIST	Response Evaluation Criteria in Solid Tumors
TTP	time to progression

References

1. Bray, F.; Ferlay, J.; Soerjomataram, I.; Siegel, R.L.; Torre, L.A.; Jemal, A. Global cancer statistics 2018: GLOBOCAN estimates of incidence and mortality worldwide for 36 cancers in 185 countries. *CA Cancer J. Clin.* **2018**, *68*, 394–424. [CrossRef] [PubMed]
2. Kim, B.H.; Park, J.W. Epidemiology of liver cancer in South Korea. *Clin. Mol. Hepatol.* **2018**, *24*, 1–9. [CrossRef]
3. Vogel, A.; Cervantes, A.; Chau, I.; Daniele, B.; Llovet, J.M.; Meyer, T.; Nault, J.C.; Neumann, U.; Ricke, J.; Sangro, B.; et al. Hepatocellular carcinoma: ESMO Clinical Practice Guidelines for diagnosis, treatment and follow-up. *Ann. Oncol.* **2018**, *29*, iv238–iv255. [CrossRef] [PubMed]
4. Pinter, M.; Peck-Radosavljevic, M. Review article: Systemic treatment of hepatocellular carcinoma. *Aliment. Pharmacol. Ther.* **2018**, *48*, 598–609. [CrossRef] [PubMed]
5. Llovet, J.M.; Ricci, S.; Mazzaferro, V.; Hilgard, P.; Gane, E.; Blanc, J.F.; de Oliveira, A.C.; Santoro, A.; Raoul, J.L.; Forner, A.; et al. Sorafenib in advanced hepatocellular carcinoma. *N. Engl. J. Med.* **2008**, *359*, 378–390. [CrossRef]

6. El-Khoueiry, A.B.; Sangro, B.; Yau, T.; Crocenzi, T.S.; Kudo, M.; Hsu, C.; Kim, T.Y.; Choo, S.P.; Trojan, J.; Welling, T.H.R.; et al. Nivolumab in patients with advanced hepatocellular carcinoma (CheckMate 040): An open-label, non-comparative, phase 1/2 dose escalation and expansion trial. *Lancet* **2017**, *389*, 2492–2502. [CrossRef]
7. Zhu, A.X.; Finn, R.S.; Edeline, J.; Cattan, S.; Ogasawara, S.; Palmer, D.; Verslype, C.; Zagonel, V.; Fartoux, L.; Vogel, A.; et al. Pembrolizumab in patients with advanced hepatocellular carcinoma previously treated with sorafenib (KEYNOTE-224): A non-randomised, open-label phase 2 trial. *Lancet Oncol.* **2018**, *19*, 940–952. [CrossRef]
8. Kudo, M.; Finn, R.S.; Qin, S.; Han, K.H.; Ikeda, K.; Piscaglia, F.; Baron, A.; Park, J.W.; Han, G.; Jassem, J.; et al. Lenvatinib versus sorafenib in first-line treatment of patients with unresectable hepatocellular carcinoma: A randomised phase 3 non-inferiority trial. *Lancet* **2018**, *391*, 1163–1173. [CrossRef]
9. Bruix, J.; Qin, S.; Merle, P.; Granito, A.; Huang, Y.H.; Bodoky, G.; Pracht, M.; Yokosuka, O.; Rosmorduc, O.; Breder, V.; et al. Regorafenib for patients with hepatocellular carcinoma who progressed on sorafenib treatment (RESORCE): A randomised, double-blind, placebo-controlled, phase 3 trial. *Lancet* **2017**, *389*, 56–66. [CrossRef]
10. Kambhampati, S.; Bauer, K.E.; Bracci, P.M.; Keenan, B.P.; Behr, S.C.; Gordan, J.D.; Kelley, R.K. Nivolumab in patients with advanced hepatocellular carcinoma and Child-Pugh class B cirrhosis: Safety and clinical outcomes in a retrospective case series. *Cancer* **2019**, *125*, 3234–3241. [CrossRef]
11. Scheiner, B.; Kirstein, M.M.; Hucke, F.; Finkelmeier, F.; Schulze, K.; von Felden, J.; Koch, S.; Schwabl, P.; Hinrichs, J.B.; Waneck, F.; et al. Programmed cell death protein-1 (PD-1)-targeted immunotherapy in advanced hepatocellular carcinoma: Efficacy and safety data from an international multicentre real-world cohort. *Aliment. Pharmacol. Ther.* **2019**, *49*, 1323–1333. [CrossRef] [PubMed]
12. Finkelmeier, F.; Czauderna, C.; Perkhofer, L.; Ettrich, T.J.; Trojan, J.; Weinmann, A.; Marquardt, J.U.; Vermehren, J.; Waidmann, O. Feasibility and safety of nivolumab in advanced hepatocellular carcinoma: Real-life experience from three German centers. *J. Cancer Res. Clin. Oncol.* **2019**, *145*, 253–259. [CrossRef] [PubMed]
13. Albillos, A.; Lario, M.; Álvarez-Mon, M. Cirrhosis-associated immune dysfunction: Distinctive features and clinical relevance. *J. Hepatol.* **2014**, *61*, 1385–1396. [CrossRef] [PubMed]
14. O'Brien, A.J.; Fullerton, J.N.; Massey, K.A.; Auld, G.; Sewell, G.; James, S.; Newson, J.; Karra, E.; Winstanley, A.; Alazawi, W.; et al. Immunosuppression in acutely decompensated cirrhosis is mediated by prostaglandin E2. *Nat. Med.* **2014**, *20*, 518–523. [CrossRef] [PubMed]
15. Choe, W.H.; Baik, S.K. Prostaglandin E2 -mediated immunosuppression and the role of albumin as its modulator. *Hepatology* **2015**, *61*, 1080–1082. [CrossRef] [PubMed]
16. Albillos, A.; Hera Ad Ade, L.; Reyes, E.; Monserrat, J.; Muñoz, L.; Nieto, M.; Prieto, A.; Sanz, E.; Alvarez-Mon, M. Tumour necrosis factor-alpha expression by activated monocytes and altered T-cell homeostasis in ascitic alcoholic cirrhosis: Amelioration with norfloxacin. *J. Hepatol.* **2004**, *40*, 624–631. [CrossRef] [PubMed]
17. Tian, Z.; Chen, Y.; Gao, B. Natural killer cells in liver disease. *Hepatology* **2013**, *57*, 1654–1662. [CrossRef]
18. Finn, R.S.; Ryoo, B.Y.; Merle, P.; Kudo, M.; Bouattour, M.; Lim, H.Y.; Breder, V.; Edeline, J.; Chao, Y.; Ogasawara, S.; et al. Pembrolizumab As Second-Line Therapy in Patients with Advanced Hepatocellular Carcinoma in KEYNOTE-240: A Randomized, Double-Blind, Phase III Trial. *J. Clin. Oncol.* **2020**, *38*, 193–202. [CrossRef]
19. Yau, T.; Park, J.; Finn, R.; Cheng, A.-L.; Mathurin, P.; Edeline, J.; Kudo, M.; Han, K.-H.; Harding, J.; Merle, P. CheckMate 459: A randomized, multi-center phase III study of nivolumab (NIVO) vs sorafenib (SOR) as first-line (1L) treatment in patients (pts) with advanced hepatocellular carcinoma (aHCC). In Proceedings of the European Society for Medical Oncology (ESMO) Congress, Barcelona, Spain, 27 September–1 October 2019; pp. v874–v875.
20. Finn, R.S.; Qin, S.; Ikeda, M.; Galle, P.R.; Ducreux, M.; Kim, T.Y.; Kudo, M.; Breder, V.; Merle, P.; Kaseb, A.O.; et al. Atezolizumab plus Bevacizumab in Unresectable hepatocellular carcinoma. *N. Engl. J. Med.* **2020**, *382*, 1894–1905. [CrossRef] [PubMed]
21. McNamara, M.G.; Slagter, A.E.; Nuttall, C.; Frizziero, M.; Pihlak, R.; Lamarca, A.; Tariq, N.; Valle, J.W.; Hubner, R.A.; Knox, J.J.; et al. Sorafenib as first-line therapy in patients with advanced Child-Pugh B hepatocellular carcinoma-a meta-analysis. *Eur. J. Cancer* **2018**, *105*, 1–9. [CrossRef] [PubMed]

22. Kudo, M.; Matilla, A.; Santoro, A.; Melero, I.; Gracian, A.C.; Acosta-Rivera, M.; Choo, S.P.; El-Khoueiry, A.B.; Kuromatsu, R.; El-Rayes, B.F. Checkmate-040: Nivolumab (NIVO) in patients (pts) with advanced hepatocellular carcinoma (aHCC) and Child-Pugh B (CPB) status. In Proceedings of the Liver Metting 2018 by Ameircan Association for the Study of Liver Diseases (AASLD), San Francisco, CA, USA, 13 November 2018.
23. Lee, P.C.; Chao, Y.; Chen, M.H.; Lan, K.H.; Lee, C.J.; Lee, I.C.; Chen, S.C.; Hou, M.C.; Huang, Y.H. Predictors of response and survival in immune checkpoint inhibitor-treated Unresectable hepatocellular carcinoma. *Cancers* **2020**, *12*, 182. [CrossRef] [PubMed]
24. Zucman-Rossi, J.; Villanueva, A.; Nault, J.C.; Llovet, J.M. Genetic landscape and biomarkers of hepatocellular carcinoma. *Gastroenterology* **2015**, *149*, 1226–1239.e1224. [CrossRef] [PubMed]
25. Yau, T.; Hsu, C.; Kim, T.Y.; Choo, S.P.; Kang, Y.K.; Hou, M.M.; Numata, K.; Yeo, W.; Chopra, A.; Ikeda, M.; et al. Nivolumab in advanced hepatocellular carcinoma: Sorafenib-experienced Asian cohort analysis. *J. Hepatol.* **2019**, *71*, 543–552. [CrossRef] [PubMed]
26. Marrero, J.A.; Kulik, L.M.; Sirlin, C.B.; Zhu, A.X.; Finn, R.S.; Abecassis, M.M.; Roberts, L.R.; Heimbach, J.K. Diagnosis, staging, and management of hepatocellular carcinoma: 2018 practice guidance by the american association for the study of liver diseases. *Hepatology* **2018**, *68*, 723–750. [CrossRef]
27. Lencioni, R.; Llovet, J.M. Modified RECIST (mRECIST) assessment for hepatocellular carcinoma. *Semin. Liver Dis.* **2010**, *30*, 52–60. [CrossRef] [PubMed]

© 2020 by the authors. Licensee MDPI, Basel, Switzerland. This article is an open access article distributed under the terms and conditions of the Creative Commons Attribution (CC BY) license (http://creativecommons.org/licenses/by/4.0/).

Review

Immune Phenotype and Immune Checkpoint Inhibitors for the Treatment of Human Hepatocellular Carcinoma

Naoshi Nishida * and **Masatoshi Kudo**

Department of Gastroenterology and Hepatology, Kindai University Faculty of Medicine; 377-2 Ohno-Higashi, Osaka-Sayama 589-8511, Japan; m-kudo@med.kindai.ac.jp
* Correspondence: naoshi@med.kindai.ac.jp; Tel.: +81-72-366-0221

Received: 1 April 2020; Accepted: 15 May 2020; Published: 18 May 2020

Abstract: Immunotherapies are promising approaches for treating hepatocellular carcinomas (HCCs) refractory to conventional therapies. However, a recent clinical trial of immune checkpoint inhibitors (ICIs) revealed that anti-tumor responses to ICIs are not satisfactory in HCC cases. Therefore, it is critical to identify molecular markers to predict outcome and develop novel combination therapies that enhance the efficacy of ICIs. Recently, several attempts have been made to classify HCC based on genome, epigenome, and transcriptome analyses. These molecular classifications are characterized by unique clinical and histological features of HCC, as well immune phenotype. For example, HCCs exhibiting gene expression patterns with proliferation signals and stem cell markers are associated with the enrichment of immune infiltrates in tumors, suggesting immune-proficient characteristics for this type of HCC. However, the presence of activating mutations in β-catenin represents a lack of immune infiltrates and refractoriness to ICIs. Although the precise mechanism that links the immunological phenotype with molecular features remains controversial, it is conceivable that alterations of oncogenic cellular signaling in cancer may lead to the expression of immune-regulatory molecules and result in the acquisition of specific immunological microenvironments for each case of HCC. Therefore, these molecular and immune characteristics should be considered for the management of HCC using immunotherapy.

Keywords: hepatocellular carcinoma; molecular classification; immune phenotype; immune checkpoint inhibitor; stem cell marker; oncogenic signal; β-catenin; genetic alteration

---

1. Introduction

Hepatocellular carcinoma (HCC) remains one of the leading causes of cancer-related morbidity worldwide and generally emerges from a background of chronic liver inflammation [1]. Recent advancements in molecular target therapy have contributed to improvements in the prognosis of HCC patients, even those with advanced disease [2]. However, most cases of HCC show a tolerance or become refractory to molecular target agents during its clinical course [3,4]. On the other hand, immunotherapies are considered to be a promising approach for HCC patients even in those refractory to conventional therapies [5], and several immune components may play a role in the development and progression of this disease [6]. Nevertheless, phase III clinical trials of immune checkpoint monotherapies in patients with HCC have failed to show superiority to control groups for overall survival (OS) and progression-free survival (PFS) [7,8].

Several attempts have been made to subclassify HCC based on genetic and epigenetic alterations [9–12]. It has also been reported that the molecular subclass of HCC sometimes reflects the immune milieu of tumors [13]. For example, an association between molecular alterations of HCC and the expression of immune checkpoint molecules has been reported [14], and alteration of oncogenic signals due to mutations may lead to altered expression of immune modulators [15]. Therefore, a profound understanding of the molecular subclasses that affect the immune status of tumors may provide valuable insight for the rational development of combination therapies using immune checkpoint inhibitors (ICIs). In this review, we focus on this important issue and introduce findings from recent studies regarding molecular classifications and immune phenotype of HCC. Furthermore, we discuss the development of novel combination therapies that may further improve the efficacy of ICIs in these refractory tumors.

2. Molecular Classification and Immune Phenotype of HCC

2.1. Oncogenic Signal Activation in HCC

Recent deep sequencing technology has led to the revelation of a complex landscape of genetic alterations in the HCC genome [16–18]. Although the majority of the alterations are considered to be passenger mutations that do not affect the immortalization or growth of HCC cells, there are several putative driver mutations that act as gain-of-function or loss-of-function mutations involved in critical signaling pathways [19]. Generally, HCCs develop in livers with chronic damage, such as that caused by hepatitis B virus (HBV), hepatitis C virus (HCV), alcoholic liver disease, and non-alcoholic fatty liver disease (NAFLD). Reportedly, mutations of *CTNNB1* are associated with alcohol intake, while *TP53* mutations are more frequently detected in HBV-positive HCCs than those with other risk factors [20]. However, genes carrying mutations are heterogeneous, regardless of etiology of this type of tumor. Activating mutations in *CTNNB1*, inactivating mutations in *TP53*, and activating mutations in *telomerase reverse transcriptase* (*TERT*) are the most frequently detected mutations in HCC [21–26]. Other genetic alterations that lead to constitutive activation of specific growth signals are relatively rare [27]. On the other hand, a considerable percentage of tumors exhibit complex patterns of genetic alterations that lead to the activation and inactivation of various signaling pathways [20]. Guichard et al. classified mutations of HCC based on the signaling pathway involved, such as Wnt/β-catenin, p53/cell cycle control, chromatin remodeling, phosphoinositide 3-kinase (PI3K)/Ras signaling, and oxidative stress and endoplasmic reticulum stress pathway [16]. In addition, constitutive activation of telomerase, which is responsible for the immortalization of cancer cells, may also act as a driver of HCC carcinogenesis [23,24].

2.2. Molecular Subclass and Tumor Characteristics

Although the genetic alterations and gene expression among individual HCCs are heterogeneous, several studies have classified HCC based on the patterns of these molecular alterations [9,10]. Genetic changes and expression may affect the phenotype of HCC and may be associated with tumor characteristics such as biological behavior [28]. Boyault et al. performed comprehensive analyses of gene expression and classified HCCs using hierarchical clustering analysis. They characterized HCC subclasses according to mutations, chromosomal alterations, copy number of HBV genomes, and DNA methylation of the promoters of *CDH1* and *CDKN2A*. Accordingly, HCCs are subclassified into six groups, with each subclass demonstrating unique molecular characteristics and clinical features [9]. Group 1–3 (G1–G3) tumors are associated with chromosomal instability and amplification and overexpression of cell-cycle/proliferation-related genes, such as *FGF19/CCND1* on 11q13 [9,29,30]. Among these, G1 is characterized by a low copy number of HBV and the expression of genes activated in fetal liver. HCCs of G2 have a high copy number of HBV and mutations in *PIK3CA* and *TP53*. Furthermore, activation of the PI3K-Akt pathway is prominent in both G1 and G2 HCC tumors. Tumors of G3 tend to carry *TSC1/TSC2* mutations. On the other hand, HCCs classified as G4–G6 exhibit low levels of chromosomal alterations. The G4 subtype contains various tumor types with mutations in *TCF1*, while G5 and G6 are strongly correlated to mutations in *CTNNB1*, leading to activation of the Wnt/β-catenin pathway. The *CTNNB1* mutations are frequently accompanied with hypermethylation in the promoter of multiple tumor suppressor genes, especially in HCV-positive and aged patients [22,31]. It has been reported that specific clinical features are associated with different subclasses, such as young age, female, African, and high α-fetoprotein (AFP) with G1, hemochromatosis with G2, and the presence of satellite nodule with G6 [9]. Hoshida et al. also reported an association of molecular features with more aggressive and less-aggressive HCCs, where the aggressive types represented the activation of E2F transcription factor 1 (E2F1) and inactivation of *TP53* [10]. As E2F1 mediates both cell-cycle progression and p53-dependent apoptosis, it is conceivable that the combination of E2F1 activation and p53 inactivation is likely to result in the acceleration of cell cycle progression and tumor growth. These investigators also identified two subclasses of aggressive HCCs (S1 and S2) based on molecular features. The subclass S1 is characterized by activation of the transforming growth factor (TGF)-β pathway and expression of Wnt target genes in the absence of *CTNNB1* mutations. On the other hand, the subclass S2 demonstrates MYC and AKT activation and overexpression of AFP and insulin-like growth factor 2 (IGF2) and is accompanied by the downregulation of interferon (IFN)-related genes. High serum AFP levels, expression of epithelial cell adhesion molecule (EpCAM), and vascular invasion are also frequently observed in S2 HCCs. Expression of stem/biliary markers, such as cytokeratin 19 (CK19), is similarly enriched in both S1 and S2 subclasses. Tumors belonging to subclass S3 are characterized by a less-aggressive phenotype and the retention of mature liver function, as exemplified by the upregulation of genes involved in metabolism, detoxification, and protein synthesis [10]. The activating mutation of *CTNNB1* is primarily observed in S3, which is enriched in the G5 and G6 subclasses of Boyault et al. [9].

On the other hand, associations between molecular alteration and clinicopathological characteristics are also reported. Calderaro et al. described the histological features of HCCs that carry *CTNNB1* and *TP53* mutations [32]. *CTNNB1* and *TP53* mutations appear to be mutually exclusive. HCCs with *CTNNB1* mutations are generally large, well-differentiated, and show microtrabecular or pseudoglandular histological patterns, cholestatic tendencies, and a lack of inflammatory infiltrates. On the other hand, *TP53* mutations are associated with poorly differentiated HCCs with a compact pattern, multinucleated and pleomorphic cells, and frequent vascular invasion. These investigators also clarified several molecular characteristics of specific HCC subtypes, including scirrhous subtypes of HCCs that showed *TSC1/TSC2* mutations, epithelial-to-mesenchymal transition, and expression of genes related to progenitor cells [32]. The steatohepatitic subtype of HCC is characterized by activation of the interleukin (IL)-6/JAK/STAT pathway with wild-type *CTNNB1*, *TERT*, and *TP53*. Interestingly, such phenotypic features are closely linked to the G1–G6 subgroups proposed by Boyault et al. with the association of progenitor phenotype to G1, macrotrabecular massive subtype and macrovascular invasion to G3, steatohepatitic subtype to G4, and cholestasis and lack of inflammatory infiltrates to G5 and G6 (Figure 1) [9].

Molecular classification	G1	G2	G3	G4	G5	G6	Periportal type
	Poor differentiation			Good differentiation			
Pathological feature		vascular invasion	Macrotrabecular massive Macrovascular invasion	Steatohepatitic Immune Infiltrates	Cholestasis Lack of immune infiltrates		
Immunohisto-chemistry	CK19, EpCAM Phospho-ERK	CK19, EpCAM	CK19	CRP	β-catenin (nucleus) Glutamine synthetase Lack of ARID1		
Immune Phenotype	Immune-low	Immune-high (T cell exhaustion)	Immune-mid/high (T cell exhaustion)	Immune-high (T cell activation)	Immune-low		
Gene expression signature /Alteration of cellular signaling	Progenitor pattern			Inflammation pattern	Hepatocyte-like pattern		
	Stem cell feature		ECM	IL6-JAK-STAT activation	Wnt/β-catenin activation		
		Mitotic cell cycle					
	AKT, MYC activation		TGF-β, WNT activation		Perivenous gene signature		Periportal gene signature
	Developmental and imprinting genes, IGF2		Cell cycle, nucleus pore		Stress and immune response	Amino acid metabolism, E-cadherin↓	HNF4A-driven gene expression
mutation	TP53 mutation, 11q13 amplification (FGF19/CCND1)			HIF1A	CTNNB1 mutation		
	AXIN1, RPS6KA3	AXIN1, PIK3CA, ATM	TSC1/TSC2				
methylation	Global DNA hypomethylation				Extensive methylation on the promoter of TSGs		
Chromosomal status	unstable				stable		
Clinical feature	female, Africa, young, high AFP HBV-low copy number	high AFP HBV-high copy number	Hemochromatosis high AFP	Without satellite nodule and vascular invasion	Satellite nodule		Smaller tumor Lowest potential for early recurrence

Figure 1. Molecular classification, clinicopathological characteristics, and immune phenotype of human hepatocellular carcinoma (HCC). Subclasses shown as G1–G6 were described by Boyault et al. [9]. The associations between the molecular subclass and pathological characteristics were reported by Calderaro et al. [32]. The classification shown as "periportal type" was described by Desert et al. [33]. Immune phenotype in this figure (immune-high, -med, and -low) was proposed by Kurebayashi et al. [34]. The bold denotes the representative findings of molecular, clinical, and pathological features. ECM: extracellular matrix. HNF4A: hepatocyte nuclear factor 4A. HIF1A: hypoxia inducible factor A. TSGs: tumor suppressor genes.

Desert et al. further subclassified the non-proliferative phenotype of HCCs that demonstrate a low potential of recurrence [33]. The transcriptomic data revealed two subclasses of non-proliferative HCCs, the periportal-type (wild-type β-catenin) and perivenous-type (mutant β-catenin). HCCs of the periportal-type show activation of a hepatocyte nuclear factor 4A-driven gene, low expression of a metastasis-specific gene, and low frequency of *TP53* mutations. Clinically, such tumors are characterized by early-stage tumors that lack macrovascular invasion. The periportal-type of HCCs represent the gene expression profile, like the S3 signature described by Hoshida et al. Although this type of HCC does not carry mutations in *CTNNB1*, such cases do exhibit a better prognosis than those of the perivenous-type. On the other hand, the perivenous-type tumors have *CTNNB1* mutations that are frequently observed in HCCs, categorized as G5 and G6 (Figure 1) [33].

2.3. Immune Phenothype of HCC

There are several studies that have clarified the association between immune status and clinical characteristics of HCC, particularly for its relation to the prognosis and the response to the treatment [35–38]. In addition, recent reports have shown a link between molecular subclass and immune phenotype of HCCs [27,34,39]. Expression of programmed cell death-ligand 1 (PD-L1) in HCC cells is reportedly associated with clinical parameters related to tumor aggressiveness, such as high serum α-fetoprotein levels, satellite nodules, vascular invasion, and poorly differentiated phenotype, as well as molecular features associated with advanced tumor [38,40]. It is also known that PD-L1 expression is more frequently detected in HCCs that express stem/biliary cell markers CK19 and Sal-like protein 4 (SALL4) [34,38,40]. This suggests that PD-L1 expression is associated with the progenitor subtype of HCCs, such as HCCs classified as G1. Furthermore, PD-L1 expression in tumor infiltrates also correlates with aggressive tumor characteristics [38].

2.3.1. Classification of HCC Based on the Gene Expression Pattern and Immune Milieu

Sia et al. found that approximately 25% of HCCs they evaluated were classified as "immune-specific class" based on gene expression profiling [39]. Furthermore, they found that this phenotype consists of two immune phenotype subclasses, active and exhausted immune subclasses, according to gene expression profiles of tumor, stromal, and immune cells. The HCCs belonging to the active immune subtype, which is related to better survival, shows enriched gene expression related to antitumor immune response, such as the expression of interferon-related and adaptive immune response genes. In contrast, the exhausted immune subtype enriched with HCCs belonging to the S1 subclass described by Hoshida et al. [10], exhibits gene expression characterized by activation of a potent immunoregulatory cytokine signal, such as transforming growth factor-β (TGF-β), which is known to regulate stroma interactions and angiogenesis, induce T-cell exhaustion, and promote M2 macrophages. Through methylome analyses, it has been suggested that immune subclasses have unique DNA methylation signatures that determine the immune response to HCC [39]. Meanwhile, another study demonstrated that alteration of genes involved in the activation of Wnt/β-catenin signaling results in poorer disease control, shorter PFS, and lower OS with respect to treatment of patients with ICIs [41]. Therefore, the presence of activating mutations involved in Wnt/β-catenin signaling is associated with innate resistant to ICIs. Reportedly, HCCs with a *CTNNB1* mutation show significantly lower enrichment scores for several immune signatures, in particular T cells, and also demonstrate overexpression of *protein tyrosine kinase 2 (PTK2)*, which may lead to immune exclusion [39]. De Galarreta et al. showed that β-catenin-driven tumors are resistant to anti-programed cell death-1 (PD-1) therapy in a mouse model where expression of chemokine (C-C motif) ligand 5 (CCL5) restores immune surveillance [42]. Therefore, an activating mutation in β-catenin may be a negative predictive marker for patients with HCC treated with ICIs.

Immune microenvironment of HCC was classified into three distinct subtypes based on immuohistochemical analyses of the immune regulatory molecules [34]. The subtypes include immune-high, immune-mid, and immune-low groups. HCC classified as the immune-high subtype show increased infiltrations of B cells, plasma cells, and T cells. Consistent with previous reports, the immune-high subtype is characterized by poorly differentiated HCC, positive for CK19 and/or Sal-like protein 4 (SALL4), and enrichment of tumors belonging to S1 and G2 subclasses (Figure 1) [34,38]. It is also confirmed that patients with HCC belonging to the immune-high subtype have better prognosis, even in cases of patients with high-grade tumor. Another study reported that HCCs with immune cell stroma exhibit distinct clinical features of dense $CD8^+$ and EBV-positive $CD20^+$ tumor infiltrating lymphocytes (TILs) and have good prognosis [43]. This type of HCC is characterized by the lack of *CTNNB1* mutations, global hypermethylation, expression of PD-1 and PD-L1 in tumor infiltrating lymphocytes (TILs), and expression of PD-L1 in tumors.

Taken together, HCC cases with the immune-high subtype, which is enriched of the tumors with progenitor/proliferative gene expression pattern, especially in the S1 and G2, may also be candidates for treatment with ICIs because this type of HCC generally shows immune infiltrates and express PD-L1 in the tumor tissues. However, a majority of inflammatory infiltrates in tumor show exhausted phenotype with expression of genes involved in Wnt/TGF-β signaling and M2 macrophage [39,40,44,45], and additional agents that alter the immune milieu should be required for the treatment of this subclass. On the other hand, HCCs with expression of adaptive immune response genes, such as IFN-γ, granzyme B, CD8A, and T-cell receptor G, may show a considerable response to ICIs [30,39]. Generally, HCCs with hepatocyte-like/non-proliferative gene expression pattern lack the activation of PD-1/PD-L1 signaling as well as gene expression related to immune infiltrates in the tumor [34,40,41].

2.3.2. Characteristics of Inflammatory Infiltrates in HCC Tissues

Expression of immune suppressive receptors in immune infiltrates are associated with shorter survival of patients with HCC. For example, T-cell immunoglobulin and mucin domain 3 (TIM-3) expression in tumor-associated macrophages (TAMs) strongly correlates with higher tumor grade and poor patients' survival [46], whereas TGF-β induces TIM-3 expression and an alternative activation of macrophages. In addition to TIM-3, expression of another immune suppressive receptor, lymphocyte activation gene-3 (LAG-3), is also increased in immune infiltrates of HCC tissues, suggesting that PD-1, TIM-3, and LAG-3 may cooperate and are implicated in inducing anti-tumor immune tolerance [14,47].

Meanwhile, Zheng et al. characterized the molecular and functional properties of T cells from HCC specimens, adjacent non-tumorous tissues, and peripheral blood using single cell sequencing [44]. In HCCs, T-cell enrichment with clonal expansion of $CD8^+$ T-cell populations with exhausted phenotype is observed according to the sequencing of T-cell receptors in TILs. These investigators found that layilin is upregulated in activated $CD8^+$ cytotoxic T-cells and regulatory T-cells (Tregs) in HCC and these cells play a role in repression of $CD8^+$ T-cell functions. Heterogeneity among the populations of exhausted tumor-infiltrating $CD8^+$ T-cells has also been reported. TILs with high expression of PD-1 show higher expression levels of genes that regulate T-cell exhaustion compared to TILs that only moderately express PD-1 [47]. Consistent with another report, cells that express high levels of PD-1 also express TIM-3 and LAG-3 and produce low levels of cytokines necessary for cytotoxic effects of T cells, such as interferon-γ (IFN-γ) and tumor necrosis factor (TNF). In addition, the expression pattern of PD-1 in $CD8^+$ TILs characterize the two subgroups of HCCs. HCC tumors with PD-1-high $CD8^+$ TILs are more aggressive than those without PD-1-high cells. PD-L1 combined positive score (CPS) can be a biomarker used to predict a favorable response to PD-1/PD-L1 blockade [48]. CPS represents PD-L1 expression in both the tumor and intra-tumor inflammatory cells and is significantly higher in cases with PD-1-high $CD8^+$ TILs than those with PD-1-low. Furthermore, incubation of PD-1-high $CD8^+$ T-cells from HCCs with anti-PD-1 and anti-TIM-3 or anti-LAG-3 antibodies restore cell proliferation and the production of IFN-γ and TNF-α in response to anti-CD3. Therefore, HCC cases with high expression of PD-1 in $CD8^+$ TILs may be good candidates for treatment with a combined immune checkpoint blockade [40,49].

In addition to the exhaustion of $CD8^+$ T cells, several stromal cells, such as myeloid-derived suppressor cells (MDSCs), tumor-associated macrophages (TAMs), Tregs, type 2 helper T (Th2) cells, and cancer-associated fibroblasts (CAFs), act in concert in refractoriness to immunotherapy in HCC patients [6,50]. Hypoxia in tumor tissues stimulates the induction of vascular endothelial growth factor (VEGF) in cancer cells and contributes to the recruitment of immune suppressive stromal cells through the binding of VEGF to its receptor on MDSCs, TAMs, Tregs, and CAFs [6,51].

So far, infiltration of MDSCs and Tregs is known to be associated with HCC progression and worse outcome of the patients [52,53]. Increase of arginase 1 in MDSC lead to the depletion of arginine, which impairs the function of immune cells [54]. TGF-β and IL-10 from MDSC stimulate Tregs and suppress natural killer (NK) cells [55]. The M2 polarization of macrophages is induced through the secretion of IL-10 from MDSCs, which result in the downregulation of IL-12 in TAMs. High IL-10 and low IL-12 levels further stimulate the induction of Th2 cells and TAM. TGF-β from MDSCs suppress $CD4^+$ and $CD8^+$ T cells and NK cells. It also induces immune suppressive receptors on T cells and TAMs [46]. TGF-β and IL-10 signaling, along with the stimulation of VEGF signal, play a role for further activation of Tregs [56]. CAFs and endothelial cells can also be players for anti-tumor immunosuppression. Prostaglandin E2 and indoleamine 2,3-dioxygenase (IDO) from CAF lead to the NK cell dysfunction [57]. Endothelial cells in cancer tissues, reportedly, produce the C-X-C motif chemokine ligand 12, resulting in the recruitment of MDSC [58]. Activation of endothelial cells also contribute to the TGF-β-mediated Treg induction.

3. Effective Application of Immune Checkpoint Inhibitors for HCC Cases

3.1. HCC Response to Immune Checkpoint Inhibitors

Although several phase II clinical trials of ICI monotherapies have shown favorable outcomes for the use of ICIs in patients with HCC [48,59], a phase III study failed to demonstrate positive results as the first-line treatment with respect to OS and PFS compared to the multi-kinase inhibitor sorafenib [7], and as the second-line treatment after sorafenib compared to best supportive care (Table 1) [8]. However, there are molecular features that may be associated with response to ICIs. For example, the HCC with microsatellite instability is reported to show good response to treatment with pembrolizumab [60]. The presence of *CTNNB1* variants is associated with the activation of Wnt/β-catenin signaling as well as a lack of immune infiltrates in HCC tumors, which are predictors of a poor response to ICIs in patients with HCC [41]. On the other hand, HCC subtypes with high inflammatory infiltrates, such as HCC of the G2 subclass, may be expected for respond to ICIs [34], although additional agents for combination therapy may be required for a good response [40]. Immunohistochemistry-based markers such as CPS may predict the anti-tumor response to ICIs [48]. However, tumor specimens are required in order to perform the immunohistochemical analysis, which are sometimes difficult to obtain in clinical settings. On the other hand, molecular markers based on genetic alterations of tumor cells based on liquid biopsy may be applicable in which DNA from peripheral blood is used for analysis. From this point of view, the development of a mutation-based molecular marker may prove to be a promising approach for identifying responders for ICIs among HCC patients. However, immune infiltrates of tumor tissues frequently express multiple immune checkpoint molecules that are likely to result in refractoriness to immune checkpoint monotherapies [14,34,40]. Therefore, additional agents for combined immune checkpoint blockades should be required to assure improved response rates.

Table 1. Clinical trials and outcomes of immune checkpoint monotherapies in HCC.

Clinical Trial ID	Trial Name	Agents [1]	Setting [2]	Key Outcome [3]
Phase I/II				
NCT01658878	CheckMate 040	**Nivolumab**	dose-escalation, n = 48, dose-expansion, n = 214	ORR: 20% [4] DCR: 64%, (37%) [5] OS: 13.2 months (8.6–NE) [6]
NCT02702414	KEYNOTE-224	**Pembrolizumab**	second-line n = 104	ORR: 17% [7] DCR: 62% OS: 12.9 months (9.7–15.5)
Phase III				
NCT03383458	CheckMate 9DX	**Nivolumab** versus placebo	adjuvant, randomized, double-blinded (n = 530)	RFS
NCT02576509	CheckMate 459	**Nivolumab** versus Sorafenib	first-line, randomized, open label, n = 743	Median OS: 16.4 months in the nivolumab group and 14.7 months in the sorafenib group. [8] Median PFS: 3.7 months for nivolumab and 3.8 months for sorafenib. ORR: 15% in the nivolumab group and 7% in the sorafenib group.
NCT03412773	Rationale-301	**Tislelizumab** versus sorafenib	first-line, randomized, open label, (n = 674)	OS
NCT02702401	KEYNOTE-240	**Pembrolizumab** versus placebo	second-line, randomized, double-blinded, n = 413	Median OS: 13.9 months in the pembrolizumab group and 10.6 months in the placebo group; HR 0.781, p = 0.0238. Median PFS: 3.0 months for pembrolizumab and 2.8 months for placebo; HR 0.781, p = 0.0022. [9] ORR: 18.3%, DCR: 62.2%

[1] Bold denotes immune checkpoint inhibitors. [2] n, number of the patients analyzed in the study. The number in the parenthesis shows the number of the planned enrollment. [3] Bold denotes the primary outcome measures of the study. Duration of responses and survival are shown as median values. The numbers in the parenthesis show 95% confidential interval (CI). [4] El-Khoueiy et al. Lancet 2017; 389: 2492–2502 [59]. [5] Disease control with stable disease for ≥6 months. [6] Median overall survival of the sorafenib progressor without viral hepatitis in the dose-expansion cohort. [7] Zhu et al. Lancet Oncol 2018; 19: 940–952 [48]. [8] Yau et al. The European Society for Medical Oncology (ESMO) 2019 congress (# LBA38). [9] Finn et al. J Clin Oncol 2019; 38: 193–202 [8]. The 95% CI of median OS: 11.6 to 16.0 months in the pembrolizumab group and 8.3 to 13.5 months in the placebo group (hazard ratio, HR, 0.781; 95% CI, 0.611 to 0.998; p = 0.0238). The 95% CI of median PFS was 2.8 to 4.1 months for pembrolizumab and 1.6 to 3.0 months for placebo (HR, 0.718; 95% CI, 0.570 to 0.904; p = 0.0022). OS and PFS did not reach statistical significance per specified criteria in this study. ORR, objective response rate; DCR, disease control rate; OS, overall survival; NE, not estimated; RSF, recurrence-free survival; PFS, progression-free survival.

3.2. Combined Immune Checkpoint Blockade Based on Inflammatory Infiltrate Characteristics of HCC

As shown above, several studies have analyzed the expression of immune suppressive receptors and ligands in inflammatory infiltrates [14,34,40,45]. Generally, inflammatory cells in HCC express several immunosuppressive molecules, suggesting that such immune cells are functionally compromised. For example, expression of PD-1, TIM-3, LAG-3, and CTLA4 is significantly higher on $CD8^+$ and $CD4^+$ T-cells in HCC tissue than those in non-tumor tissues or peripheral blood, and dendric cells (DCs), monocytes, and B cells in tumors express ligands for these receptors [45]. In addition, tumor-associated antigen (TAA)-specific $CD8^+$ TILs express higher levels of PD-1, TIM-3, and LAG-3 compared to that of other $CD8^+$ TILs. Importantly, antibodies against PD-L1, TIM-3, or LAG-3 restore responses of HCC-derived T cells to tumor antigens, and treatment with combinations of these antibodies demonstrate additive effects in the restoration of T-cell function response to TAA [45]. On the other hand, Brown et al. reported the resistance of tumor cells to ICIs through the upregulation of IDO in patients with HCC [61]. Both anti-CTLA4 and anti-PD-1 antibodies induce IDO and the combination of ICIs with 1-methyl-D-tryptophan, an inhibitor of IDO, is able to suppress tumor growth of HCC in a mouse model. Therefore, anti-PD-1 therapy combined with anti-TIM-3, anti-LAG-3, or IDO inhibitor may be worth consideration for patients with HCCs that have exhausted immune infiltrates (Figure 2a). In addition to the phase III combined immune checkpoint blockade using anti-PD-1/PD-L1 and anti-CTLA-4 antibodies, currently, phase I/II clinical trials for the combinations of anti-PD-1 and anti-TIM-3 antibodies (ClinicalTrials.gov NCT03680508), anti-PD-1 and anti-LAG-3 antibodies (NCT03250832), and anti-PD-1 antibody and IDO inhibitors (NCT03695250) are ongoing (Table 2).

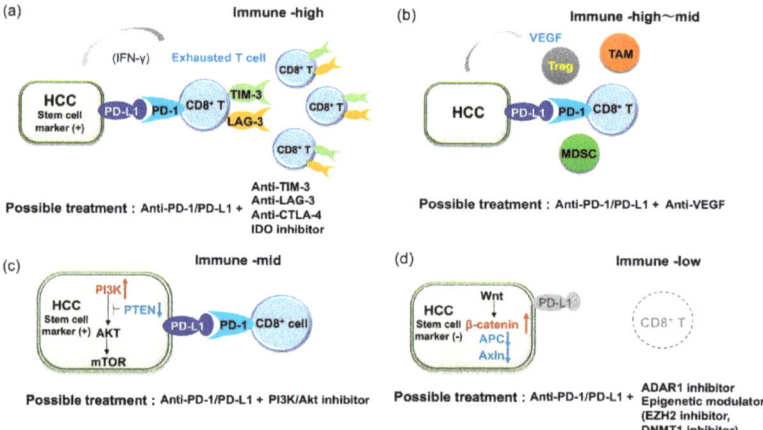

Figure 2. Illustrative figures of expected combination therapies for HCC patient refractory to immune checkpoint monotherapies. (**a**) In cases with expression of PD-L1 in HCC but multiple co-inhibitory receptors on tumor infiltrates, dual blockade of PD-1/PD-L1 and anti-TIM-3 or anti-LAG-3 should be required. (**b**) Because VEGF is known to play an important role for induction of immune suppressive molecules and cells, dual blockade of PD-1/PD-L1 and VEGF axis should be effective. (**c**) In cases with expression of PD-L1 and activating mutation in the PI3K-mTOR pathway in HCC, dual blockade of PD-1/PD-L1 and the PI3K-mTOR pathway might be effective. Notably, both anti-PD-1/PD-L1 and anti-PI3K-mTOR agents could target cancer stem cells (CSCs). (**d**) In cases with a lack of $CD8^+$ T cell infiltration in tumor (activating mutation in the β-catenin pathway is common in this type), ADAR1 inhibitor and epigenetic modulator might induce the recruitment of $CD8^+$ T cells into tumor and contribute to the induction of anti-tumor immunity.

Table 2. Clinical trials and outcomes of combined immune checkpoint blockade in HCC.

Clinical Trial ID	Trial Name	Agents [1]	Setting [2]	Key Outcome [3]
Phase I/II				
NCT01658878	CheckMate 040	**Nivolumab + Ipilimumab**	n = 50	**ORR: 32%** [4] DCR: 54% OS: 22.8 months (9.4–NE) DOR: 17.5 months (4.6–30.5)
NCT02519348		**Durvalumab ± Tremelimumab**	n = 40	**ORR: 25%** [5] DCR: 57.5%
NCT03680508		**TSR-002 + TSR-042** (Dostarlimab)	first-line, (n = 42)	**ORR**
NCT03250832		**TSR-033 + TSR-042**	dose escalation and dose expansion cohorts (n = 200)	**AEs** for dose escalation cohort **ORR** for dose expansion cohort
NCT03695250		**BMS986205 + Nivolumab**	first- or second-line, (n = 23)	**AEs and ORR**
Phase III				
NCT04039607	CheckMate9DW	**Nivolumab + Ipilimumab** versus Sorafenib/Lenvatinib	first-line, randomized, open label, (n = 1084)	**OS**
NCT03298451	HIMARAYA	**Durvalumab ± Tremelimumab** versus Sorafenib	first-line, randomized, open label, (n = 1310)	**OS**

[1] Bold denotes immune checkpoint inhibitors. [2] n, number of the patients analyzed in the study. The number in the parenthesis shows the number of the planned enrollment. [3] Bold denotes the primary outcome measures of the study. Duration of responses and survival are shown as median values. The numbers in the parenthesis show 95% confidential interval. [4] Yau et al. J Clin Oncol. 2019; 37 (supplement abstract 4012). [5] Kelley et al. J Clin Oncol 2017; 35 (supplement abstract 4073). DOR, duration of response; AEs, adverse events.

3.3. Combined Blockade of PD-1/PD-L1 and VEGF Axis

Because HCC is known as a hypervascular tumor where the development of tumor vessels plays an important role in its pathogenesis [62,63], several ongoing clinical studies are evaluating the combination of anti-angiogenic agents and ICIs (Table 3) [64]. Multiple agents that target VEGF and its receptor (VEGFR) are proven to be effective in the treatment of HCC, including the anti-VEGFR2 antibody, ramucirumab [65]. In addition, anti-angiogenic agents are believed to alter the immunosuppressive microenvironment in HCC [6]. It has been reported that anti-angiogenesis normalizes the leaky vascular network induced by VEGF, where the lack of adhesion molecules on endothelial cells may impair the extravasation of T cells [62,66] and induce an immune proficient condition. VEGF play a role in the recruitment of Tregs into tumor tissues and M2 polarization of macrophages via the increase of IL-4 and IL-10. VEGF is also critical for inhibition of the maturation of dendric cells (DCs) by activating NF-κB, production of IDO in tumor cell and macrophage, T-cell exhaustion by inducing PD-1, LAG-3 and TIM-3, accumulation of myeloid-derived suppressor cells (MDSCs), and inhibition of natural killer cell activity [6,67]. Therefore, a combination of ICIs with anti-VEGF agents should be effective (Figure 2b) [67–69], although the dosage that best improves the therapeutic effect of ICIs needs to be defined in individual agents [70]. Accordingly, dual blockade of the VEGF/VEGFR and PD-1/PD-L1 axes in patients with advanced HCC using the anti-PD-L1 antibody atezolizumab and the anti-VEGF-A antibody bevacizumab, or the anti-PD-1 antibody camrelizumab and the VRGFR2-TKI apatinib results in considerable ORR (Table 3) [64]. In addition, other combinations modulating immune microenvironment, such as the combination of anti-PD-1 antibody with an inhibitor of TGF-β receptor, is also under the early phase clinical trial (Table 3: NCT02423343).

Table 3. Clinical trials and outcomes of the combination therapies with immune checkpoint inhibitors and molecular targeted agents.

Clinical Trial ID	Trial Name	Agents [1]	Setting [2]	Key Outcome [3]
Phase I/II				
NCT03299946	CaboNivo	Cabozantinib + **Nivolumab**	neoadjuvant, (n = 15)	AEs and number of patients who complete the treatment.
NCT03006926		Lenvatinib + **Pembrolizumab**	first-line, (dose-escalation, dose-expansion), n = 30 (n = 97)	ORR: 53.3% (34.3–71.7), DOR: 8.3 months (3.8–11.0) [4] DCR = 90.0%; 73.5–97.9, PFS: 9.7 months 7.7–NE, OS: 14.6 months 9.9–NE.
NCT03289533	VEGF Liver 100	**Avelumab** + Axitinib	AFP ≥400 ng/mL, n = 22	AE ORR: 13.6% (2.9–34.9) [5] DCR: 68.2 (45.1–86.1) PFS: 5.5 months (1.9–7.4) OS: 12.7 months (0.0–NE) DOR: 5.5 months (3.7–7.3)
NCT03418922		Lenvatinib + **Nivolumab**	first-line, (n = 30)	DLT, AEs
NCT02715531	GO30140	**Atezolizumab** + Bevacizumab	n = 73	ORR: 27% [6] PFS: 7.5 months (0.4–23.9+)
NCT01658878	CheckMate 040	Cabozantinib + **Nivolumab ± Ipilimumab**	first or second-line, (dose-escalation, dose-expansion), (n = 1097, across all cohorts)	safety, tolerability, ORR
NCT03170960	COSMIC-021	Cabozantinib + **Atezolizumab**	first-line, (dose-escalation and dose-expansion), (n = 1732, across all cohorts)	MTD, ORR
NCT03347292		Regorafenib + **Pembrolizumab**	first-line, (dose-escalation and dose-expansion, n = 57)	TEAE, DLT
NCT03539822	CAMILLA	Cabozantinib + **Durvalumab**	second-line, (n = 30)	MTD
NCT03475953	REGOMUNE	Regorafenib + **Avelumab**	Second-line, (n = 212)	Recommended phase II dose, ORR
NCT02572687		Ramucirumab + **Durvalumab**	Second-line and AFP ≥ 1.5x ULN, n = 28	DLTs ORR: 11% [7] PFS: 4.4 months (1.6–5.7) OS: 10.8 months (5.1–18.4)
NCT3463876	RESCUE	**SHR-121 (Camrelizumab)** + Apatinib	n = 18 (n = 40)	ORR: 38.9% [8] DCR: 83.3% PFS: 7.2 months (2.6–NE)
NCT02423343		Galunisertib (TGFβ receptor I inhibitor) + **Nivolumab**	second-line and AFP ≥ 200 ng/mL, (dose escalation and cohort expansion, n = 75)	MTD
Phase III				
NCT03847428	EMERALD-2	**Durvalumab** ± Bevacizumab versus placebo	adjuvant, randomized, double-blinded, (n = 888)	RFS
NCT03434379	IMbrave150	**Atezolizumab** + Bevacizumab versus sorafenib	first-line, randomized, open label, n = 501	OS: not reached for Atezolizumab + bevacizumab vs 13.2 months for sorafenib; HR 0.58, p = 0.006 [9] PFS: 6.8 months for Atezolizumab + bevacizumab versus 4.3 months for sorafenib; HR 0.59, p < 0.0001 ORR: 27%
NCT03713593	LEAP-002	Lenvatinib + **Pembrolizumab** versus Lenvatinib	first-line, randomized, double-blinded, (n = 750)	OS, PFS
NCT03755791	COSMIC-312	Cabozantinib + **Atezolizumab** versus Sorafenib versus Cabozantinib	first-line, randomized, open label, (n = 740)	OS, PFS

[1] Bold denotes immune checkpoint inhibitors. [2] n, number of the patients analyzed in the study. The number in the parenthesis shows the number of the planned enrollment. [3] Bold denotes the primary outcome measures of the study. Duration of responses and survival are shown as median values. The numbers in the parenthesis show 95% confidential interval. [4] Ikeda et al. The American Association for Cancer Research (AACR) annual meeting 2019 (abstract #18). [5] Mudo et al. J. Clin Oncol 2019; 37 (supplement. abstract 4072). [6] Pishvaian et al. ESMO 2018 congress (# LBA26). [7] Bang et al. J Clin Oncol 2019; 37 (supplement. abstract). [8] Xu et al. J Clin Oncol 2018; 36 (supplement. abstract 4075). [9] Cheng et al. ESMO Asia2019 congress (# LBA3). DLT, dose-limiting toxicity; MTD, maximum tolerated dose; TEAEs, treatment-emergent adverse event.

3.4. Immune Checkpoint Inhibitors of Cancer Stem Cells

As previously reported, PD-L1 is expressed in the progenitor subtype of HCCs [34,38]. We also found a significant increase of PD-L1 expression in CK19-positive and/or SALL4-positive HCCs compared to those not expressing such markers [40]. Interestingly, genetic alterations involved in the PI3K-Akt pathway are more frequently detected in PD-L1-positive tumors than in PD-L1-negative tumors [40]. Inactivation of phosphatase and tensin homolog deleted from chromosome 10 (PTEN), which is known to suppress PI3K, leads to the expression of PD-L1 in glioma [71]. More importantly, a recent report suggests that an inactivating mutation of *PTEN* and activating mutation of *PI3KCA* are associated with CK19 expression in HCC [72], where expression of PD-L1 is common. As activation of the PI3K-Akt pathway is a characteristic of cancer stem cells (CSCs) [73], genetic alterations and constitutive activation of this pathway may give rise to the overexpression of PD-L1 and induce stem cell features in HCCs. From this perspective, blockade of the PD-1/PD-L1 axis may be effective for HCC with stem cell-like characteristics, which is resistant to conventional therapies. However, we have also found that infiltration of $CD8^+$ cells is not as prominent in PD-L1-positive HCCs with mutations in the PI3K-Akt pathway compared to those without the mutations. Constitutive activation of the PI3K-Akt pathway in HCC might induce PD-L1 expression, even in a non-inflamed background, where a lack of $CD8^+$ T-cells could be an obstacle for sufficient action of anti-PD-1/PD-L1 monotherapy. On the other hand, it is also suggested that the PI3K-Akt pathway is frequently activated in CSCs and PI3K inhibitors preferentially target CSCs [73]. As the expression of stem cell markers in HCC is associated with PD-L1 expression and since anti-PD-1/PD-L1 antibody might also target CSCs, a dual blockade of the PD1/PD-L1 axis and PI3K-Akt pathway may be an option for treating patients with HCC showing stem cell features (Figure 2c) [74].

3.5. Current Limitation of Immune Checkpoint Inhibitors and Challenge for HCC with Lack of Immune Infiltrates

HCC patients with dense lymphocyte infiltration reportedly show a marked reduction of response rate after curative resection of tumor, suggesting that TILs are critical for anti-tumor immune response [75]. From this point of view, it is conceivable that "immune cold tumor" with lack of immune infiltrates should be refractory to ICIs [66]. Ishizuka et al. reported that loss-of-function of the RNA-editing enzyme adenosine deaminase acting on RNA (ADAR1) overcomes immune checkpoint blockade resistance caused by inactivation of antigen presentation by tumor cells [76]. This restoration of sensitivity to immunotherapy may occur without recognition of TAA by $CD8^+$ T-cells. As ADAR1 is able to act as an oncogene and its overexpression plays a role in the carcinogenesis of HCC [77], intervention of ADAR1 activity may also be a promising approach as an effective immunotherapy in patients with HCC refractory to ICIs due to the lack of $CD8^+$ TILs (Figure 2d).

On the other hand, results from methylome analyses of cancer tissues suggest that epigenetic alterations in HCC may affect the anti-tumor immune response. Hong et al. investigated the role of epigenetic therapy on enhancing immunotherapy responses in HCC [78]. Treatment of HCC cell lines with inhibitors of enhancer of zeste homolog 2 (EZH2) and DNA methyltransferase 1 (DNMT1) improved the induction of Th1 chemokines and HCC-related antigens upon treatment with anti-PD-L1 antibody. Furthermore, using an in vivo model, they found that the combination of PD-1/PD-L1 blockade with an epigenetic modulator improves the trafficking of $CD8^+$ T-cells into tumor tissues and promotes tumor regression. Therefore, epigenetic modulation may reactivate the epigenetically repressed chemokine responsible for T-cell trafficking and induce neoantigens as immune targets. Thus, the combination of epigenetic therapy with ICIs might also be applicable to cases with refractory HCC (Figure 2d). Schonfeld et al. showed that polymorphism in the protein arginine methyltransferase 1 (PRMT1) was associated with protein expression and modulated the expression of PD-L1 and PL-L2 in HCC cells [79], suggesting that intervention of PRMT1 activity could also restore the response to immune checkpoint inhibitors in some patients.

For the development of biomarkers that predict the tumor response to immunotherapy, it is critical to improve the outcome of the treatment. Previous reports point out that tumors with active

IFN-γ signaling show immune classes that can be candidates for immunotherapy [30,39]. In addition, expression of PD-L1 in tumor cells and tumor infiltrates (CPS) was reportedly associated with tumor response in HCC cases [48]. Detection of activating mutation in *CTNNB1* should also be informative to know immune cold phenotype and lack of response to ICIs in HCC [41]. On the other hand, Feun et al. indicated that baseline plasma TGF-β level could be a predictive biomarker for the response to pembrolizumab [80], and clinical trials of combined blockade of PD-1/PD-L1 and TGF-β axis are ongoing (Table 3). Dong et al. analyzed multiple tumors of the same patients for genetic structure, neoantigens, T cell receptor repertoires, and immune infiltrates, and found that only a few tumors were under the control of immunosurveillance and the majority carry a variety of immune escape mechanisms, even in a single case [81]. From this point of view, precise analysis of immune phenotype of HCC should contribute to the establishment of personalized immunotherapy in HCC cases.

4. Conclusions

Several studies have demonstrated the efficacy of immune checkpoint inhibitors in HCC, even in tumors that are resistant to conventional therapies. However, only a small subset of HCCs show an anti-tumor response to immune checkpoint monotherapy [5,48,59,82]. Therefore, understanding the immunological microenvironment of HCC is crucial since the response to anti-PD-1 therapy may be determined by the immune status of the tumor [13,15,83]. As the mutational signature of HCC may affect its immunophenotype thorough the induction of immune regulatory molecules and cells, the data presented here may be informative in the development of effective combination therapies using ICIs for treating patients with HCC, especially those who are refractory to conventional therapies.

Author Contributions: N.N. wrote the original version of the manuscript, and N.N. and M.K. revised the manuscript. All authors have read and agreed to the published version of the manuscript.

Funding: This work was supported in part by a Grant-in-Aid for Scientific Research from the Japan Society for the Promotion of Science (KAKENHI: 16K09382, N. Nishida, and 18K07922, M. Kudo) and a grant from the Smoking Research Foundation (N. Nishida).

Conflicts of Interest: The authors declare no conflict of interest.

References

1. Villanueva, A. Hepatocellular Carcinoma. *N. Engl. J. Med.* **2019**, *380*, 1450–1462. [CrossRef] [PubMed]
2. Kudo, M. Systemic therapy for hepatocellular carcinoma: Latest advances. *Cancers* **2018**, *10*, 412. [CrossRef] [PubMed]
3. Nishida, N.; Kitano, M.; Sakurai, T.; Kudo, M. Molecular mechanism and prediction of sorafenib chemoresistance in human hepatocellular carcinoma. *Dig. Dis.* **2015**, *33*, 771–779. [CrossRef] [PubMed]
4. Nishida, N.; Arizumi, T.; Hagiwara, S.; Ida, H.; Sakurai, T.; Kudo, M. MicroRNAs for the prediction of early response to sorafenib treatment in human hepatocellular carcinoma. *Liver Cancer* **2017**, *6*, 113–125. [CrossRef]
5. Nishida, N.; Kudo, M. Immune checkpoint blockade for the treatment of human hepatocellular carcinoma. *Hepatol. Res.* **2018**, *48*, 622–634. [CrossRef]
6. Prieto, J.; Melero, I.; Sangro, B. Immunological landscape and immunotherapy of hepatocellular carcinoma. *Nat. Rev. Gastroenterol. Hepatol.* **2015**, *12*, 681–700. [CrossRef]
7. Yau, T.; Park, J.W.; Finn, R.S.; Cheng, A.; Mathurin, P.; Edeline, J.; Kudo, M.; Han, K.; Harding, J.J.; Merle, P.; et al. CheckMate 459: A randomized, multi-center phase 3 study of Nivolumab (NIVO) vs Sorafenib (SOR) as first-line (1L) treatment in patients (pts) with advanced hepatocellular carcinoma (aHCC). *Ann. Oncol.* **2019**, *30* (Suppl. 5), v851–v934. [CrossRef]
8. Finn, R.S.; Ryoo, B.Y.; Merle, P.; Kudo, M.; Bouattour, M.; Lim, H.Y.; Breder, V.; Edeline, J.; Chao, Y.; Ogasawara, S.; et al. Pembrolizumab as second-line therapy in patients with advanced hepatocellular carcinoma in KEYNOTE-240: A randomized, double-blind, phase III trial. *J. Clin. Oncol.* **2020**, *38*, 193–202. [CrossRef]

9. Boyault, S.; Rickman, D.S.; de Reynies, A.; Balabaud, C.; Rebouissou, S.; Jeannot, E.; Herault, A.; Saric, J.; Belghiti, J.; Franco, D.; et al. Transcriptome classification of HCC is related to gene alterations and to new therapeutic targets. *Hepatology* **2007**, *45*, 42–52. [CrossRef]
10. Hoshida, Y.; Toffanin, S.; Lachenmayer, A.; Villanueva, A.; Minguez, B.; Llovet, J.M. Molecular classification and novel targets in hepatocellular carcinoma: Recent advancements. *Semin. Liver Dis.* **2010**, *30*, 35–51. [CrossRef]
11. Nishida, N.; Kudo, M. Recent advancements in comprehensive genetic analyses for human hepatocellular carcinoma. *Oncology* **2013**, *84* (Suppl. 1), 93–97. [CrossRef]
12. Nishida, N.; Kudo, M.; Nishimura, T.; Arizumi, T.; Takita, M.; Kitai, S.; Yada, N.; Hagiwara, S.; Inoue, T.; Minami, Y.; et al. Unique association between global DNA hypomethylation and chromosomal alterations in human hepatocellular carcinoma. *PLoS ONE* **2013**, *8*, e72312. [CrossRef] [PubMed]
13. Nishida, N.; Kudo, M. Oncogenic signal and tumor microenvironment in hepatocellular carcinoma. *Oncology* **2017**, *93* (Suppl. 1), 160–164. [CrossRef] [PubMed]
14. Yarchoan, M.; Xing, D.; Luan, L.; Xu, H.; Sharma, R.B.; Popovic, A.; Pawlik, T.M.; Kim, A.K.; Zhu, Q.; Jaffee, E.M.; et al. Characterization of the immune microenvironment in hepatocellular carcinoma. *Clin. Cancer Res.* **2017**, *23*, 7333–7339. [CrossRef] [PubMed]
15. Nishida, N.; Kudo, M. Immunological microenvironment of hepatocellular carcinoma and its clinical implication. *Oncology* **2017**, *92* (Suppl. 1), 40–49. [CrossRef] [PubMed]
16. Guichard, C.; Amaddeo, G.; Imbeaud, S.; Ladeiro, Y.; Pelletier, L.; Maad, I.B.; Calderaro, J.; Bioulac-Sage, P.; Letexier, M.; Degos, F.; et al. Integrated analysis of somatic mutations and focal copy-number changes identifies key genes and pathways in hepatocellular carcinoma. *Nat. Genet.* **2012**, *44*, 694–698. [CrossRef]
17. Fujimoto, A.; Totoki, Y.; Abe, T.; Boroevich, K.A.; Hosoda, F.; Nguyen, H.H.; Aoki, M.; Hosono, N.; Kubo, M.; Miya, F.; et al. Whole-Genome sequencing of liver cancers identifies etiological influences on mutation patterns and recurrent mutations in chromatin regulators. *Nat. Genet.* **2012**, *44*, 760–764. [CrossRef]
18. Nishida, N.; Kudo, M.; Nagasaka, T.; Ikai, I.; Goel, A. Characteristic patterns of altered DNA methylation predict emergence of human hepatocellular carcinoma. *Hepatology* **2012**, *56*, 994–1003. [CrossRef]
19. Nishida, N.; Goel, A. Genetic and epigenetic signatures in human hepatocellular carcinoma: A systematic review. *Curr. Genom.* **2011**, *12*, 130–137. [CrossRef]
20. Schulze, K.; Imbeaud, S.; Letouze, E.; Alexandrov, L.B.; Calderaro, J.; Rebouissou, S.; Couchy, G.; Meiller, C.; Shinde, J.; Soysouvanh, F.; et al. Exome sequencing of hepatocellular carcinomas identifies new mutational signatures and potential therapeutic targets. *Nat. Genet.* **2015**, *47*, 505–511. [CrossRef]
21. Nishida, N.; Fukuda, Y.; Kokuryu, H.; Toguchida, J.; Yandell, D.W.; Ikenega, M.; Imura, H.; Ishizaki, K. Role and mutational heterogeneity of the p53 gene in hepatocellular carcinoma. *Cancer Res.* **1993**, *53*, 368–372. [PubMed]
22. Nishida, N.; Nishimura, T.; Nagasaka, T.; Ikai, I.; Goel, A.; Boland, C.R. Extensive methylation is associated with beta-catenin mutations in hepatocellular carcinoma: Evidence for two distinct pathways of human hepatocarcinogenesis. *Cancer Res.* **2007**, *67*, 4586–4594. [CrossRef]
23. Nakayama, J.; Tahara, H.; Tahara, E.; Saito, M.; Ito, K.; Nakamura, H.; Nakanishi, T.; Tahara, E.; Ide, T.; Ishikawa, F. Telomerase activation by hTRT in human normal fibroblasts and hepatocellular carcinomas. *Nat. Genet.* **1998**, *18*, 65–68. [CrossRef] [PubMed]
24. Nault, J.C.; Mallet, M.; Pilati, C.; Calderaro, J.; Bioulac-Sage, P.; Laurent, C.; Laurent, A.; Cherqui, D.; Balabaud, C.; Zucman-Rossi, J. High frequency of telomerase reverse-transcriptase promoter somatic mutations in hepatocellular carcinoma and preneoplastic lesions. *Nat. Commun.* **2013**, *4*, 2218. [CrossRef] [PubMed]
25. Shibata, T.; Aburatani, H. Exploration of liver cancer genomes. *Nat. Rev. Gastroenterol. Hepatol.* **2014**, *11*, 340–349. [CrossRef] [PubMed]
26. Zucman-Rossi, J.; Villanueva, A.; Nault, J.C.; Llovet, J.M. Genetic landscape and biomarkers of hepatocellular carcinoma. *Gastroenterology* **2015**, *149*, 1226–1239. [CrossRef] [PubMed]
27. Cancer Genome Atlas Research Network. Comprehensive and integrative genomic characterization of hepatocellular carcinoma. *Cell* **2017**, *169*, 1327–1341. [CrossRef]
28. Nishida, N.; Nishimura, T.; Kaido, T.; Minaga, K.; Yamao, K.; Kamata, K.; Takenaka, M.; Ida, H.; Hagiwara, S.; Minami, Y.; et al. Molecular scoring of hepatocellular carcinoma for predicting metastatic recurrence and requirements of systemic chemotherapy. *Cancers* **2018**, *10*, 367. [CrossRef]

29. Nishida, N.; Fukuda, Y.; Komeda, T.; Kita, R.; Sando, T.; Furukawa, M.; Amenomori, M.; Shibagaki, I.; Nakao, K.; Ikenaga, M.; et al. Amplification and overexpression of the cyclin D1 gene in aggressive human hepatocellular carcinoma. *Cancer Res.* **1994**, *54*, 3107–3110.
30. Rebouissou, S.; Nault, J.C. Advances in molecular classification and precision oncology in hepatocellular carcinoma. *J. Hepatol.* **2020**, *72*, 215–229. [CrossRef]
31. Nishida, N.; Nagasaka, T.; Nishimura, T.; Ikai, I.; Boland, C.R.; Goel, A. Aberrant methylation of multiple tumor suppressor genes in aging liver, chronic hepatitis, and hepatocellular carcinoma. *Hepatology* **2008**, *47*, 908–918. [CrossRef]
32. Calderaro, J.; Couchy, G.; Imbeaud, S.; Amaddeo, G.; Letouze, E.; Blanc, J.F.; Laurent, C.; Hajji, Y.; Azoulay, D.; Bioulac-Sage, P.; et al. Histological subtypes of hepatocellular carcinoma are related to gene mutations and molecular tumour classification. *J. Hepatol.* **2017**, *67*, 727–738. [CrossRef] [PubMed]
33. Desert, R.; Rohart, F.; Canal, F.; Sicard, M.; Desille, M.; Renaud, S.; Turlin, B.; Bellaud, P.; Perret, C.; Clement, B.; et al. Human hepatocellular carcinomas with a periportal phenotype have the lowest potential for early recurrence after curative resection. *Hepatology* **2017**, *66*, 1502–1518. [CrossRef] [PubMed]
34. Kurebayashi, Y.; Ojima, H.; Tsujikawa, H.; Kubota, N.; Maehara, J.; Abe, Y.; Kitago, M.; Shinoda, M.; Kitagawa, Y.; Sakamoto, M. Landscape of immune microenvironment in hepatocellular carcinoma and its additional impact on histological and molecular classification. *Hepatology* **2018**, *68*, 1025–1041. [CrossRef]
35. Gao, Q.; Wang, X.Y.; Qiu, S.J.; Yamato, I.; Sho, M.; Nakajima, Y.; Zhou, J.; Li, B.Z.; Shi, Y.H.; Xiao, Y.S.; et al. Overexpression of PD-L1 significantly associates with tumor aggressiveness and postoperative recurrence in human hepatocellular carcinoma. *Clin. Cancer Res.* **2009**, *15*, 971–979. [CrossRef] [PubMed]
36. Umemoto, Y.; Okano, S.; Matsumoto, Y.; Nakagawara, H.; Matono, R.; Yoshiya, S.; Yamashita, Y.; Yoshizumi, T.; Ikegami, T.; Soejima, Y.; et al. Prognostic impact of programmed cell death 1 ligand 1 expression in human leukocyte antigen class I-positive hepatocellular carcinoma after curative hepatectomy. *J. Gastroenterol.* **2015**, *50*, 65–75. [CrossRef]
37. Gabrielson, A.; Wu, Y.; Wang, H.; Jiang, J.; Kallakury, B.; Gatalica, Z.; Reddy, S.; Kleiner, D.; Fishbein, T.; Johnson, L.; et al. Intratumoral CD3 and CD8 T-cell densities associated with relapse-free survival in HCC. *Cancer Immunol. Res.* **2016**, *4*, 419–430. [CrossRef]
38. Calderaro, J.; Rousseau, B.; Amaddeo, G.; Mercey, M.; Charpy, C.; Costentin, C.; Luciani, A.; Zafrani, E.S.; Laurent, A.; Azoulay, D.; et al. Programmed death ligand 1 expression in hepatocellular carcinoma: Relationship with clinical and pathological features. *Hepatology* **2016**, *64*, 2038–2046. [CrossRef]
39. Sia, D.; Jiao, Y.; Martinez-Quetglas, I.; Kuchuk, O.; Villacorta-Martin, C.; Castro de Moura, M.; Putra, J.; Camprecios, G.; Bassaganyas, L.; Akers, N.; et al. Identification of an immune-specific class of hepatocellular carcinoma, based on molecular features. *Gastroenterology* **2017**, *153*, 812–826. [CrossRef]
40. Nishida, N.; Sakai, K.; Morita, M.; Aoki, T.; Takita, M.; Hagiwara, S.; Komeda, Y.; Takenaka, M.; Minami, Y.; Ida, H.; et al. Association between genetic and immunological background of hepatocellular carcinoma and expression of programmed cell death-1. *Liver Cancer* **2020**, in press. [CrossRef]
41. Harding, J.J.; Nandakumar, S.; Armenia, J.; Khalil, D.N.; Albano, M.; Ly, M.; Shia, J.; Hechtman, J.F.; Kundra, R.; El Dika, I.; et al. Prospective genotyping of hepatocellular carcinoma: Clinical implications of next-generation sequencing for matching patients to targeted and immune therapies. *Clin. Cancer Res.* **2019**, *25*, 2116–2126. [CrossRef] [PubMed]
42. Ruiz de Galarreta, M.; Bresnahan, E.; Molina-Sanchez, P.; Lindblad, K.E.; Maier, B.; Sia, D.; Puigvehi, M.; Miguela, V.; Casanova-Acebes, M.; Dhainaut, M.; et al. β-Catenin activation promotes immune escape and resistance to Anti-PD-1 therapy in hepatocellular carcinoma. *Cancer Discov.* **2019**, *9*, 1124–1141. [CrossRef] [PubMed]
43. Kang, H.J.; Oh, J.H.; Chun, S.M.; Kim, D.; Ryu, Y.M.; Hwang, H.S.; Kim, S.Y.; An, J.; Cho, E.J.; Lee, H.; et al. Immunogenomic landscape of hepatocellular carcinoma with immune cell stroma and EBV-positive tumor-infiltrating lymphocytes. *J. Hepatol.* **2019**, *71*, 91–103. [CrossRef] [PubMed]
44. Zheng, C.; Zheng, L.; Yoo, J.K.; Guo, H.; Zhang, Y.; Guo, X.; Kang, B.; Hu, R.; Huang, J.Y.; Zhang, Q.; et al. Landscape of infiltrating T cells in liver cancer revealed by single-cell sequencing. *Cell* **2017**, *169*, 1342–1356. [CrossRef]
45. Zhou, G.; Sprengers, D.; Boor, P.P.C.; Doukas, M.; Schutz, H.; Mancham, S.; Pedroza-Gonzalez, A.; Polak, W.G.; de Jonge, J.; Gaspersz, M.; et al. Antibodies against immune checkpoint molecules restore functions of tumor-infiltrating T cells in hepatocellular carcinomas. *Gastroenterology* **2017**, *153*, 1107–1119. [CrossRef]

46. Yan, W.; Liu, X.; Ma, H.; Zhang, H.; Song, X.; Gao, L.; Liang, X.; Ma, C. Tim-3 fosters HCC development by enhancing TGF-β-mediated alternative activation of macrophages. *Gut* **2015**, *64*, 1593–1604. [CrossRef]
47. Kim, H.D.; Song, G.W.; Park, S.; Jung, M.K.; Kim, M.H.; Kang, H.J.; Yoo, C.; Yi, K.; Kim, K.H.; Eo, S.; et al. Association between expression level of PD1 by tumor-infiltrating CD8(+) T cells and features of hepatocellular carcinoma. *Gastroenterology* **2018**, *155*, 1936–1950. [CrossRef]
48. Zhu, A.X.; Finn, R.S.; Edeline, J.; Cattan, S.; Ogasawara, S.; Palmer, D.; Verslype, C.; Zagonel, V.; Fartoux, L.; Vogel, A.; et al. Pembrolizumab in patients with advanced hepatocellular carcinoma previously treated with sorafenib (KEYNOTE-224): A non-randomised, open-label phase 2 trial. *Lancet Oncol.* **2018**, *19*, 940–952. [CrossRef]
49. Chang, B.; Shen, L.; Wang, K.; Jin, J.; Huang, T.; Chen, Q.; Li, W.; Wu, P. High number of PD-1 positive intratumoural lymphocytes predicts survival benefit of cytokine-induced killer cells for hepatocellular carcinoma patients. *Liver Int.* **2018**, *38*, 1449–1458. [CrossRef]
50. Roth, G.S.; Decaens, T. Liver immunotolerance and hepatocellular carcinoma: Patho-Physiological mechanisms and therapeutic perspectives. *Eur. J. Cancer* **2017**, *87*, 101–112. [CrossRef]
51. Faivre, S.; Rimassa, L.; Finn, R.S. Molecular therapies for HCC: Looking outside the box. *J. Hepatol.* **2020**, *72*, 342–352. [CrossRef]
52. Arihara, F.; Mizukoshi, E.; Kitahara, M.; Takata, Y.; Arai, K.; Yamashita, T.; Nakamoto, Y.; Kaneko, S. Increase in CD14 + HLA-DR-/low myeloid-derived suppressor cells in hepatocellular carcinoma patients and its impact on prognosis. *Cancer Immunol. Immunother.* **2013**, *62*, 1421–1430. [CrossRef]
53. Chen, K.J.; Lin, S.Z.; Zhou, L.; Xie, H.Y.; Zhou, W.H.; Taki-Eldin, A.; Zheng, S.S. Selective recruitment of regulatory T cell through CCR6-CCL20 in hepatocellular carcinoma fosters tumor progression and predicts poor prognosis. *PLoS ONE* **2011**, *6*, e24671. [CrossRef] [PubMed]
54. Hoechst, B.; Ormandy, L.A.; Ballmaier, M.; Lehner, F.; Kruger, C.; Manns, M.P.; Greten, T.F.; Korangy, F. A new population of myeloid-derived suppressor cells in hepatocellular carcinoma patients induces CD4(+)CD25(+)Foxp3(+) T cells. *Gastroenterology* **2008**, *135*, 234–243. [CrossRef] [PubMed]
55. Li, H.; Han, Y.; Guo, Q.; Zhang, M.; Cao, X. Cancer-Expanded myeloid-derived suppressor cells induce anergy of NK cells through membrane-bound TGF-β1. *J. Immunol.* **2009**, *182*, 240–249. [CrossRef] [PubMed]
56. Quezada, S.A.; Peggs, K.S.; Simpson, T.R.; Allison, J.P. Shifting the equilibrium in cancer immunoediting: From tumor tolerance to eradication. *Immunol. Rev.* **2011**, *241*, 104–118. [CrossRef] [PubMed]
57. Li, T.; Yang, Y.; Hua, X.; Wang, G.; Liu, W.; Jia, C.; Tai, Y.; Zhang, Q.; Chen, G. Hepatocellular carcinoma-associated fibroblasts trigger NK cell dysfunction via PGE2 and IDO. *Cancer Lett.* **2012**, *318*, 154–161. [CrossRef]
58. Chen, Y.; Huang, Y.; Reiberger, T.; Duyverman, A.M.; Huang, P.; Samuel, R.; Hiddingh, L.; Roberge, S.; Koppel, C.; Lauwers, G.Y.; et al. Differential effects of sorafenib on liver versus tumor fibrosis mediated by stromal-derived factor 1 α/C-X-C receptor type 4 axis and myeloid differentiation antigen-positive myeloid cell infiltration in mice. *Hepatology* **2014**, *59*, 1435–1447. [CrossRef]
59. El-Khoueiry, A.B.; Sangro, B.; Yau, T.; Crocenzi, T.S.; Kudo, M.; Hsu, C.; Kim, T.Y.; Choo, S.P.; Trojan, J.; Welling, T.H.R.; et al. Nivolumab in patients with advanced hepatocellular carcinoma (CheckMate 040): An open-label, non-comparative, phase 1/2 dose escalation and expansion trial. *Lancet* **2017**, *389*, 2492–2502. [CrossRef]
60. Kawaoka, T.; Ando, Y.; Yamauchi, M.; Suehiro, Y.; Yamaoka, K.; Kosaka, Y.; Fuji, Y.; Uchikawa, S.; Morio, K.; Fujino, H.; et al. Incidence of microsatellite instability-high hepatocellular carcinoma among Japanese patients and response to pembrolizumab. *Hepatol. Res.* **2020**. [CrossRef]
61. Brown, Z.J.; Yu, S.J.; Heinrich, B.; Ma, C.; Fu, Q.; Sandhu, M.; Agdashian, D.; Zhang, Q.; Korangy, F.; Greten, T.F. Indoleamine 2,3-dioxygenase provides adaptive resistance to immune checkpoint inhibitors in hepatocellular carcinoma. *Cancer Immunol. Immunother.* **2018**, *67*, 1305–1315. [CrossRef]
62. Jain, R.K. Normalization of tumor vasculature: An emerging concept in antiangiogenic therapy. *Science* **2005**, *307*, 58–62. [CrossRef] [PubMed]
63. Mossenta, M.; Busato, D.; Baboci, L.; Cintio, F.D.; Toffoli, G.; Bo, M.D. New insight into therapies targeting angiogenesis in hepatocellular carcinoma. *Cancers* **2019**, *11*, 1086. [CrossRef] [PubMed]
64. Kudo, M. Combination cancer immunotherapy with molecular targeted agents/anti-CTLA-4 antibody for hepatocellular carcinoma. *Liver Cancer* **2019**, *8*, 1–11. [CrossRef] [PubMed]

65. Zhu, A.X.; Kang, Y.K.; Yen, C.J.; Finn, R.S.; Galle, P.R.; Llovet, J.M.; Assenat, E.; Brandi, G.; Pracht, M.; Lim, H.Y.; et al. Ramucirumab after sorafenib in patients with advanced hepatocellular carcinoma and increased alpha-fetoprotein concentrations (REACH-2): A randomised, double-blind, placebo-controlled, phase 3 trial. *Lancet Oncol.* **2019**, *20*, 282–296. [CrossRef]
66. Teng, M.W.; Ngiow, S.F.; Ribas, A.; Smyth, M.J. Classifying cancers based on T-cell infiltration and PD-L1. *Cancer Res.* **2015**, *75*, 2139–2145. [CrossRef]
67. Yi, M.; Jiao, D.; Qin, S.; Chu, Q.; Wu, K.; Li, A. Synergistic effect of immune checkpoint blockade and anti-angiogenesis in cancer treatment. *Mol. Cancer* **2019**, *18*, 60. [CrossRef]
68. Shigeta, K.; Datta, M.; Hato, T.; Kitahara, S.; Chen, I.X.; Matsui, A.; Kikuchi, H.; Mamessier, E.; Aoki, S.; Ramjiawan, R.R.; et al. Dual Programmed death receptor-1 and vascular endothelial growth factor receptor-2 blockade promotes vascular normalization and enhances antitumor immune responses in hepatocellular carcinoma. *Hepatology* **2019**, 1247–1261. [CrossRef]
69. Nishida, N. Clinical implications of the dual blockade of the PD-1/PD-L1 and vascular endothelial growth factor axes in the treatment of hepatocellular carcinoma. *Hepatobiliary Surg. Nutr.* **2020**, in press. [CrossRef]
70. Cheng, A.L.; Hsu, C.; Chan, S.L.; Choo, S.P.; Kudo, M. Challenges of combination therapy with immune checkpoint inhibitors for hepatocellular carcinoma. *J. Hepatol.* **2020**, *72*, 307–319. [CrossRef]
71. Parsa, A.T.; Waldron, J.S.; Panner, A.; Crane, C.A.; Parney, I.F.; Barry, J.J.; Cachola, K.E.; Murray, J.C.; Tihan, T.; Jensen, M.C.; et al. Loss of tumor suppressor PTEN function increases B7-H1 expression and immunoresistance in glioma. *Nat. Med.* **2007**, *13*, 84–88. [CrossRef] [PubMed]
72. Chen, D.; Li, Z.; Cheng, Q.; Wang, Y.; Qian, L.; Gao, J.; Zhu, J.Y. Genetic alterations and expression of PTEN and its relationship with cancer stem cell markers to investigate pathogenesis and to evaluate prognosis in hepatocellular carcinoma. *J. Clin. Pathol.* **2019**, *72*, 588–596. [CrossRef] [PubMed]
73. Zhou, H.; Yu, C.; Kong, L.; Xu, X.; Yan, J.; Li, Y.; An, T.; Gong, L.; Gong, Y.; Zhu, H.; et al. B591, a novel specific pan-PI3K inhibitor, preferentially targets cancer stem cells. *Oncogene* **2019**, *38*, 3371–3386. [CrossRef] [PubMed]
74. Li, H.; Li, X.; Liu, S.; Guo, L.; Zhang, B.; Zhang, J.; Ye, Q. Programmed cell death-1 (PD-1) checkpoint blockade in combination with a mammalian target of rapamycin inhibitor restrains hepatocellular carcinoma growth induced by hepatoma cell-intrinsic PD-1. *Hepatology* **2017**, *66*, 1920–1933. [CrossRef] [PubMed]
75. Wada, Y.; Nakashima, O.; Kutami, R.; Yamamoto, O.; Kojiro, M. Clinicopathological study on hepatocellular carcinoma with lymphocytic infiltration. *Hepatology* **1998**, *27*, 407–414. [CrossRef] [PubMed]
76. Ishizuka, J.J.; Manguso, R.T.; Cheruiyot, C.K.; Bi, K.; Panda, A.; Iracheta-Vellve, A.; Miller, B.C.; Du, P.P.; Yates, K.B.; Dubrot, J.; et al. Loss of ADAR1 in tumours overcomes resistance to immune checkpoint blockade. *Nature* **2019**, *565*, 43–48. [CrossRef]
77. Chan, T.H.; Lin, C.H.; Qi, L.; Fei, J.; Li, Y.; Yong, K.J.; Liu, M.; Song, Y.; Chow, R.K.; Ng, V.H.; et al. A disrupted RNA editing balance mediated by ADARs (Adenosine DeAminases that act on RNA) in human hepatocellular carcinoma. *Gut* **2014**, *63*, 832–843. [CrossRef]
78. Hong, Y.K.; Li, Y.; Pandit, H.; Li, S.; Pulliam, Z.; Zheng, Q.; Yu, Y.; Martin, R.C.G. Epigenetic modulation enhances immunotherapy for hepatocellular carcinoma. *Cell. Immunol.* **2019**, *336*, 66–74. [CrossRef]
79. Schonfeld, M.; Zhao, J.; Komatz, A.; Weinman, S.A.; Tikhanovich, I. The polymorphism rs975484 in the protein arginine methyltransferase-1 gene modulates expression of immune checkpoint genes in hepatocellular carcinoma. *J. Biol. Chem.* **2020**, *295*, 7126–7137. [CrossRef]
80. Feun, L.G.; Li, Y.Y.; Wu, C.; Wangpaichitr, M.; Jones, P.D.; Richman, S.P.; Madrazo, B.; Kwon, D.; Garcia-Buitrago, M.; Martin, P.; et al. Phase 2 study of pembrolizumab and circulating biomarkers to predict anticancer response in advanced, unresectable hepatocellular carcinoma. *Cancer* **2019**, *125*, 3603–3614. [CrossRef]
81. Dong, L.Q.; Peng, L.H.; Ma, L.J.; Liu, D.B.; Zhang, S.; Luo, S.Z.; Rao, J.H.; Zhu, H.W.; Yang, S.X.; Xi, S.J.; et al. Heterogeneous immunogenomic features and distinct escape mechanisms in multifocal hepatocellular carcinoma. *J. Hepatol.* **2020**, *72*, 896–908. [CrossRef] [PubMed]

82. Nishida, N.; Kudo, M. Role of immune checkpoint blockade in the treatment for human hepatocellular carcinoma. *Dig. Dis.* **2017**, *35*, 618–622. [CrossRef] [PubMed]
83. Hou, J.; Zhang, H.; Sun, B.; Karin, M. The immunobiology of hepatocellular carcinoma in humans and mice: Basic concepts and therapeutic implications. *J. Hepatol.* **2020**, *72*, 167–182. [CrossRef] [PubMed]

© 2020 by the authors. Licensee MDPI, Basel, Switzerland. This article is an open access article distributed under the terms and conditions of the Creative Commons Attribution (CC BY) license (http://creativecommons.org/licenses/by/4.0/).

Review

Scientific Rationale for Combined Immunotherapy with PD-1/PD-L1 Antibodies and VEGF Inhibitors in Advanced Hepatocellular Carcinoma

Masatoshi Kudo

Department of Gastroenterology and Hepatology, Kindai University Faculty of Medicine, Osaka-Sayama 589-8511, Japan; m-kudo@med.kindai.ac.jp; Tel.: +81-72-366-0221 (ext. 3149); Fax: +81-72-367-2880

Received: 19 March 2020; Accepted: 17 April 2020; Published: 27 April 2020

Abstract: A successful phase III trial for the combination of atezolizumab and bevacizumab (the IMbrave150 trial) in advanced hepatocellular carcinoma has recently been reported. This is groundbreaking because nivolumab and pembrolizumab, both programmed cell death-1 (PD-1) antibodies, have failed to show efficacy as first- and second-line therapeutics, respectively, in phase III clinical trials. Immunotherapy with a combination of atezolizumab and bevacizumab resulted in better survival than treatment with sorafenib for the first time since sorafenib was approved in 2007. The high efficacy of the combination of PD-1/programmed death ligand 1 (PD-L1) and vascular endothelial growth factor (VEGF) antibodies is not only due to their additive effects on tumor growth, but also to their reprogramming of the immunosuppressive microenvironment into an immunostimulatory microenvironment. These results were confirmed in a phase Ib trial that showed significantly longer progression-free survival in the atezolizumab plus bevacizumab group than in patients that received atezolizumab alone. These results demonstrate that immunotherapy with a combination of PD-1/PD-L1 and VEGF inhibitors is effective and may result in a reprogramming of the tumor microenvironment. The results of an ongoing phase III trial of a PD-1 antibody in combination with the VEGF receptor tyrosine kinase inhibitor (TKI) are highly anticipated.

Keywords: hepatocellular carcinoma; immune checkpoint inhibitor; PD-1 antibody; PD-L1 antibody; anti-VEGF inhibitor

1. Introduction

At the European Society for Medical Oncology (ESMO) Asia in November 2019, the positive results of the IMbrave150 study, a trial which compared the effects of the combination of atezolizumab and bevacizumab with those of sorafenib [1], drew attention to the possibility of immunotherapy with a combination of programmed cell death-1 (PD-1)/programmed death ligand 1 (PD-L1) and vascular endothelial growth factor (VEGF) inhibitors. This review outlines the scientific rationale for the therapeutic combination of PD-1/PD-L1 and VEGF antibodies, proof-of-concept results of the phase Ib trial, and results of other phase Ib trials for similar combination strategies.

2. The Rationale Underlying the Combination of PD-1/PD-L1 and VEGF Inhibitors

At tumor sites, VEGF released by hypoxic cancer cells and vascular endothelial cells promotes tumor growth, invasion, and metastasis by increasing neovascularization [2]. Simultaneously, VEGF enhances the mobilization and proliferation of various cells, including regulatory T cells (Tregs), and the release of immunosuppressive cytokines [2,3]. It also enhances the mobilization of tumor-associated macrophages (TAMs) and their polarization to an M2 phenotype. Tregs and TAMs promote tumor growth through the release of VEGF and angiopoietin-2, among other mechanisms [4]. VEGF can also activate myeloid-derived suppressor cells (MDSCs), which in turn release more VEGF [4]. Furthermore, VEGF inhibits dendritic cell maturation and antigen presentation in the priming phase. Thus, VEGF reduces the proliferation and activation of naive CD8+ cells by suppressing dendritic cell activity even in the presence of neoantigens [4] (Figure 1). VEGF-induced Tregs, TAMs, and MDSCs reduce the proliferation and function of CD8+ cells. VEGF also prevents antigen-activated CD8+ cells from infiltrating the tumor tissue through its effects on tumor angiogenesis. In addition, VEGF creates a microenvironment that inhibits the function of T cells in the tumor during the effector phase of the immune response [4]. Furthermore, immunosuppressive cells (Tregs, TAMs, and MDSCs) promote immune escape by releasing immunosuppressive cytokines, including interleukin (IL)-10 and transforming growth factor beta (TGF-β), and by inhibiting dendritic cell maturation and activation, NK cell activation, and T cell activation and proliferation [2–25] (Figure 1). The cancer immunity cycle begins with the uptake and presentation of neoantigens released from necrotic tumor cells by dendritic cells. This is followed by seven steps: (1) tumor antigen release, (2) tumor antigen uptake and presentation by dendritic cells, (3) T cell priming and activation, (4) T cell migration to the tumor, (5) T cell invasion of the tumor, (6) cancer cell recognition by T cells, and (7) attack on tumor cells by T cells, which leads to cancer cell death and release of additional tumor antigens [5] (Figure 2). VEGF promotes immune escape at almost every step of the cancer immunity cycle [6–9]. Furthermore, hepatic interstitial cells such as Kupffer cells, liver endothelial cells, and hepatic stellate cells are involved in maintaining immune tolerance in the healthy liver and may contribute to the immunosuppressive microenvironment in hepatocellular carcinoma [26].

The administration of molecular targeted drugs that inhibit VEGF activity, such as multi-kinase inhibitors that inhibit VEGF receptors, leads to an increase in antigen presentation by dendritic cells [8]. These drugs also promote T cell activation in the priming phase [8] and improve the migration of T cells from the lymph nodes to the tumor site by normalizing the tumor vasculature [15]. In addition, these drugs have been found to suppress the generation of Tregs, TAMs, and MDSCs at the tumor site, and to negatively regulate the expression of immunosuppressive cytokines such as TGF-β and IL-10 [10]. VEGF inhibitors therefore reprogram the immunosuppressive tumor microenvironment into an immunostimulatory environment [6,8]. The administration of PD-1/PD-L1 antibodies under such conditions enhances the antitumor activity of T cells (Figures 3 and 4). As described above, the combination of VEGF and PD-1/PD-L1 inhibitors promotes antitumor immunity according to the four Rs. First, a reversal of the VEGF-mediated inhibition of dendritic cell maturation results in the effective priming and activation of T cells (Recognition) [9]. Second, anti-VEGF antibodies normalize the tumor vasculature and promote the effective infiltration of T cells into the tumor (Recruitment) [15]. Third, anti-VEGF antibodies inhibit the activity of MDSCs, Tregs, and TAMs, leading to the reprogramming of the immunosuppressive microenvironment into an immunostimulatory microenvironment (Reprogramming) [6]. Fourth, PD-1/PD-L1 antibodies enhance the ability of T cells to attack tumor cells (Restoration) (Figure 3). These four Rs lead to efficient cancer immunity and tumor growth inhibition. Proteins released by the killed tumor cells are taken up by dendritic cells, and then processed into tumor antigen peptides that are presented on major histocompatibility complex (MHC) class I molecules, leading to a progression through the cancer immunity cycle and further tumor attacks [5] (Figure 2). As described above, normalization of the VEGF-suppressed tumor microenvironment with molecular targeted agents against VEGF leads to the efficient attack on tumors by activated T cells [5–25,27] (Figures 2 and 4). In addition, non-clinical study of lenvatinib, a tyrosine kinase inhibitor (TKI), showed that the inhibition of VEGF activity reduced TAMs and Tregs in the tumor microenvironment, leading to a decrease in TGF-β and IL-10, a decreased expression of T cell exhaustion markers such as PD-1 and TIM-3, and an increased expression of immunostimulatory cytokines such as IL-12 [28–31]. These findings form the rationale for a trial of the combination of TKIs and anti-PD-1/PD-L1 antibodies.

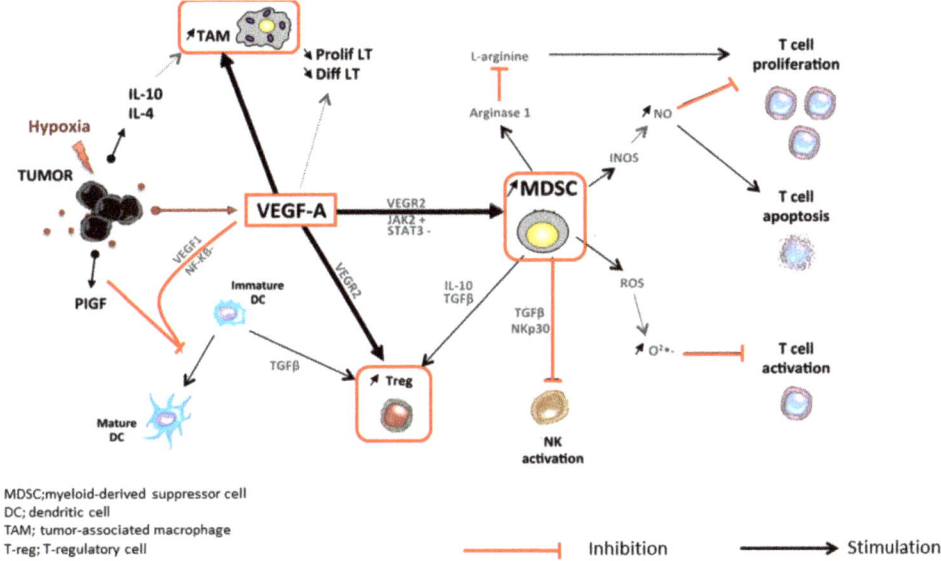

Figure 1. Immune suppressive microenvironment induced by VEGF (modified from ref. [4] with permission).

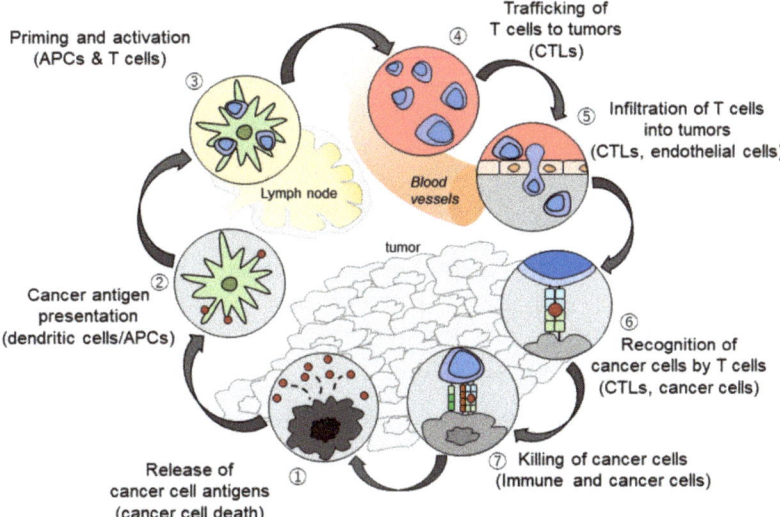

Figure 2. The Cancer-Immunity Cycle (modified from ref. [5] with permission).

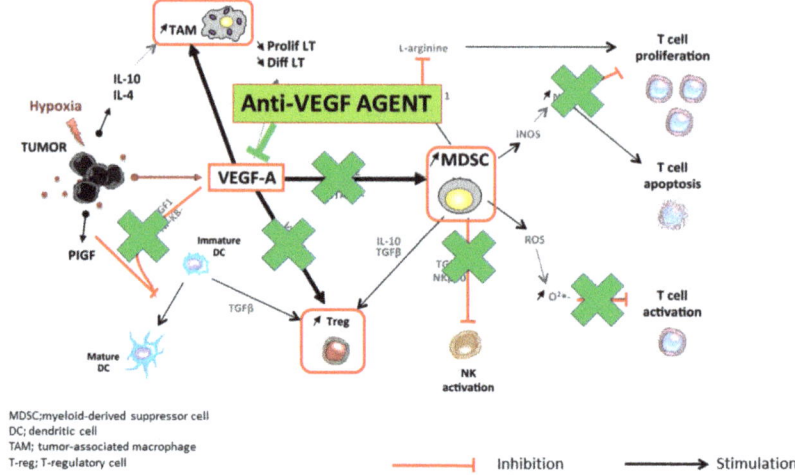

Figure 3. Anti-VEGF antibody reprograms the tumor microenvironment from immune suppressive to immune permissive (modified from ref. [4] with permission).

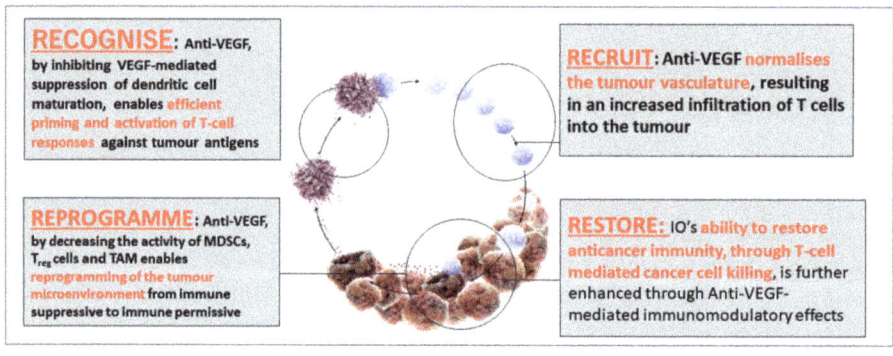

Figure 4. Scientific rationale of Immune-checkpoint Inhibitors plus Anti-VEGF: 4 Roles of anti-VEGF inhibitors in Cancer Immunity cycle, Recognise, Recruitment, Reprogramme, and Restore (original Figure).

3. Classification of the Tumor Microenvironment and Determination of Immunotherapeutic Strategies

Cancers are classified into four types based on the presence of tumor-infiltrating CD8+ T cells and the expression of PD-L1 [32] (Figure 5). Type I tumors contain tumor-infiltrating lymphocytes and express PD-L1. Type I cancers generally show an adequate response to monotherapy with immune checkpoint inhibitors. By contrast, type IV tumors lack PD-L1 expression, although they do contain tumor-infiltrating lymphocytes. Type IV tumors are not responsive to immune checkpoint inhibitors because the immunosuppressive tumor microenvironment inhibits the proliferation and activity of CD8+ cells in these tumors. In type I, there is an initial antitumor immune response, in which perforin, granzyme, and interferon gamma (IFN-γ) are released by activated CD8+ cells, resulting in an immune attack on the cancer cells [32]. However, IFN-γ binds to IFN-γ receptors on the cancer cell surface and upregulates the expression of PD-L1 through the Janus kinase (JAK)-signal transducer and activator of transcription (STAT) signaling pathway [31]. This leads to immune escape, whereby cancer cells evade the attack by activated CD8+ cells. Therefore, type I cancers are responsive to monotherapy with PD-1/PD-L1 antibodies. By contrast, type IV tumors do not show an initial local immune response, even though CD8+ cells are present and the tumor expression of PD-L1 is low. These tumors are never attacked by CD8+ cells because T cell activity is inhibited by the immunosuppressive microenvironment. Therefore, induction of IFN-γ and PD-L1 expression is not observed [28,32]. As expected, such cancers are not responsive to anti-PD-1/PD-L1 antibody monotherapy due to the absence of immune escape through the PD-1/PD-L1 axis. Thus, PD-1 antibody monotherapy is not predicted to be effective in cancers without PD-L1 expression, even if there are large numbers of tumor-infiltrating lymphocytes. In such tumors, anti-VEGF antibodies or inhibitors may reprogram the immunosuppressive microenvironment into an immunostimulatory microenvironment by targeting Tregs, TAMs, and MDSCs, leading to an attack by antigen-specific T cells. This, in turn, would lead to the induction of PD-L1 on the cancer cell surface by IFN-γ. In this scenario, PD-1/PD-L1 antibodies could inhibit immune escape through the PD-1/PD-L1 axis [28,32]. Therefore, this combination therapy could be effective in tumors that are unresponsive to anti-PD-1/PD-L1 monotherapy. Dramatic tumor inhibition could therefore result from the concomitant administration of PD-1/PD-L1 antibodies and VEGF antibodies or TKIs in type IV tumors (Figures 3 and 4) [32]. However, in Type II and III tumors, where no tumor-infiltrating lymphocytes are present, another strategy to increase immunogenicity may be necessary.

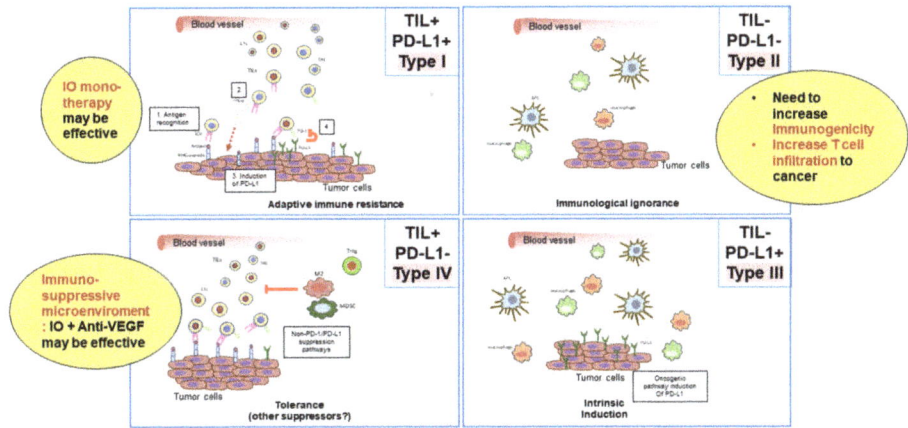

Figure 5. Cancer is classified into 4 types depending on immune microenvironment (TIL: CD8+ cell and PD-L1 expression) (Type I-IV) (modified from ref. [32] with permission).

4. The Results of a Phase Ib Trial of the Combination of Atezolizumab and Bevacizumab (Clinical Trials.Gov Identifier NCT02715531)

4.1. The Use of the Combination of Atezolizumab (a PD-L1 Antibody) and Bevacizumab (a VEGF Antibody) in Unresectable Hepatocellular Carcinoma (Arm A)

Arm A of NCT02715531 was a single-arm phase Ib study of the combination of atezolizumab (a PD-L1 antibody) and bevacizumab (a VEGF antibody) in unresectable hepatocellular carcinoma. Updated results from the 104 unresectable hepatocellular carcinoma patients in Arm A were presented at the annual meeting of the European Society for Medical Oncology (ESMO) in Barcelona, in the fall of 2019 [33]. Fifty-three percent of patients had macroscopic vascular invasion (MVI), of whom 88% were hepatocellular carcinoma patients with highly advanced extrahepatic spread (EHS). Although these were highly advanced cases, evaluation by an independent imaging facility (IRF) based on Response Evaluation Criteria in Solid Tumors (RECIST, version 1.1) showed an overall response rate (ORR) of 36% (95% confidence interval [CI], 26–46%). The ORR based on the modified RECIST (mRECIST) was 39%. The percentage of patients achieving a complete response (CR) based on RECIST 1.1 was

12%. Moreover, the partial response (PR) rate and disease control rate (DCR) were 24% and 71%, respectively. The median duration of response was not reached (95% CI, 11.8–not estimated [NE]). There were 20 patients (54%) with a duration of response ≥ 9 months and 11 patients (30%) with long-term responses (duration of response ≥ 12 months).

In addition, the progression-free survival (PFS) and overall survival (OS) were extremely good (PFS, 7.3 months [95% CI, 5.4–9.9]; OS, 17.1 months [95% CI, 13.8–not reached]). The result is very promising considering the fact that 53%, 88%, and 36% of patients had MVI, EHS with or without MVI, and alpha-fetoprotein (AFP) > 400 ng/mL, respectively.

4.2. Randomized Controlled Arm Comparing the Combination of Atezolizumab Plus Bevacizumab Versus Atezolizumab Alone (Arm F)

Arm F of the study compared PFS in unresectable hepatocellular carcinoma between the combination of atezolizumab (1200 mg) and bevacizumab (15 mg/kg) (every 3 weeks), and atezolizumab alone (1200 mg) as a first-line therapy. This was a proof-of-concept study to determine whether the favorable outcomes observed in Arm A were due to atezolizumab alone or to the combined effect of bevacizumab plus atezolizumab. Importantly, the ORR of the combination of atezolizumab and bevacizumab was slightly higher (20%) than that of atezolizumab alone (17%), which is consistent with data from other trials on the ORR of immune checkpoint inhibitors alone (about 15–18.3% [34–39]). In fact, the median PFS was 5.6 months (95% CI, 3.6–7.4) for atezolizumab plus bevacizumab, and 3.4 months (95% CI, 1.9–5.2) for atezolizumab alone. The hazard ratio was 0.55 (95% CI, 0.40–0.74; $p = 0.0108$). These data clearly showed the beneficial effect of bevacizumab on atezolizumab therapy. The PFS of atezolizumab plus bevacizumab in Arm F (5.6 months) was shorter than that in Arm A (7.3 months). However, this result may be due to the fact that the median follow-up period of Arm F was shorter (6.6 months vs. 12.4 months). With extended follow-up, the PFS in Arm F may have been equivalent to that of Arm A. In any case, the results of Arm F clearly supported the hypothesis that bevacizumab reprograms the immunosuppressive microenvironment into an immunostimulatory environment, enhancing the efficacy of atezolizumab (Figure 4).

5. Results of Phase Ib Studies of Other Combinations of PD-1/PD-L1 Antibodies and VEGF Inhibitors

In addition to the trial of atezolizumab and bevacizumab described above, other studies are examining the efficacy of combined PD-1/PD-L1 and VEGF inhibition. One such study, the LEAP-002 study, is a phase III clinical trial of pembrolizumab and lenvatinib [40,41]. This trial is ongoing and the results are highly anticipated. In addition, multiple other clinical trials of immune checkpoint inhibitors and VEGF inhibitors have been completed (Table 1). The number of patients who received pembrolizumab and lenvatinib ($n = 67$) was lower than the number of patients who received atezolizumab and bevacizumab in Arm A of the phase Ib trial described above ($n = 104$). The ORR (40.3%), DCR (85.1%), PFS (9.7 months), and OS (20.4 months) of the combination of pembrolizumab and lenvatinib were higher than those of the combination of atezolizumab and bevacizumab [42]. Furthermore, the efficacy of the combination of nivolumab and lenvatinib (evaluated by an independent imaging committee based on RECIST 1.1), which was recently reported at the annual meeting of the American Society of Clinical Oncology, Gastrointestinal Cancers (ASCO GI), was higher than that of the other two combination therapies (ORR, 54.2%; DCR, 91.7%; PFS, 7.4 months; and OS, not reached) [43]. Of course, it is not adequate to compare the results of single-arm trials with different patient populations, small sample sizes, and short observation periods. However, the results are very promising. The ORR and PFS of the combination of camerelizumab and apatinib were 38.9% and 7.2 months, respectively [44]. However, there have been no updated reports on this combination. Moreover, the reported results of the combination of avelumab and axitinib [45] were slightly inferior to those of other combination therapies (ORR, 13.6%; PFS, 5.5 months; and OS, 12.7 months, based on RECIST 1.1). Therefore, at present, the most promising ongoing trial is the LEAP-002 study [40,41]. The decision whether or not to proceed to phase III trials of the combination of nivolumab and lenvatinib has currently drawn attention. In any case, the efficacy of all other combinations of anti-PD-1/PD-L1 antibodies and TKIs or anti-VEGF antibodies, except for the combination of avelumab and axitinib, is higher than that of nivolumab (a PD-1 antibody) alone (ORR, 15%; DCR, 55%; PFS, 3.7 months; and OS, 16.4 months) [34] or pembrolizumab alone (ORR, 18.3%; DCR, 62.2%; PFS, 3.0 months; OS, 13.9 months) [36]. Therefore, combined immunotherapy is expected to shift the paradigm as a first-line treatment option in advanced hepatocellular carcinoma [41,46].

Table 1. Efficacy of Immune Checkpoint Inhibitors and Combination Immunotherapy with VEGF Antibodies/Tyrosine Kinase Inhibitors in Phase 1b Trials according to RECIST 1.1.

Efficacy	Anti-PD-1 Monotherapy (Phase 3 Trial)		Anti-PD-1/PD-L1 plus TKI/Anti-VEGF (Phase 1b Trial)				
	Nivolumab [34] ($n = 214$)	Pembrolizumab [36] ($n = 278$)	Atezolizumab + bevacizumab [33] ($n = 104$)	Pembrolizumab + Lenvatinib [42] ($n = 67$)	Camrelizumab + apatinib [44] ($n = 18$)	Avelumab + axitinib [45] ($n = 22$)	Nivolumab + Lenvatinib [43] ($n = 24$)
ORR (95% CI)	15%	18.3% (14.0–23.4)	36% (26–46)	40.3% (28.5–53.0)	38.9%	13.6% (2.9–34.9)	54.2% (32.8–74.4)
DCR (95% CI)	55%	62.2%	71%	85.1% (74.3–92.6)	83.3%	68.2% (45.1–86.1)	91.7% (73.0–99.0)
PFS, months (95% CI)	3.7 (3.1–3.9)	3.0 (2.8–4.1)	7.4 (5.6–10.7)	9.7 (5.3–13.8)	7.2 (2.6–NE)	5.5 (1.9–7.4)	7.4 (3.7–NE)
OS, months (95% CI)	16.4 (13.9–18.4)	13.9 (11.6–16.0)	17.1 (13.8–NE)	20.4 (11.0–NE)	NR	12.7 (8.0–NE)	NR
DOR, months (M)	23.3 (3.1–34.5+)	13.8 (1.5–23.6)	NE (11.7–NE)	11.0 (5.6–11.0)	NA	5.5 (3.7–7.3)	NA

DCR, disease control rate; DOR, duration of response; NA, not available; NE; not evaluable; NR, not reached; ORR, objective response rate (RECIST 1.1); OS, overall survival; PFS, progression-fee survival. TKI, tyrosine kinase inhibitor.

6. Conclusions

This article described the scientific rationale for the combination of PD-1/PD-L1 antibodies plus VEGF inhibitors, and discussed the results of a phase Ib trial of this combination. We also described the results of Arm F of a randomized phase Ib trial of the combination of atezolizumab and bevacizumab, a combination that also achieved positive results in the phase III IMbrave150 study. The results of the phase Ib trial (Arm F) and the success of the phase III IMbrave150 study suggest that the tumor microenvironment was changed by bevacizumab, enabling greater responses to the immune checkpoint blockade, as hypothesized. In addition to the improvement in PFS, in the phase III IMbrave150 study, the OS was also improved, which was an unexpected finding [1]. These results are paradigm-changing as well as practice-changing. This study suggested that the immunosuppressive tumor microenvironment was successfully reprogrammed into an immunostimulatory microenvironment that was responsive to an immune checkpoint blockade. Therefore, the promising results that have been reported with combinations of anti-PD-1/PD-L1 antibodies and VEGF inhibitors (bevacizumab or TKIs) may be due to a normalization of the tumor microenvironment. In addition to the combination of atezolizumab and bevacizumab, therapies with other combinations targeting the same pathways (Table 1), especially the combinations of penbrolizumab and lenvatinib (the LEAP-002 study) and atezolizumab and cabozantinib (the COSMIC-312 trial), are highly promising (Figure 6 and Table 2) [1,34,36,47–65]. Furthermore, other phase III trials of combinations with CTLA-4 inhibitors [66] (durvalumab plus tremelimumab [HIMALAYA study] and nivolumab plus ipilimumab [the CheckMate 9DW study]) are currently being conducted (Figure 1 and Table 2). In the era of combination immunotherapy, the treatment of hepatocellular carcinoma, including the proper use of molecular targeted drugs after progression on immunotherapy [67,68], has entered a period of a major paradigm shift.

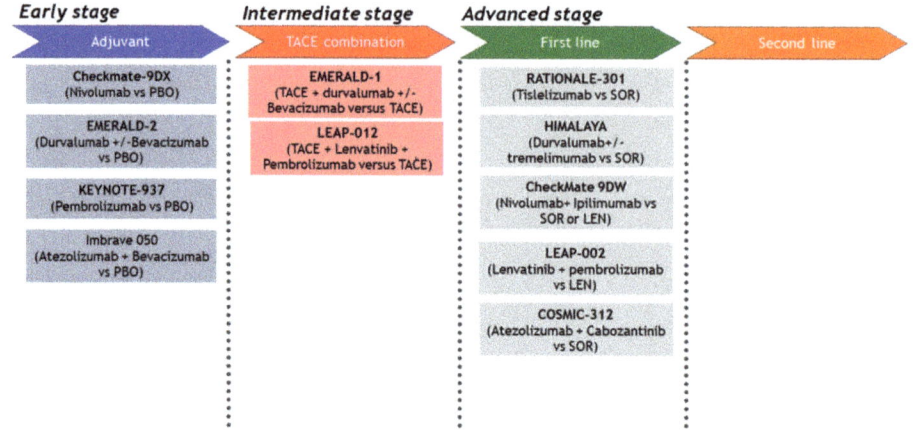

Figure 6. Ongoing Phase III trials in HCC (original Figure).

Table 2. Phase III Clinical Trials of Advanced Stage HCC.

Target Population		Design	Trial Name	Result	Presentation	Publication	1st Author
Advanced	First line	1. Sorafenib vs. Sunitinib	SUN1170	Negative	ASCO 2011	JCO 2013	Cheng AL [47]
		2. Sorafenib ± Erlotinib	SEARCH	Negative	ESMO 2012	JCO 2015	Zhu AX [48]
		3. Sorafenib vs. Brivanib	BRISK-FL	Negative	AASLD 2012	JCO 2013	Johnson PJ [49]
		4. Sorafenib vs. Linifanib	LiGHT	Negative	ASCO-GI 2013	JCO 2015	Cainap C [50]
		5. Sorafenib ± Doxorubicin	CALGB 80802	Negative	ASCO-GI 2016		
		6. Sorafenib ±- HAIC	SILIUS	Negative	EASL 2016	Lancet GH 2018	Kudo M [51]
		7. Sorafenib ± Y90	SARAH	Negative	EASL 2017	Lancet-O 2017	Vilgrain V [52]
		8. Sorafenib ± Y90	SIRveNIB	Negative	ASCO 2017	JCO 2018	Chow PKH [53]
		9. Sorafenib vs. Lenvatinib	REFLECT	Positive	ASCO 2017	Lancet 2018	Kudo M [54]
		10. Sorafenib vs. Nivolumab	CheckMate-459	Negative	ESMO 2019		Yau T [34]
		11. Sorafenib ± Y90	SORAMIC	Negative	EASL 2018	J Hepatol 2019	Ricke J [55]
		12. Sorafenib vs. Atezolizumab + Bevacizumab	IMbrave150	Positive	ESMO-Asia 2019		Cheng AL [1]
		13. Sorafenib vs. Durvalumab +Tremelimumab vs. Durva	HIMALAYA	Ongoing			
		14. Sorafenib vs. Tislelizumab	Rationale301	Ongoing			
		15. Lenvatinib ± Pembrolizumab	LEAP002	Ongoing			
		16. Lenvatinib or Sorafenib vs. Nivolumab + Ipilimumab	CheckMate 9DW	Ongoing			
		17. Sorafenib vs. Atezolizumab + Cabozantinib	COSMIC-312	Ongoing			
	Second line	1. Brivanib vs. Placebo	BRISK-PS	Negative	EASL 2012	JCO 2013	Llovet JM [56]
		2. Everolimus vs. Placebo	EVOLVE-1	Negative	ASCO-GI 2014	JAMA 2014	Zhu AX [57]
		3. Ramucirumab vs. Placebo	REACH	Negative	ESMO 2014	Lancet-O 2015	Zhu AX [58]
		4. S-1 vs. Placebo	S-CUBE	Negative	ASCO 2015	Lancet GH 2017	Kudo M [59]
		5. ADI-PEG 20 vs. Placebo	NA	Negative	ASCO 2016	Ann Oncol 2018	Abou-Alfa GK [60]
		6. Regorafenib vs. Placebo	RESORCE	Positive	WCGC 2016	Lancet 2017	Bruix J [61]
		7. Tivantinib vs. Placebo	METIV-HCC	Negative	ASCO 2017	Lancet-O 2018	Rimassa L [62]
		8. Tivantinib vs. Placebo	JET-HCC	Negative	ESMO 2017		
		9. DT# vs. Placebo	ReLive	Negative	ILCA 2017	Lancet Gastroenterol Hepatol	Merle P [63]
		10. Cabozantinib vs. Placebo	CELESTIAL	Positive	ASCO-GI 2018	NEJM 2018	Abou-Alfa G [64]
		11. Ramucirumab vs. Placebo	REACH-2	Positive	ASCO 2018	Lancet-O 2019	Zhu AX [65]
		12. Pembrolizumab vs. Placebo	KEYNOTE-240	Negative	ASCO 2019	JCO 2020	Finn RS [36]

Red: Positive trials, Blue: Ongoing trials, Black: Negative trials.

Funding: This research received no external funding.

Conflicts of Interest: Masatoshi Kudo has received grants from Taiho Pharmaceuticals, Chugai Pharmaceuticals, Otsuka, Takeda, Sumitomo Dainippon-Sumitomo, Daiichi Sankyo, AbbVie, Astellas Pharma, and Bristol-Myers Squibb. He has also received grants and personal fees from Merck Sharpe and Dohme (MSD), Eisai, and Bayer, and is an adviser for MSD, Eisai, Bayer, Bristol-Myers Squibb, Eli Lilly, and ONO Pharmaceuticals.

References

1. Cheng, A.L.; Qin, S.; Ikeda, M.; Galle, P.R.; Ducreux, M.; Zhu, A.X.; Kim, T.Y.; Kudo, M.; Breder, V.; Merle, P.; et al. IMbrave150: Efficacy and safety results from a ph III study evaluating atezolizumab (atezo) + bevacizumab (bev) vs sorafenib (Sor) as first treatment (tx) for patients (pts) with unresectable hepatocellular carcinoma (HCC). *Ann. Oncol.* **2019**, *30* (Suppl. 9), ix186–ix187. [CrossRef]
2. Fukumura, D.; Kloepper, J.; Amoozgar, Z.; Duda, D.G.; Jain, R.K. Enhancing cancer immunotherapy using antiangiogenics: Opportunities and challenges. *Nat. Rev. Clin. Oncol.* **2018**, *15*, 325–340. [CrossRef]
3. Chouaib, S.; Messai, Y.; Couve, S.; Escudier, B.; Hasmim, M.; Noman, M.Z. Hypoxia promotes tumor growth in linking angiogenesis to immune escape. *Front. Immunol.* **2012**, *3*, 21. [CrossRef] [PubMed]
4. Voron, T.; Marcheteau, E.; Pernot, S.; Colussi, O.; Tartour, E.; Taieb, J.; Terme, M. Control of the immune response by pro-angiogenic factors. *Front. Oncol.* **2014**, *4*, 70. [CrossRef] [PubMed]
5. Chen, D.S.; Mellman, I. Oncology meets immunology: The cancer-immunity cycle. *Immunity* **2013**, *39*, 1–10. [CrossRef]
6. Hegde, P.S.; Wallin, J.J.; Mancao, C. Predictive markers of anti-VEGF and emerging role of angiogenesis inhibitors as immunotherapeutics. *Semin. Cancer Biol.* **2018**, *52*, 117–124. [CrossRef]
7. Ferrara, N.; Hillan, K.J.; Gerber, H.P.; Novotny, W. Discovery and development of bevacizumab, an anti-VEGF antibody for treating cancer. *Nat. Rev. Drug Discov.* **2004**, *3*, 391–400. [CrossRef]
8. Gabrilovich, D.I.; Chen, H.L.; Girgis, K.R.; Cunningham, H.T.; Meny, G.M.; Nadaf, S.; Kavanaugh, D.; Carbone, D.P. Production of vascular endothelial growth factor by human tumors inhibits the functional maturation of dendritic cells. *Nat. Med.* **1996**, *2*, 1096–1103. [CrossRef]
9. Gabrilovich, D.; Ishida, T.; Oyama, T.; Ran, S.; Kravtsov, V.; Nadaf, S.; Carbone, D.P. Vascular endothelial growth factor inhibits the development of dendritic cells and dramatically affects the differentiation of multiple hematopoietic lineages in vivo. *Blood* **1998**, *92*, 4150–4166. [CrossRef]
10. Elovic, A.E.; Ohyama, H.; Sauty, A.; McBride, J.; Tsuji, T.; Nagai, M.; Weller, P.F.; Wong, D.T. IL-4-dependent regulation of TGF-alpha and TGF-beta1 expression in human eosinophils. *J. Immunol.* **1998**, *160*, 6121–6127.
11. Guermonprez, P.; Valladeau, J.; Zitvogel, L.; Thery, C.; Amigorena, S. Antigen presentation and T cell stimulation by dendritic cells. *Annu. Rev. Immunol.* **2002**, *20*, 621–667. [CrossRef] [PubMed]
12. Villadangos, J.A.; Schnorrer, P. Intrinsic and cooperative antigen-presenting functions of dendritic-cell subsets in vivo. *Nat. Rev. Immunol.* **2007**, *7*, 543–555. [CrossRef] [PubMed]
13. Griffioen, A.W.; Damen, C.A.; Blijham, G.H.; Groenewegen, G. Tumor angiogenesis is accompanied by a decreased inflammatory response of tumor-associated endothelium. *Blood* **1996**, *88*, 667–673. [CrossRef] [PubMed]
14. Griffioen, A.W.; Damen, C.A.; Martinotti, S.; Blijham, G.H.; Groenewegen, G. Endothelial intercellular adhesion molecule-1 expression is suppressed in human malignancies: The role of angiogenic factors. *Cancer Res.* **1996**, *56*, 1111–1117. [PubMed]
15. Goel, S.; Duda, D.G.; Xu, L.; Munn, L.L.; Boucher, Y.; Fukumura, D.; Jain, R.K. Normalization of the vasculature for treatment of cancer and other diseases. *Physiol. Rev.* **2011**, *91*, 1071–1121. [CrossRef] [PubMed]
16. Motz, G.T.; Santoro, S.P.; Wang, L.P.; Garrabrant, T.; Lastra, R.R.; Hagemann, I.S.; Lal, P.; Feldman, M.D.; Benencia, F.; Coukos, G. Tumor endothelium FasL establishes a selective immune barrier promoting tolerance in tumors. *Nat. Med.* **2014**, *20*, 607–615. [CrossRef]
17. Hodi, F.S.; Lawrence, D.; Lezcano, C.; Wu, X.; Zhou, J.; Sasada, T.; Zeng, W.; Giobbie-Hurder, A.; Atkins, M.B.; Ibrahim, N.; et al. Bevacizumab plus ipilimumab in patients with metastatic melanoma. *Cancer Immunol. Res.* **2014**, *2*, 632–642. [CrossRef]
18. Wallin, J.J.; Bendell, J.C.; Funke, R.; Sznol, M.; Korski, K.; Jones, S.; Hernandez, G.; Mier, J.; He, X.; Hodi, F.S.; et al. Atezolizumab in combination with bevacizumab enhances antigen-specific T-cell migration in metastatic renal cell carcinoma. *Nat. Commun.* **2016**, *7*, 12624. [CrossRef]

19. Gabrilovich, D.I.; Nagaraj, S. Myeloid-derived suppressor cells as regulators of the immune system. *Nat. Rev. Immunol.* **2009**, *9*, 162–174. [CrossRef]
20. Huang, C.; Li, J.; Song, L.; Zhang, D.; Tong, Q.; Ding, M.; Bowman, L.; Aziz, R.; Stoner, G.D. Black raspberry extracts inhibit benzo(a)pyrene diol-epoxide-induced activator protein 1 activation and VEGF transcription by targeting the phosphotidylinositol 3-kinase/Akt pathway. *Cancer Res.* **2006**, *66*, 581–587. [CrossRef]
21. Ko, S.Y.; Guo, H.; Barengo, N.; Naora, H. Inhibition of ovarian cancer growth by a tumor-targeting peptide that binds eukaryotic translation initiation factor 4E. *Clin. Cancer Res. Off. J. Am. Assoc. Cancer Res.* **2009**, *15*, 4336–4347. [CrossRef] [PubMed]
22. Kusmartsev, S.; Eruslanov, E.; Kubler, H.; Tseng, T.; Sakai, Y.; Su, Z.; Kaliberov, S.; Heiser, A.; Rosser, C.; Dahm, P.; et al. Oxidative stress regulates expression of VEGFR1 in myeloid cells: Link to tumor-induced immune suppression in renal cell carcinoma. *J. Immunol.* **2008**, *181*, 346–353. [CrossRef] [PubMed]
23. Herbst, R.S.; Soria, J.C.; Kowanetz, M.; Fine, G.D.; Hamid, O.; Gordon, M.S.; Sosman, J.A.; McDermott, D.F.; Powderly, J.D.; Gettinger, S.N.; et al. Predictive correlates of response to the anti-PD-L1 antibody MPDL3280A in cancer patients. *Nature* **2014**, *515*, 563–567. [CrossRef] [PubMed]
24. Chen, X.; Zeng, Q.; Wu, M.X. Improved efficacy of dendritic cell-based immunotherapy by cutaneous laser illumination. *Clin. Cancer Res. Off. J. Am. Assoc. Cancer Res.* **2012**, *18*, 2240–2249. [CrossRef] [PubMed]
25. Zou, W.; Chen, L. Inhibitory B7-family molecules in the tumour microenvironment. *Nat. Rev. Immunol.* **2008**, *8*, 467–477. [CrossRef] [PubMed]
26. Tiegs, G.; Lohse, A.W. Immune tolerance: What is unique about the liver. *J Autoimmun.* **2010**, *34*, 1–6. [CrossRef]
27. Oyama, T.; Ran, S.; Ishida, T.; Nadaf, S.; Kerr, L.; Carbone, D.P.; Gabrilovich, D.I. Vascular endothelial growth factor affects dendritic cell maturation through the inhibition of nuclear factor-kappa B activation in hemopoietic progenitor cells. *J. Immunol.* **1998**, *160*, 1224–1232.
28. Kato, Y.; Tabata, K.; Kimura, T.; Yachie-Kinoshita, A.; Ozawa, Y.; Yamada, K.; Ito, J.; Tachino, S.; Hori, Y.; Matsuki, M.; et al. Lenvatinib plus anti-PD-1 antibody combination treatment activates CD8+ T cells through reduction of tumor-associated macrophage and activation of the interferon pathway. *PLoS ONE* **2019**, *14*, e0212513. [CrossRef]
29. Kudo, M. Combination Cancer Immunotherapy in Hepatocellular Carcinoma. *Liver Cancer* **2018**, *7*, 20–27. [CrossRef]
30. Kudo, M. Targeted and immune therapies for hepatocellular carcinoma: Predictions for 2019 and beyond. *World J. Gastroenterol.* **2019**, *25*, 789–807. [CrossRef]
31. Kudo, M. Combination Cancer Immunotherapy with Molecular Targeted Agents/Anti-CTLA-4 Antibody for Hepatocellular Carcinoma. *Liver Cancer* **2019**, *8*, 1–11. [CrossRef] [PubMed]
32. Teng, M.W.; Ngiow, S.F.; Ribas, A.; Smyth, M.J. Classifying Cancers Based on T-cell Infiltration and PD-L1. *Cancer Res.* **2015**, *75*, 2139–2145. [CrossRef] [PubMed]
33. Hsu, C.H.; Lee, M.S.; Lee, K.H.; Numata, K.; Stein, S.; Verret, W.; Hack, S.; Spahn, J.; Liu, B.; Huang, C.; et al. Randomised efficacy and safety results for atezolizumab + bevacizumab in patients with previously untreated, unresectable hepatocellular carcinoma. *Ann. Oncol.* **2019**, *30* (Suppl. 9), ix187. [CrossRef]
34. Yau, T.; Park, J.W.; Finn, R.S.; Cheng, A.L.; Mathurin, P.; Edeline, J.; Kudo, M.; Han, K.H.; Harding, J.J.; Merle, P.; et al. CheckMate 459: A randomized, multi-center phase III study of nivolumab vs sorafenib as first-line treatment in patients with advanced hepatocellular carcinoma. *Ann. Oncol.* **2019**, *30* (Suppl. 5), v874–v875. [CrossRef]
35. Zhu, A.X.; Finn, R.S.; Edeline, J.; Cattan, S.; Ogasawara, S.; Palmer, D.; Verslype, C.; Zagonel, V.; Fartoux, L.; Vogel, A.; et al. Pembrolizumab in patients with advanced hepatocellular carcinoma previously treated with sorafenib (KEYNOTE-224): A non-randomised, open-label phase 2 trial. *Lancet Oncol.* **2018**, *19*, 940–952. [CrossRef]
36. Finn, R.S.; Ryoo, B.Y.; Merle, P.; Kudo, M.; Bouattour, M.; Lim, H.Y.; Breder, V.; Edeline, J.; Chao, Y.; Ogasawara, S.; et al. Pembrolizumab As Second-Line Therapy in Patients With Advanced Hepatocellular Carcinoma in KEYNOTE-240: A Randomized, Double-Blind, Phase III Trial. *J. Clin. Oncol. Off. J. Am. Soc. Clin. Oncol.* **2020**, *38*, 193–202. [CrossRef]
37. Kudo, M. Pembrolizumab for the Treatment of Hepatocellular Carcinoma. *Liver Cancer* **2019**, *8*, 143–154. [CrossRef]
38. Kudo, M. Immune checkpoint blockade in hepatocellular carcinoma: 2017 update. *Liver Cancer* **2017**, *6*, 1–12. [CrossRef]

39. El-Khoueiry, A.B.; Sangro, B.; Yau, T.; Crocenzi, T.S.; Kudo, M.; Hsu, C.; Kim, T.Y.; Choo, S.P.; Trojan, J.; Welling, T.H.R.; et al. Nivolumab in patients with advanced hepatocellular carcinoma (CheckMate 040): An open-label, non-comparative, phase 1/2 dose escalation and expansion trial. *Lancet* **2017**, *389*, 2492–2502. [CrossRef]
40. Llovet, J.M.; Kudo, M.; Cheng, A.L.; Finn, R.S.; Galle, P.R.; Kaneko, S.; Meyer, T.; Qin, S.; dutcus, C.E.; Chen, E.; et al. First-Line Combination Therapy With Lenvatinib Plus Pembrolizumab for Patients with Advanced Hepatocellular Carcinoma: Phase 3 LEAP-002 Study. In Proceedings of the ILCA 13th Annual Conference, Chicago, IL, USA, 20–22 September 2019.
41. Kudo, M. Immuno-Oncology Therapy for Hepatocellular Carcinoma: Current Status and Ongoing Trials. *Liver Cancer* **2019**, *8*, 221–238. [CrossRef]
42. Llovet, J.M.; Finn, R.S.; Ikeda, M.; Sung, M.W.; Baron, A.D.; Kudo, M.; Okusaka, T.; Kobayashi, M.; Kumada, H.; Kaneko, S.; et al. A phase 1b trial of lenvatinib plus pembrolizumab in unresectable hepatocellular carcinoma: Upated results. *Ann. Oncol.* **2019**, *30* (Suppl. 5), v253–v324. [CrossRef]
43. Kudo, M.; Ikeda, K.; Motomura, K.; Okusaka, T.; Kato, N.; Dutcus, C.E.; Hisai, T.; Suzuki, M.; Ikezawa, H.; Iwata, T.; et al. A Phase 1b Study of Lenvatinib Plus Nivolumab in Patients With Unresectable Hepatocellular Carcinoma. In Proceedings of the ASCO-GI, San Francisco, CA, USA, 23–25 January 2020.
44. Xu, J.M.; Zhang, Y.; Jia, R.; Wang, Y.; Liu, R.; Zhang, G.; Zhao, C.; Zhang, Y.; Zhou, J.; Wang, Q. Anti-programmed death-1 antibody SHR-1210 (S) combined with apatinib (A) for advanced hepatocellular carcinoma (HCC), gastric cancer (GC) or esophagogastric junction (EGJ) cancer refractory to standard therapy: A phase 1 trial. *J. Clin. Oncol.* **2018**, *36*, 4075. [CrossRef]
45. Kudo, M.; Motomura, K.; Wada, Y.; Inaba, Y.; Sakamoto, Y.; Kurosaki, M.; Umeyama, Y.; Kamei, Y.; Yoshimitsu, J.; Fujii, Y.; et al. First-line avelumab + axitinib in patients with advanced hepatocellular carcinoma: Results from a phase 1b trial (VEGF Liver 100). *J. Clin. Oncol.* **2019**, *37* (Suppl. 15). [CrossRef]
46. Cheng, A.L.; Hsu, C.; Chan, S.L.; Choo, S.P.; Kudo, M. Challenges of combination therapy with immune checkpoint inhibitors for hepatocellular carcinoma. *J. Hepatol.* **2020**, *72*, 307–319. [CrossRef] [PubMed]
47. Cheng, A.L.; Kang, Y.K.; Lin, D.Y.; Park, J.W.; Kudo, M.; Qin, S.; Chung, H.C.; Song, X.; Xu, J.; Poggi, G.; et al. Sunitinib versus sorafenib in advanced hepatocellular cancer: results of a randomized phase III trial. *J. Clin. Oncol.* **2013**, *31*, 4067–4075. [CrossRef]
48. Zhu, A.X.; Rosmorduc, O.; Evans, T.R.; Ross, P.J.; Santoro, A.; Carrilho, F.J.; Bruix, J.; Qin, S.; Thuluvath, P.J.; Llovet, J.M.; et al. SEARCH: a phase III, randomized, double-blind, placebo-controlled trial of sorafenib plus erlotinib in patients with advanced hepatocellular carcinoma. *J. Clin. Oncol.* **2015**, *33*, 559–566. [CrossRef] [PubMed]
49. Johnson, P.J.; Qin, S.; Park, J.W.; Poon, R.T.; Raoul, J.L.; Philip, P.A.; Hsu, C.H.; Hu, T.H.; Heo, J.; Xu, J.; et al. Brivanib versus sorafenib as first-line therapy in patients with unresectable, advanced hepatocellular carcinoma: results from the randomized phase III BRISK-FL study. *J. Clin. Oncol.* **2013**, *31*, 3517–3524. [CrossRef]
50. Cainap, C.; Qin, S.; Huang, W.T.; Chung, I.J.; Pan, H.; Cheng, Y.; Kudo, M.; Kang, Y.K.; Chen, P.J.; Toh, H.C.; et al. Linifanib versus Sorafenib in patients with advanced hepatocellular carcinoma: results of a randomized phase III trial. *J. Clin. Oncol.* **2015**, *33*, 172–179. [CrossRef]
51. Kudo, M.; Ueshima, K.; Yokosuka, O.; Ogasawara, S.; Obi, S.; Izumi, N.; Aikata, H.; Nagano, H.; Hatano, E.; Sasaki, Y.; et al. Sorafenib plus low-dose cisplatin and fluorouracil hepatic arterial infusion chemotherapy versus sorafenib alone in patients with advanced hepatocellular carcinoma (SILIUS): a randomised, open label, phase 3 trial. *Lancet Gastroenterol. Hepatol.* **2018**, *3*, 424–432. [CrossRef]
52. Vilgrain, V.; Pereira, H.; Assenat, E.; Guiu, B.; Ilonca, A.D.; Pageaux, G.P.; Sibert, A.; Bouattour, M.; Lebtahi, R.; Allaham, W.; et al. Efficacy and safety of selective internal radiotherapy with yttrium-90 resin microspheres compared with sorafenib in locally advanced and inoperable hepatocellular carcinoma (SARAH): an open-label randomised controlled phase 3 trial. *Lancet Oncol.* **2017**, *18*, 1624–1636. [CrossRef]
53. Chow, P.K.H.; Gandhi, M.; Tan, S.B.; Khin, M.W.; Khasbazar, A.; Ong, J.; Choo, S.P.; Cheow, P.C.; Chotipanich, C.; Lim, K.; et al. SIRveNIB: Selective Internal Radiation Therapy Versus Sorafenib in Asia-Pacific Patients With Hepatocellular Carcinoma. *J. Clin. Oncol.* **2018**, *36*, 1913–1921. [CrossRef] [PubMed]
54. Kudo, M.; Finn, R.S.; Qin, S.; Han, K.H.; Ikeda, K.; Piscaglia, F.; Baron, A.; Park, J.W.; Han, G.; Jassem, J.; et al. Lenvatinib versus sorafenib in first-line treatment of patients with unresectable hepatocellular carcinoma: a randomised phase 3 non-inferiority trial. *Lancet* **2018**, *391*, 1163–1173. [CrossRef]

55. Ricke, J.; Klumpen, H.J.; Amthauer, H.; Bargellini, I.; Bartenstein, P.; de Toni, E.N.; Gasbarrini, A.; Pech, M.; Peck-Radosavljevic, M.; Popovic, P.; et al. Impact of combined selective internal radiation therapy and sorafenib on survival in advanced hepatocellular carcinoma. *J. Hepatol.* **2019**, *71*, 1164–1174. [CrossRef] [PubMed]
56. Llovet, J.M.; Decaens, T.; Raoul, J.L.; Boucher, E.; Kudo, M.; Chang, C.; Kang, Y.K.; Assenat, E.; Lim, H.Y.; Boige, V.; et al. Brivanib in patients with advanced hepatocellular carcinoma who were intolerant to sorafenib or for whom sorafenib failed: results from the randomized phase III BRISK-PS study. *J. Clin. Oncol.* **2013**, *31*, 3509–3516. [CrossRef] [PubMed]
57. Zhu, A.X.; Kudo, M.; Assenat, E.; Cattan, S.; Kang, Y.K.; Lim, H.Y.; Poon, R.T.; Blanc, J.F.; Vogel, A.; Chen, C.L.; et al. Effect of everolimus on survival in advanced hepatocellular carcinoma after failure of sorafenib: the EVOLVE-1 randomized clinical trial. *JAMA* **2014**, *312*, 57–67. [CrossRef] [PubMed]
58. Zhu, A.X.; Park, J.O.; Ryoo, B.Y.; Yen, C.J.; Poon, R.; Pastorelli, D.; Blanc, J.F.; Chung, H.C.; Baron, A.D.; Pfiffer, T.E.; et al. Ramucirumab versus placebo as second-line treatment in patients with advanced hepatocellular carcinoma following first-line therapy with sorafenib (REACH): a randomised, double-blind, multicentre, phase 3 trial. *Lancet Oncol.* **2015**, *16*, 859–870. [CrossRef]
59. Kudo, M.; Moriguchi, M.; Numata, K.; Hidaka, H.; Tanaka, H.; Ikeda, M.; Kawazoe, S.; Ohkawa, S.; Sato, Y.; Kaneko, S.; et al. S-1 versus placebo in patients with sorafenib-refractory advanced hepatocellular carcinoma (S-CUBE): a randomised, double-blind, multicentre, phase 3 trial. *Lancet Gastroenterol. Hepatol.* **2017**, *2*, 407–417. [CrossRef]
60. Abou-Alfa, G.K.; Qin, S.; Ryoo, B.Y.; Lu, S.N.; Yen, C.J.; Feng, Y.H.; Lim, H.Y.; Izzo, F.; Colombo, M.; Sarker, D.; et al. Phase III randomized study of second line ADI-PEG 20 plus best supportive care versus placebo plus best supportive care in patients with advanced hepatocellular carcinoma. *Ann. Oncol.* **2018**, *29*, 1402–1408. [CrossRef]
61. Bruix, J.; Qin, S.; Merle, P.; Granito, A.; Huang, Y.H.; Bodoky, G.; Pracht, M.; Yokosuka, O.; Rosmorduc, O.; Breder, V.; et al. Regorafenib for patients with hepatocellular carcinoma who progressed on sorafenib treatment (RESORCE): a randomised, double-blind, placebo-controlled, phase 3 trial. *Lancet* **2017**, *389*, 56–66. [CrossRef]
62. Rimassa, L.; Assenat, E.; Peck-Radosavljevic, M.; Pracht, M.; Zagonel, V.; Mathurin, P.; Rota Caremoli, E.; Porta, C.; Daniele, B.; Bolondi, L.; et al. Tivantinib for second-line treatment of MET-high, advanced hepatocellular carcinoma (METIV-HCC): a final analysis of a phase 3, randomised, placebo-controlled study. *Lancet Oncol.* **2018**, *19*, 682–693. [CrossRef]
63. Merle, P.; Blanc, J.F.; Phelip, J.M.; Pelletier, G.; Bronowicki, J.P.; Touchefeu, Y.; Pageaux, G.; Gerolami, R.; Habersetzer, F.; Nguyen-Khac, E.; et al. Doxorubicin-loaded nanoparticles for patients with advanced hepatocellular carcinoma after sorafenib treatment failure (RELIVE): a phase 3 randomised controlled trial. *Lancet Gastroenterol. Hepatol.* **2019**, *4*, 454–465. [CrossRef]
64. Abou-Alfa, G.K.; Meyer, T.; Cheng, A.L.; El-Khoueiry, A.B.; Rimassa, L.; Ryoo, B.Y.; Cicin, I.; Merle, P.; Chen, Y.; Park, J.W.; et al. Cabozantinib in Patients with Advanced and Progressing Hepatocellular Carcinoma. *N. Engl. J. Med.* **2018**, *379*, 54–63. [CrossRef] [PubMed]
65. Zhu, A.X.; Kang, Y.K.; Yen, C.J.; Finn, R.S.; Galle, P.R.; Llovet, J.M.; Assenat, E.; Brandi, G.; Pracht, M.; Lim, H.Y.; et al. Ramucirumab after sorafenib in patients with advanced hepatocellular carcinoma and increased alpha-fetoprotein concentrations (REACH-2): a randomised, double-blind, placebo-controlled, phase 3 trial. *Lancet Oncol.* **2019**, *20*, 282–296. [CrossRef]
66. Kudo, M. Scientific rationale for combination immunotherapy of hepatocellular carcinoma with anti-PD-1/PD-L1 and anti-CTLA-4 antibodies. *Liver Cancer* **2019**, *8*, 413–426. [CrossRef]
67. Bouattour, M.; Mehta, N.; He, A.R.; Cohen, E.I.; Nault, J.C. Systemic Treatment for Advanced Hepatocellular Carcinoma. *Liver Cancer* **2019**, *8*, 341–358. [CrossRef] [PubMed]
68. Rimassa, L.; Pressiani, T.; Merle, P. Systemic Treatment Options in Hepatocellular Carcinoma. *Liver Cancer* **2019**, *8*, 427–446. [CrossRef]

© 2020 by the author. Licensee MDPI, Basel, Switzerland. This article is an open access article distributed under the terms and conditions of the Creative Commons Attribution (CC BY) license (http://creativecommons.org/licenses/by/4.0/).

Review

Recent Advances in Immunotherapy for Hepatocellular Carcinoma

Shigeharu Nakano †, Yuji Eso †, Hirokazu Okada, Atsushi Takai, Ken Takahashi * and Hiroshi Seno

Department of Gastroenterology and Hepatology, Graduate School of Medicine, Kyoto University, Kyoto 606-8501, Japan; shnakano@kuhp.kyoto-u.ac.jp (S.N.); yujieso@kuhp.kyoto-u.ac.jp (Y.E.); okada_hiro@kuhp.kyoto-u.ac.jp (H.O.); atsushit@kuhp.kyoto-u.ac.jp (A.T.); seno@kuhp.kyoto-u.ac.jp (H.S.)
* Correspondence: takaken@kuhp.kyoto-u.ac.jp; Tel.: +81-75-751-4302
† These authors contributed equally to this work.

Received: 8 March 2020; Accepted: 24 March 2020; Published: 25 March 2020

Abstract: Hepatocellular carcinoma (HCC) is one of the leading causes of cancer-related death since most patients are diagnosed at advanced stage and the current systemic treatment options using molecular-targeted drugs remain unsatisfactory. However, the recent success of cancer immunotherapies has revolutionized the landscape of cancer therapy. Since HCC is characterized by metachronous multicentric occurrence, immunotherapies that induce systemic and durable responses could be an appealing treatment option. Despite the suppressive milieu of the liver and tumor immunosurveillance escape mechanisms, clinical studies of checkpoint inhibitors in patients with advanced HCC have yielded promising results. Here, we provide an update on recent advances in HCC immunotherapies. First, we describe the unique tolerogenic properties of hepatic immunity and its interaction with HCC and then review the status of already or nearly available immune checkpoint blockade-based therapies as well as other immunotherapy strategies at the preclinical or clinical trial stage.

Keywords: hepatocellular carcinoma; immunotherapy; immune checkpoint inhibitor; PD-1; CTLA-4; combination therapy

1. Introduction

Hepatocellular carcinoma (HCC) is the most common type of primary liver cancer and poses a serious health problem worldwide [1]. Although various surveillance systems and treatment strategies have been developed and are recommended by guidelines, including surgical resection, radiofrequency ablation (RFA), transarterial chemoembolization (TACE), systemic therapy, and liver transplantation, the prognosis of HCC remains poor due to high levels of high intra- and extra-hepatic recurrence and metastasis [2,3]. Systemic therapies using molecular-targeted agents (MTAs) have been considered efficient and are recommended for patients with advanced-stage HCC [2,4]; however, the regimens currently available are often unsatisfactory. Therefore, a novel approach that uses a different mechanism to these conventional therapies is required to improve the prognosis of HCC.

The recent development of cancer immunotherapies using immune checkpoint inhibitors (ICIs) targeting cytotoxic T-lymphocyte-associated protein-4 (CTLA-4) and anti-programmed cell death protein-1 (PD-1) has dramatically changed the landscape of cancer therapy and was awarded the Nobel Prize in 2018. Several monoclonal antibodies (mAbs) targeting CTLA-4, PD-1, or its ligand programmed cell death-ligand 1 (PD-L1) have now been approved by the FDA for various types of cancers [5]. The liver is a tolerogenic organ [6] that is relevant to successful allograft acceptance after transplantation. Thus, the development of antitumor immunity against HCC might be speculated to be synergistically impeded by this tolerogenic nature of the liver and the immunosuppressive tumor microenvironment of HCC. However, the potential of cancer immunotherapy to induce systemic and durable antitumor responses may make it an ideal therapeutic option for HCC characterized by metachronous multicentric occurrence. Indeed, several ICI therapies targeting PD-1/PD-L1 and CTLA-4 have already demonstrated promising activity against HCC and manageable safety in clinical trials, thus have been approved by the FDA. Combination ICI-based strategies have also shown promising results, while other classes of immunotherapies have begun to emerge and are being tested in preclinical and clinical studies.

In this review, we first provide an overview of the unique intrinsic immunotolerant environment of the liver and the immune evasion mechanisms of HCC, and then review recent advances in different immunotherapy approaches and their combinations for treating HCC.

2. Tolerogenic Liver Immune Environment and HCC Immune Evasion Mechanisms

The liver is a tolerogenic organ in which a unique immune environment prevents the overactivation of the immune system to antigens derived from food and bacterial products in the portal flow [6]. Immune tolerance in the liver is induced by non-parenchymal cells. Kupffer cells (KCs) are liver-resident macrophages that play a role in pathogen clearance mediated by innate immune activation [7]. However, under physiological conditions, KCs induce tolerance by impairing T cell activation or preferentially expanding regulatory T cells (Tregs) by secreting immunosuppressive factors such as IL-10, TGF-β, and prostaglandin E2 [8,9]. Liver sinusoidal endothelial cells (LSECs), which act as antigen-presenting cells (APCs) and form a cellular barrier between the liver parenchyma and sinusoid [10], are characterized by low co-stimulatory molecule levels, high immune checkpoint molecule levels, and immunosuppressive cytokine production, all of which impede their potential for T cell activation and induce immune tolerance [11,12]. Hepatic dendritic cells (DCs) mediate the induction of T cell tolerance rather than their activation [13], presumably, as they are under the influence of IL-10 and TGF-β secreted by KCs and LSECs [14]. In addition to these non-parenchymal cells, hepatocytes also function as APCs by directly interacting with and presenting antigens to naïve T cells; however, hepatocytes predispose T cells towards tolerance because they lack co-stimulatory molecule expression [15]. Together, these immunosuppressive features of the liver might impede the development of antitumor immunity.

HCC evades host immunosurveillance via multiple mechanisms; for instance, HCC cells silence the expression of tumor antigens or antigen presentation-related molecules so that cytotoxic T cells (CTLs) cannot recognize tumor cells [16,17]. HCC cells also escape immunosurveillance by expressing immune checkpoint molecules such as PD-L1 and producing various immunoinhibitory molecules, including TGF-β, IL-10, indoleamine 2, 3-dioxygenase, arginase, and adenosine [18,19]. Immunosuppressive stromal cells are also a critical component of immune dysregulation. Myeloid-derived suppressor cells (MDSCs) are a heterogeneous population of immature myeloid cells that inhibit T cell activation via iNOS, ROS, and increased arginase activity, and induce Treg expansion by producing IL-10 and TGF-β [20]. Moreover, the frequency of MDSCs in HCC patients has been reported to correlate with tumor progression [21]. Macrophages are generally categorized as having an M1 or M2 state; M1 macrophages display an antitumor phenotype by producing high and low levels of IL-12 and IL-10, respectively, whereas M2 macrophages exhibit a tumor supportive phenotype with opposite cytokine profiles. During HCC progression, hepatic macrophages are skewed from an M1 phenotype to an M2 phenotype characteristic of tumor-associated macrophages (TAMs), which act as immune suppressor cells and support tumor growth by promoting angiogenesis and tumor invasion [22]. Tregs can also impede immune surveillance against HCC due to their immunosuppressive functions; indeed, they have been shown to densely infiltrate the tumor site in patients with HCC, with the number of intratumoral Tregs acting as an independent prognostic factor of overall survival (OS) and disease-free survival (DFS) in those patients [23].

Even under these immunosuppressive conditions, several studies have shown that antitumor immunity exists in patients with HCC. For instance, T cells specific for four different tumor-associated antigens (TAAs) were detected in both the tumor tissue and peripheral blood of patients with HCC, with the breadth of T cell response correlating with survival [24]. Another study found that the intratumoral density of activated CTLs in patients with HCC after resection was associated with OS and that the intratumoral balance between CTLs and Tregs was associated with OS and DFS [25]. These observations suggest that the immunogenic potential of HCC could be controlled by optimized immunotherapy.

3. PD-1/PD-L1 and CTLA-4-Blockade Therapies

3.1. Basic Immunobiology of PD-1 and CTLA-4

Immune checkpoint molecules—among which, PD-1 and CTLA-4 are the best studied—play essential roles in preventing T cell overactivation by interacting with APCs and other cell types. PD-1 is a member of the CD28 family that is expressed on activated T cells, B cells, and myeloid cells and negatively regulates the immune system. The engagement of PD-1 by its ligand PD-L1 leads to the transmission of suppressive signals into T cells and the induction of peripheral tolerance [26]. In the liver, PD-L1 is constitutively expressed on liver non-parenchymal cells such as LSECs and KCs [27]; however, PD-L1 is aberrantly expressed in various tumors, including HCC tumor cells, allowing them to escape from host immune surveillance. Indeed, it has been demonstrated that tumor PD-L1 expression is associated with HCC prognosis after curative surgical treatment, suggesting that the PD-1/PD-L1 pathway is an immune escape mechanism in HCC [19]. Another member of the CD28 family, CTLA-4, is induced on naïve T cells by antigen activation but is constitutively expressed on Tregs [28]. CTLA-4 binds to CD80 and CD86 more tightly than CD28, which provides a positive signal required for T cell activation; therefore, CTLA-4 induces peripheral tolerance by counteracting CD28-mediated costimulatory signals [28]. Importantly, the expression of CTLA-4 on Tregs depletes APCs of CD80 and CD86, leaving them with a reduced ability to prime naïve T cells [28]. The intensive study of PD-1- and CTLA-4-mediated immunosuppression culminated in the dramatic success of cancer immunotherapies [29] and many clinical trials of ICI mono- and combination therapies targeting PD-1/PD-L1 and CTLA-4 in HCC have now been conducted.

3.2. ICI Monotherapies Directed Against PD-1 and CTLA-4

Many clinical trials have been conducted for ICI monotherapies in HCC (Table 1) and the first to be approved by the FDA was the anti-PD-1 mAb nivolumab. A phase I/II trial of nivolumab in patients with advanced HCC (CheckMate-040) showed promising results. In the dose-expansion phase in which a total of 214 patients in 4 cohorts were enrolled, the objective response rate (ORR) was 20%, the disease control rate (DCR) was 64%, and progression free survival (PFS) was 4.1 months [30]. Since adverse events (AEs) were fairly mild [30], nivolumab was approved by the FDA in September 2017 as a second-line treatment for unresectable HCC after sorafenib failure, based on subgroup analysis in CheckMate-040 [4]. However, a phase III trial (CheckMate-459) evaluating nivolumab versus sorafenib as first-line treatments in patients with unresectable HCC revealed that the trial did not achieve statistical significance for its primary OS endpoint as per the prespecified analysis [31]. The CheckMate-9DX trial is currently evaluating adjuvant nivolumab versus a placebo in HCC patients at high risk of recurrence after curative hepatic resection or ablation.

Pembrolizumab is another anti-PD-1 mAb that was granted accelerated approval by the FDA in May 2017 for patients with unresectable or metastatic microsatellite instability-high (MSI-H) or mismatch repair deficient (dMMR) solid tumors that continued to progress after conventional treatment, based on the data from five clinical trials [32]. A phase II trial (KEYNOTE-224) revealed the potential of pembrolizumab against HCC after sorafenib failure, with an ORR of 17% with one complete response (CR), a DCR of 61%, and AEs (>grade 3) reported in 26% of patients [33]. Based on this data, pembrolizumab was granted accelerated approval by the FDA in November 2018 as a second-line treatment after sorafenib. A phase III trial (KEYNOTE-240) comparing pembrolizumab to a placebo as a second-line treatment demonstrated that pembrolizumab was associated with a longer median OS and PFS; however, these findings were not deemed statistically significant according to the prespecified statistical plan [34]. Two further phase III trials are currently ongoing: KEYNOTE-394 is evaluating pembrolizumab versus a placebo and best supportive care in Asian patients with systemically treated advanced HCC, while KEYNOTE-937 is evaluating pembrolizumab versus a placebo as an adjuvant therapy in HCC patients after curative treatment.

In addition, the anti-PD-L1 mAb Durvalumab was tested in a phase I/II trial (NCT01693562) of patients with advanced HCC who had been previously treated with sorafenib, achieving an OS rate of 10.3% in 39 patients [35]. The investigational IgG4 anti-PD-1 Ab, tislelizumab (BGB-A317), was designed to bind minimally to FcγR on macrophages in order to abrogate antibody-dependent phagocytosis, which is a potential mechanism of anti-PD-1 therapy resistance. Tislelizumab has demonstrated a good preliminary safety profile and antitumor activity in a phase I trial and a phase III trial (RATIONALE 301) of tislelizumab versus sorafenib as a first-line treatment in patients with unresectable HCC is currently underway [36].

The anti-CTLA-4 mAb, tremelimumab, has been tested in a small phase II pilot trial (NCT01008358) of HCV-infected patients with advanced HCC, demonstrating partial response (PR) and stable disease (SD) rates of 17.6 and 58.8%, respectively. Moreover, the treatment was well tolerated, and no patients needed steroids due to severe immune-related AEs (irAEs) [37].

Table 1. Summary of clinical trials of ICI monotherapy for HCC.

Trial identifier	Target	Drugs	Phase	N	Patient Group	ORR	DCR	PFS (Median, mo)	OS (Median, mo)
NCT01658878 (CheckMate040) [30]	PD-1	Nivolumab	I/II	214*	Naive/Pre-treated	20.0%	64.0%	4	NR
NCT02576509 (CheckMate459) [31]	PD-1	Nivolumab vs. Sorafenib	III	743	Naïve	15% vs. 7%	N/A	3.7 vs. 3.8	16.4 vs. 14.7
NCT03383458 (CheckMate 9DX)	PD-1	Nivolumab vs. Placebo	III	530	Adjuvant	N/A	N/A	N/A	N/A
NCT02702414 (KEYNOTE-224) [33]	PD-1	Pembrolizumab	II	104	Pre-treated	17.0%	61.0%	4.9	12.9
NCT02702401 (KEYNOTE-240) [34]	PD-1	Pembrolizumab vs. Placebo	III	413	Pre-treated	18.3% vs. 4.4%	62.2% vs. 53.3%	3.0 vs. 2.8	13.9 vs. 10.6
NCT03062358 (KEYNOTE-394)	PD-1	Pembrolizumab vs. Placebo	III	N/A	Pre-treated	N/A	N/A	N/A	N/A
NCT03867084 (KEYNOTE-937)	PD-1	Pembrolizumab vs. Placebo	III	N/A	Adjuvant	N/A	N/A	N/A	N/A
NCT03412773 (RATIONALE-301) [36]	PD-1	Tislelizumab vs. Sorafenib	III	N/A	Naïve	N/A	N/A	N/A	N/A
NCT01693562 [35]	PD-L1	Durvalumab	I/II	39	Pre-treated	10.3%	33.3%	NA	13.2
NCT01008358 [37]	CTLA-4	Tremelimumab	II	20	Pre-treated	17.6%	76.4%	6.48	8.2

N, number of patients; N/A; not available; NR, not reached; * dose-expansion phase.

4. ICI-Based Combination Therapy

Although ICI monotherapy regimens have shown benefits in some HCC patients with generally acceptable AE profiles, their response rates (approximately 20%) have been unsatisfactory, presumably due to the immunosuppressive properties of the liver and HCC tumor microenvironment. To achieve enhanced therapeutic efficacy, several types of combination strategy are currently being explored (Table 2).

Table 2. Summary of clinical trials of ICI combination therapy for HCC.

Trial Identifier	Target	Drugs	Phase	N	Patient Group	ORR	DCR	PFS (Median, mo)	OS (Median, mo)
ICI + ICI									
NCT01658878 (CheckMate040) [38]	PD-1 + CTLA-4	Nivolumab + Ipilumumab	II	148	Pre-treated	31% (5%CR)	49.0%	NA	22.8 (arm A)
NCT04039607 (CheckMate 9DW)	PD-1 + CTLA-4	Nivolumab + Ipilimumab vs. Sorafenib/lenvatinib	III	1084	Naïve	N/A	N/A	N/A	N/A
NCT02519348 [39]	PD-L1 + CTLA-4	Durvalumab + Tremelimumab	I/II	40	Naïve/Pre-treated	15.0%	57.5% at 4 mo	NA	NA
NCT03298451 (HIMALAYA)	PD-L1 + CTLA-4	Durvalumab + Tremelimumab vs. Sorafenib	III	1310	Naïve	N/A	N/A	N/A	N/A
ICI + MTA									
NCT03006926 (KEYNOTE-524) [40]	PD-1 + MTA	Pembrolizumab + Lenvatinib	Ib	30	Naïve	36.7%	90.0%	9.7 (TTP)	14.6
NCT03713593 (LEAP-002) [41]	PD-1 + MTA	Pembrolizumab + Lenvatinib vs. Lenvatinib	III	750	Naïve	N/A	N/A	N/A	N/A
NCT03434379 (IMbrave150) [42]	PD-L1 + MTA	Atezolizumab + Bevacizumab vs. Sorafenib	III	501	Naïve	33% vs. 13%	NA	6.8 vs.4.3	NR vs. 13.2
NCT04102098 (IMbrave050)	PD-L1 + MTA	Atezolizumab + Bevacizumab vs. Placebo	III	662	Adjuvant	N/A	N/A	N/A	N/A
NCT03847428 (EMERALD-2)	PD-L1 + MTA	Durvalumab + Bevacizumab vs. Bevacizumab	III	888	Adjuvant	N/A	N/A	N/A	N/A
NCT03764293	PD-1 + MTA	SHR-1210 + Apatinib vs. Sorafenib	III	510	Naïve	N/A	N/A	N/A	N/A

Table 2. Cont.

Trial Identifier	Target	Drugs	Phase	N	Patient Group	ORR	DCR	PFS (Median, mo)	OS (Median, mo)
ICI + MTA									
NCT03755791 (COSMIC-312)	PD-L1 + MTA	Atezolizumab + Cabozantinib vs. Sorafenib	III	740	Naive	N/A	N/A	N/A	N/A
NCT03794440 (ORIENT-32)	PD-1 + MTA	Sintilimab + Bevacizumab biosimilar vs. Sorafenib	III	566	Naive	N/A	N/A	N/A	N/A
ICI + Chemo									
NCT03605706	PD-1 + chemotherapeutic agents	SHR-1210 + FOLFOX4 regimen vs. Sorafenib or FOLFOX4 regimen	III	448	Naive	N/A	N/A	N/A	N/A
ICI + ablation									
NCT01853618 [43]	CTLA-4	Tremelimumab + ablation	I/II	32	Advanced	26.0%	85.0%	7.4 (TTP)	12.3
ICI + TACE									
NCT03778957 (EMERALD-1)	PD-L1	Durvalumab + TACE or Durvalumab + Bevacizumab +TACE vs. TACE alone	III	600	Locoregional(Naive)	N/A	N/A	N/A	N/A
ICI + Radiation									
NCT03316872	PD-1 + Radiation	Pembrolizumab + Radiation (SBRT)	II	30	Pre-treated	N/A	N/A	N/A	N/A
NCT03099564	PD-1 + radioembolization	Pembrolizumab + Y90 radioembolization	I	30	Locoregional	N/A	N/A	N/A	N/A
NCT03033446	PD-1 + radioembolization	Nivolumab + Y90 radioembolization	II	40	Advanced	N/A	N/A	N/A	N/A

MTA, molecular-targeted agent; SBRT, stereotactic body radiotherapy; TACE, transarterial chemoembolization; Y90, Yttrium-90; Apatinib, VEGFR2 inhibitor; Cabozantinib, multi kinase inhibitors of MET, VEGFR2, FLT3, c-KIT and RET; N, number of patients; N/A, not available; NR, not reached.

4.1. Combination of ICIs with Other ICIs or Immunostimulatory Agents

Anti-PD-1/PD-L1 and anti-CTLA-4 mAb combination strategies have been evaluated in various types of cancers. In November 2019, the FDA granted breakthrough therapy designation for nivolumab in combination with the anti-CTLA-4 inhibitor ipilimumab for patients with advanced HCC who had previously been treated with sorafenib based on data from the phase I/II CheckMate-040 study of nivolumab plus ipilimumab [38]. The study demonstrated that nivolumab + ipilimumab achieved clinically meaningful responses and had an acceptable safety profile compared to nivolumab monotherapy (ORR: 31% and 14%, respectively), with a median OS of 22.8 months in the nivolumab + ipilimumab group. Another trial (CheckMate-9DW) evaluating nivolumab + ipilimumab versus standard care (sorafenib or lenvatinib) in patients with advanced HCC who have received no prior systemic therapy is currently ongoing.

Durvalumab + tremelimumab has been evaluated in a phase I/II study of patients with advanced HCC, with an ORR of 17.5% and 7/40 evaluable patients showing PR [39]. The combination was well tolerated and showed no unexpected safety signals; therefore, a randomized phase III HIMALAYA study is currently evaluating the efficacy and safety of the durvalumab + tremelimumab combination and durvalumab monotherapy versus sorafenib as a first-line treatment for patients with unresectable HCC and no prior systemic therapy. In January 2020, the FDA granted durvalumab + tremelimumab orphan drug designation for treating patients with HCC.

Other immune checkpoint molecules, such as LAG3 and TIM-3, can also be targeted and combined with PD-1/PD-L1 or CTLA-4 blockade. For instance, phase I basket trials are currently evaluating the dual immune checkpoint blockade of LAG-3 and PD-1 (NCT03005782) and dual TIM-3 and PD-L1 blockade (NCT03099109) in patients with HCC. In addition, combination strategies involving agonistic antibodies that target costimulatory molecules such as 4-1BB, CD40, and OX40 appear to be promising. In a preclinical study, triple combination therapies targeting 4-1BB, OX40, and PD-L1 demonstrated prolonged survival in HCC-bearing mice, providing proof of concept for this combination [44]. A phase I/II basket trial is underway to evaluate the combination of agonistic anti-OX40 Abs with nivolumab and ipilimumab in patients with HCC (NCT03241173).

4.2. Combination of ICI and Non-Immunological Systemic Therapies

Several clinical trials are currently investigating combinations of ICIs and molecular-targeted therapies. For instance, anti-VEGF therapy has been demonstrated to not only normalize immunosuppressive tumor vasculature but also activate DCs and decrease Tregs and MDSCs [45]. In addition, a recent study demonstrated that anti-VEGF therapy rescues effector T cells from exhaustion by downregulating the transcription factor TOX [46]. Therefore, anti-VEGF therapies utilizing multi tyrosine kinase inhibitors (lenvatinib) or anti-VEGFR monoclonal antibodies (bevacizumab) appear to be quite promising in combination with ICIs.

In July 2019, the FDA granted breakthrough therapy designation for pembrolizumab in combination with lenvatinib for the first-line treatment of patients with unresectable HCC who are amenable to locoregional treatment, based on the results of a phase Ib trial (KEYNOTE-524) [40]. Consequently, a phase III trial (LEAP-002) evaluating lenvatinib + pembrolizumab vs. lenvatinib + placebo as a first-line therapy for advanced HCC is currently ongoing [41]. Recently, the results of a phase III trial (IMbrave 150) evaluating the anti-PD-L1 mAb atezolizumab + bevacizumab versus sorafenib monotherapy for patients with unresectable HCC without prior systemic therapy were presented at ESMO Asia Congress 2019, revealing that the atezolizumab + bevacizumab combination significantly improved OS and PFS compared to sorafenib [42]. Another phase III study (IMbrave 050) is currently comparing the same combination with active surveillance in HCC patients at high risk of recurrence after curative treatment, while a phase III trial (EMERALD-2) is also evaluating the durvalumab + bevacizumab combination or durvalumab alone in the same adjuvant setting. Several phase III trials are also evaluating other combinations of ICIs and MTAs: SHR-1210 + apatinib

(NCT03764293), atezolizumab + cabozantinib (NCT03755791/COSMIC-312), and sintilimab (anti-PD-1) + bevacizumab biosimilar (NCT03794440/ORIENT-32).

Chemotherapeutic drugs are generally considered to be immunosuppressive agents due to their toxicity against immune cells; however, they may also be a promising partner to ICIs as they cause immunogenic cell death, allowing the release of tumor antigens and danger-associated molecular patterns from the dead tumor and enhancing the immune response [47]. In addition, some anticancer drugs downregulate Tregs and MDSCs, further promoting tumor eradication [48]. Therefore, a phase III trial evaluating SHR-1210 (anti-PD-1 Ab) + FOLFOX4 as first-line therapy in patients with advanced HCC is currently underway (NCT03605706).

4.3. Combination of ICIs and Non-Immunological Locoregional Therapies

Standard locoregional therapies for HCC can trigger effector T cell responses via the release of tumor-specific antigens from dead tumor cells; therefore, the combined use of locoregional therapies such as RFA, TACE, and radiation could improve the effectiveness of immunotherapies against HCC. The combination of tremelimumab + RFA was tested in a phase I/II trial (NCT01853618) of patients with advanced HCC, with PR and SD noted in five (26%) and 12 (63%) of the 19 evaluable patients, with a median time to progression (TTP) and OS of 7.4 and 12.3 months, respectively. Moreover, pathological evaluation revealed that the accumulation of intratumoral CD8+ T cells in patients had a clinical benefit [43]. TACE has been suggested to exert immunostimulatory effects as the number of α-fetoprotein (AFP)-specific T cells was observed to increase after TACE [49]. Therefore, a phase III trial (EMERALD-1) is currently evaluating TACE in combination with durvalumab and bevacizumab in patients with multiple HCCs (NCT03778957). Radiation with dual checkpoint blockade reportedly induces optimal responses in melanoma, with a previous preclinical study of melanoma demonstrating that anti-CTLA-4 increases the CTL:Treg ratio while anti-PD-L1 rescues T cell exhaustion. Moreover, radiation expanded the T cell receptor (TCR) repertoire, thereby enhancing the antitumor activity of dual checkpoint blockade [50]. Thus, these results provide proof of concept for combining ICIs and radiation to treat HCC and phase II trials are currently underway to evaluate pembrolizumab in combination with stereotactic body radiation therapy (SBRT) (NCT03316872) or Y90 (NCT03099564), and nivolumab with Y90 (NCT03033446) to treat HCC.

5. Exploring ICI Biomarkers

Considering the success of ICIs, it is necessary to identify predictive biomarkers for patients that will respond better to ICIs, particularly since PD-L1 expression on tumor cells does not correlate with the response to anti-PD1 therapy in patients with HCC [30]. MSI, the result of dMMR, was the first predictive biomarker for PD-1 inhibitors to be approved by the FDA [51]. MSI-H colon cancers display favorable responses to ICIs; however, MSI-H appears to be a rare event in HCC [52]. Recently, next-generation sequencing has identified Wnt/CTNNB1 mutations as possible biomarkers for predicting ICI resistance in patients with advanced HCC; however, next-generation sequencing is too complex and costly to use in clinical practice [53]. Therefore, the development of clinically and economically feasible biomarkers is a crucial yet unmet requirement in this field.

6. Non-ICI Immunotherapies

While ICIs release the brake on cancer immunity to unleash dysfunctional antitumor CTLs, there are other "active" immunotherapies that accelerate cancer immunity, such as cancer vaccines, oncolytic virotherapy, and cell-based therapy.

6.1. Cancer Vaccines

The two main cancer vaccine strategies are DC vaccines and peptide vaccines. DCs are potent APCs that can promote tumor-specific T cell responses. In DC vaccines, DCs are loaded with tumor antigens ex vivo and administered to patients as a cellular vaccine. In a preclinical mouse model,

DC vaccines pulsed with tumor cell lysate effectively eradicated tumors and displayed histological evidence of intratumoral lymphocyte infiltration [54]. Unfortunately, clinical trials using DCs pulsed with tumor antigen peptides [55] or tumor cell lysate [56,57] have only demonstrated marginal activity in patients with advanced HCC thus far.

Peptide vaccines for HCC utilize shared TAAs, including AFP, glypican-3 (GPC3), and telomerase reverse transcriptase (TERT). A phase I trial of an AFP-derived peptide vaccine in 15 patients with HCC found that the vaccine was well tolerated, with CR in one patient (AFP-specific CTL response) and SD in eight patients [58]. GPC3 is another antigen that is highly expressed in HCC. In a phase I trial of 33 patients, the GPC3 peptide vaccine was well tolerated with one patient showing PR and 19 showing SD. Importantly, the GPC3 peptide vaccine induced a GPC3-specific CTL response which correlated with OS. [59]. The same group later demonstrated that PD-1 blockade augmented the efficacy of the GPC3 vaccine by increasing the number of vaccine-induced CTLs [60]. A phase II trial of a TERT-derived peptide vaccine (GV1001) in combination with low dose cyclophosphamide showed no effective antitumor response or prolonged TTP [61]. Overall, low-level clinical responses have been observed for DC- and TAA-based peptide vaccines so far; therefore, further trials should examine their combination with immunotherapy.

Neoantigen vaccines are a new cancer vaccine strategy that utilizes tumor neoantigens, which are the products of non-synonymous tumor-specific mutations and are expected to be an ideal therapeutic vaccine as they can achieve a full personalization. First, tumor mutations are analyzed by next-generation sequencing and then candidate neoantigen peptides are predicted on the basis of HLA-binding algorithms [62]. Results from phase I clinical trials testing neoantigen vaccine in advanced melanoma are quite encouraging [63,64]. The cancer vaccine development for hepatocellular carcinoma (HEPAVAC) project, which aims to produce "off-the-shelf" shared antigen-based vaccines for HCC, also includes the actively personalized vaccine (APVAC) protocol based on patient-specific neoantigens [65].

6.2. Oncolytic Virotherapy

Oncolytic virotherapy is a novel approach for cancer immunotherapy [66] that utilizes JX-594 (also known as Pexa-Vec), a vaccinia virus designed to preferentially replicate in and lyse tumor cells, thereby causing the release of antigens from the dead tumor cells and triggering antitumor immunity. This antitumor immunity can be further stimulated by inserting the human granulocyte-macrophage colony stimulating factor transgene into JX-594 [67], with a phase I trial showing that JX-594 has a good safety profile in patients with primary or metastatic liver cancer [68]. A randomized phase II trial has also been conducted to evaluate the safety and antitumor efficacy of JX-594 in patients with advanced HCC, finding that intratumoral JX-594 injection was well tolerated at both low and high doses. Moreover, tumor regression was observed in injected and non-injected tumors, with one CR and three PRs, and OS was significantly longer in patients that received the high dose than the low dose (median 14.1 and 6.7 months, respectively) [69]. A randomized open label phase III trial comparing sorafenib alone and JX-594 + sorafenib in patients with advanced HCC is currently underway (NCT02562755).

6.3. Cell-Based Immunotherapy

Cell-based immunotherapy, also known as adoptive cell transfer (ACT), is also a promising strategy that has been explored extensively. For HCC, cytokine-induced killer cells (CIKs), TCR-engineered T cells, and chimeric antigen receptor T cells (CAR-T) are the major strategies. CIKs are a mixture of heterogeneous immune cells generated by the ex vivo expansion of peripheral blood mononuclear cells in the presence of IL-2, IFN-γ, and anti-CD3 mAbs. CIKs consist of NKT cells, NK cells, and CTLs [70] and display strong cytolytic activity against tumor cells independently of MHC restriction [70]. A randomized phase II trial in treatment-naïve patients with HCC demonstrated that CIK therapy prolonged OS and PFS [71], while a multicenter open-label randomized phase III trial in patients with

HCC after curative treatment demonstrated that CIK therapy prolonged recurrence-free survival and OS [72].

TCR-engineered T cells are generated by integrating cloned tumor antigen-specific TCR into T cells, circumventing the technical difficulties of TIL therapy wherein TILs must be isolated from tumor tissue and expanded ex vivo before being infused back into patients. In mouse models, TCR-engineered T cells recognizing AFP and GPC3 have been reported to control liver tumor growth [73,74], while phase I trials are currently evaluating genetically modified T cells expressing AFP-specific TCRs in patients with advanced HCC (NCT03132792) and autologous TCR-engineered T cell therapy targeting MAGEA1 in solid tumors such as HCC (NCT03441100).

The essential structure of CARs consists of an extracellular single-chain antibody domain that recognizes tumor antigens and an intracellular domain that transmits activation and proliferation signals into cells [75]. Antigen recognition allows CAR-T cells to eliminate cancer cells in an MHC restriction-independent manner, thus solving the problem of tumor immune escape via MHC downregulation [75]. In xenograft mouse models, CAR-T cells targeting GPC3 have been shown to eradicate GPC3-positive HCC [76], while a phase I trial of anti-GPC3 CAR-T cells with or without lymphodepletion treatment has been conducted in six patients with relapsed or refractory GPC3 positive HCC. PR and SD were observed in one and three patients, respectively, with no dose-limiting toxicity identified and only one serious AE of grade 3 fever was reported [77]. In addition, early clinical trials are currently examining CAR-T cells targeting AFP (NCT03349255), MUC-1 (NCT03198546), and EpCAM (NCT03013712).

7. Conclusions

HCC is a serious global health problem because current regimens have limited efficacy in HCC patients, particularly at an advanced disease stage. Cancer immunotherapy has been a significant breakthrough in cancer treatment in recent years and there has been growing interest regarding its application in HCC. As reviewed here, several classes of immunotherapy have emerged for HCC—among which, ICIs targeting PD-1/PD-L1 and CTLA4 hold the greatest promise. However, many studies evaluating ICI-based therapies and other therapeutic strategies are in progress. There are positive and negative factors that should be taken into account for developing successful immunotherapy for HCC (Table 3). Most importantly, it should be designed to counteract the unique immunosuppressive environment of the liver itself in addition to HCC. Therefore, a deeper understanding of the mechanisms underlying HCC immunology will allow the rational design of optimal therapies that coordinate the activation of both innate and adaptive immunity. Research efforts should also be directed toward identifying predictive biomarkers to avoid inappropriate treatment or overtreatment, particularly since current immunotherapies can display limited efficacy in a minority of patients, serious irAEs, and high financial cost.

Table 3. Positive and negative factors for developing successful immunotherapy for HCC.

	Positive factors		Negative Factors
1	Immunotherapy can induce not only systemic but also durable responses by immunological memory, both of which are advantageous for controlling HCC that is characterized by metachronous multicentric occurrence.	1	Paucity of biomarkers predicting responders and non-responders.
2	The presence of tumor-infiltrating lymphocytes (TILs) in HCC suggests the potential of hosts to induce endogenous tumor immunity.	2	Tolerogenic nature of hepatic immunity and immunosuppressive tumor microenvironment of HCC.
3	Several ICIs have already demonstrated manageable safety and promising activity in clinical trials.	3	Response rates of ICI monotherapy are not satisfactory.

Author Contributions: Conceptualization, S.N., Y.E. and K.T.; writing—original draft preparation, S.N., Y.E. (equal contribution with S.N.), H.O., A.T., K.T. and H.S.; writing—review and editing, K.T. All authors have read and agreed to the published version of the manuscript.

Funding: This work is supported by Japan Society for the Promotion of Science KAKENHI 17K09421 (K.T.).

Conflicts of Interest: The authors declare no conflict of interest.

References

1. Bray, F.; Ferlay, J.; Soerjomataram, I.; Siegel, R.L.; Torre, L.A.; Jemal, A. Global cancer statistics 2018: GLOBOCAN estimates of incidence and mortality worldwide for 36 cancers in 185 countries. *CA Cancer J. Clin.* **2018**, *68*, 394–424. [CrossRef] [PubMed]
2. European Association for the Study of the Liver. EASL Clinical Practice Guidelines: Management of hepatocellular carcinoma. *J. Hepatol.* **2018**, *69*, 182–236. [CrossRef]
3. Zheng, Z.; Liang, W.; Wang, D.; Schroder, P.M.; Ju, W.; Wu, L.; Shang, Y.; Guo, Z.; He, X. Adjuvant chemotherapy for patients with primary hepatocellular carcinoma: A meta-analysis. *Int. J. Cancer* **2015**, *136*, E751–E759. [CrossRef]
4. Eso, Y.; Marusawa, H. Novel approaches for molecular targeted therapy against hepatocellular carcinoma. *Hepatol. Res.* **2018**, *48*, 597–607. [CrossRef] [PubMed]
5. Hargadon, K.M.; Johnson, C.E.; Williams, C.J. Immune checkpoint blockade therapy for cancer: An overview of FDA-approved immune checkpoint inhibitors. *Int. Immunopharmacol.* **2018**, *62*, 29–39. [CrossRef] [PubMed]
6. Crispe, I.N. Liver antigen-presenting cells. *J. Hepatol.* **2011**, *54*, 357–365. [CrossRef] [PubMed]
7. Knolle, P.A.; Gerken, G. Local control of the immune response in the liver. *Immunol. Rev.* **2000**, *174*, 21–34. [CrossRef]
8. Krenkel, O.; Tacke, F. Liver macrophages in tissue homeostasis and disease. *Nat. Rev. Immunol.* **2017**, *17*, 306–321. [CrossRef]
9. Breous, E.; Somanathan, S.; Vandenberghe, L.H.; Wilson, J.M. Hepatic regulatory T cells and Kupffer cells are crucial mediators of systemic T cell tolerance to antigens targeting murine liver. *Hepatology* **2009**, *50*, 612–621. [CrossRef]
10. Limmer, A.; Knolle, P.A. Liver sinusoidal endothelial cells: A new type of organ-resident antigen-presenting cell. *Arch. Immunol. Ther. Exp.* **2001**, *49*, S7–S11.
11. Diehl, L.; Schurich, A.; Grochtmann, R.; Hegenbarth, S.; Chen, L.; Knolle, P.A. Tolerogenic maturation of liver sinusoidal endothelial cells promotes B7-homolog 1-dependent CD8+ T cell tolerance. *Hepatology* **2008**, *47*, 296–305. [CrossRef]
12. von Oppen, N.; Schurich, A.; Hegenbarth, S.; Stabenow, D.; Tolba, R.; Weiskirchen, R.; Geerts, A.; Kolanus, W.; Knolle, P.; Diehl, L. Systemic antigen cross-presented by liver sinusoidal endothelial cells induces liver-specific CD8 T-cell retention and tolerization. *Hepatology* **2009**, *49*, 1664–1672. [CrossRef] [PubMed]
13. Bamboat, Z.M.; Stableford, J.A.; Plitas, G.; Burt, B.M.; Nguyen, H.M.; Welles, A.P.; Gonen, M.; Young, J.W.; DeMatteo, R.P. Human liver dendritic cells promote T cell hyporesponsiveness. *J. Immunol.* **2009**, *182*, 1901–1911. [CrossRef] [PubMed]
14. Lau, A.H.; Thomson, A.W. Dendritic cells and immune regulation in the liver. *Gut* **2003**, *52*, 307–314. [CrossRef] [PubMed]
15. Holz, L.E.; Benseler, V.; Bowen, D.G.; Bouillet, P.; Strasser, A.; O'Reilly, L.; d'Avigdor, W.M.; Bishop, A.G.; McCaughan, G.W.; Bertolino, P. Intrahepatic murine CD8 T-cell activation associates with a distinct phenotype leading to Bim-dependent death. *Gastroenterology* **2008**, *135*, 989–997. [CrossRef] [PubMed]
16. Matsui, M.; Machida, S.; Itani-Yohda, T.; Akatsuka, T. Downregulation of the proteasome subunits, transporter, and antigen presentation in hepatocellular carcinoma, and their restoration by interferon-gamma. *J. Gastroenterol. Hepatol.* **2002**, *17*, 897–907. [CrossRef]
17. Dunn, G.P.; Bruce, A.T.; Ikeda, H.; Old, L.J.; Schreiber, R.D. Cancer immunoediting: From immunosurveillance to tumor escape. *Nat. Immunol.* **2002**, *3*, 991–998. [CrossRef]
18. Prieto, J.; Melero, I.; Sangro, B. Immunological landscape and immunotherapy of hepatocellular carcinoma. *Nat. Rev. Gastroenterol. Hepatol.* **2015**, *12*, 681–700. [CrossRef]

19. Gao, Q.; Wang, X.Y.; Qiu, S.J.; Yamato, I.; Sho, M.; Nakajima, Y.; Zhou, J.; Li, B.Z.; Shi, Y.H.; Xiao, Y.S.; et al. Overexpression of PD-L1 significantly associates with tumor aggressiveness and postoperative recurrence in human hepatocellular carcinoma. *Clin. Cancer Res.* **2009**, *15*, 971–979. [CrossRef]
20. Meirow, Y.; Kanterman, J.; Baniyash, M. Paving the Road to Tumor Development and Spreading: Myeloid-Derived Suppressor Cells are Ruling the Fate. *Front. Immunol.* **2015**, *6*, 523. [CrossRef]
21. Arihara, F.; Mizukoshi, E.; Kitahara, M.; Takata, Y.; Arai, K.; Yamashita, T.; Nakamoto, Y.; Kaneko, S. Increase in CD14+HLA-DR -/low myeloid-derived suppressor cells in hepatocellular carcinoma patients and its impact on prognosis. *Cancer Immunol. Immunother.* **2013**, *62*, 1421–1430. [CrossRef] [PubMed]
22. Degroote, H.; Van Dierendonck, A.; Geerts, A.; Van Vlierberghe, H.; Devisscher, L. Preclinical and Clinical Therapeutic Strategies Affecting Tumor-Associated Macrophages in Hepatocellular Carcinoma. *J. Immunol. Res.* **2018**, *2018*, 7819520. [CrossRef] [PubMed]
23. Chen, K.J.; Lin, S.Z.; Zhou, L.; Xie, H.Y.; Zhou, W.H.; Taki-Eldin, A.; Zheng, S.S. Selective recruitment of regulatory T cell through CCR6-CCL20 in hepatocellular carcinoma fosters tumor progression and predicts poor prognosis. *PLoS ONE* **2011**, *6*, e24671. [CrossRef] [PubMed]
24. Flecken, T.; Schmidt, N.; Hild, S.; Gostick, E.; Drognitz, O.; Zeiser, R.; Schemmer, P.; Bruns, H.; Eiermann, T.; Price, D.A.; et al. Immunodominance and functional alterations of tumor-associated antigen-specific CD8+ T-cell responses in hepatocellular carcinoma. *Hepatology* **2014**, *59*, 1415–1426. [CrossRef] [PubMed]
25. Gao, Q.; Qiu, S.J.; Fan, J.; Zhou, J.; Wang, X.Y.; Xiao, Y.S.; Xu, Y.; Li, Y.W.; Tang, Z.Y. Intratumoral balance of regulatory and cytotoxic T cells is associated with prognosis of hepatocellular carcinoma after resection. *J. Clin. Oncol.* **2007**, *25*, 2586–2593. [CrossRef] [PubMed]
26. Okazaki, T.; Honjo, T. PD-1 and PD-1 ligands: From discovery to clinical application. *Int. Immunol.* **2007**, *19*, 813–824. [CrossRef]
27. Iwai, Y.; Terawaki, S.; Ikegawa, M.; Okazaki, T.; Honjo, T. PD-1 inhibits antiviral immunity at the effector phase in the liver. *J. Exp. Med.* **2003**, *198*, 39–50. [CrossRef]
28. Fritz, J.M.; Lenardo, M.J. Development of immune checkpoint therapy for cancer. *J. Exp. Med.* **2019**, *216*, 1244–1254. [CrossRef]
29. Sharma, P.; Allison, J.P. The future of immune checkpoint therapy. *Science* **2015**, *348*, 56–61. [CrossRef]
30. El-Khoueiry, A.B.; Sangro, B.; Yau, T.; Crocenzi, T.S.; Kudo, M.; Hsu, C.; Kim, T.-Y.; Choo, S.-P.; Trojan, J.; Welling, T.H.; et al. Nivolumab in patients with advanced hepatocellular carcinoma (CheckMate 040): An open-label, non-comparative, phase 1/2 dose escalation and expansion trial. *Lancet* **2017**, *389*, 2492–2502. [CrossRef]
31. Yau, T.; Park, J.W.; Finn, R.S.; Cheng, A.L.; Mathurin, P.; Edeline, J.; Kudo, M.; Han, K.H.; Harding, J.J.; Merle, P.; et al. LBA38_PR-CheckMate 459: A randomized, multi-center phase III study of nivolumab (NIVO) vs. sorafenib (SOR) as first-line (1L) treatment in patients (pts) with advanced hepatocellular carcinoma (aHCC). *Ann. Oncol.* **2019**, *30*, v874–v875. [CrossRef]
32. Prasad, V.; Kaestner, V.; Mailankody, S. Cancer Drugs Approved Based on Biomarkers and Not Tumor Type-FDA Approval of Pembrolizumab for Mismatch Repair-Deficient Solid Cancers. *JAMA Oncol.* **2018**, *4*, 157–158. [CrossRef] [PubMed]
33. Zhu, A.X.; Finn, R.S.; Edeline, J.; Cattan, S.; Ogasawara, S.; Palmer, D.; Verslype, C.; Zagonel, V.; Fartoux, L.; Vogel, A.; et al. Pembrolizumab in patients with advanced hepatocellular carcinoma previously treated with sorafenib (KEYNOTE-224): A non-randomised, open-label phase 2 trial. *Lancet Oncol.* **2018**, *19*, 940–952. [CrossRef]
34. Finn, R.S.; Ryoo, B.Y.; Merle, P.; Kudo, M.; Bouattour, M.; Lim, H.Y.; Breder, V.; Edeline, J.; Chao, Y.; Ogasawara, S.; et al. Pembrolizumab as Second-Line Therapy in Patients with Advanced Hepatocellular Carcinoma in KEYNOTE-240: A Randomized, Double-Blind, Phase III Trial. *J. Clin. Oncol.* **2020**, *38*, 193–202. [CrossRef]
35. Wainberg, Z.A.; Segal, N.H.; Jaeger, D.; Lee, K.-H.; Marshall, J.; Antonia, S.J.; Butler, M.; Sanborn, R.E.; Nemunaitis, J.J.; Carlson, C.A.; et al. Safety and clinical activity of durvalumab monotherapy in patients with hepatocellular carcinoma (HCC). *J. Clin. Oncol.* **2017**, *35*, 4071. [CrossRef]
36. Qin, S.; Finn, R.S.; Kudo, M.; Meyer, T.; Vogel, A.; Ducreux, M.; Macarulla, T.M.; Tomasello, G.; Boisserie, F.; Hou, J.; et al. RATIONALE 301 study: Tislelizumab versus sorafenib as first-line treatment for unresectable hepatocellular carcinoma. *Future Oncol.* **2019**, *15*, 1811–1822. [CrossRef]

37. Sangro, B.; Gomez-Martin, C.; de la Mata, M.; Iñarrairaegui, M.; Garralda, E.; Barrera, P.; Riezu-Boj, J.I.; Larrea, E.; Alfaro, C.; Sarobe, P.; et al. A clinical trial of CTLA-4 blockade with tremelimumab in patients with hepatocellular carcinoma and chronic hepatitis C. *J. Hepatol.* **2013**, *59*, 81–88. [CrossRef]
38. Yau, T.; Kang, Y.K.; Kim, T.Y.; El-Khoueiry, A.B.; Santoro, A.; Sangro, B.; Melero, I.; Kudo, M.; Hou, M.M.; Matilla, A.; et al. Nivolumab (NIVO)+ ipilimumab (IPI) combination therapy in patients (pts) with advanced hepatocellular carcinoma (aHCC): Results from CheckMate 040. *J. Clin. Oncol.* **2019**. [CrossRef]
39. Kelley, R.; Abou-Alfa, G.; Bendell, J. Phase I/II study of durvalumab and tremelimumab in patients with unresectable hepatocellular carcinoma (HCC): Phase I safety and efficacy analyses. *J. Clin. Oncol.* **2017**, *35*, 4073. [CrossRef]
40. Ikeda, M.; Sung, M.W.; Kudo, M.; Kobayashi, M.; Baron, A.D.; Finn, R.S.; Kaneko, S.; Zhu, A.X.; Kubota, T.; Kralijevic, S.; et al. Abstract CT061: A Phase Ib trial of lenvatinib (LEN) plus pembrolizumab (PEMBRO) in unresectable hepatocellular carcinoma (uHCC): Updated results. *Cancer Res.* **2019**, *79*, CT061. [CrossRef]
41. Llovet, J.M.; Kudo, M.; Cheng, A.-L.; Finn, R.S.; Galle, P.R.; Kaneko, S.; Meyer, T.; Qin, S.; Dutcus, C.E.; Chen, E.; et al. Lenvatinib (len) plus pembrolizumab (pembro) for the first-line treatment of patients (pts) with advanced hepatocellular carcinoma (HCC): Phase 3 LEAP-002 study. *J. Clin. Oncol.* **2019**, *37*. [CrossRef]
42. Cheng, A.L.; Qin, S.; Ikeda, M.; Galle, P.; Ducreux, M.; Zhu, A.; Kim, T.Y.; Kudo, M.; Breder, V.; Merle, P.; et al. LBA3-IMbrave150: Efficacy and safety results from a ph III study evaluating atezolizumab (atezo) + bevacizumab (bev) vs. sorafenib (Sor) as first treatment (tx) for patients (pts) with unresectable hepatocellular carcinoma (HCC). *Ann. Oncol.* **2019**, *30*, ix186–ix187. [CrossRef]
43. Duffy, A.G.; Ulahannan, S.V.; Makorova-Rusher, O.; Rahma, O.; Wedemeyer, H.; Pratt, D.; Davis, J.L.; Hughes, M.S.; Heller, T.; ElGindi, M.; et al. Tremelimumab in combination with ablation in patients with advanced hepatocellular carcinoma. *J. Hepatol.* **2017**, *66*, 545–551. [CrossRef] [PubMed]
44. Morales-Kastresana, A.; Sanmamed, M.F.; Rodriguez, I.; Palazon, A.; Martinez-Forero, I.; Labiano, S.; Hervas-Stubbs, S.; Sangro, B.; Ochoa, C.; Rouzaut, A.; et al. Combined immunostimulatory monoclonal antibodies extend survival in an aggressive transgenic hepatocellular carcinoma mouse model. *Clin. Cancer Res.* **2013**, *19*, 6151–6162. [CrossRef]
45. Fukumura, D.; Kloepper, J.; Amoozgar, Z.; Duda, D.G.; Jain, R.K. Enhancing cancer immunotherapy using antiangiogenics: Opportunities and challenges. *Nat. Rev. Clin. Oncol.* **2018**, *15*, 325–340. [CrossRef]
46. Kim, C.G.; Jang, M.; Kim, Y.; Leem, G.; Kim, K.H.; Lee, H.; Kim, T.S.; Choi, S.J.; Kim, H.D.; Han, J.W.; et al. VEGF-A drives TOX-dependent T cell exhaustion in anti-PD-1-resistant microsatellite stable colorectal cancers. *Sci. Immunol.* **2019**, *4*. [CrossRef]
47. Pol, J.; Vacchelli, E.; Aranda, F.; Castoldi, F.; Eggermont, A.; Cremer, I.; Sautes-Fridman, C.; Fucikova, J.; Galon, J.; Spisek, R.; et al. Trial Watch: Immunogenic cell death inducers for anticancer chemotherapy. *Oncoimmunology* **2015**, *4*, e1008866. [CrossRef]
48. Zheng, Y.; Dou, Y.; Duan, L.; Cong, C.; Gao, A.; Lai, Q.; Sun, Y. Using chemo-drugs or irradiation to break immune tolerance and facilitate immunotherapy in solid cancer. *Cell. Immunol.* **2015**, *294*, 54–59. [CrossRef]
49. Ayaru, L.; Pereira, S.P.; Alisa, A.; Pathan, A.A.; Williams, R.; Davidson, B.; Burroughs, A.K.; Meyer, T.; Behboudi, S. Unmasking of alpha-fetoprotein-specific CD4(+) T cell responses in hepatocellular carcinoma patients undergoing embolization. *J. Immunol.* **2007**, *178*, 1914–1922. [CrossRef]
50. Twyman-Saint Victor, C.; Rech, A.J.; Maity, A.; Rengan, R.; Pauken, K.E.; Stelekati, E.; Benci, J.L.; Xu, B.; Dada, H.; Odorizzi, P.M.; et al. Radiation and dual checkpoint blockade activate non-redundant immune mechanisms in cancer. *Nature* **2015**, *520*, 373–377. [CrossRef]
51. Le, D.T.; Durham, J.N.; Smith, K.N.; Wang, H.; Bartlett, B.R.; Aulakh, L.K.; Lu, S.; Kemberling, H.; Wilt, C.; Luber, B.S.; et al. Mismatch repair deficiency predicts response of solid tumors to PD-1 blockade. *Science* **2017**, *357*, 409–413. [CrossRef] [PubMed]
52. Goumard, C.; Desbois-Mouthon, C.; Wendum, D.; Calmel, C.; Merabtene, F.; Scatton, O.; Praz, F. Low Levels of Microsatellite Instability at Simple Repeated Sequences Commonly Occur in Human Hepatocellular Carcinoma. *Cancer Genom. Proteom.* **2017**, *14*, 329–339. [CrossRef]
53. Harding, J.J.; Nandakumar, S.; Armenia, J.; Khalil, D.N.; Albano, M.; Ly, M.; Shia, J.; Hechtman, J.F.; Kundra, R.; El Dika, I.; et al. Prospective Genotyping of Hepatocellular Carcinoma: Clinical Implications of Next-Generation Sequencing for Matching Patients to Targeted and Immune Therapies. *Clin. Cancer Res.* **2019**, *25*, 2116–2126. [CrossRef] [PubMed]

54. Lee, W.C.; Wang, H.C.; Jeng, L.B.; Chiang, Y.J.; Lia, C.R.; Huang, P.F.; Chen, M.F.; Qian, S.; Lu, L. Effective treatment of small murine hepatocellular carcinoma by dendritic cells. *Hepatology* **2001**, *34*, 896–905. [CrossRef] [PubMed]
55. Tada, F.; Abe, M.; Hirooka, M.; Ikeda, Y.; Hiasa, Y.; Lee, Y.; Jung, N.C.; Lee, W.B.; Lee, H.S.; Bae, Y.S.; et al. Phase I/II study of immunotherapy using tumor antigen-pulsed dendritic cells in patients with hepatocellular carcinoma. *Int. J. Oncol.* **2012**, *41*, 1601–1609. [CrossRef]
56. Iwashita, Y.; Tahara, K.; Goto, S.; Sasaki, A.; Kai, S.; Seike, M.; Chen, C.L.; Kawano, K.; Kitano, S. A phase I study of autologous dendritic cell-based immunotherapy for patients with unresectable primary liver cancer. *Cancer Immunol. Immunother.* **2003**, *52*, 155–161. [CrossRef]
57. El Ansary, M.; Mogawer, S.; Elhamid, S.A.; Alwakil, S.; Aboelkasem, F.; Sabaawy, H.E.; Abdelhalim, O. Immunotherapy by autologous dendritic cell vaccine in patients with advanced HCC. *J. Cancer Res. Clin. Oncol.* **2013**, *139*, 39–48. [CrossRef]
58. Nakagawa, H.; Mizukoshi, E.; Kobayashi, E.; Tamai, T.; Hamana, H.; Ozawa, T.; Kishi, H.; Kitahara, M.; Yamashita, T.; Arai, K.; et al. Association Between High-Avidity T-Cell Receptors, Induced by alpha-Fetoprotein-Derived Peptides, and Anti-Tumor Effects in Patients with Hepatocellular Carcinoma. *Gastroenterology* **2017**, *152*, 1395–1406.e10. [CrossRef]
59. Sawada, Y.; Yoshikawa, T.; Nobuoka, D.; Shirakawa, H.; Kuronuma, T.; Motomura, Y.; Mizuno, S.; Ishii, H.; Nakachi, K.; Konishi, M.; et al. Phase I trial of a glypican-3-derived peptide vaccine for advanced hepatocellular carcinoma: Immunologic evidence and potential for improving overall survival. *Clin. Cancer Res.* **2012**, *18*, 3686–3696. [CrossRef]
60. Sawada, Y.; Yoshikawa, T.; Shimomura, M.; Iwama, T.; Endo, I.; Nakatsura, T. Programmed death-1 blockade enhances the antitumor effects of peptide vaccine-induced peptide-specific cytotoxic T lymphocytes. *Int. J. Oncol.* **2015**, *46*, 28–36. [CrossRef]
61. Greten, T.F.; Forner, A.; Korangy, F.; N'Kontchou, G.; Barget, N.; Ayuso, C.; Ormandy, L.A.; Manns, M.P.; Beaugrand, M.; Bruix, J. A phase II open label trial evaluating safety and efficacy of a telomerase peptide vaccination in patients with advanced hepatocellular carcinoma. *BMC Cancer* **2010**, *10*, 209. [CrossRef]
62. Li, L.; Goedegebuure, S.P.; Gillanders, W.E. Preclinical and clinical development of neoantigen vaccines. *Ann. Oncol.* **2017**, *28*, xii11–xii17. [CrossRef]
63. Ott, P.A.; Hu, Z.; Keskin, D.B.; Shukla, S.A.; Sun, J.; Bozym, D.J.; Zhang, W.; Luoma, A.; Giobbie-Hurder, A.; Peter, L.; et al. An immunogenic personal neoantigen vaccine for patients with melanoma. *Nature* **2017**, *547*, 217–221. [CrossRef]
64. Sahin, U.; Derhovanessian, E.; Miller, M.; Kloke, B.P.; Simon, P.; Lower, M.; Bukur, V.; Tadmor, A.D.; Luxemburger, U.; Schrors, B.; et al. Personalized RNA mutanome vaccines mobilize poly-specific therapeutic immunity against cancer. *Nature* **2017**, *547*, 222–226. [CrossRef]
65. Buonaguro, L. New vaccination strategies in liver cancer. *Cytokine Growth Factor Rev.* **2017**, *36*, 125–129. [CrossRef]
66. Chiocca, E.A. Oncolytic viruses. *Nat. Rev. Cancer* **2002**, *2*, 938–950. [CrossRef]
67. Parato, K.A.; Breitbach, C.J.; Le Boeuf, F.; Wang, J.; Storbeck, C.; Ilkow, C.; Diallo, J.S.; Falls, T.; Burns, J.; Garcia, V.; et al. The oncolytic poxvirus JX-594 selectively replicates in and destroys cancer cells driven by genetic pathways commonly activated in cancers. *Mol. Ther.* **2012**, *20*, 749–758. [CrossRef]
68. Park, B.H.; Hwang, T.; Liu, T.C.; Sze, D.Y.; Kim, J.S.; Kwon, H.C.; Oh, S.Y.; Han, S.Y.; Yoon, J.H.; Hong, S.H.; et al. Use of a targeted oncolytic poxvirus, JX-594, in patients with refractory primary or metastatic liver cancer: A phase I trial. *Lancet Oncol.* **2008**, *9*, 533–542. [CrossRef]
69. Heo, J.; Reid, T.; Ruo, L.; Breitbach, C.J.; Rose, S.; Bloomston, M.; Cho, M.; Lim, H.Y.; Chung, H.C.; Kim, C.W.; et al. Randomized dose-finding clinical trial of oncolytic immunotherapeutic vaccinia JX-594 in liver cancer. *Nat. Med.* **2013**, *19*, 329–336. [CrossRef]
70. Gao, X.; Mi, Y.; Guo, N.; Xu, H.; Xu, L.; Gou, X.; Jin, W. Cytokine-Induced Killer Cells as Pharmacological Tools for Cancer Immunotherapy. *Front. Immunol.* **2017**, *8*, 774. [CrossRef]
71. Yu, X.; Zhao, H.; Liu, L.; Cao, S.; Ren, B.; Zhang, N.; An, X.; Yu, J.; Li, H.; Ren, X. A randomized phase II study of autologous cytokine-induced killer cells in treatment of hepatocellular carcinoma. *J. Clin. Immunol.* **2014**, *34*, 194–203. [CrossRef]

72. Lee, J.H.; Lim, Y.S.; Yeon, J.E.; Song, T.J.; Yu, S.J.; Gwak, G.Y.; Kim, K.M.; Kim, Y.J.; Lee, J.W.; Yoon, J.H. Adjuvant immunotherapy with autologous cytokine-induced killer cells for hepatocellular carcinoma. *Gastroenterology* **2015**, *148*, 1383–1391.e6. [CrossRef]
73. Sun, L.; Guo, H.; Jiang, R.; Lu, L.; Liu, T.; He, X. Engineered cytotoxic T lymphocytes with AFP-specific TCR gene for adoptive immunotherapy in hepatocellular carcinoma. *Tumour Biol.* **2016**, *37*, 799–806. [CrossRef]
74. Dargel, C.; Bassani-Sternberg, M.; Hasreiter, J.; Zani, F.; Bockmann, J.H.; Thiele, F.; Bohne, F.; Wisskirchen, K.; Wilde, S.; Sprinzl, M.F.; et al. T Cells Engineered to Express a T-Cell Receptor Specific for Glypican-3 to Recognize and Kill Hepatoma Cells In Vitro and in Mice. *Gastroenterology* **2015**, *149*, 1042–1052. [CrossRef]
75. June, C.H.; O'Connor, R.S.; Kawalekar, O.U.; Ghassemi, S.; Milone, M.C. CAR T cell immunotherapy for human cancer. *Science* **2018**, *359*, 1361–1365. [CrossRef]
76. Gao, H.; Li, K.; Tu, H.; Pan, X.; Jiang, H.; Shi, B.; Kong, J.; Wang, H.; Yang, S.; Gu, J.; et al. Development of T cells redirected to glypican-3 for the treatment of hepatocellular carcinoma. *Clin. Cancer Res.* **2014**, *20*, 6418–6428. [CrossRef]
77. Zhai, B.; Shi, D.; Gao, H.; Qi, X.; Jiang, H.; Zhang, Y.; Chi, J.; Ruan, H.; Wang, H.; Ru, Q.C.; et al. A phase I study of anti-GPC3 chimeric antigen receptor modified T cells (GPC3 CAR-T) in Chinese patients with refractory or relapsed GPC3+ hepatocellular carcinoma (r/r GPC3+ HCC). *J. Clin. Oncol.* **2017**, *35*, 3049. [CrossRef]

© 2020 by the authors. Licensee MDPI, Basel, Switzerland. This article is an open access article distributed under the terms and conditions of the Creative Commons Attribution (CC BY) license (http://creativecommons.org/licenses/by/4.0/).

Article

A Disintegrin and Metalloproteinase 9 (ADAM9) in Advanced Hepatocellular Carcinoma and Their Role as a Biomarker During Hepatocellular Carcinoma Immunotherapy

Sooyeon Oh [1], YoungJoon Park [2], Hyun-Jung Lee [3], Jooho Lee [4,*], Soo-Hyeon Lee [5], Young-Seok Baek [5], Su-Kyung Chun [6], Seung-Min Lee [3], Mina Kim [4], Young-Eun Chon [4], Yeonjung Ha [4], Yuri Cho [7], Gi Jin Kim [2], Seong-Gyu Hwang [4] and KyuBum Kwack [2]

1. Chaum Life Center, CHA University School of Medicine, Seoul 06062, Korea
2. Department of Biomedical Science, College of Life Science, CHA University, Seongnam 13488, Korea
3. Center for Research & Development, CHA Advanced Research Institute, Seongnam 13488, Korea
4. Department of Gastroenterology, CHA Bundang Medical Center, CHA University School of Medicine, Seongnam 13496, Korea
5. Immunotherapy Development Team, R & D Division, CHA Biolab, Seongnam 13488, Korea
6. Department of Food Science and Biotechnology, College of Life Science, CHA University, Seongnam 13488, Korea
7. Department of Gastroenterology, CHA Gangnam Medical Center, CHA University School of Medicine, Seoul 06135, Korea
* Correspondence: ljh0505@cha.ac.kr; Tel.: +82-31-780-1811

Received: 28 January 2020; Accepted: 19 March 2020; Published: 21 March 2020

Abstract: The chemotherapeutics sorafenib and regorafenib inhibit shedding of MHC class I-related chain A (MICA) from hepatocellular carcinoma (HCC) cells by suppressing a disintegrin and metalloprotease 9 (ADAM9). MICA is a ligand for natural killer (NK) group 2 member D (NKG2D) and is expressed on tumor cells to elicit attack by NK cells. This study measured *ADAM9* mRNA levels in blood samples of advanced HCC patients ($n = 10$). In newly diagnosed patients ($n = 5$), the plasma *ADAM9* mRNA level was significantly higher than that in healthy controls (3.001 versus 1.00, $p < 0.05$). Among four patients treated with nivolumab therapy, two patients with clinical response to nivolumab showed significant decreases in fold changes of serum *ADAM9* mRNA level from 573.98 to 262.58 and from 323.88 to 85.52 ($p < 0.05$); however, two patients with no response to nivolumab did not. Using the Cancer Genome Atlas database, we found that higher expression of *ADAM9* in tumor tissues was associated with poorer survival of HCC patients (log-rank $p = 0.00039$), while *ADAM10* and *ADAM17* exhibited no such association. In addition, *ADAM9* expression showed a positive correlation with the expression of inhibitory checkpoint molecules. This study, though small in sample size, clearly suggested that *ADAM9* mRNA might serve as biomarker predicting clinical response and that the ADAM9-MICA-NKG2D system can be a good therapeutic target for HCC immunotherapy. Future studies are warranted to validate these findings.

Keywords: hepatocellular carcinoma; a disintegrin and metalloprotease 9; nivolumab; natural killer; immunotherapy

1. Introduction

Hepatocellular carcinoma (HCC) is the fifth most prevalent cancer and the third leading cause of cancer associated mortality worldwide [1,2]. Advanced stage HCC accompanied with portal vein invasion, distant metastasis, or lymph node metastasis is hard to treat due to the underlying liver cirrhosis, frequent recurrence, or multiple occurrence [3,4]. Currently, the tyrosine kinase inhibitors (TKI) sorafenib and lenvatinib are the first-line treatments in advanced HCC [5–8]. Regorafenib, nivolumab, cabozantinib, and ramucirumab were approved as second line therapies following their recent successes in several clinical trials [9–15]. Despite these breakthroughs, many patients still do not survive advanced HCC.

Among the recent breakthroughs in HCC treatments, immunotherapy is particularly notable. As a leading therapeutic in immunotherapy targeting programmed cell death 1 protein (PD-1), nivolumab demonstrated a 20% objective response rate and a 64% disease control rate in HCC patients who progressed on sorafenib [16]. Adoptive cellular immunotherapy using natural killer (NK) cells, cytokine induced killer (CIK) cells, or dendritic cells have also been studied [3]. Some adoptive cellular immunotherapy treatment modalities exhibited survival benefit in HCC patients [17]. These encouraging results were attributed to the immunological characteristics of the liver, which shelters a large pool of immune cells belonging to both the innate and acquired immune systems. The interactions between HCC and immune cells play a major role in tumor escape from immune surveillance resulting in HCC progression [18]. Inadequate co-stimulation, failure of tumor associated antigens (TAA) processing and presentation by dendritic cells, along with the suppression of effector T and NK cells are proposed mechanisms by which the HCC tumor cells evade the host immune system [19]. At the same time, HCC is indicated as a hot tumor that is characterized by increased expression of checkpoint molecules such as PD-1 and PD-1 ligand 1 (PD-L1) proteins, a large pool of tumor infiltrating immune cells, and high tumor mutational burdens, giving rise to ample amount of TAAs [20]. These immunological features enable HCC to benefit from immunotherapy. Thus, harnessing these immune characters through combination immunotherapy is proposed as the next important step in treatment of advanced HCC [4,13–15,21,22].

In this regard, a disintegrin and metalloproteases 9 (*ADAM9*) pathway in relation to MHC class I-related chain A (MICA) may provide a strategic ground for combination immunotherapy. Cell membrane-bound MICA (mMICA) is a ligand for NK group 2 member D (NKG2D), a stimulatory receptor on NK cells. The mMICA expressed in human HCC cells signals NK cells and other immune cells to kill the transformed hepatoma cells [23,24]. However, ADAM9, a matrix metalloproteinase (MMP) expressed by cancer cells, cleaves mMICA, releasing soluble MICA (sMICA). This sMICA acts as a decoy, weakening the cytotoxic immunity provided by NK cells and CD8+ T cells. This mMICA shedding by *ADAM9* protease is identified as a mechanism of HCC escape from the host immune surveillance. Fortunately, sorafenib and regorafenib inhibit the expression of *ADAM9* mRNA [23,24], restoring the host immunity against HCC and generating a room for synergistic action by adoptive cell therapy with NK cells or CD8+ T cells. Thus, the ADAM9-MICA-NKG2D system may provide a strategic target for a novel chemoimmunotherapy combining adoptive NK cell therapy and sorafenib or regorafenib [24].

In this pilot observational study, we aimed to characterize *ADAM9* mRNA expression in blood samples of advanced HCC patients according to their clinical courses. To support our findings, we probed the role of *ADAM9* as a prognostic biomarker for HCC using the Cancer Genome Atlas (TCGA) database. Furthermore, we present the case of a patient who achieved complete remission with regorafenib and autologous NK cell combination immunotherapy.

2. Results

2.1. Patient Characteristics

This study was conducted in CHA Bundang Hospital between January 2017 and November 2019. Advanced HCC patients eligible for this study, who were to be treated with sorafenib, regorafenib, or nivolumab as standard-of-care therapy and met our inclusion and exclusion criteria, were invited to the present study. A total of 10 patients participated in this study. The demographic and clinical details of each participant are listed in Table 1.

Table 1. Demographic details and clinical characteristics of study participants ($n = 10$).

Subject No.	Age	Sex	Etiology	Antiviral Therapy	Serum AFP (ng/mL)	Child-Pugh Class	Tumor Size (cm)	PV Invasion (Vp) *	mUICC Stage	Extra-Hepatic Metastasis	Type of Therapy
#1	58	M	HBV	TDF	97,387	A	8	4	IVb	Yes	Sorafenib
#2	61	M	HBV	ETV	>200,000	B	22	2	Ivb	Yes	Sorafenib
#3	61	M	HBV	ETV	190	B	5.2	3	Ivb	Yes	Sorafenib
#4	55	M	HBV	TDF	71.7	A	3	0	III	No	TACE, Sorafenib
#5	58	M	HBV	ETV	14.7	A	10	2	Ivb	Yes	Sorafenib, Nivolumab
#6	59	M	HBV	ETV	82.4	A	4	2	Iva	No	Sorafenib, Nivolumab + NK cell therapy, Regorafenib + NK cell therapy
#7	45	M	HBV	TDF	154.7	B	11	4	Iva	No	Sorafenib, Nivolumab
#8	44	F	HCV	DAC/SUN	6519.4	A	4.5	0	Ivb	Yes	Sorafenib, Regorafenib, Nivolumab
#9	58	F	HBV	TDF	66	A	9	2	Ivb	Yes	Sorafenib, Nivolumab Regorafenib
#10	76	M	NASH	none	4594.1	A	5.2	1	Ivb	Yes	Sorafenib, Nivolumab

Abbreviations: No, number; M, male; F, female; AFP, alpha-fetoprotein; PV, portal vein; mUICC, modified Union for International Cancer Control; HBV, hepatitis B virus; HCV, hepatitis C virus; NASH, nonalcoholic steatohepatitis; TACE, transarterial chemoembolization; NK, natural killer; TDF, tenofovir disoproxil fumarate; ETV, entecavir; DAC, daclatasvir; SUN, asunaprevir. * The extent of portal vein invasion (Vp) by tumor thrombosis was documented according to the Liver Cancer Study Group of Japan classification: Vp0 = no portal vein invasion, Vp1 = segmental portal vein invasion, Vp2 = right anterior/posterior portal vein, Vp3 = right/left portal vein and Vp4 = main trunk [25].

The 10 participants comprised eight chronic hepatitis B (CHB), one chronic hepatitis C (CHC) and one non-viral HCC patients. At the time of enrollment, five participants (Subject No. (#) 1 to #5) were newly diagnosed and treatment-naïve; they were subjected to sorafenib as first-line therapy. In addition, four patients (Subject #7 to #10) had failed first-line sorafenib therapy and were subjected to nivolumab therapy, whereas one patient (Subject #6) was enrolled after the patient had already reached near complete remission.

The baseline characteristics of the participants and their HCC are summarized in Table S1. The mean age was 57.4 years, and 80% of the participants were male. The modified Union for International Cancer Control (mUICC) TNM stage was III in 1 patient (10%), IVa in 2 patients (20%), and IVb in 7 patients (70%). The Barcelona Clinic Liver Cancer (BCLC) stage was B in 1 patient (10%) and C in 9 patients (90%). Most patients (70%) had well-preserved liver function (Child–Pugh class A).

2.2. Overexpression of Plasma ADAM9 mRNA in Untreated HCC Patients

The plasma levels of *ADAM9* mRNA were tested in the five newly diagnosed and treatment-naïve HCC patients (Table 1, Subject #1 to #5) and expressed as fold changes compared to the healthy control group ($n = 5$, 100% female, mean age 34.2 years). The mean value of pre-treatment plasma *ADAM9* mRNA levels in the HCC patients was significantly higher than that in the healthy controls (3.001 ± 0.279 vs. 1.00 ± 0.005, $p < 0.05$) (Figure 1).

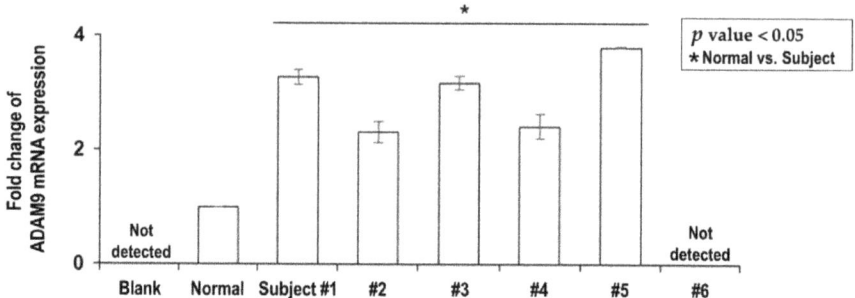

Figure 1. Expression level of *ADAM9* mRNA in plasma samples of treatment naïve advanced HCC patients and one HCC patient with near complete remission. Each patient (Subject #1 to #5) with newly diagnosed and treatment naïve HCC had significantly elevated expression of *ADAM9* mRNA in plasma compared with normal controls (3.001 ± 0.279 vs. 1.00 ± 0.005, $p < 0.05$). The HCC patient who had reached near complete remission (Subject #6) did not express *ADAM9* mRNA in plasma at all.

2.3. Decreased ADAM9 mRNA Expression Correlated with Response to Nivolumab

In four patients who were to receive nivolumab as second- or third-line treatment (Table 1, Subject #7 to #10), *ADAM9* mRNA expression in serum was significantly elevated compared to that in the healthy controls (Figure 2; pre-nivolumab mean, 323.39 ± 88.67 vs. 1.00 ± 0.000003, $p < 0.05$). To investigate the functional relevance between the change in *ADAM9* mRNA expression and clinical response, the serum levels of *ADAM9* mRNA were followed during the nivolumab therapy.

After three and four cycles of nivolumab therapy, respectively, Subjects #7 and #8 showed progressive disease (PD) on follow-up computed tomography (CT) scans. In these patients, serum levels of *ADAM9* mRNA in the follow-up were not significantly different from the pre-treatment levels ($p > 0.05$) (Figure 2A,B). Both patients succumbed to HCC progression and liver failure within 6 months since the start of nivolumab therapy.

In contrast, Subjects #9 and #10 started to reveal tumor regression after four cycles of nivolumab therapy. Their clinical courses are presented in Figures S1 and S2. Later, both patients exhibited partial response (PR) and survived longer than 6 months after the nivolumab therapy. In these patients, serum levels of *ADAM9* mRNA decreased significantly. In Subject #9, the serum *ADAM9* mRNA dropped from the pre-treatment level of 573.98 ± 5.16 to 523.85 ± 7.07 ($p < 0.05$) after two cycles, and further dropped to 262.58 ± 20.13 ($p < 0.05$) after 4 cycles of nivolumab therapy (Figure 2C). In Subject #10, it dropped from 323.88 ± 10.67 to 85.52 ± 5.59 ($p < 0.05$, Figure 2D) after two cycles of nivolumab therapy.

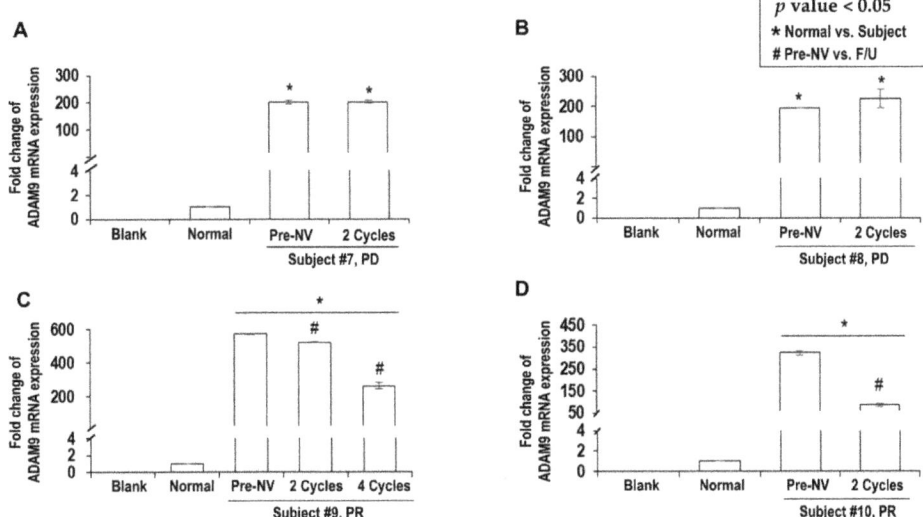

Figure 2. Expression level changes of serum *ADAM9* mRNA in advanced HCC patients treated with nivolumab. The number of nivolumab cycles completed prior to follow-up blood sampling is denoted on the X-axis. After 2 cycles of nivolumab therapy, non-responders did not show significant change in *ADAM9* mRNA (**A**, Subject #7; **B**, Subject #8). In contrast, responders exhibited a significant decrease in *ADAM9* mRNA (**C**, Subject #9, from 573.98 ± 5.16 to 523.85 ± 7.0 ($p < 0.05$) after 2 cycles, and further down to 262.58 ± 20.13 ($p < 0.05$) after 4 cycles; **D**, Subject #10, from 323.88 ± 10.67 to 85.52 ± 5.59 ($p < 0.05$) after 2 cycles). Abbreviations: NV, nivolumab; F/U, follow-up; PD, progressive disease; PR, partial response.

2.4. Immunophenotype Changes following Nivolumab Therapy

Lymphocyte immunophenotypes were tested before and after three cycles of nivolumab therapy for Subject #7 and four cycles for the rest (Table 2 and Figures S3–S6). Before nivolumab therapy, NK cells in Subjects #8 and #9 and cytotoxic T cells in Subject #9 were depleted, and these were therefore not detected for inhibitory checkpoint markers. In subject #8 and #10, PD-1 or T cell immunoglobulin- and mucin-domain-containing molecule-3 (TIM-3) positive cytotoxic T cells decreased significantly, but this change was not correlated with response to nivolumab. In contrast, TIM-3 positive helper T cells decreased in responders, but increased in non-responders.

Table 2. Lymphocyte immunophenotypes of the 4 patients who received nivolumab therapy (%).

Cell Type	Phenotype Marker	Non-Responders				Responders			
		Subject #7		Subject #8		Subject #9		Subject #10	
		Pre-NV	3 Cycles	Pre-NV	4 Cycles	Pre-NV	4 Cycles	Pre-NV	4 Cycles
T cells	$CD3^+$	63.62	52.98	74.53	84.57	6.18	19.02	45.01	53.65
Helper T cells	$CD3^+CD4^+$	27.46	26.36	59.91	52.43	5.51	10.58	30.57	35.58
Cytotoxic T cells	$CD3^+CD8^+$	34.15	20.03	13.95	30.7	0.73	7.41	14.57	18.75
B cells	$CD19^+$	10.15	7.27	1.88	1.06	3.21	60.7	19.1	5.84
NK cells	$CD3^-CD56^+$	10.77	5	0.21	5.52	0.61	7.49	19.87	18.88
NKT cells	$CD3^+CD56^+$	4	2.37	2.03	2.59	0.23	0.78	6.68	4.63

Table 2. Cont.

Cell Type	Phenotype Marker	Non-Responders				Responders			
		Subject #7		Subject #8		Subject #9		Subject #10	
		Pre-NV	3 Cycles	Pre-NV	4 Cycles	Pre-NV	4 Cycles	Pre-NV	4 Cycles
Cytotoxic T cells	CD3$^+$CD8$^+$PD-1$^+$	3.95	6.48	22.34	2.87	ND	0.12	24.65	7.99
	CD3$^+$CD8$^+$TIM3$^+$	21.05	52.11	15.06	5.3	ND	45.32	11.57	6.73
	CD3$^+$CD8$^+$LAG3$^+$	6.58	16.47	38.57	37.64	ND	3.33	29.95	30.63
	CD3$^+$CD8$^+$BTLA$^+$	14.47	11.32	29.87	54.32	ND	6.7	2.7	1.79
Helper T cells	CD3$^+$CD4$^+$PD-1$^+$	0	1.22	15.44	2.15	7.69	7.34	13.09	3.3
	CD3$^+$CD4$^+$TIM3$^+$	19.81	37.3	3.42	11.89	69.74	45.52	10.2	4.01
	CD3$^+$CD4$^+$LAG3$^+$	0.94	16.51	35.16	25.04	8.21	2.42	6.93	4.53
	CD3$^+$CD4$^+$BTLA$^+$	13.16	9.4	24.81	29.19	0.51	2	1.94	11.14
NK cells	CD3$^-$CD56$^+$PD-1$^+$	8.7	1.24	ND	1.58	ND	1.19	4.39	3.94
	CD3$^-$CD56$^+$TIM-3$^+$	4.35	22.39	ND	21.45	ND	33.6	25.45	14.6
	CD3$^-$CD56$^+$LAG3$^+$	0	2.74	ND	14.51	ND	5.06	46.1	25.15
	CD3$^-$CD56$^+$BTLA$^+$	47.83	26.18	ND	9.15	ND	3.45	1.63	1.47

Blood samples for the follow-up were acquired after three cycles of nivolumab therapy for Subject #7 and four cycles for the rest. Abbreviations: NV, nivolumab; HCC, hepatocellular carcinoma; NK, natural killer; NKT, natural killer-T; CD, cluster differentiation; ND, not detected; PD-1, programmed cell death 1; TIM-3, T cell immunoglobulin- and mucin-domain-containing molecule 3; LAG-3, lymphocyte activation gene 3; BTLA, B and T lymphocyte attenuator.

2.5. Serum ADAM9 mRNA Expression was Completely Suppressed in Complete Response of HCC

Plasma *ADAM9* mRNA level was completely suppressed in one patient (Table 1 and Figure 1, Subject #6), who achieved complete response (CR). The clinical treatment course and the outcomes are detailed in Figures S7 and S8. This patient was a 59-year-old man with CHB HCC and had tumor recurrence after surgical resection. Since the HCC progressed on sorafenib and radiotherapy, nivolumab was administered in combination with activated autologous NK cell therapy (2–6 × 10^9 cells/100 mL, intravenous infusion every 4 weeks). With serum AFP level increasing continuously, disease progression at the molecular level was suspected that nivolumab was switched to the third-line chemotherapy, regorafenib. To allow synergistic action of NK cells upon suppression of *ADAM9* protease and sMICA by regorafenib [24], the patient continued to receive the NK cell therapy every 4 weeks up to six times and then every 8 weeks up to a total of 15 times. At 6 months after beginning regorafenib with concomitant NK cell therapy, the patient achieved nearly complete regression of HCC. At this point, we acquired the patient's blood sample for analysis and found that *ADAM9* mRNA was not detectable at all in his plasma (Figure 1). From then on, immunophenotype changes were also checked and a decrease in inhibitory checkpoint molecules was observed (Table S2).

2.6. ADAM9 was Associated with HCC Prognosis in TCGA Database

To evaluate the effect of *ADAM9* expression on HCC prognosis, we performed *in-silico* analyses of 370 HCC patients from the TCGA database. Kaplan-Meier plots revealed that the group with *ADAM9* expression the higher than the median had a significantly poorer overall survival rate (Log-rank test $p = 3.9 \times 10^{-4}$) (Figure 3A). In addition, *ADAM9* was significantly upregulated in primary tumor tissues of HCC ($n = 370$) compared with adjacent normal liver tissues ($n = 50$) (*t*-test $p = 4.6 \times 10^{-6}$) (Figure 3B). Unlike *ADAM9*, other ADAM family genes *ADAM10* and *ADAM17* neither differed in their expression levels between HCC tumor tissues and adjacent normal liver tissues, nor showed significant correlation with survival analysis (Figures S9 and S10).

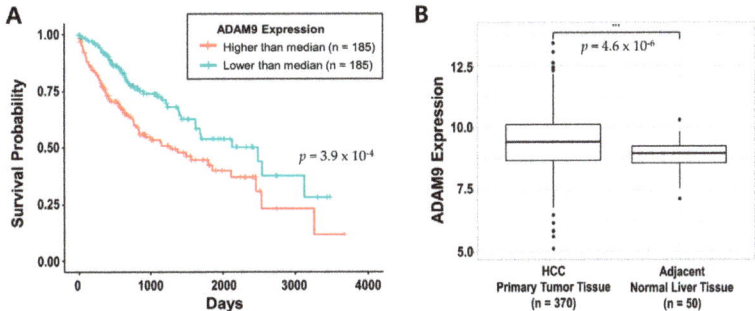

Figure 3. Effect of *ADAM9* expression on HCC prognosis in the TCGA database. (**A**) Kaplan–Meier plot of HCC patients (n = 370) according to *ADAM9* expressions level higher or lower than median (n = 185 for each group). (**B**) Box-plot comparing *ADAM9* expression between HCC primary tumor (n = 370) and adjacent normal liver tissue (n = 50). Abbreviations: HCC, hepatocellular carcinoma; TCGA, The Cancer Genome Atlas.

2.7. *ADAM9 Expression is Positively Correlated with PD-1, TIM-3 and BTLA*

The correlation between *ADAM9* expression and expression of immune checkpoint molecules (*PD-1, TIM-3, lymphocyte activation gene-3 (LAG-3)* and *B and T lymphocyte attenuator (BTLA))* in HCC patients (n = 370) from TCGA database was analyzed. *ADAM9* expression was positively correlated with the expression of *PD-1, TIM-3*, and *BTLA* but not with that of *LAG-3* (Figure 4). *TIM-3* had the strongest positive correlation with *ADAM9* (Correlation coefficient r = 0.37 and $p = 1.3 \times 10^{-13}$) (Figure 4B).

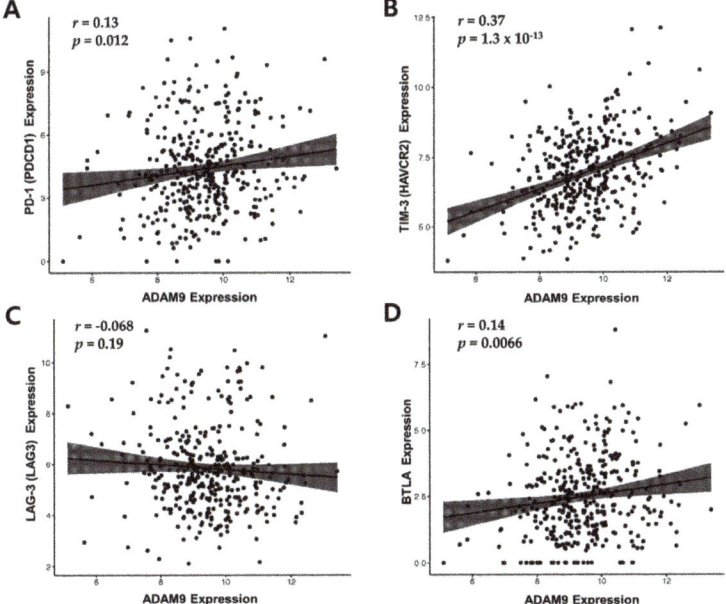

Figure 4. Correlation between *ADAM9* expression and four immune checkpoint genes *PD-1* (**A**), *TIM-3* (**B**), *LAG-3* (**C**), and *BTLA* (**D**). Abbreviations: r, Pearson's correlation coefficient; PD-1, programmed cell death 1; TIM-3, T cell immunoglobulin- and mucin-domain-containing molecule 3; LAG-3, lymphocyte activation gene 3; BTLA, B and T lymphocyte attenuator.

3. Discussion

In the present study, we found that the *ADAM9* blood mRNA level was significantly elevated in HCC patients compared to healthy controls. Furthermore, the magnitude of this elevation was much greater in the patients with previous treatment failure than in the newly diagnosed treatment-naïve patients. These findings suggested that *ADAM9* expression increase as HCC progresses, and that the immune evasion mechanisms involving *ADAM9* protease probably aggravate as HCC progresses. In addition, serum *ADAM9* mRNA levels decreased significantly in the two patients (one CHB and one non-viral HCC) who responded to nivolumab. Furthermore, *ADAM9* mRNA was not detected at all in the plasma of one HCC patient who achieved CR with regorafenib and NK cell combination immunotherapy. As such, the present study demonstrated, for the first time, the functional relevance of blood *ADAM9* mRNA levels with clinical response to HCC treatments. Since the sample size was too small in our study, we probed the TCGA data to find supporting evidence and found that higher *ADAM9* expression level was significantly associated with poor prognosis of HCC. In contrast, *ADAM10* and *ADAM17*, which were also reportedly related to MICA shedding [26], showed no such association.

In our study, the change of serum *ADAM9* mRNA was easily correlated with the clinical response to nivolumab, while the observed lymphocyte immunophenotype changes were intriguing but not enough to draw a solid conclusion. Namely, Subject #8 did show some favorable changes in her lymphocyte immunophenotypes (mainly PD-1+ or TIM-3+ cytotoxic T cells) but tumor progression was still seen. Another interesting finding was the change of TIM-3 positive helper T cell; responders (Subjects #9 and #10) showed a decrease while non-responders (Subjects #7 and #8) showed an increase. In addition, *ADAM9* was strongly correlated with *TIM-3* in the TCGA database. These findings altogether suggest that key features of lymphocyte immunophenotype changes that can predict treatment response early on may exist. Our study indicated that the proportion of TIM-3 positive helper T cells may be a good candidate marker, and that unleashing helper T cell from exhaustion may be more important than unleashing cytotoxic T cells. Future studies are needed to elucidate the interplay between the ADAM9-MICA-NKG2D system and lymphocyte immunophenotypes, and to find the relevance between such factors and clinical outcome.

ADAMs belong to the zinc protease superfamily, and they are usually transmembrane proteins [27]. Containing disintegrin and metalloprotease domains, ADAMs take part in multiple cellular functions including cell adhesion and migration, proteolysis of the extracellular matrix and shedding of membrane proteins [27,28]. Several studies have indicated that ADAMs are involved in tumor development and progression of HCC [23,24,27–40]. Though the pathogenesis of HCC is multifactorial, it is largely due to hepatitis B (HBV) or C virus (HCV) infection and alcoholic or non-alcoholic fatty liver disease. These underlying liver conditions result in chronic inflammation and liver fibrosis, causing continuous remodeling of the extracellular matrix [27]. ADAMs are involved in this inflammatory process that leads to development of HCC [27]. Among several ADAMs associated with HCC, those related to MICA shedding are noteworthy as MICA is a critical part of cytotoxic cellular immunity. Our study revealed that only *ADAM9* was significantly associated with prognosis of HCC while others, *ADAM10* and *ADAM17*, were not. Regarding *ADAM9*, previous studies demonstrated that transcriptional suppression of *ADAM9* led to inhibition of proliferation and invasion activities of HCC cell lines [23,24,36–40]. Inhibition of *ADAM9* protease also showed similar results [41]. Some of these results were backed by the increased mMICA expression and subsequently increased susceptibility of HCC cells to NK cells [23,24,40]. On the other hand, treatment with interleukin (IL)-1β on HCC cell lines increased the expression of *ADAM9* and sMICA, and the IL-1β-treated HCC cells became more resistant to the cytolytic activity of NK cells [42].

MICA is a ligand of NKG2D, and it triggers NK cell or CD8+ T cell-mediated cytokine release and cytotoxicity towards the target cells [24]. MICA expression is induced in response to various types of stress such as heat, DNA damage, and viral infection [43]. The *ADAM9* protease that interferes with the MICA-NKG2D system was overexpressed in human HCC tissues [24]. In vitro, *ADAM9* knockdown increased mMICA expression, decreased sMICA production, and increased the cytolytic activity of NK cells against HCC cells [23]. Sorafenib suppresses the expression of ADAM9. Therefore, sorafenib treatment in vitro showed the same results as the *ADAM9* knockdown [23]. This important phenomenon was reproduced by another group. This time regorafenib was tested. Regorafenib suppressed the expression of *ADAM9* and ADAM10 to a greater extent than sorafenib [24]. This suppression led to mMICA accumulation and inhibition of sMICA production [24]. The authors suggested that regorafenib is superior to sorafenib as regorafenib suppresses not only *ADAM9* but also ADAM10 [24].

Beyond these experimental findings, the involvement of the MICA-NKG2D system was described in HCC patients. Both mMICA and sMICA showed elevated expression in human HCC [44–46]. Serum levels of sMICA were significantly elevated in chronic liver disease and HCC patients compared to healthy controls [44]. Furthermore, there was a stepwise increase in the level of sMICA as liver disease progressed from chronic hepatitis to cirrhosis, low-grade HCC and high-grade HCC [44]. In CHC patients, serum IL-1β levels were positively correlated with serum sMICA level, and serum IL-1β levels were significantly higher in CHC patients with HCC than those without HCC [42]. On the other hand, sMICA decrease could suggest favorable response to HCC treatment. For example, serum sMICA levels decreased, and NKG2D expression on NK cells and CD8+ T cells increased significantly after transcatheter arterial embolization therapy [44]. This could be related to our finding that *ADAM9* mRNA decreased in HCC patients who responded to treatments. Interestingly, there were divergent findings on the level of sMICA and its association with incidence or prognosis of HCC. In HCC with CHB, higher sMICA was correlated with vascular invasion and poor prognosis and was associated with a G allele of single nucleotide polymorphism (SNP) *rs2596542*, a risk allele for HBV-induced HCC ($p = 0.029$, odds ratio = 1.19) [47]. In contrast, A allele of SNP *rs2596542* was significantly associated with the higher risk of HCV-induced HCC ($p = 4.21 \times 10^{-13}$, odds ratio = 1.39) as well as low levels of sMICA [48]. These findings seemingly contradict each other, but could be explained by the fact that there may be an interplay between HCC etiology, i.e., the type of hepatitis virus or genetic factors and ADAM9-MICA-NKG2D system.

HCC occurrence in HBV infection can be partly ascribed to the perturbation of signaling pathways by HBV-encoded X protein (HBx) incorporated into the human genome [49]. HBx was associated with enhanced expression of MMPs such as *ADAM9* and ADAM10 [50,51]. More MMPs expressed could mean more MICA shedding. Therefore, this viral factor explains the association between the higher sMICA levels and the poorer prognosis of HCC with CHB [43,47]. In the absence of such interactions between virus and MMPs, the level of sMICA may, more or less, linearly follow that of mMICA expression, as mMICA is the source of sMICA [48]. In this case, low sMICA levels reflect low mMICA expression resulted from host genetic factor, which may indicate weakened cytotoxic immune surveillance, as was observed in HCV-induced HCC [48]. In addition, hepatitis B surface antigen is known to inhibit MICA expression via induction of cellular micro RNAs [52].

Anti-viral treatment for HBV and HCV was much improved by the development of new antiviral agents. This notwithstanding, the development of HCC still remains a concern even after HBV suppression [53,54] or HCV eradication [55]. In these patients, NK cells were usually observed to have depressed function with decreased capacity of interferon-γ production, impaired IL-15 production, or decreased expression of NK cell activation receptors including NKG2D [51,56,57]. Representing 30–50% of all hepatic lymphocytes, NK cells increase up to 90% in patients with hepatic malignancy. Therefore, ADAM9-MICA-NKG2D system may also serve as a good target for HCC prevention strategy in patients chronically infected with HBV or HCV. In this regard, it is encouraging that one group found an approved drug for anti-alcoholism, disulfiram, and showed that it effectively restored mMICA expression by inhibiting ADAM10 and did not have unfavorable off-target effects [58].

One of the unmet needs in this era of immunologic treatments for HCC or other cancers is a biomarker that can predict treatment response well in advance. It would be very useful if the marker could dictate the best cancer immunotherapy course. Despite the proven efficacy of nivolumab in HCC, expression of a known immune marker, PD-L1, in tumor tissues, fell short of predicting treatment response [16]. Our and previous studies have suggested that *ADAM9* mRNA may fill this unmet need [23,24,43]. *ADAM9* transcript elevation might suggest the existence of an *ADAM9* associated immune evasion mechanism in tumors. Decrease or increase of *ADAM9* mRNA after 2–4 cycles of treatment may help predict treatment response or failure early on. Finally, a decrease of *ADAM9* mRNA may indicate a candidate who can expect synergistic effects by adding adoptive cell therapy. Adoptive cell therapy composed of NK or CD8+ T cells have shown efficacy in treatment of HCC [3,17,22,59–61]. However, the patient population who had survival benefits was mostly restricted to those with minimal tumor burden after treatments with curative modality [17,22,62]. Therefore, there remains significant room for improvement of the adoptive cell therapy in patients with high burden of HCC [17,22,62]. This combination strategy may greatly enhance the survival of advanced HCC patients, as was shown in Subject #6. Thus, *ADAM9* mRNA may help select the patients who may benefit from such combination immunotherapy.

There were limitations in our study. The population size was too small. We did not investigate whether the findings of our study could be applied for other agents approved for advanced HCC or not. To complement such weaknesses, we probed the TCGA data and retrieved encouraging results. This pilot study demonstrates that *ADAM9* mRNA is associated with clinical response to HCC treatment and that there is an important link between the MICA-NKG2D system and prognosis of HCC, by showing how *ADAM9* mRNA expression changes over time as HCC patients received nivolumab therapy. Thus, *ADAM9* mRNA has potential as a biomarker to predict the clinical response of HCC patients, and the ADAM9-MICA-NKG2D system may be a good therapeutic target in HCC immunotherapy. Restoration of MICA-NKG2D signaling by suppressing *ADAM9* mRNA and its influence on the tumor microenvironment and its potential as a prognostic marker of human HCC patients should be investigated in future studies. The knowledge garnered will help us to formulate better strategies to manage the advanced HCC patients.

4. Materials and Methods

4.1. Patients

This study was conducted in CHA Bundang Hospital between January 2017 and October 2019. HCC patients who were to be treated with sorafenib, regorafenib or nivolumab as the standard-of-care therapy were eligible for this study. Patients who were 19 years of age or older and were willing to participate in this study were enrolled. For enrollment, the patient must have met the inclusion criteria which included presentation of HCC, diagnosed histologically or radiologically, in accordance with the guidelines of the American Association for the Study of Liver Diseases or the European Association for the Study of the Liver [5,10]. Other inclusion criteria were Child-Pugh score ≤ 7, BCLC stage B or C, Eastern Cooperative Oncology Group (ECOG) performance status score of 0 or 1, adequate bone marrow (hemoglobin ≥ 9 g/dL, granulocyte count >1000/mm^3 and platelet count ≥ 40,000/mm^3), adequate liver function (serum aspartate aminotransferase and alanine aminotransferase < 5 times of upper normal limit, bilirubin ≤ 3 mg/dL, prothrombin time international normalized ratio (PT INR) < 1.5) and adequate renal function (serum creatinine < 1.5 times the upper normal limit). In addition, HCC lesions must be measurable by CT or magnetic resonance imaging (MRI) and there should be at least one lesion ≥ 1.0 cm in maximum diameter. Exclusion criteria were evidence of malignant tumor of other than HCC, uncontrolled ascites, hepatic encephalopathy, major bleeding event within 30 days, severe bacterial infection, infection with HIV, major cardiac disease, anticoagulation therapy, pregnant or breast-feeding woman, dysphagia precluding drug administration, and other contraindications for systemic HCC therapy. Both written and oral consent was obtained prior to sample collection. The study was approved by the institutional review board (IRB) of the CHA Bundang Medical Center (IRB protocol: 2016-03-039-019).

4.2. Study Design and Protocol

This study was a prospective observational clinical research study. After enrollment, the first blood samples were drawn within 3 weeks before and 4 days after the commencement of chemotherapy with sorafenib, regorafenib or nivolumab. Afterwards, follow-up blood samples were acquired at the 5th, 10th and 20th week or when chemotherapy was terminated. Blood samples were drawn into two plain bottles (BD Vacutainer®, 5 mL), two EDTA bottles (BD Vacutainer®, 3 mL) and one heparin bottle (BD Vacutainer®, 10 mL). Immune cells acquired from the EDTA and heparin bottles were immediately subjected to examination of immunophenotypes and functional assays. The plasma samples in the heparin bottles and serum samples in the plain bottles were stored at −80 °C for further *ADAM9* mRNA quantification.

Patients' age, sex, etiology of HCC and Child–Pugh score were recorded. Clinical and TNM stages were classified according to the BCLC clinical stage and the mUICC stages, respectively [5,63]. Tumor response (CR, PR, stable disease, and PD) was evaluated in accordance with the modified Response Evaluation Criteria in Solid Tumors (mRECIST) [64] every 10–12 weeks for more than 1 year via dynamic liver CT or MRI. Complete blood cell count, blood chemistry (liver function, renal function, metabolic function, electrolyte), PT INR, and tumor markers (AFP and PIVKA-II) were regularly checked as part of standard care.

4.3. mRNA Isolation and Real-Time PCR

To evaluate the serum or plasma *ADAM9* mRNA expression levels, quantitative real-time reverse transcription-polymerase chain reaction (RT-PCR) was performed using a CFX Connect real-time system (Bio-Rad, Hercules, CA, USA). Total RNA was isolated from serum or plasma samples using the miRNeasy Serum/Plasma Kit (Qiagen, Hilden, Germany). For normal controls, we used banking serum and plasma from women ($n = 5$, mean age 34.2 years, range 29–41 years) who delivered at term ($\geq$35 gestational weeks) because we had difficulty with collection of normal serum. The collection and use of these samples for research purposes was approved by the IRB of CHA Hospital (Seoul, Korea) (IRB protocol: 2006-12). The mRNA levels of *ADAM9* and glyceraldehyde phosphate dehydrogenase (GAPDH) were determined by quantitative real-time RT-PCR using SYBR green mastermix (Roche, Basel, Switzerland) according to the manufacturer's instructions. PCR reactions were performed by denaturation at 95 °C for 10 min followed by 45 cycles of amplification at 95 °C for 10 s, 55 °C for 15 s, and then melting curves were performed after PCR amplification under the following conditions: 60 °C to 95 °C with a temperature transition rate of 1 °C/s. The sequence of the PCR primers used for detection of *ADAM9* and *GAPDH* were as follows: *ADAM9* forward primer, 5'-GGAAACTGCCTT CTTAATATTCCAAA-3', *ADAM9* reverse primer, 5'-CCCAGCGTC CACCAACTTAT-3', *GAPDH* forward primer, 5'- CTCCTCTTCGGCAGCACA-3', *GAPDH* reverse primer, 5'- AACGCTTCACCTA ATTTG CG T -3'. *GAPDH* mRNA from each sample was quantified as an endogenous control of internal RNA. All experiments were performed in duplicate.

4.4. Flow Cytometric Analysis

HCC patient's blood in the heparin bottle was transferred to a 15 mL conical tube containing 5 mL filcol-paque plus (GE Healthcare, Chicago, IL, USA) and centrifuged at 2000 rpm for 10 min. Plasma was transferred to a 5 mL cryovial (Corning Inc, Corning, NY, USA) and stored at −80 °C. The buffy coat layer was transferred to a new 15 mL conical tube washed 2 times with PBS. Peripheral blood mononuclear cells (PBMCs) were stained with anti-CD3-eFluor 506 (ebioscience, San Diego, CA, USA), anti-CD3-eFluor 450 (ebioscience, USA), anti-CD4-eFluor 450 (ebioscience, USA), anti-CD4-PE (ebioscience), anti-CD8-Alexa Fluor 700 (ebioscience), anti-CD19-PE (ebioscience), anti-CD56-Alexa Fluor 700 (ebioscience), anti-CD16-PerCP/Cy5.5 (BioLegend, San Diego, CA, USA), anti-NKG2D-FITC (ebioscience), anti-CD158b-PerCP/Cy5.5 (BioLegend), anti-TIM3-FITC (ebioscience), anti-LAG-3-PE (BioLegend), anti-BTLA-APC/Cy7 (BioLegend), anti-PD-1-APC (BioLegend), anti-CTLA4-APC

(Miltenyi Biotec, Bergisch Gladbach, Germany), anti-mouse IgG1 kappa isotype control-FITC (ebioscience, USA), anti-mouse IgG1 kappa isotype control-PE (ebioscience), anti-mouse IgG1 kappa isotype control-PerCP/Cy5.5 (BioLegend), REA control-APC (Miltenyi Biotec), anti-mouse IgG1 kappa isotype control-Alexa Fluor 700 (ebioscience), anti-mouse IgG1 kappa isotype control-APC/Cy7 (BioLegend), anti-mouse IgG1 kappa isotype control-eFluor 506 (ebioscience), anti-mouse IgG1 kappa isotype control-eFluor 450 (ebioscience). Stained cells were analyzed with CytoFLEX flow cytometry (Beckman Coulter, Brea, CA, USA) and resulting data were analyzed using Kaluza version 1.5a analysis software (Beckman Coulter).

4.5. Statistical Analysis

Data are expressed as the mean (± standard deviation) or frequencies (percentages), as appropriate. The statistical significance of differences between the groups was determined by Student t test or two-sample t test. A p-value of <0.05 was considered statistically significant. Statistical analyses were performed with SPSS 22.0 (SPSS Inc., Chicago, IL, USA).

4.6. In-Silico Analysis with TCGA Database

We downloaded transcriptomic, survival and clinical data of HCC patients (indexed as LIHC) from the Xena TCGA database hub (https://xenabrowser.net). The transcriptomic data included 370 patients and was generated by the University of North Carolina TCGA genome characterization center. The survival data includes information on overall survival. Statistical analyses were performed with t-test, Pearson's correlation analysis, Cox-regression and log-rank analysis. All statistical analyses with TCGA dataset were performed with Python (Version 2.7.10) and R-studio (Version 1.1.456).

5. Conclusions

ADAM9 mRNA was overexpressed in blood samples of patients with advanced HCC. Decreased levels of *ADAM9* mRNA in the blood was significantly associated with clinical response to HCC treatment with nivolumab. Also, using the TCGA database, we found that higher expression of *ADAM9* in HCC tumor tissues is associated with poor survival of HCC patients. Therefore, *ADAM9* mRNA has a potential as a biomarker predicting clinical response, and the ADAM9-MICA-NKG2D system may serve as a therapeutic target of HCC immunotherapy.

Supplementary Materials: The following are available online at http://www.mdpi.com/2072-6694/12/3/745/s1, Table S1: Baseline patient and tumor characteristics of the study population (n = 10), Table S2: Changes in the proportions of immune checkpoint molecules and immune cells in the HCC patient (Subject #6), who showed complete tumor response following regorafenib and NK cell combination therapy, Figures S1 and S2: Clinical course and serum AFP levels of Subjects #9 and #10, Figures S3–S6: Lymphocyte distribution and inhibitory checkpoint molecule expression before and after nivolumab therapy in Subjects #7-#10. Figure S7: Clinical course and serum AFP levels of Subject #6, Figure S8: Completely resolved HCC after regorafenib and autologous NK cell combined therapy in one patient (Subject #6), Figure S9–S10: Effect of *ADAM10* and *ADAM17* expression on HCC prognosis.

Author Contributions: S.O. and J.L. contributed to the study design, data collection, study analysis, manuscript writing, critical review of the manuscript, and final approval of the manuscript submission. S.O., Y.P., J.L., H.-J.L., S.-H.L., Y.-S.B., K.K. and G.J.K. performed the experiments, analyzed the data, and wrote the manuscript. S.-M.L. and S.-K.C. assisted in performing the experiments. Y.C., Y.-E.C., Y.H., M.K., S.-G.H. and K.K. assisted in analyzing and interpreting the data. S.O., J.L., K.K. and G.J.K. contributed to data interpretation, critical review of the manuscript, and final approval of the manuscript submission. All authors have read and agreed to the published version of the manuscript.

Funding: This research was supported by a grant from the "Gyeonggi-Incheon study group of the Korean Association for the Study of the Liver."

Acknowledgments: The authors thank 'Immunotherapy Development Team of CHA Biolab (Seongnam, Korea)' for their technical assistance.

Conflicts of Interest: The authors declare no conflict of interest.

References

1. Chon, Y.E.; Park, H.; Hyun, H.K.; Ha, Y.; Kim, M.N.; Kim, B.K.; Lee, J.H.; Kim, S.U.; Kim, D.Y.; Ahn, S.H.; et al. Development of a New Nomogram Including Neutrophil-to-Lymphocyte Ratio to Predict Survival in Patients with Hepatocellular Carcinoma Undergoing Transarterial Chemoembolization. *Cancers* **2019**, *11*, 509. [CrossRef] [PubMed]
2. Ferlay, J.; Soerjomataram, I.; Dikshit, R.; Eser, S.; Mathers, C.; Rebelo, M.; Parkin, D.M.; Forman, D.; Bray, F. Cancer incidence and mortality worldwide: Sources, methods and major patterns in GLOBOCAN 2012. *Int. J. Cancer* **2015**, *136*, E359–E386. [CrossRef] [PubMed]
3. Lee, J.-H.; Oh, S.-Y.; Kim, J.Y.; Nishida, N. Cancer immunotherapy for hepatocellular carcinoma. *Hepatoma. Res.* **2018**, *4*, 51. Available online: https://hrjournal.net/article/view/2776 (accessed on 20 March 2020). [CrossRef]
4. Nishida, N.; Kudo, M. Immune checkpoint blockade for the treatment of human hepatocellular carcinoma. *Hepatol. Res.* **2018**, *48*, 622–634. [CrossRef]
5. European Association for the Study of the Liver. EASL Clinical Practice Guidelines: Management of hepatocellular carcinoma. *J. Hepatol.* **2018**, *69*, 182–236. [CrossRef]
6. Kudo, M.; Finn, R.S.; Qin, S.; Han, K.H.; Ikeda, K.; Piscaglia, F.; Baron, A.; Park, J.W.; Han, G.; Jassem, J.; et al. Lenvatinib versus sorafenib in first-line treatment of patients with unresectable hepatocellular carcinoma: A randomised phase 3 non-inferiority trial. *Lancet* **2018**, *391*, 1163–1173. [CrossRef]
7. Saffo, S.; Taddei, T.H. Systemic Management for Advanced Hepatocellular Carcinoma: A Review of the Molecular Pathways of Carcinogenesis, Current and Emerging Therapies, and Novel Treatment Strategies. *Dig. Dis. Sci.* **2019**, *64*, 1016–1029. [CrossRef]
8. Ueshima, K.; Nishida, N.; Hagiwara, S.; Aoki, T.; Minami, T.; Chishina, H.; Takita, M.; Minami, Y.; Ida, H.; Takenaka, M.; et al. Impact of Baseline ALBI Grade on the Outcomes of Hepatocellular Carcinoma Patients Treated with Lenvatinib: A Multicenter Study. *Cancers* **2019**, *11*, 952. [CrossRef]
9. Daher, S.; Massarwa, M.; Benson, A.A.; Khoury, T. Current and Future Treatment of Hepatocellular Carcinoma: An Updated Comprehensive Review. *J. Clin. Transl. Hepatol.* **2018**, *6*, 69–78. [CrossRef]
10. Heimbach, J.K.; Kulik, L.M.; Finn, R.S.; Sirlin, C.B.; Abecassis, M.M.; Roberts, L.R.; Zhu, A.X.; Murad, M.H.; Marrero, J.A. AASLD guidelines for the treatment of hepatocellular carcinoma. *Hepatology* **2018**, *67*, 358–380. [CrossRef]
11. Kulik, L.; El-Serag, H.B. Epidemiology and Management of Hepatocellular Carcinoma. *Gastroenterology* **2019**, *156*, 477–491.e471. [CrossRef] [PubMed]
12. Bteich, F.; Di Bisceglie, A.M. Current and Future Systemic Therapies for Hepatocellular Carcinoma. *Gastroenterol. Hepatol.* **2019**, *15*, 266–272.
13. Liu, X.; Qin, S. Immune Checkpoint Inhibitors in Hepatocellular Carcinoma: Opportunities and Challenges. *Oncologist* **2019**, *24*, S3–S10. [CrossRef] [PubMed]
14. Kudo, M. Combination Cancer Immunotherapy with Molecular Targeted Agents/Anti-CTLA-4 Antibody for Hepatocellular Carcinoma. *Liver Cancer* **2019**, *8*, 1–11. [CrossRef] [PubMed]
15. Marrero, J.A.; Kulik, L.M.; Sirlin, C.B.; Zhu, A.X.; Finn, R.S.; Abecassis, M.M.; Roberts, L.R.; Heimbach, J.K. Diagnosis, Staging, and Management of Hepatocellular Carcinoma: 2018 Practice Guidance by the American Association for the Study of Liver Diseases. *Hepatology* **2018**, *68*, 723–750. [CrossRef] [PubMed]
16. El-Khoueiry, A.B.; Sangro, B.; Yau, T.; Crocenzi, T.S.; Kudo, M.; Hsu, C.; Kim, T.Y.; Choo, S.P.; Trojan, J.; Welling, T.H.R.; et al. Nivolumab in patients with advanced hepatocellular carcinoma (CheckMate 040): An open-label, non-comparative, phase 1/2 dose escalation and expansion trial. *Lancet* **2017**, *389*, 2492–2502. [CrossRef]
17. Lee, J.H.; Lee, J.H.; Lim, Y.S.; Yeon, J.E.; Song, T.J.; Yu, S.J.; Gwak, G.Y.; Kim, K.M.; Kim, Y.J.; Lee, J.W.; et al. Adjuvant immunotherapy with autologous cytokine-induced killer cells for hepatocellular carcinoma. *Gastroenterology* **2015**, *148*, 1383–1391.e1386. [CrossRef]
18. Schmidt, N.; Neumann-Haefelin, C.; Thimme, R. Cellular immune responses to hepatocellular carcinoma: Lessons for immunotherapy. *Dig. Dis.* **2012**, *30*, 483–491. [CrossRef]
19. Kudo, M. Immune Checkpoint Inhibition in Hepatocellular Carcinoma: Basics and Ongoing Clinical Trials. *Oncology* **2017**, *92* (Suppl. 1), 50–62. [CrossRef]
20. Chen, D.S.; Mellman, I. Elements of cancer immunity and the cancer-immune set point. *Nature* **2017**, *541*, 321–330. [CrossRef]

21. Li, L.; Li, W.; Wang, C.; Yan, X.; Wang, Y.; Niu, C.; Zhang, X.; Li, M.; Tian, H.; Yao, C.; et al. Adoptive transfer of natural killer cells in combination with chemotherapy improves outcomes of patients with locally advanced colon carcinoma. *Cytotherapy* **2018**, *20*, 134–148. [CrossRef] [PubMed]
22. Oh, S.; Lee, J.H.; Kwack, K.; Choi, S.W. Natural Killer Cell Therapy: A New Treatment Paradigm for Solid Tumors. *Cancers* **2019**, *11*, 1534. [CrossRef] [PubMed]
23. Kohga, K.; Takehara, T.; Tatsumi, T.; Ishida, H.; Miyagi, T.; Hosui, A.; Hayashi, N. Sorafenib inhibits the shedding of major histocompatibility complex class I-related chain A on hepatocellular carcinoma cells by down-regulating a disintegrin and metalloproteinase 9. *Hepatology* **2010**, *51*, 1264–1273. [CrossRef] [PubMed]
24. Arai, J.; Goto, K.; Stephanou, A.; Tanoue, Y.; Ito, S.; Muroyama, R.; Matsubara, Y.; Nakagawa, R.; Morimoto, S.; Kaise, Y.; et al. Predominance of regorafenib over sorafenib: Restoration of membrane-bound MICA in hepatocellular carcinoma cells. *J. Gastroenterol. Hepatol.* **2018**, *33*, 1075–1081. [CrossRef] [PubMed]
25. Kudo, M.; Izumi, N.; Ichida, T.; Ku, Y.; Kokudo, N.; Sakamoto, M.; Takayama, T.; Nakashima, O.; Matsui, O.; Matsuyama, Y. Report of the 19th follow-up survey of primary liver cancer in Japan. *Hepatol. Res.* **2016**, *46*, 372–390. [CrossRef] [PubMed]
26. Waldhauer, I.; Goehlsdorf, D.; Gieseke, F.; Weinschenk, T.; Wittenbrink, M.; Ludwig, A.; Stevanovic, S.; Rammensee, H.G.; Steinle, A. Tumor-associated MICA is shed by ADAM proteases. *Cancer Res.* **2008**, *68*, 6368–6376. [CrossRef]
27. Mazzocca, A.; Giannelli, G.; Antonaci, S. Involvement of ADAMs in tumorigenesis and progression of hepatocellular carcinoma: Is it merely fortuitous or a real pathogenic link? *Biochim. Biophys. Acta* **2010**, *1806*, 74–81. [CrossRef]
28. Seals, D.F.; Courtneidge, S.A. The ADAMs family of metalloproteases: Multidomain proteins with multiple functions. *Genes Dev.* **2003**, *17*, 7–30. [CrossRef]
29. Xia, C.; Zhang, D.; Li, Y.; Chen, J.; Zhou, H.; Nie, L.; Sun, Y.; Guo, S.; Cao, J.; Zhou, F.; et al. Inhibition of hepatocellular carcinoma cell proliferation, migration, and invasion by a disintegrin and metalloproteinase-17 inhibitor TNF484. *J. Res. Med. Sci.* **2019**, *24*, 26. [CrossRef]
30. Shiu, J.S.; Hsieh, M.J.; Chiou, H.L.; Wang, H.L.; Yeh, C.B.; Yang, S.F.; Chou, Y.E. Impact of ADAM10 gene polymorphisms on hepatocellular carcinoma development and clinical characteristics. *Int. J. Med. Sci.* **2018**, *15*, 1334–1340. [CrossRef]
31. Li, Y.; Ren, Z.; Wang, Y.; Dang, Y.Z.; Meng, B.X.; Wang, G.D.; Zhang, J.; Wu, J.; Wen, N. ADAM17 promotes cell migration and invasion through the integrin beta1 pathway in hepatocellular carcinoma. *Exp. Cell Res.* **2018**, *370*, 373–382. [CrossRef]
32. Honda, H.; Takamura, M.; Yamagiwa, S.; Genda, T.; Horigome, R.; Kimura, N.; Setsu, T.; Tominaga, K.; Kamimura, H.; Matsuda, Y.; et al. Overexpression of a disintegrin and metalloproteinase 21 is associated with motility, metastasis, and poor prognosis in hepatocellular carcinoma. *Sci. Rep.* **2017**, *7*, 15485. [CrossRef]
33. Liu, Y.; Zhang, W.; Liu, S.; Liu, K.; Ji, B.; Wang, Y. miR-365 targets ADAM10 and suppresses the cell growth and metastasis of hepatocellular carcinoma. *Oncol. Rep.* **2017**, *37*, 1857–1864. [CrossRef]
34. Li, S.Q.; Wang, D.M.; Zhu, S.; Ma, Z.; Li, R.F.; Xu, Z.S.; Han, H.M. The important role of ADAM8 in the progression of hepatocellular carcinoma induced by diethylnitrosamine in mice. *Hum. Exp. Toxicol.* **2015**, *34*, 1053–1072. [CrossRef]
35. Liu, S.; Zhang, W.; Liu, K.; Ji, B.; Wang, G. Silencing ADAM10 inhibits the in vitro and in vivo growth of hepatocellular carcinoma cancer cells. *Mol. Med. Rep.* **2015**, *11*, 597–602. [CrossRef]
36. Dong, Y.; Wu, Z.; He, M.; Chen, Y.; Chen, Y.; Shen, X.; Zhao, X.; Zhang, L.; Yuan, B.; Zeng, Z. ADAM9 mediates the interleukin-6-induced Epithelial-Mesenchymal transition and metastasis through ROS production in hepatoma cells. *Cancer Lett.* **2018**, *421*, 1–14. [CrossRef]
37. Hu, D.; Shen, D.; Zhang, M.; Jiang, N.; Sun, F.; Yuan, S.; Wan, K. MiR-488 suppresses cell proliferation and invasion by targeting *ADAM9* and lncRNA HULC in hepatocellular carcinoma. *Am. J. Cancer Res.* **2017**, *7*, 2070–2080.
38. Wan, D.; Shen, S.; Fu, S.; Preston, B.; Brandon, C.; He, S.; Shen, C.; Wu, J.; Wang, S.; Xie, W.; et al. miR-203 suppresses the proliferation and metastasis of hepatocellular carcinoma by targeting oncogene *ADAM9* and oncogenic long non-coding RNA HULC. *Anti-Cancer Agents Med. Chem.* **2016**, *16*, 414–423. [CrossRef]
39. Zhou, C.; Liu, J.; Li, Y.; Liu, L.; Zhang, X.; Ma, C.Y.; Hua, S.C.; Yang, M.; Yuan, Q. microRNA-1274a, a modulator of sorafenib induced a disintegrin and metalloproteinase 9 (ADAM9) down-regulation in hepatocellular carcinoma. *FEBS Lett.* **2011**, *585*, 1828–1834. [CrossRef]

40. Kohga, K.; Tatsumi, T.; Takehara, T.; Tsunematsu, H.; Shimizu, S.; Yamamoto, M.; Sasakawa, A.; Miyagi, T.; Hayashi, N. Expression of CD133 confers malignant potential by regulating metalloproteinases in human hepatocellular carcinoma. *J. Hepatol.* **2010**, *52*, 872–879. [CrossRef]
41. Itabashi, H.; Maesawa, C.; Oikawa, H.; Kotani, K.; Sakurai, E.; Kato, K.; Komatsu, H.; Nitta, H.; Kawamura, H.; Wakabayashi, G.; et al. Angiotensin II and epidermal growth factor receptor cross-talk mediated by a disintegrin and metalloprotease accelerates tumor cell proliferation of hepatocellular carcinoma cell lines. *Hepatol. Res.* **2008**, *38*, 601–613. [CrossRef]
42. Kohga, K.; Tatsumi, T.; Tsunematsu, H.; Aono, S.; Shimizu, S.; Kodama, T.; Hikita, H.; Yamamoto, M.; Oze, T.; Aketa, H.; et al. Interleukin-1beta enhances the production of soluble MICA in human hepatocellular carcinoma. *Cancer Immunol. Immunother. CII* **2012**, *61*, 1425–1432. [CrossRef]
43. Goto, K.; Kato, N. MICA SNPs and the NKG2D system in virus-induced HCC. *J. Gastroenterol.* **2015**, *50*, 261–272. [CrossRef]
44. Kohga, K.; Takehara, T.; Tatsumi, T.; Ohkawa, K.; Miyagi, T.; Hiramatsu, N.; Kanto, T.; Kasugai, T.; Katayama, K.; Kato, M.; et al. Serum levels of soluble major histocompatibility complex (MHC) class I-related chain A in patients with chronic liver diseases and changes during transcatheter arterial embolization for hepatocellular carcinoma. *Cancer Sci.* **2008**, *99*, 1643–1649. [CrossRef]
45. Jinushi, M.; Takehara, T.; Tatsumi, T.; Kanto, T.; Groh, V.; Spies, T.; Kimura, R.; Miyagi, T.; Mochizuki, K.; Sasaki, Y.; et al. Expression and role of MICA and MICB in human hepatocellular carcinomas and their regulation by retinoic acid. *Int. J. Cancer* **2003**, *104*, 354–361. [CrossRef]
46. Groh, V.; Rhinehart, R.; Secrist, H.; Bauer, S.; Grabstein, K.H.; Spies, T. Broad tumor-associated expression and recognition by tumor-derived gamma delta T cells of MICA and MICB. *Proc. Natl. Acad. Sci. USA* **1999**, *96*, 6879–6884. [CrossRef]
47. Kumar, V.; Yi Lo, P.H.; Sawai, H.; Kato, N.; Takahashi, A.; Deng, Z.; Urabe, Y.; Mbarek, H.; Tokunaga, K.; Tanaka, Y.; et al. Soluble MICA and a MICA variation as possible prognostic biomarkers for HBV-induced hepatocellular carcinoma. *PLoS ONE* **2012**, *7*, e44743. [CrossRef]
48. Kumar, V.; Kato, N.; Urabe, Y.; Takahashi, A.; Muroyama, R.; Hosono, N.; Otsuka, M.; Tateishi, R.; Omata, M.; Nakagawa, H.; et al. Genome-wide association study identifies a susceptibility locus for HCV-induced hepatocellular carcinoma. *Nat. Genet.* **2011**, *43*, 455–458. [CrossRef]
49. Feitelson, M.A.; Bonamassa, B.; Arzumanyan, A. The roles of hepatitis B virus-encoded X protein in virus replication and the pathogenesis of chronic liver disease. *Expert Opin. Ther. Targets* **2014**, *18*, 293–306. [CrossRef]
50. Ou, D.P.; Tao, Y.M.; Tang, F.Q.; Yang, L.Y. The hepatitis B virus X protein promotes hepatocellular carcinoma metastasis by upregulation of matrix metalloproteinases. *Int. J. Cancer* **2007**, *120*, 1208–1214. [CrossRef]
51. Lunemann, S.; Malone, D.F.; Hengst, J.; Port, K.; Grabowski, J.; Deterding, K.; Markova, A.; Bremer, B.; Schlaphoff, V.; Cornberg, M.; et al. Compromised function of natural killer cells in acute and chronic viral hepatitis. *J. Infect. Dis.* **2014**, *209*, 1362–1373. [CrossRef]
52. Wu, J.; Zhang, X.J.; Shi, K.Q.; Chen, Y.P.; Ren, Y.F.; Song, Y.J.; Li, G.; Xue, Y.F.; Fang, Y.X.; Deng, Z.J.; et al. Hepatitis B surface antigen inhibits MICA and MICB expression via induction of cellular miRNAs in hepatocellular carcinoma cells. *Carcinogenesis* **2014**, *35*, 155–163. [CrossRef]
53. Chotiyaputta, W.; Lok, A.S. Hepatitis B virus variants. *Nat. Rev. Gastroenterol. Hepatol.* **2009**, *6*, 453–462. [CrossRef]
54. Abu-Amara, M.; Feld, J.J. Does antiviral therapy for chronic hepatitis B reduce the risk of hepatocellular carcinoma? *Semin. Liver Dis.* **2013**, *33*, 157–166. [CrossRef]
55. Morgan, R.L.; Baack, B.; Smith, B.D.; Yartel, A.; Pitasi, M.; Falck-Ytter, Y. Eradication of hepatitis C virus infection and the development of hepatocellular carcinoma: A meta-analysis of observational studies. *Ann. Intern. Med.* **2013**, *158*, 329–337. [CrossRef]
56. Li, Y.; Wang, J.J.; Gao, S.; Liu, Q.; Bai, J.; Zhao, X.Q.; Hao, Y.H.; Ding, H.H.; Zhu, F.; Yang, D.L.; et al. Decreased peripheral natural killer cells activity in the immune activated stage of chronic hepatitis B. *PLoS ONE* **2014**, *9*, e86927. [CrossRef]
57. Jinushi, M.; Takehara, T.; Tatsumi, T.; Kanto, T.; Groh, V.; Spies, T.; Suzuki, T.; Miyagi, T.; Hayashi, N. Autocrine/paracrine IL-15 that is required for type I IFN-mediated dendritic cell expression of MHC class I-related chain A and B is impaired in hepatitis C virus infection. *J. Immunol.* **2003**, *171*, 5423–5429. [CrossRef]

58. Goto, K.; Arai, J.; Stephanou, A.; Kato, N. Novel therapeutic features of disulfiram against hepatocellular carcinoma cells with inhibitory effects on a disintegrin and metalloproteinase 10. *Oncotarget* **2018**, *9*, 18821–18831. [CrossRef]
59. Takayama, T.; Makuuchi, M.; Sekine, T.; Terui, S.; Shiraiwa, H.; Kosuge, T.; Yamazaki, S.; Hasegawa, H.; Suzuki, K.; Yamagata, M.; et al. Distribution and therapeutic effect of intraarterially transferred tumor-infiltrating lymphocytes in hepatic malignancies. A preliminary report. *Cancer* **1991**, *68*, 2391–2396. [CrossRef]
60. Kawata, A.; Une, Y.; Hosokawa, M.; Wakizaka, Y.; Namieno, T.; Uchino, J.; Kobayashi, H. Adjuvant chemoimmunotherapy for hepatocellular carcinoma patients. Adriamycin, interleukin-2, and lymphokine-activated killer cells versus adriamycin alone. *Am. J. Clin. Oncol.* **1995**, *18*, 257–262. [CrossRef]
61. Takayama, T.; Sekine, T.; Makuuchi, M.; Yamasaki, S.; Kosuge, T.; Yamamoto, J.; Shimada, K.; Sakamoto, M.; Hirohashi, S.; Ohashi, Y.; et al. Adoptive immunotherapy to lower postsurgical recurrence rates of hepatocellular carcinoma: A randomised trial. *Lancet* **2000**, *356*, 802–807. [CrossRef]
62. Liu, D.; Staveley-O'Carroll, K.F.; Li, G. Immune-based therapy clinical trials in hepatocellular carcinoma. *J. Clin. Cell. Immunol.* **2015**, *6*, 376. [CrossRef]
63. Ueno, S.; Tanabe, G.; Nuruki, K.; Hamanoue, M.; Komorizono, Y.; Oketani, M.; Hokotate, H.; Inoue, H.; Baba, Y.; Imamura, Y.; et al. Prognostic performance of the new classification of primary liver cancer of Japan (4th edition) for patients with hepatocellular carcinoma: A validation analysis. *Hepatol. Res.* **2002**, *24*, 395–403. [CrossRef]
64. Lencioni, R.; Llovet, J.M. Modified RECIST (mRECIST) assessment for hepatocellular carcinoma. *Semin. Liver Dis.* **2010**, *30*, 52–60. [CrossRef]

 © 2020 by the authors. Licensee MDPI, Basel, Switzerland. This article is an open access article distributed under the terms and conditions of the Creative Commons Attribution (CC BY) license (http://creativecommons.org/licenses/by/4.0/).

Article

Predictors of Response and Survival in Immune Checkpoint Inhibitor-Treated Unresectable Hepatocellular Carcinoma

Pei-Chang Lee [1,2,3], Yee Chao [4,*], Ming-Huang Chen [4], Keng-Hsin Lan [1,3], Chieh-Ju Lee [3,5], I-Cheng Lee [2,3,5], San-Chi Chen [2,4,5], Ming-Chih Hou [2,3] and Yi-Hsiang Huang [2,3,5,*]

1. Institute of Pharmacology, School of Medicine, National Yang-Ming University, Taipei 11221, Taiwan; tympanum3688@gmail.com (P.-C.L.); khlan@vghtpe.gov.tw (K.-H.L.)
2. Faculty of Medicine, School of Medicine, National Yang-Ming University, Taipei 11221, Taiwan; iclee@vghtpe.gov.tw (I.-C.L.); scchen16@vghtpe.gov.tw (S.-C.C.); mchou@vghtpe.gov.tw (M.-C.H.)
3. Division of Gastroenterology and Hepatology, Department of Medicine, Taipei Veterans General Hospital, Taipei 11217, Taiwan; ssbugi@gmail.com
4. Department of Oncology, Taipei Veterans General Hospital, Taipei 11217, Taiwan; mhchen9@vghtpe.gov.tw
5. Institute of Clinical medicine, School of Medicine, National Yang-Ming University, Taipei 11221, Taiwan
* Correspondence: ychao@vghtpe.gov.tw (Y.C.); yhhuang@vghtpe.gov.tw (Y.-H.H.); Tel.: +886-2-28712121 (ext. 7506) (Y.-H.H.)

Received: 9 December 2019; Accepted: 8 January 2020; Published: 11 January 2020

Abstract: Immune checkpoint inhibitors (ICIs) with nivolumab and pembrolizumab are promising agents for advanced hepatocellular carcinoma (HCC) but lack of effective biomarkers. We aimed to investigate the potential predictors of response and factors associated with overall survival (OS) for ICI treatment in unresectable HCC patients. Ninety-five patients who received nivolumab or pembrolizumab for unresectable HCC were enrolled for analyses. Radiologic evaluation was based on RECIST v1.1. Factors associated with outcomes were analyzed. Of 90 patients with evaluable images, the objective response rate (ORR) was 24.4%. Patients at Child–Pugh A or received combination treatment had higher ORR. Early alpha-fetoprotein (AFP) >10% reduction (within 4 weeks) was the only independent predictor of best objective response (odds ratio: 7.259, $p = 0.001$). For patients with baseline AFP ≥10 ng/mL, significantly higher ORR (63.6% vs. 10.2%, $p < 0.001$) and disease control rate (81.8% vs. 14.3%, $p < 0.001$) were observed in those with early AFP reduction than those without. In addition, early AFP reduction and albumin-bilirubin (ALBI) grade or Child–Pugh class were independent factors associated with OS in different models. In conclusion, a 10-10 rule of early AFP response can predict objective response and survival to ICI treatment in unresectable HCC. ALBI grade and Child–Pugh class determines survival by ICI treatment.

Keywords: alpha fetoprotein response; immune checkpoint inhibitor; unresectable hepatocellular carcinoma

1. Introduction

Hepatocellular carcinoma (HCC) is the fifth most common cancer and the second leading cause of cancer related death worldwide that constitutes a major global health problem [1,2]. Despite improvement in surveillance and hepatitis B vaccination, hepatitis C treatment, a large number of patients still present with unresectable, advanced-stage disease and require systemic therapy [2]. Sorafenib has long been the first and the only effective systemic treatment for advanced HCC [3,4]. Recently, several positive results from the phase 2/3 trials of first or second line settings enable HCC patients access to more treatment options [5–7].

Manipulation of immune checkpoints by targeted antibodies, such as anti-programmed cell death-1 (PD-1) antibody, has recently emerged as an effective anticancer strategy for many types of cancers including HCC [8]. Nivolumab and pembrolizumab, the anti-PD-1 antibodies, are FDA conditionally approved immune checkpoint inhibitors (ICIs) for HCC as a second line treatment after sorafenib failure [5,6,9]. Based on the multi-cohort phase 1/2 trial CheckMate-040, phase 2 trial keynote-224, and phase 3 trial of Keynote-240 [5–7], only 14–18% of HCC patients could get a tumor response by nivolumab or pembrolizumab. Traditionally, PD-L1 expression level is a determinant marker of response in lung cancer, gastric cancer, head and neck cancer, and urothelial cancer [10–12]. However, previous CheckMate-040 and Keynote-224 studies could not show a significant association between PD-L1 expression level and tumor response in HCC [5,6]. As ICI treatment is expensive and has potential risk of immune-related adverse events, a baseline or early biomarker can help physicians to encourage suitable patients to maintain the treatment [2]. However, so far, it still remains an unmet medical need as there is no well-identified biomarker for HCC immunotherapy. In this study, we aimed to identify potential predictors of treatment response and overall survival (OS) in patients treated with ICI for unresectable HCC.

2. Results

2.1. Demographic Characteristics of the Study Cohort

Upon enrollment, most patients were within Child–Pugh class A (72.6%); but more than half of them were classified beyond albumin-bilirubin (ALBI) grade 1 (71.6%). A total of 78.9% of the patients were at BCLC stage C, and the maximal tumor size was 5.2 cm (IQR, 2.3–8.8). The median alpha-fetoprotein (AFP) level was 865.6 ng/mL, and 15.8% of the patients had low AFP level (<10 ng/mL). In addition, 41.1% received ICI as first-line systemic therapy, while 58.9% had experienced sorafenib failure. Among 95 patients, 13 received combination therapy with ICIs and tyrosine kinase inhibitors (six with sorafenib, six with lenvatinib, and one with regorafenib). Four and three patients developed grade 2 immunotherapy-related pneumonitis and hepatitis, respectively. Six patients suffered from grade 1/2 skin reactions. The detailed baseline characteristics are presented in Table 1.

Table 1. Characteristics and outcomes of hepatocellular carcinoma (HCC) patients treated with immune checkpoint inhibitors.

Characteristics	$n = 95$
Age, y	65.5 (57.2–72.9)
Sex (male), n (%)	73 (76.8)
HBsAg-positive, n (%)	62 (65.3)
Anti-HCV-positive, n (%)	21 (22.1)
Max. tumor size, cm	5.2 (2.3–8.8)
Tumor >50% liver volume, n (%)	30 (31.6)
Multiple tumors, n (%)	89 (93.7)
Extrahepatic metastasis, n (%)	48 (50.5)
Portal vein invasion, n (%)	51 (53.7)
AFP, ng/mL	609.7 (37.5–4832.3)
<10 ng/mL, n (%)	15 (15.8)
10–400 ng/mL, n (%)	27 (28.4)
≥400 ng/mL, n (%)	53 (55.8)
BCLC stage B/C, n (%)	20/75 (21.1/78.9)
Prothrombin time, INR	1.10 (1.05–1.23)
Platelet count, K/cumm	145 (102–218)
ALT, U/L	39 (25–61)
AST, U/L	57 (35–97)
Total bilirubin, mg/dL	1.03 (0.55–1.52)

Table 1. Cont.

Characteristics	n = 95
Albumin, g/dL	3.6 (3.2–4.0)
Neutrophil-lymphocyte ratio	4.16 (2.89–6.85)
Presence of ascites, n (%)	37 (38.9)
Child–Pugh score	6 (5–7)
Child–Pugh class A/B/C, n (%)	69/23/3 (72.6/24.2/3.2)
ALBI grade 1/2/3, n (%)	27/58/10 (28.4/61.1/10.5)
First line systemic therapy, n (%)	39 (41.1)
Prior therapy to ICI, n (%)	
Surgical resection	35 (36.8)
RFA/PEIT/MWA	31/9/1 (32.6/9.5/1.1)
TACE/RT/TARE (Y-90)	55/23/5 (57.9/24.2/5.3)
Sorafenib	56 (58.9)
Nivolumab/Pembrolizumab, n (%)	92/3 (96.8/3.2)
Combined ICI with TKI, n (%)	13 (13.7)
Immune-related AEs	
Skin reactions/Pneumonitis/Hepatitis	6/4/3 (6.3/4.2/3.2)
Post PD treatment, n (%)	
TACE/RT/TARE (Y-90)	9/8/2 (9.5/8.4/2.1)
Regorafenib/Lenvatinib/Carbozantinib	8/16/2 (8.4/16.8/2.1)
Ramucirumab	4 (4.2)
Sorafenib/Traditional CT	7/6 (7.4/6.3)
Death	47 (49.5)

The data are expressed as median (interquartile range) unless marked with number (percentage) in behind. Abbreviations: AEs, adverse events; AFP, alpha fetoprotein; ALBI grade, albumin-bilirubin grade; ALT, alanine aminotransferase; AST, aspartate aminotransferase; BCLC stage, Barcelona-Clinic liver cancer stage; CI, confidence interval; CT, chemotherapy; HBsAg, hepatitis B surface antigen; HCV, hepatitis C; ICI, immune checkpoint inhibitor; INR, international normalized ratio; MWA, microwave ablation; PD, progressive disease; PEIT, percutaneous ethanol injection in tumor; RFA, radiofrequency ablation; RT, radiotherapy; TACE, transarterial chemoembolization; TARE (Y-90), transarterial radioembolization (Yttrium-90); TKI, tyrosine kinase inhibitors.

2.2. Treatment Response to ICI Therapy

The median duration of ICI treatment was 10.4 weeks (IQR, 4.8–22.3) with a median of five cycles (ranged 1–35) administered. As presented in Table 2, the disease control rate (DCR) was 36.7%, including six complete response (CR), 16 partial responses (PR), and 11 stable diseases. The best objective response rate (ORR) was 26.9% and 20.0% between patients at Child–Pugh A and B, respectively. Combination treatment had a significantly higher ORR than ICI monotherapy (46.2% vs. 20.8%, $p = 0.049$). The median time to response was 63 days (IQR, 48–75) after a median five cycles of ICI treatment (IQR, 4–6); and the median duration of response was not yet reached for responders (16/22 kept ongoing with response). Noteworthily, three Child–Pugh B patients whose tumors controlled well by ICI notably improved their liver reserve to Child–Pugh A after treatment.

In univariate analysis, AFP >10% reduction within the first 4 weeks of treatment, baseline ALT level, as well as combination treatment were associated with best objective response. In multivariate analysis, early AFP response was the only independent predictor of best objective response to ICI treatment (odds ratio: 7.259, $p = 0.001$) (Table 3). Besides, early AFP reduction was also associated with best disease control by ICI therapy (Table S1).

Table 2. Treatment response to immune checkpoint inhibitors.

Evaluable Response	All Patients (n = 95)	Child–Pugh A (n = 69)	Child–Pugh B (n = 23)	Child–Pugh C (n = 3)	Combination Treatment (n = 13)	Monotherapy (n = 82)
Best Response, n (%)						
Complete response	6 (6.7)	5 (7.5)	1 (5.0)	0	1 (7.7)	5 (6.5)
Partial response	16 (17.8)	13 (19.4)	3 (15.0)	0	5 (38.5)	11 (14.3)
Stable disease	11 (12.2)	10 (14.9)	1 (5.0)	0	1 (7.7)	10 (13.0)
Progressive disease	57 (63.3)	39 (58.2)	15 (75.0)	3 (100.0)	6 (46.2)	51 (66.2)
Non-assessable	5	2	3	0	0	5
Objective response rate	22 (24.4)	18 (26.9)	4 (20.0)	0	6 (46.2)	16 (20.8)
Disease control rate	33 (36.7)	28 (41.8)	5 (25.0)	0	7 (53.8)	26 (33.8)
For Responders						
Time to response (days)	63 (48–75)	64 (52–76)	52 (21–72)	–	57 (43–73)	63 (55–77)
Duration of response (months)	Not yet reached (16 ongoing)	Not yet reached (13 ongoing)	Not yet reached (three ongoing)	–	Not yet reached (five ongoing)	Not yet reached (11 ongoing)

Table 3. Factors associated with best objective response in 90 patients with evaluable responses.

Characteristics		Univariate Analysis			Multivariate Analysis		
		OR	95% CI	p Value	OR	95% CI	p Value
Age, y	>60 vs. ≤60	0.447	0.167–1.192	0.108			
Sex	Male vs. Female	0.691	0.228–2.092	0.514			
HBsAg-positive	Yes vs. No	1.651	0.573–4.756	0.353			
Anti-HCV-positive	Yes vs. No	0.722	0.213–2.446	0.601			
Tumor size, cm	>7 vs. ≤7	0.754	0.271–2.094	0.588			
Tumor number	Multiple vs. single	0.625	0.106–3.670	0.625			
Tumor shape	Infiltrative vs. nodular	2.250	0.813–6.227	0.118			
Tumor/Liver volume	>50% vs. ≤50%	0.900	0.308–2.633	0.847			
Portal vein invasion	Yes vs. No	1.131	0.431–2.969	0.802			
Main portal vein invasion	Yes vs. No PVI	1.046	0.278–3.932	0.947			
Portal branches invasion	Yes vs. No PVI	1.295	0.441–3.803	0.638			
Extrahepatic metastasis	Yes vs. No	0.580	0.219–1.537	0.273			
BCLC stage	Stage C vs. B	1.385	0.409–4.689	0.601			
AFP, ng/mL	>400 vs. ≤400	0.789	0.301–2.068	0.630			
AFP, ng/mL	<10 vs. ≤10	0.737	0.188–2.894	0.662			
NLR	>2.5 vs. ≤2.5	1.529	0.390–5.992	0.542			
Prothrombin time, INR	>1.2 vs. ≤1.2	1.211	0.422–3.470	0.722			
Platelet count	>100K vs. ≤100K	0.821	0.275–2.447	0.723			
ALT, U/L	> 40 vs. ≤40	0.294	0.097–0.888	0.030	0.384	0.109–1.349	0.135
AST, U/L	> 40 vs. ≤40	0.465	0.172–1.255	0.131			
Ascites	Yes vs. No	0.536	0.186–1.539	0.246			
Child–Pugh class	Class B, C vs. A	0.537	0.172–1.914	0.366			
ALBI grade	Grade 2,3 vs. 1	0.520	0.190–1.422	0.203			
Prior Sorafenib treatment	Yes vs. No	1.011	0.380–2.687	0.982			
Combined treatment *	Yes vs. No	3.813	1.083–13.419	0.037	2.522	0.572–11.111	0.222
AFP reduction at fourth week †	Yes vs. No	7.437	2.545–21.735	<0.001	7.259	2.359–22.337	0.001
IO related AEs	Yes vs. No	0.916	0.228–3.678	0.901			

Abbreviations: AEs, adverse events; AFP, alpha fetoprotein; ALBI grade, albumin-bilirubin grade; ALT, alanine aminotransferase; AST, aspartate aminotransferase; BCLC stage, Barcelona-Clinic liver cancer stage; CI, confidence interval; HBV, hepatitis B; HCV, hepatitis C; INR, international normalized ratio; IO, immunotherapy; OR, odds ratio; NLR, neutrophil-lymphocyte ratio. * Combined treatment: combined immune checkpoint inhibitors with tyrosine kinase inhibitors, including sorafenib, lenvatinib, and regorafenib. † AFP reduction at fourth week: AFP reduced >10% from baseline serum level.

2.3. Association between Tumor Response and Early AFP Response

As 10% AFP reduction might not be meaningful for HCCs with baseline level less than 10 ng/mL, the AFP response was further categorized by baseline AFP level. For patients with baseline AFP ≥10 ng/mL, significantly higher ORR (63.6% vs. 10.2%, $p < 0.001$) and DCR (81.8% vs. 14.3%, $p < 0.001$) were observed in those with early AFP reduction than those without. However, such association was not observed in patients with baseline AFP level <10 ng/mL (Figure 1).

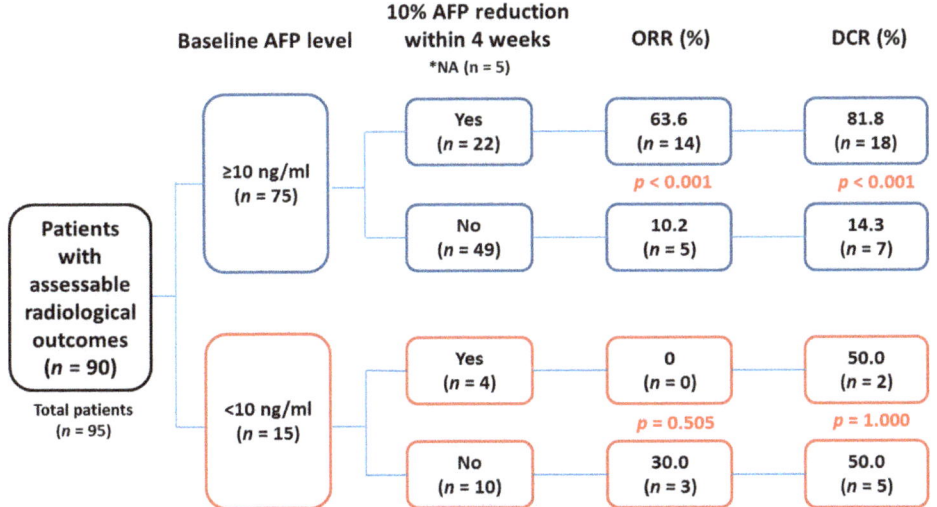

Figure 1. The association between tumor response and early alpha-fetoprotein (AFP) reduction categorized by AFP ≥ or <10 ng/mL. NA (not assessed): total of five patients did not have an AFP value within 4 weeks of treatment that could not be assessed for early AFP response.

2.4. Response in HCC Patients with Available PD-L1 Level and Evaluable Images

Of 18 patients whose tumor specimens were assessed for PD-L1 expression, three patients had TPS ≥ 1% or CPS ≥ 1%, and all of them achieved partial response to ICI treatment. In the other 15 patients with low expression of PD-L1 (<1%), 60.0% developed PD (p = 0.206) (Figure S1).

2.5. Uni- and Multivariate Analysis for Factors Associated with OS for All HCC Patients

During a median follow-up period of 5.2 (IQR, 3.2–12.5) months, 47 deaths occurred. The median overall survival was 11.9 months (95% C.I. 5.6–18.2). As shown in Figure 2, patients with objective tumor response had significantly better OS than those that developed PD (median OS: not yet reached vs. 6.1 months). Besides, patients with early AFP reduction >10% also had significantly better OS than non-responders (median OS: 24.7 vs. 5.6 months, p = 0.014; Figure 3). In addition, Child–Pugh A vs. B/C (median OS: 24.7 vs. 3.8 and 0.6 months, p < 0.001) (Figure 4A), and ALBI grade 1 vs. 2/3 (median OS: not yet reached, vs. 5.6 and 3.2 months; p < 0.001) (Figure 4B) were associated with OS. No significant survival difference was reported according to prior sorafenib treatment (Figure S2).

As declared in Table 4, early AFP reduction (hazard ratio (HR): 0.234, p = 0.001) and Child–Pugh A (HR: 0.238, p = 0.002) were the independent predictors to better OS in patients received ICI treatment (Multivariate analysis model 1). Similarly, early AFP response (HR: 0.243, p = 0.001) and ALBI grade 1 (HR: 0.220, p = 0.002) were also good survival predictors in the model 2. After including tumor response into analysis, presence of tumor response, serum AST level, and good liver reserves were identified as independent survival predictors (Table S2).

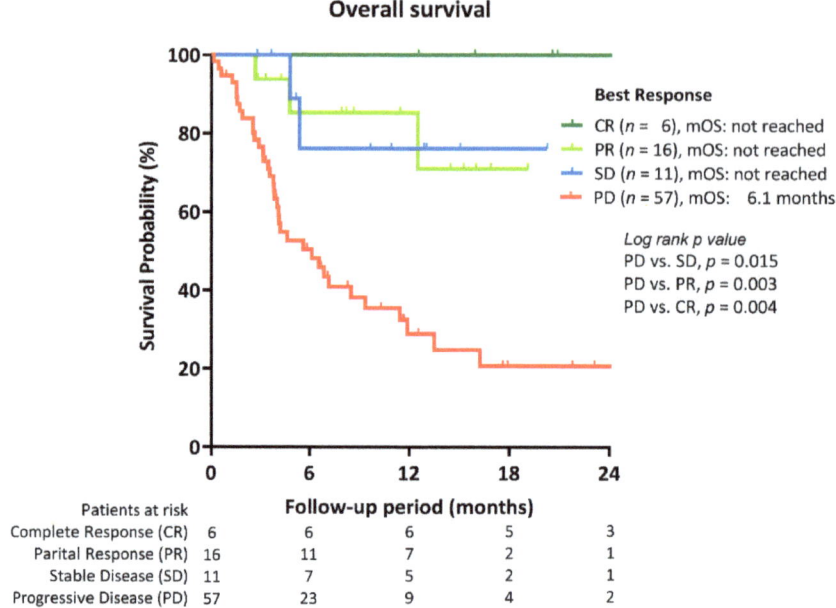

Figure 2. Overall survival (OS) of unresectable HCC according to treatment response to immune checkpoint inhibitors.

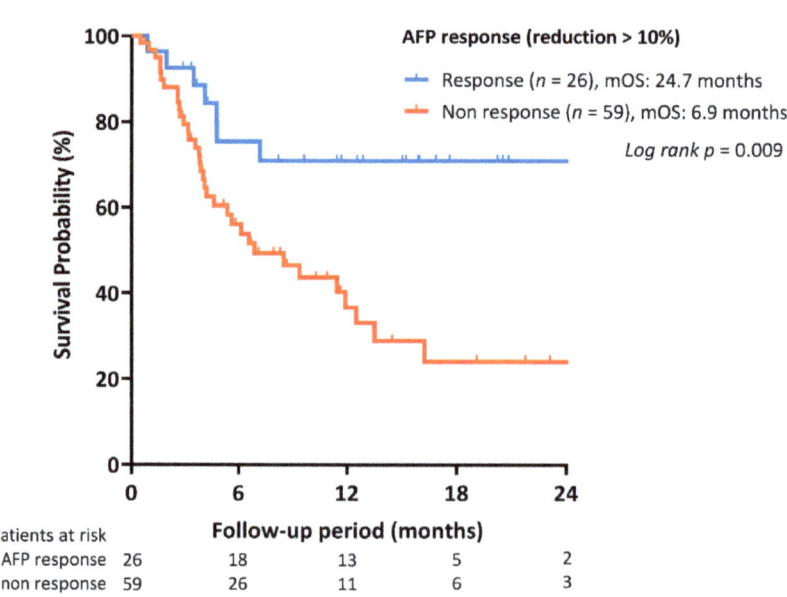

Figure 3. OS of HCC patients according to AFP reduction within 4 weeks treatment of immune checkpoint inhibitors.

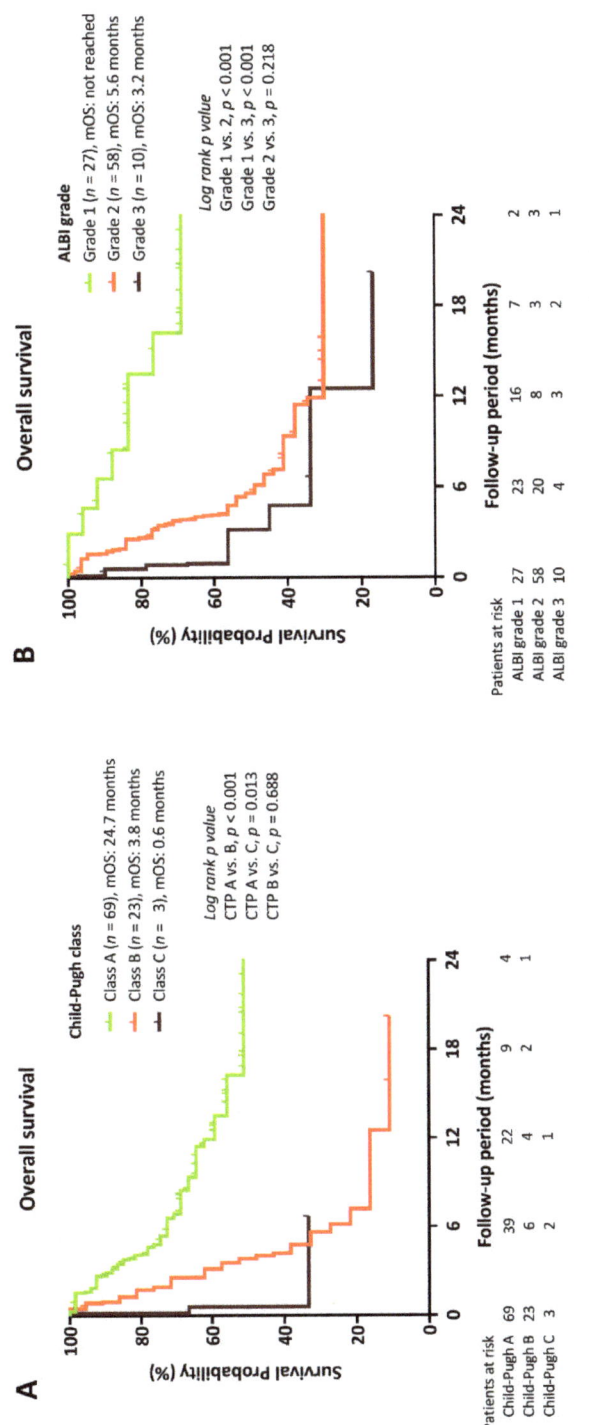

Figure 4. OS of HCC patients stratified by (**A**) Child–Pugh classification, and (**B**) Albumin-bilirubin (ALBI) grade.

Table 4. Factors associated with overall survival in 95 patients treated with immune checkpoint inhibitors.

Characteristics		Univariate			Multivariate (Model 1) #			Multivariate (Model 2) #		
		HR	95% CI	p	HR	95% CI	p	HR	95% CI	p
Age, y	>60 vs. ≤60	1.252	0.676–2.318	0.476			NA			NA
Sex	Male vs. Female	0.632	0.337–1.186	0.153			NA			NA
HBsAg-positive	Yes vs. No	1.020	0.555–1.874	0.950			NA			NA
Anti-HCV-positive	Yes vs. No	1.393	0.729–2.661	0.315			NA			NA
Tumor size, cm	>7 vs. ≤7	2.450	1.362–4.409	0.003			NS			NS
Tumor number	Multiple vs. single	3.709	0.510–26.946	0.195			NA			NA
Tumor/Liver volume	>50% vs. ≤50%	2.425	1.323–4.444	0.004			NS			NS
Portal vein invasion	Yes vs. No	1.829	1.008–3.321	0.047			NS			NS
Extrahepatic metastasis	Yes vs. No	1.444	0.804–2.591	0.219			NA			NA
BCLC stage	Stage C vs. B	1.854	0.828–4.154	0.134			NA			NA
AFP, ng/mL	>400 vs. ≤400	2.039	1.102–3.773	0.023			NS			NS
AFP, ng/mL	<10 vs. ≤10	0.255	0.079–0.826	0.023			NS			NS
NLR	>2.5 vs. ≤2.5	1.010	0.467–2.185	0.981			NA			NA
Prothrombin time, INR	>1.2 vs. ≤1.2	1.585	0.842–2.983	0.154			NS			NS
Platelet count	>100K vs. ≤100K	0.928	0.479–1.799	0.825			NA			NA
ALT, U/L	>40 vs. ≤40	2.463	1.370–4.428	0.003			NS			NS
AST, U/L	>40 vs. ≤40	4.762	2.015–11.255	<0.001			NS			NS
Ascites	Yes vs. No	2.782	1.551–4.989	0.001			NA			NA
Child-Pugh class	Class A vs. B	0.260	0.143–0.472	<0.001	0.289	0.134–0.624	0.002			NA
ALBI grade	Grade1 vs. 2/3	0.189	0.079–0.453	<0.001			NA	0.220	0.084–0.576	0.002
Prior Sorafenib treatment	Yes vs. No	0.952	0.528–1.717	0.870			NA			NA
Combined treatment *	Yes vs. No	0.408	0.125–1.331	0.137			NS			NS
AFP reduction at fourth week †	Yes vs. No	0.372	0.172–0.809	0.013	0.234	0.096–0.569	0.001	0.243	0.104–0.565	0.001
Immunotherapy related AEs	Yes vs. No	0.746	0.294–1.893	0.537			NA			NA

Abbreviations: ALBI grade, albumin-bilirubin grade; AEs, adverse events; AFP, alpha fetoprotein; AL(S)T, alanine(aspartate) aminotransferase; BCLC stage, AEs, adverse events; AFP, alpha fetoprotein; ALBI grade, albumin-bilirubin grade; ALT, alanine aminotransferase; AST, aspartate aminotransferase; BCLC stage, Barcelona-Clinic liver cancer stage; CI, confidence interval; HBsAg, hepatitis B surface antigen; HCV, hepatitis C; INR, international normalized ratio; NLR, neutrophil-lymphocyte ratio. * Combined treatment: combined immune checkpoint inhibitors with tyrosine kinase inhibitors, including sorafenib, lenvatinib, and regorafenib. † AFP reduction at fourth week: AFP reduced >10% from baseline serum level. # Model 1 enrolled significant parameters in univariate analysis into multivariate analysis, except ascites and ALBI grade. Model 2 enrolled significant parameters in univariate analysis into multivariate analysis, except Child-Pugh class.

3. Discussion

This is the largest real-world cohort from Asian patients with unresectable HCC treated by ICIs until now. A better ORR (24.4%) was observed than previous studies, but stable disease was fewer [7,8,13]. Impressively, early AFP response within 4 weeks of treatment was identified as the independent predictor to objective response in our patients. Besides, better liver reserves (Child–Pugh class A or ALBI grade 1) and early AFP response were also good predictors of survival.

The predictive role of AFP reduction in HCC response to various treatments has been reported [14–16]. In sorafenib-treated HCC, a decline of AFP >20% from baseline level after 4 to 8 weeks of treatment was suggested as a surrogate marker to predict treatment response and survival benefits [17]. In an extended analysis of CheckMate-040, however, the authors failed to find biomarkers predicting treatment response to nivolumab [18]. In a recent real-life experience of ICI-treated HCC, no factor was identified to associate with response, either [19]. Early AFP reduction >20% within the first 4 weeks of ICI treatment was recently reported in relation to treatment efficacy for patients with baseline AFP > 20 ng/mL [20]. In this study, we proposed a novel 10-10 rule to early predict ICI response based on baseline AFP level ≥10 ng/mL, and 10% reduction within 4 weeks of treatment. A >10% reduction of AFP (ORR: 63.6% vs. 10.2%, $p < 0.001$; DCR: 81.8% vs. 14.3%, $p < 0.001$) performed a better discriminative ability in tumor response than AFP reduction >20% (ORR: 64.7% vs. 14.8%, $p < 0.001$; DCR: 82.4% vs. 20.4%, $p < 0.001$) or >30% (ORR: 61.5% vs. 19.0%, $p = 0.001$; DCR: 84.6% vs. 24.1%, $p < 0.001$). These findings suggested the 10-10 rule can serve as guidance for ICI treatment in advanced HCC.

Unlike the real-world report of sorafenib-treated HCC with inferior survival benefit in Eastern population [21], the median OS of our patients was similar to the data of CheckMate-040 and a recent multicenter real-world study [7,19]. Our results were also in line with the report from Asian cohort of CheckMate-040 with comparable ORR and survival [22]. The prognosis of advanced HCC depends not only on tumor burden, but also on liver reserve [21,23,24]. Consistent with the survival-predictive ability of ALBI grade in sorafenib-failed HCC [21,25], our data confirmed ALBI grade as an independent survival predictor in patients received ICI treatment. The ORR was 12.2% for Child–Pugh B HCC in CheckMate-040 [26]. In a recent case series enrolled 18 patients with advanced HCC and Child–Pugh B cirrhosis, the ORR was 17%, and the median OS was 5.9 months [27]. In this study, we declared a comparable ORR as 20.0% (three PR and one CR) in Child-Pugh B patients but presented with a shorter OS. Although only five of our Child–Pugh B patients had disease controlled by ICI treatment; notably, three of them improved their liver function to Child–Pugh A along with excellent survival benefits (the median OS was not yet reached). Inconsistent with prior statements indicated that patients with poor liver reserves may not get benefit from oncological management [13,28], these recent findings suggested that Child–Pugh B patients could still get benefit from ICI treatment.

Synergic benefits of combination therapy to advance-staged HCC have been explored recently [29]. Current ongoing clinical trials suggested that combination treatment with lenvatinib plus pembrolizumab, atezolizumab plus bevacizumab, or nivolumab plus ipilimumab had promising ORR higher than 30%, or even 60% [30–32]. As recently reported in ESMO Asia 2019, the phase 3 IMbrave 150 has demonstrated significant improvements with atezolizumab and bevacizumab over sorafenib in OS and RFS for unresectable HCC (ORR 27%, DCR 74% by RECIST 1.1). In this study, a significantly better ORR (46.2%) was noted in patients that received combination therapy compared with ICI monotherapy. However, it did not independently predict objective response to ICI treatment in the multivariate analysis. Further investigation is still needed to clarify the role of combined treatment in management of HCC.

PD-L1 expression by either tumor cells or intratumoral inflammatory cells is related to HCC aggressiveness and might account for the response of immune checkpoint inhibitor [33]. Although numerically a higher ORR was observed in patients whose PD-L1 expression level was ≥1% in previous studies [7,8], the difference did not reach statistical difference. In our data, all patients with ≥1% TPS or CPS had PR to ICI treatment; whereas, most patients with <1% PD-L1 expression presented with PD.

These findings suggested that PD-L1 expressions might play some role in the response to ICI treatment in HCC.

The adverse effects (AEs) of immune checkpoint inhibitor are different from toxicities caused by chemotherapy. In general, the immunotherapy related AEs was low in our cohort and we did not observe a high incidence of immunotherapy related AEs in Child–Pugh B patients.

There are several limitations in this study. First, this is a retrospective study that only enrolled patients in single hospital. However, our hospital is the main leading tertiary medical center in Taiwan. The information bias would be ameliorated by regular tumor reassessment by contrast-enhanced image and clinical evaluation. Besides, it is so far the largest real-life Asian ICI-treated HCC cohort; and is the first study demonstrated the 10-10 role of AFP to predict response. Second, the level of PD-L1 expression was only performed in few patients, although our pivotal results were similar to previous studies with improved prediction to treatment efficacy [7,8]. Third, most of our patients (73.5%) had chronic hepatitis B as the underlying hepatic disease. Our results should be applied to other populations with caution.

4. Materials and Methods

4.1. Patients

From May 2017 to August 2019, 95 patients had received nivolumab or pembrolizumab treatment for unresectable HCC in Taipei Veterans General Hospital and were retrospectively enrolled in this study. None of them enrolled in previous or ongoing ICI clinical trials. Among them, 90 patients with evaluable image studies following the treatment before the cut-off date of data were recruited for further assessment of treatment response. Of the five subjects not available for assessment, four patients died before the first radiological evaluation and one patient was lost to follow-up. The diagnosis was according to the AALSD treatment guidelines for HCC [34]. ICIs were prescribed to these patients because of treatment failure or intolerable adverse events to sorafenib, deteriorated liver reserves beyond Child-Pugh class A so that was unable to apply sorafenib according to the reimbursement criteria of National Health Insurance in Taiwan [21], or patients who experienced ineffective transarterial chemoembolization for their intermediate-staged HCC. The study was approved by the Institutional Review Board of the Taipei Veterans General Hospital (IRB numbers: 2017-09-007CC, 2019-07-007AC, and 2019-08-006B). All alive patients have signed informed consent; and the informed consent of others was waived by IRB because of retrospective design.

4.2. Treatment and Outcome Assessment

ICIs were administered according to the recommended dosing and safety information (2–3 mg/kg, every 2 weeks for nivolumab and every 3 weeks for pembrolizumab). The safety assessment and grading was performed based on the National Cancer Institute Common Terminology Criteria for Adverse Events (NCI CTCAE; version 4.03). Besides, clinical evaluation with Child-Pugh class, albumin-bilirubin grade [35,36], hemogram, serum chemistry, and alpha-fetoprotein (AFP) level were performed every 2 to 3 weeks during the treatment. An early AFP response was defined as >10% reduction from baseline level within 4 weeks of treatment.

Clinical tumor response was assessed by RECIST version 1.1 based on contrast-enhanced abdominal computed tomography scan or magnetic resonance imaging [7,37]. The image examinations were carried out every 6–8 weeks during ICIs treatment. The OS was measured from the date of starting ICIs until the date of death; and the time to response was the interval between ICIs initiation and occurrence of first objective response.

4.3. PD-L1 Expression Analysis

PD-L1 expression was measured by immunohistochemistry pharmDx assay (Agilent Technologies, Santa Clara, CA, USA) on archive HCC tissues for 18 patients. The anti-PD-L1 28-8 antibody was used for nivolumab-treated HCC, and anti-PD-L1 22C3 antibody was applied for pembrolizumab-treated HCC [7,8]. Expression levels were reported by tumor proportion score (TPS) and/or combined positive score (CPS), respectively [7,8].

4.4. Biochemical Tests

Serum biochemistry tests were measured by systemic multi-autoanalyzer (Technicon SMAC, Technicon Instruments Corp., Tarrytown, NY, USA). Serum AFP levels were measured by chemiluminescent microparticle immunoassay (ARCHITECT AFP assay, Abbott Ireland Diagnostics Division, Sligo, Ireland) with clinically reportable range from 1 to 1,998,000 ng/mL.

4.5. Statistical Analysis

Continuous variables were expressed as median (interquartile ranges—IQR), while categorical variables were analyzed as frequency and percentages. The Pearson chi-square analysis or Fisher's exact test was used to compare categorical variables, while the Student's t-test or Mann–Whitney U test was applied for continuous variables. Survival was estimated by the Kaplan–Meier method and compared by the log-rank test. Additionally, Cox's proportional-hazard model was used to identify prognostic factors for survivals. To avoid the effect of collinearity, ALBI grade and BCLC or Child-Pugh class were not included in the same multivariate model. For all analyses, $p < 0.05$ was considered statistically significant. All statistical analyses were performed using the Statistical Package for Social Sciences (SPSS 17.0 for Windows, SPSS Inc., Chicago, IL, USA).

5. Conclusions

The 10-10 rule of early AFP response can predict objective response and survival to ICI treatment in unresectable HCC. Besides, good liver reserves confer better survival among these patients. These findings help to provide effective on-treatment guidance of ICI treatment for HCC patients.

Supplementary Materials: The following are available online at http://www.mdpi.com/2072-6694/12/1/182/s1, Figure S1: Best response in 18 HCC patients with measured PD-L1 level and evaluable response; Figure S2: Differentiated overall survival (OS) of unresectable hepatocellular carcinoma treated with immune checkpoint inhibitors. Table S1: Factors associated with best disease control in 90 patients with evaluable response; Table S2: Factors associated with overall survival in 95 patients treated with immune checkpoint inhibitors (including tumor response in analyses).

Author Contributions: Conceptualization, Y.-H.H.; data curation, P.-C.L., Y.C., M.-H.C., K.-H.L., C.-J.L., I.-C.L. and S.-C.C.; formal analysis, P.-C.L., C.-J.L. and M.-C.H.; writing—original draft preparation, P.-C.L. and Y.-H.H.; writing—review and editing, M.C.H. and Y.C.; supervision, Y.C. and Y.H.H. All authors have read and agreed to the published version of the manuscript.

Funding: The work was partly supported by the Taipei Veterans General Hospital, Taipei, Taiwan (grant numbers V109C-048); and Ministry of Science and Technology, Taiwan (grant numbers MOST108-2628-B-075-006).

Conflicts of Interest: The authors declare no conflict of interest.

References

1. Ferlay, J.; Soerjomataram, I.; Dikshit, R.; Eser, S.; Mathers, C.; Rebelo, M.; Parkin, D.M.; Forman, D.; Bray, F. Cancer incidence and mortality worldwide: Sources, methods and major patterns in GLOBOCAN 2012. *Int. J. Cancer* **2015**, *136*, E359–E386. [CrossRef]
2. Galle, P.R.; Forner, A.; Llovet, J.M.; Mazzaferro, V.; Piscaglia, F.; Raoul, J.L.; Schirmacher, P.; Vilgrain, V. EASL Clinical Practice Guidelines: Management of hepatocellular carcinoma. *J. Hepatol.* **2018**, *69*, 182–236. [CrossRef]
3. Llovet, J.M.; Ricci, S.; Mazzaferro, V.; Hilgard, P.; Gane, E.; Blanc, J.F.; de Oliveira, A.C.; Santoro, A.; Raoul, J.L.; Forner, A.; et al. Sorafenib in advanced hepatocellular carcinoma. *N. Engl. J. Med.* **2008**, *359*, 378–390. [CrossRef]

4. Cheng, A.L.; Kang, Y.K.; Chen, Z.; Tsao, C.J.; Qin, S.; Kim, J.S.; Luo, R.; Feng, J.; Ye, S.; Yang, T.S.; et al. Efficacy and safety of sorafenib in patients in the Asia-Pacific region with advanced hepatocellular carcinoma: A phase III randomised, double-blind, placebo-controlled trial. *Lancet Oncol.* **2009**, *10*, 25–34. [CrossRef]
5. El-Khoueiry, A.B.; Sangro, B.; Yau, T.; Crocenzi, T.S.; Kudo, M.; Hsu, C.; Kim, T.Y.; Choo, S.P.; Trojan, J.; Welling, T.H.R.; et al. Nivolumab in patients with advanced hepatocellular carcinoma (CheckMate 040): An open-label, non-comparative, phase 1/2 dose escalation and expansion trial. *Lancet* **2017**, *389*, 2492–2502. [CrossRef]
6. Zhu, A.X.; Finn, R.S.; Edeline, J.; Cattan, S.; Ogasawara, S.; Palmer, D.; Verslype, C.; Zagonel, V.; Fartoux, L.; Vogel, A.; et al. Pembrolizumab in patients with advanced hepatocellular carcinoma previously treated with sorafenib (KEYNOTE-224): A non-randomised, open-label phase 2 trial. *Lancet Oncol.* **2018**, *19*, 940–952. [CrossRef]
7. Finn, R.S.; Ryoo, B.Y.; Merle, P.; Kudo, M.; Bouattour, M.; Lim, H.Y.; Breder, V.; Edeline, J.; Chao, Y.; Ogasawara, S.; et al. Results of KEYNOTE-240: Phase 3 study of pembrolizumab (Pembro) vs best supportive care (BSC) for second line therapy in advanced hepatocellular carcinoma (HCC). *J. Clin. Oncol.* **2019**, *37*, 4004. [CrossRef]
8. Topalian, S.L.; Hodi, F.S.; Brahmer, J.R.; Gettinger, S.N.; Smith, D.C.; McDermott, D.F.; Powderly, J.D.; Carvajal, R.D.; Sosman, J.A.; Atkins, M.B.; et al. Safety, activity, and immune correlates of anti-PD-1 antibody in cancer. *N. Engl. J. Med.* **2012**, *366*, 2443–2454. [CrossRef] [PubMed]
9. Finkelmeier, F.; Waidmann, O.; Trojan, J. Nivolumab for the treatment of hepatocellular carcinoma. *Expert Rev. Anticancer Ther.* **2018**, *18*, 1169–1175. [CrossRef] [PubMed]
10. Garon, E.B.; Rizvi, N.A.; Hui, R.; Leighl, N.; Balmanoukian, A.S.; Eder, J.P.; Patnaik, A.; Aggarwal, C.; Gubens, M.; Horn, L.; et al. Pembrolizumab for the treatment of non-small-cell lung cancer. *N. Engl. J. Med.* **2015**, *372*, 2018–2028. [CrossRef]
11. Fuchs, C.S.; Doi, T.; Jang, R.W.; Muro, K.; Satoh, T.; Machado, M.; Sun, W.; Jalal, S.I.; Shah, M.A.; Metges, J.P.; et al. Safety and Efficacy of Pembrolizumab Monotherapy in Patients With Previously Treated Advanced Gastric and Gastroesophageal Junction Cancer: Phase 2 Clinical KEYNOTE-059 Trial. *JAMA Oncol.* **2018**, *4*, e180013. [CrossRef] [PubMed]
12. Khunger, M.; Hernandez, A.V.; Pasupuleti, V.; Rakshit, S.; Pennell, N.A.; Stevenson, J.; Mukhopadhyay, S.; Schalper, K.; Velcheti, V. Programmed Cell Death 1 (PD-1) Ligand (PD-L 1) Expression in Solid Tumors As a Predictive Biomarker of Benefit From PD-1/PD-L 1 Axis Inhibitors: A Systematic Review and Meta-Analysis. *JCO Precis. Oncol.* **2017**, *1*. [CrossRef]
13. Finkelmeier, F.; Czauderna, C.; Perkhofer, L.; Ettrich, T.J.; Trojan, J.; Weinmann, A.; Marquardt, J.U.; Vermehren, J.; Waidmann, O. Feasibility and safety of nivolumab in advanced hepatocellular carcinoma: Real-life experience from three German centers. *J. Cancer Res. Clin. Oncol.* **2019**, *145*, 253–259. [CrossRef] [PubMed]
14. Memon, K.; Kulik, L.; Lewandowski, R.J.; Wang, E.; Ryu, R.K.; Riaz, A.; Nikolaidis, P.; Miller, F.H.; Yaghmai, V.; Baker, T.; et al. Alpha-fetoprotein response correlates with EASL response and survival in solitary hepatocellular carcinoma treated with transarterial therapies: A subgroup analysis. *J. Hepatol.* **2012**, *56*, 1112–1120. [CrossRef] [PubMed]
15. Jung, J.; Yoon, S.M.; Han, S.; Shim, J.H.; Kim, K.M.; Lim, Y.S.; Lee, H.C.; Kim, S.Y.; Park, J.H.; Kim, J.H. Alpha-fetoprotein normalization as a prognostic surrogate in small hepatocellular carcinoma after stereotactic body radiotherapy: A propensity score matching analysis. *BMC Cancer* **2015**, *15*, 987. [CrossRef] [PubMed]
16. Mehta, N.; Dodge, J.L.; Roberts, J.P.; Hirose, R.; Yao, F.Y. Alpha-Fetoprotein Decrease from > 1, 000 to <500 ng/mL in Patients with Hepatocellular Carcinoma Leads to Improved Posttransplant Outcomes. *Hepatology* **2019**, *69*, 1193–1205.
17. Shao, Y.Y.; Lin, Z.Z.; Hsu, C.; Shen, Y.C.; Hsu, C.H.; Cheng, A.L. Early alpha-fetoprotein response predicts treatment efficacy of antiangiogenic systemic therapy in patients with advanced hepatocellular carcinoma. *Cancer* **2010**, *116*, 4590–4596. [CrossRef]
18. Meyer, T.; Melero, I.; Yau, T.; Hsu, C.; Kudo, M.; Choo, S.P.; Trojan, J.; Welling, T.; Kang, Y.-K.; Yeo, W.; et al. Hepatic safety and biomarker assessments in sorafenib-experienced patients with advanced hepatocellular carcinoma treated with nivolumab in the CheckMate-040 study. *J. Hepatol.* **2018**, *68* (Suppl. 1), S16. [CrossRef]

19. Scheiner, B.; Kirstein, M.M.; Hucke, F.; Finkelmeier, F.; Schulze, K.; von Felden, J.; Koch, S.; Schwabl, P.; Hinrichs, J.B.; Waneck, F.; et al. Programmed cell death protein-1 (PD-1)-targeted immunotherapy in advanced hepatocellular carcinoma: Efficacy and safety data from an international multicentre real-world cohort. *Aliment. Pharmacol. Ther.* **2019**, *49*, 1323–1333. [CrossRef]
20. Shao, Y.Y.; Liu, T.H.; Hsu, C.; Lu, L.C.; Shen, Y.C.; Lin, Z.Z.; Cheng, A.L.; Hsu, C.H. Early alpha-foetoprotein response associated with treatment efficacy of immune checkpoint inhibitors for advanced hepatocellular carcinoma. *Liver Int.* **2019**. [CrossRef]
21. Lee, P.C.; Chen, Y.T.; Chao, Y.; Huo, T.I.; Li, C.P.; Su, C.W.; Lee, M.H.; Hou, M.C.; Lee, F.Y.; Lin, H.C.; et al. Validation of the albumin-bilirubin grade-based integrated model as a predictor for sorafenib-failed hepatocellular carcinoma. *Liver Int.* **2018**, *38*, 321–330. [CrossRef] [PubMed]
22. Yau, T.; Hsu, C.; Kim, T.Y.; Choo, S.P.; Kang, Y.K.; Hou, M.M.; Numata, K.; Yeo, W.; Chopra, A.; Ikeda, M.; et al. Nivolumab in advanced hepatocellular carcinoma: Sorafenib-experienced Asian cohort analysis. *J. Hepatol.* **2019**, *71*, 543–552. [CrossRef] [PubMed]
23. Reig, M.; Rimola, J.; Torres, F.; Darnell, A.; Rodriguez-Lope, C.; Forner, A.; Llarch, N.; Rios, J.; Ayuso, C.; Bruix, J. Postprogression survival of patients with advanced hepatocellular carcinoma: Rationale for second-line trial design. *Hepatology* **2013**, *58*, 2023–2031. [CrossRef] [PubMed]
24. Iavarone, M.; Cabibbo, G.; Biolato, M.; Della Corte, C.; Maida, M.; Barbara, M.; Basso, M.; Vavassori, S.; Craxi, A.; Grieco, A.; et al. Predictors of survival in patients with advanced hepatocellular carcinoma who permanently discontinued sorafenib. *Hepatology* **2015**, *62*, 784–791. [CrossRef] [PubMed]
25. Pinato, D.J.; Yen, C.; Bettinger, D.; Ramaswami, R.; Arizumi, T.; Ward, C.; Pirisi, M.; Burlone, M.E.; Thimme, R.; Kudo, M.; et al. The albumin-bilirubin grade improves hepatic reserve estimation post-sorafenib failure: Implications for drug development. *Aliment. Pharmacol. Ther.* **2017**, *45*, 714–722. [CrossRef] [PubMed]
26. Kudo, M.; Matilla, A.; Santoro, A.; Melero, I.; Gracian, A.C.; Acosta-Rivera, M.; Choo, S.P.; El-Khoueiry, A.B.; Kuromatsu, R.; El-Rayes, B.F.; et al. Checkmate-040: Nivolumab (NIVO) in patients (pts) with advanced hepatocellular carcinoma (aHCC) and Child-Pugh B (CPB) status. *J. Clin. Oncol.* **2019**, *37*, 327. [CrossRef]
27. Kambhampati, S.; Bauer, K.E.; Bracci, P.M.; Keenan, B.P.; Behr, S.C.; Gordan, J.D.; Kelley, R.K. Nivolumab in patients with advanced hepatocellular carcinoma and Child-Pugh class B cirrhosis: Safety and clinical outcomes in a retrospective case series. *Cancer* **2019**, *125*, 3234–3241. [CrossRef]
28. Cabibbo, G.; Petta, S.; Barbara, M.; Attardo, S.; Bucci, L.; Farinati, F.; Giannini, E.G.; Negrini, G.; Ciccarese, F.; Rapaccini, G.L.; et al. Hepatic decompensation is the major driver of death in HCV-infected cirrhotic patients with successfully treated early hepatocellular carcinoma. *J. Hepatol.* **2017**, *67*, 65–71. [CrossRef]
29. Kudo, M. Targeted and immune therapies for hepatocellular carcinoma: Predictions for 2019 and beyond. *World J. Gastroenterol.* **2019**, *25*, 789–807. [CrossRef]
30. Ikeda, M.; Sung, M.W.; Kudo, M.; Kobayashi, M.; Baron, A.D.; Finn, R.S.; Kaneko, S.; Zhu, A.X.; Kubota, T.; Kraljevic, S.; et al. A phase 1b trial of lenvatinib (LEN) plus pembrolizumab (PEM) in patients (pts) with unresectable hepatocellular carcinoma (uHCC). *J. Clin. Oncol.* **2018**, *36*, 4076. [CrossRef]
31. Stein, S.; Pishvaian, M.J.; Lee, M.S.; Lee, K.-H.; Hernandez, S.; Kwan, A.; Liu, B.; Grossman, W.; Iizuka, K.; Ryoo, B.-Y. Safety and clinical activity of 1L atezolizumab + bevacizumab in a phase Ib study in hepatocellular carcinoma (HCC). *J. Clin. Oncol.* **2018**, *36*, 4074. [CrossRef]
32. Yau, T.; Kang, Y.-K.; Kim, T.-Y.; El-Khoueiry, A.B.; Santoro, A.; Sangro, B.; Melero, I.; Kudo, M.; Hou, M.-M.; Matilla, A.; et al. Nivolumab (NIVO) + ipilimumab (IPI) combination therapy in patients (pts) with advanced hepatocellular carcinoma (aHCC): Results from CheckMate 040. *J. Clin. Oncol.* **2019**, *37*, 4012. [CrossRef]
33. Calderaro, J.; Rousseau, B.; Amaddeo, G.; Mercey, M.; Charpy, C.; Costentin, C.; Luciani, A.; Zafrani, E.S.; Laurent, A.; Azoulay, D.; et al. Programmed death ligand 1 expression in hepatocellular carcinoma: Relationship With clinical and pathological features. *Hepatology* **2016**, *64*, 2038–2046. [CrossRef] [PubMed]
34. Heimbach, J.K.; Kulik, L.M.; Finn, R.S.; Sirlin, C.B.; Abecassis, M.M.; Roberts, L.R.; Zhu, A.X.; Murad, M.H.; Marrero, J.A. AASLD guidelines for the treatment of hepatocellular carcinoma. *Hepatology* **2018**, *67*, 358–380. [CrossRef]
35. Johnson, P.J.; Berhane, S.; Kagebayashi, C.; Satomura, S.; Teng, M.; Reeves, H.L.; O'Beirne, J.; Fox, R.; Skowronska, A.; Palmer, D.; et al. Assessment of liver function in patients with hepatocellular carcinoma: A new evidence-based approach-the ALBI grade. *J. Clin. Oncol.* **2015**, *33*, 550–558. [CrossRef]

36. Liu, P.H.; Hsu, C.Y.; Hsia, C.Y.; Lee, Y.H.; Chiou, Y.Y.; Huang, Y.H.; Lee, F.Y.; Lin, H.C.; Hou, M.C.; Huo, T.I. ALBI and PALBI grade predict survival for HCC across treatment modalities and BCLC stages in the MELD Era. *J. Gastroenterol. Hepatol.* **2017**, *32*, 879–886. [CrossRef]
37. Eisenhauer, E.A.; Therasse, P.; Bogaerts, J.; Schwartz, L.H.; Sargent, D.; Ford, R.; Dancey, J.; Arbuck, S.; Gwyther, S.; Mooney, M.; et al. New response evaluation criteria in solid tumours: Revised RECIST guideline (version 1.1). *Eur. J. Cancer* **2009**, *45*, 228–247. [CrossRef]

© 2020 by the authors. Licensee MDPI, Basel, Switzerland. This article is an open access article distributed under the terms and conditions of the Creative Commons Attribution (CC BY) license (http://creativecommons.org/licenses/by/4.0/).

Review
Rationale of Immunotherapy in Hepatocellular Carcinoma and Its Potential Biomarkers

David Tai [1], Su Pin Choo [1,2] and Valerie Chew [3,*]

[1] National Cancer Centre, Singapore 169610, (NCCS), Singapore; david.tai.w.m@singhealth.com.sg (D.T.); choo.su.pin@singhealth.com.sg (S.P.C.)
[2] Curie Oncology, Mount Elizabeth Novena Specialist Centre, Singapore 329563, Singapore
[3] Translational Immunology Institute (TII), SingHealth-DukeNUS Academic Medical Centre, Singapore 169856, Singapore
* Correspondence: valerie.chew.s.p@singhealth.com.sg

Received: 30 October 2019; Accepted: 29 November 2019; Published: 3 December 2019

Abstract: Hepatocellular carcinoma (HCC), the most common type of liver cancer, is derived mostly from a background of chronic inflammation. Multiple immunotherapeutic strategies have been evaluated in HCC, with some degree of success, particularly with immune checkpoint blockade (ICB). Despite the initial enthusiasm, treatment benefit is only appreciated in a modest proportion of patients (response rate to single agent ~20%). Therapy-induced immune-related adverse events (irAEs) and economic impact are pertinent considerations with ICB. It is imperative that a deeper understanding of its mechanisms of action either as monotherapy or in combination with other therapeutic agents is needed. We herein discuss the latest developments in the immunotherapeutic approaches for HCC, the potential predictive biomarkers and the rationale for combination therapies. We also outline promising future immunotherapeutic strategies for HCC patients.

Keywords: immunotherapy; biomarkers; combination immunotherapy; immune-related adverse events (irAEs); hepatocellular carcinoma (HCC)

1. Introduction

Cancer immunotherapy is a rapidly evolving field, which has revolutionized the treatment landscape in oncology this past decade [1]. Unlike conventional cancer therapies, immunotherapeutic approaches do not directly target tumor cells; instead, they target the patient's immune system or the tumor microenvironment (TME) [2]. A variety of strategies have been explored: cytokine administration, cancer vaccines, adoptive cellular therapy, and immune checkpoint blockade (ICB) [3]. Among which, ICB have been the focus of cancer immunotherapy due to its promising outcomes across multiple advanced solid malignancies, including hepatocellular carcinoma (HCC) [4,5]. HCC is the most common type of primary liver cancer. It is the sixth most common cancer type and the fourth leading cause of cancer death worldwide [6]. Survival after curative surgery remains relatively low. Five-year disease-free survival rates after resection ranges between 24% and 36%, with recurrence rates being as high as 70% [7–9]. Before the emergence of immunotherapy, therapeutic development in advanced HCC has been limited partly due to its complex and heterogeneous disease etiologies [10].

The response rate for HCC patients treated with single-agent ICB is modest, at ~15–20%. Moreover, 15–25% of these ICB-treated patients experienced grade 3/4 treatment or immune-related adverse events (TRAEs or irAEs), such as rash, pruritus, diarrhea, and an increase in aspartate aminotransferase (AST) and alanine aminotransferase (ALT) [11,12]. Therefore, a better understanding of mechanistic properties of ICB and predictive biomarkers of response and toxicities is crucial for improved treatment in HCC. This review highlights the current knowledge of immunotherapy in HCC, with particular focus on ICB and the growing understanding of biomarkers discoveries. We also endeavor to provide rationale for combination strategies with ICB and perspectives on personalized immuno-therapeutics for HCC.

2. Current Landscape of Immunotherapy in HCC

2.1. Immune Checkpoint Blockade (ICB) Therapy

The key mechanism of action for ICB is to block the immune exhaustion or inhibitory pathways induced by chronic immune response against tumor antigen, in order to reactivate the antitumor immune response [13,14]. Immune checkpoint inhibitors are monoclonal antibodies designed to target multiple checkpoint molecules, such as PD-1, CTLA-4, Tim-3, Lag-3, and VISTA, expressed primarily by T cells, as well as PD-L1, the ligand for PD-1, expressed primarily by the tumor or other immune cells [14]. PD-1, PD-L1, and CTLA-4 inhibitors are the most widely evaluated ICB therapies in clinical trials for various solid cancers, including HCC. A summary of the major clinical trials using ICB as monotherapy in HCC, their response rates, and rates of >grade 3 irAEs are provided in Table 1. Combination strategies utilizing ICB are described in greater detail below, in Section 2.4.

2.1.1. Anti-PD-1 Therapy

Two phase I/II clinical studies in HCC, CheckMate040, and Keynote224, using anti-PD-1 monoclonal antibodies nivolumab and pembrolizumab respectively, have been reported [11,12]. CheckMate040, is a multicohort phase I/II, open-label, dose escalation, and expansion trial, using nivolumab alone or in combination with ipilimumab (anti-CTLA4 monoclonal antibody). In cohort 1 and 2 of Checkmate040 (cohort 1: all 214 patients and cohort 2: 85 Asian patients), patients with advanced HCC who were treatment naïve or progressed/intolerant to sorafenib were treated with nivolumab. Keynote224 was a phase II, open-label trial that assessed the efficacy and safety of pembrolizumab in 104 patients with advanced HCC, who were previously treated with sorafenib. Nivolumab demonstrated an objective response rate (ORR) of 20% and disease control rate (DCR) of 64%, whereas pembrolizumab showed an ORR of 17% and DCR of 62% (see Table 1) [11,12]. A subsequent Asian cohort analysis from CheckMate040 demonstrated an ORR of 15% [15]. Both of these trials demonstrated superior ORR and DCR compared to historical responses of sorafenib in advanced HCC (~2% ORR in the SHARP trial) [16]. In addition, the median duration of response was up to 17 months reported in sorafenib experienced HCC patients treated with nivolumab [11], underlining the durability of control in a proportion of patients. Both CheckMate040, and Keynote224 reported moderate (15–25%) >grade 3/4 irAEs (see Table 1). The response rates of anti-PD1 therapy in HCC is, however, modest compared to other cancers, like melanoma (ORR 44%) [17] and renal cell carcinoma (ORR 25%) [18].

Despite encouraging results obtained from initial single-arm studies, two phase III trials in advanced HCC: CheckMate459 (NCT02576509), and Keynote240 (NCT02702401), using nivolumab and pembrolizumab respectively, failed to meet their predetermined primary endpoints of overall survival (OS). In Keynote240 trial, pembrolizumab, when compared to placebo in advanced HCC patients previously treated with sorafenib, did not meet the predetermined dual primary endpoints of improved OS (HR: 0.78; one-sided p = 0.0238) and PFS (HR: 0.78; one-sided p = 0.0209). Of note, however, the ORR of 18.3% was comparable to earlier studies with median duration of response of 13.8 months [19]. CheckMate459 trial, which compared nivolumab versus sorafenib as first-line treatment in patients with unresectable HCC, also did not meet its prespecified primary endpoint of OS [20]. Median OS was 16.4 months for nivolumab and 14.7 months for sorafenib (HR, 0.85 [95% CI, 0.72–1.02]; p = 0.0752). An improvement in ORR was observed with nivolumab compared with sorafenib (odds ratio (95% CI), 2.41 (1.48–3.92)) (see Table 1). Grade 3/4 treatment related adverse events were reported in 81 patients (22%) in the nivolumab arm and 179 patients (49%) in the sorafenib arm [20]. Despite both studies not meeting their primary endpoints, there was a clear trend toward improved OS in favor of ICB. Nevertheless, treatment effect of single-agent ICB appears binary with a modest proportion of patients truly deriving benefit. This underlines the need for a predictive biomarker of response as well as rational combination strategies.

2.1.2. Anti-PD-L1 Therapy

Several anti-PD-L1 monoclonal antibodies are currently under clinical trials in advanced HCC include avelumab, durvalumab, and atezolizumab. Avelumab monotherapy is currently being evaluated in a phase II study (NCT03389126). Durvalumab monotherapy was evaluated in a phase I/II trial in various solid tumors and reported an ORR of 10.3% in 39 HCC patients who declined, were intolerant, or progressed on prior sorafenib [21] (see Table 1). Atezolizumab monotherapy was compared against combination of atezolizumab and bevacizumab (anti-VEGF antibody) in advanced HCC patients in the Arm F of Phase Ib GO30140 study [22]. Median progression-free survival (PFS) was 3.4 months in the monotherapy arm, compared to 5.6 months in the combination arm (HR 0.55, $p = 0.018$) [22].

2.1.3. Anti-CTLA-4 Monoclonal Antibodies

Anti-CTLA4 antibody (Ipilimumab) was first approved by FDA in 2011 for the treatment of melanoma, following the result from the phase III trial, showing significant overall survival benefit compared to gp100 vaccine alone [23]. Another anti-CTLA4 antibody, tremelimumab, was evaluated for safety, antitumor, and antiviral activity in HCV-related HCC as monotherapy in a single-arm phase II trial (NCT01008358) [24]. An ORR of 17.6% was reported among 17 patients (see Table 1) as well as anti-HCV viral immunity [24]. Result from another phase I/II study of durvalumab and tremelimumab in patients with unresectable HCC (NCT02519348) will be announced in the near future [25].

Table 1. Immune checkpoint monotherapy clinical trials in HCC.

Study Name	Phase	Target	Treatments	Estimated Enrolment ^ (n)	ORR (%)	DCR (%)	PFS (Median, mo)	OS (Median, mo)	Adverse Effect † > Grade 3 (%)
NCT01658878 (CheckMate040) [11]	I/II	PD-1	Nivolumab	214	20%	64%	4	15.1	25%
NCT01658878 (CheckMate040-Asian cohort analysis) [15]	I/II	PD-1	Nivolumab	85	15%	49%	NA	14.9	16%
NCT02702414 (Keynote224) [12]	II	PD-1	Pembrolizumab	104	17%	62%	4.9	12.9	15%
NCT02576509 (CheckMate459) [20]	III	PD-1	Nivolumab vs. Sorafenib	743 (371 vs. 372)	15% vs. 7%	55% vs. 58%	3.7 vs. 3.8	16.4 vs. 14.7	22% vs. 49%
NCT02702401 (Keynote240) [19]	III	PD-1	Pembrolizumab vs. placebo	413 (278 vs. 135)	16.9% vs. 4.4%	62.2% vs. 53.3%	3.0 vs. 2.8	13.9 vs. 10.6	18.6% vs. 7.5%
NCT01693562 [21]	I/II	PD-L1	Durvalumab	39	10.3%	32.5% #	2.7	13.2	20%
NCT01008358 [24]	II	CTLA-4	Tremelimumab	20	17.6%	76.4%	6.48 (TTP)	8.2	45%

^, most updated from clinicaltrials.gov as of August 2019; n, number of patients; ORR, overall response rate; DCR, disease control rate; PFS, progression-free survival; OS, overall survival; mo, months; †, treatment-related adverse effects; NA, not available; #, CR+ PR + SD > 24 weeks; TTP, time to progression.

2.2. Current Knowledge on Biomarkers for ICB and Its Relevance in HCC

Predictive biomarkers of response in ICB across different cancer types have been extensively reviewed [26–28]. We summarize the key biomarkers from intratumoral tissues (tumor or TME specific tissue markers) and extratumoral tissues (from peripheral blood, serum or feces) in Table 2 and provide evidence and perspectives, where available, on HCC.

2.2.1. PD-L1 Expression

PD-L1 expression is one of the earliest and most widely used biomarkers of response to immunotherapy. PD-L1 IHC is approved by FDA as a companion diagnostic when considering use of anti-PD1 therapy in NSCLC [29,30]. Despite this, the utility of PD-L1 expression across multiple tumor types has been disparate: some with positive association [31–35], while others with no association [11,12,18,36,37] with clinical outcome. Within HCC tissues, PD-L1 was found to be expressed by the tumor cells [38] and macrophages [39], both of which were associated with poor post-resection prognosis; meanwhile, PD-1 was expressed mainly by the T cells, including regulatory T cells (Treg) [40,41]. It has also been shown that the PD-L1 expression in HCC is generally low (~10% by tumor cells) and highly heterogeneous across different anti-PD-L1 staining antibodies used [42]. Indeed, tumor PD-L1 expression was not a robust biomarker for response to anti-PD-1 therapy in both CheckMate040 and Keynote224 trials in HCC [11,12]. Reasons for contradictory results from clinical trials using PD-L1 as a biomarker include the different assays for detection, the spatial heterogeneity in expression of PD-L1, and various standards and cutoffs used in assessing positive staining [29,30,43].

Another important consideration is that nontumor host cells could also express PD-L1 and be considered as the biomarker for response to anti-PD-1/anti-PD-L1 ICB [44]. For instance, studies in melanoma [45], urothelial carcinoma [46], and HCC [12] have found that PD-L1 expression on nontumor host cells, such as TILs, was associated with response to anti-PD1 or anti-PD-L1 therapy. More recently, circulating exosomal PD-L1 was shown to correlate with clinical response to anti-PD-1 therapy, in a study conducted in patients with advanced melanoma [47]. Increased circulating exosomal PD-L1 was indicative of adaptive response by the tumor cells to T-cell reinvigoration [47]. One recent preclinical study in mice models demonstrated that, by suppressing exosomal PD-L1, antitumor immune response and memory could be induced even in the anti-PD-L1-resistant models [48]. Given the high intratumoral heterogeneity of HCC tumors as described previously [49,50], exosomal markers could serve as an attractive biomarker to predict clinical outcome to immunotherapy.

Recent research focuses also on the post-translational regulation of PD-L1 expression [51–53]. For instance, epigenetic regulation of PD-L1 protein expression by microRNA has been implicated in various cancers [51]. Maintenance of PD-L1 on the cell membrane and prevention from its lysosomal degradation by regulatory proteins, such as CMTM6, could also play an important role [54]. Additionally, given the link between inflammation and IFNγ-induced upregulation of PD-L1 expression in tumor [55,56], IFNγ signature has also been shown to be a biomarker of response to ICB in multiple cancer types [57,58]. Such data in HCC is currently lacking.

Table 2. Biomarkers predictive of response of immune checkpoint therapy.

Source	Biomarker	Assay Type	Cancer Type	Clinical Relevance	Relevance to HCC
Intra-tumoral	PD-L1 expression	Immunohistochemistry (IHC)	Multiple	Expression on tumor cells [31–35] or immune cells [12,45,46] showed positive association with response. No significant association with response [11,12,18,36,37].	No association with response [11,12]. Marginal association between PD-L1 expression on nontumor host cell (Keynote224) with response [12].
	IFNγ signature	NGS or targeted genes seq	Multiple	Predictive of response to ICB [57,58].	No direct evidence in HCC yet.CC yet.
	TMB	NGS, WES, or WGS	Multiple	Higher TMB was positively associated with improved response to ICB [59–62].	Positive association with response (mixed cancer types including HCC) [59,63]. No significant association with anti-PD-1 ICB (17 HCC patients) [64].
	Tumor transcriptomic diversity	Single-cell RNA seq	HCC/iCCA	Lower tumor transcriptomic diversity was associated with PFS and OS of liver cancer patients treated with mixed ICB [65].	
	Wnt/β-catenin pathway mutation	NGS, WES, or targeted genes seq	Melanoma/HCC	Wnt/β-catenin mutation was linked to T cell exclusion, immunosuppressive TME and resistance to ICB [66,67].	Wnt pathway mutation was related to resistance to therapy (27 HCC patients) [68].
	TILs density (hot/cold)	IHC or RNA seq	Multiple	Higher TILs density (particularly CD8+ TILs) was associated with superior clinical response [45,69–71].	An increase in CD8+ T-cell tumor infiltration and effector T-cells [72] or cytolytic T cell infiltrates [65] was associated with response to ICB in HCC.
	T-cell repertoire	RNA seq or TCR seq	Melanoma and lung cancer	TIL clonality positively correlated with response [69,73–75].	
	HLA diversity	NGS, WES	Melanoma and NSCLC	HLA-I heterozygosity was associated with improved OS after ICB [76].	
	Specific CD8+ T-cell phenotypes	Flow cytometry	Melanoma and NSCLC	Increased density of cytolytic [73], PD-1+CD8+ T cells [77] and TCF7+ memory-like CD8+ T cells [78] were positively associated with ICB response.	
	Treg	Flow cytometry, IHC, or RNA seq	Multiple	Higher frequency of Treg was linked to unresponsiveness to [79] and hyperprogression after ICB [80].	No direct evidence in HCC yet.
	Macrophages	IHC, flow cytometry, or RNA seq	Multiple	TAM [81] and MDSC [79] is associated with unresponsiveness to PD-1 ICB.	No direct evidence in HCC yet.

Table 2. Cont.

Source	Biomarker	Assay Type	Cancer Type	Clinical Relevance	Relevance to HCC
	T-cell clonality	TCR repertoire sequencing	Multiple	Pretreatment TCR diversity and on-therapy TCR clonal expansion were correlated with clinical benefit [82–84].	No direct evidence in HCC yet.
	T-cell phenotypes	Flow cytometry, CyTOF	Melanoma	Higher T-cell reinvigoration [85] and T-cell activation [86] were associated with clinical outcome after anti-PD-1 therapy.	No direct evidence in HCC yet.
	MDSCs	Flow cytometry	Melanoma	Peripheral blood level of MDSCs correlated with poor anti-CTLA-4 response [87–90].	No direct evidence in HCC yet.
	Neutrophils/leukocytes	Flow cytometry	Multiple	Higher peripheral blood neutrophil/lymphocytes ratios were associated with decreased PFS and OS after ICB treatment [91–94].	No direct evidence in HCC yet.
	Treg	Flow cytometry	Melanoma	High baseline frequency [89], on-therapy increased [95] or decreased [96] in frequency of circulating Treg was associated with disease control upon ICB.	No direct evidence in HCC yet.
Extra-tumoral	LDH	Serum LDH detection	Melanoma and NSCLC	Baseline or on-therapy change of serum LDH levels correlated with OS of ICB-treated patients [89,96-100].	No direct evidence in HCC yet.
	Exosomal PD-L1	Exosome purification and characterization	Melanoma	Increased increase in circulating exosomal PD-L1 during early stages of treatment, as an indicator of the adaptive response of the tumour cells to T cell reinvigoration, stratifies clinical responders from non-respondersIncrease in circulating exosomal PD-L1 during early stages of treatment positively correlated with clinical response to anti-PD-1 therapy [47].	No direct evidence in HCC yet.
	cfDNA	cfDNA isolation followed by WES and WGS	Multiple	Specific mutations and TMB detected from circulating cfDNA associated with response to ICB [101,102].	Hypermutated circulating tumor DNA was associated with clinical outcome in 69 ICB-treated cancer patients (includes 3 HCC patients) [101].
	Gut microbiome	PCR or 16S rRNA gene sequencing	Multiple	Specific or diversity of gut microbiome was associated with response to ICB [103–107].	No direct evidence in HCC yet.

PD-L1, programmed cell death 1 ligand; PD-1, programmed cell death 1; HCC, hepatocellular carcinoma; seq. sequencing; ICB, immune checkpoint blockade that include anti-PD-1/PD-L1/CTLA-4 unless specified; TMB, tumor mutational burden; NGS, next-generation sequencing; WES, whole exome sequencing; WGS, whole genome sequencing; TME, tumor microenvironment; iCCA, intrahepatic cholangiocarcinoma; TILs, tumor-infiltrating lymphocytes; IHC, immunohistochemistry; TCR, T cell receptor; HLA, human leukocyte antigen; NSCLC, non-small cell lung cancer; CyTOF, Cytometry by Time-of-Flight; PFS, progression-free survival; OS, overall survival; MDSCs, myeloid derived suppressor cells; Treg, regulatory T cells; LDH, lactate dehydrogenase; cfDNA, cell-free DNA.

2.2.2. Tumor Mutational Burden (TMB) and Specific Genomic Mutations

Tumor mutational burden (TMB) correlates with responses with ICB across multiple cancer types, including HCC [59–63]. One cross cancer study on TMB indicated that tumors with high TMB would also have higher expression PD-L1 on tumor cells, predicting their response to anti-PD-1/PD-L1 therapy [63]. Indeed, tumors with high TMB are associated with more neoantigens and linked to a more inflamed tumor microenvironment, higher IFNγ expression, and upregulation of PD-L1 expression [55,108]. In this study, TMB level is considered moderate for HCC, consistent with a modest response rate to ICB in HCC [63]. However, a separate small case series of 17 HCC patients treated with anti-PD-1 ICB showed no significant association between TMB and response [64]. Furthermore, DNA mismatch repair (MMR) gene deficiency, which results from a heavy mutational burden and predictive of response to immunotherapy, is infrequent in HCC [109].

Specific tumor mutations, such as Wnt alteration/β-catenin mutation, are linked to a T-cell exclusion or immunosuppressive TME and resistance to ICB in patients with advanced melanoma [66,67]. A study involving 27 HCC patients who received ICB (a mix of anti-PD1/anti-PD-L1/anti-CTLA4 or combination) found Wnt-pathway mutation to be predictive of resistance to therapy [68]. A more recent single-cell RNA sequencing study on biopsy samples taken from 19 liver cancer patients (9 HCC and 10 cholangiocarcinoma patients) treated with mixed ICB regimens showed that patients with less transcriptomically diverse tumors demonstrated a better response and survival profile [65]. This study also identified VEGFA as one of the possible mechanisms of resistance to ICB, hence providing rationale for combination of ICB with an anti-angiogenic agent [65]. However, it is not known how TMB is related to the transcriptomic diversity of tumors. Further studies are needed to clarify the relevance of TMB and specific molecular alterations (e.g., Wnt alteration/β-catenin mutation or VEGF overexpression) in relation to response to ICB.

2.2.3. Tumor-Infiltrating Lymphocytes (TILs) Density and Phenotypes

Density of TILs, particularly CD8+ T cells, connotes a better prognosis in various cancer types, including HCC [110–114]. Several studies have shown that higher TILs density, particularly for CD8+ T cells, predicts for better survival after ICB [45,69–71]. CD8+ T-cells density at the invasive margin, and not at the center of the tumor, was the most important determinant of better outcomes in melanoma patients treated with anti-PD1 ICB [69]. In addition, T-cell receptor (TCR) diversity or clonality, indicative of its ability to recognize diverse repertoire of tumor antigens, has also been shown to correlate with response to ICB [69,73–75]. As recognition of tumor antigens depends on antigen-presentation components, it is hence not surprising that HLA diversity predicts better responses to immunotherapy [76].

Apart from density and location of TILs, their phenotypes also play an important role. For instance, the cytolytic property of T cells, indicated by expression of pro-inflammatory genes perforin and granzyme, was associated with response to anti-PD-1 therapy in melanoma patients, despite no significant change in TILs density [73,77]. A study using single-cell RNA sequencing (scRNA seq) technologies to profile TILs found that the ratio of activated to exhausted CD8+ T cells in the tumor correlated with the response to ICB in melanoma patients [78]. The recent scRNA seq study in liver cancer patients treated with mixed ICBs also concurred with these findings that tumor-infiltrating cytolytic T cells play an important role in predicting response to immunotherapy [65]. Another immunoprofiling study in HCC cohort who received preoperative ICB treatment, followed by resection, showed that an increase in effector T cell was associated with complete response [72]. Both studies underlined the importance of TILs, particularly its phenotypes, as predictive biomarker of response in HCC patients. In fact, it was previously shown that only ~20% of HCC tumors were considered well infiltrated by immune cells [67,115], consistent with the reported clinical outcomes in anti-PD-1 ICB monotherapy trials.

Other immune subsets such as regulatory T cells (Treg) or macrophages also have predictive values for response to ICB. For instance, higher frequency of Treg, myeloid derived suppressor cells

(MDSCs) [79] and tumor-associated macrophages (TAM) [81] are linked to unresponsiveness to ICB. Treg has been linked to cancer hyper-progression after ICB [80], further underlining its important regulatory role in ICB response. Their roles in HCC remain to be elucidated.

2.2.4. Peripheral Immune Cells' Phenotypes

Peripheral blood is an important biological material for monitoring clinical response after ICB. As T cells are the primary targets for ICB, the pretreatment diversity of TCR repertoire is an important biomarker of response to ICB in the circulating blood [82–84]. The phenotypes of T cells have also been studied, and the ratio of reinvigorated CD8+ T cell to the tumor burden [85], as well as the activation status of both CD4+ and CD8+ T cells [86], upon treatment could predict for response to ICB in melanoma patients. Other circulating immune cells, such as immunosuppressive MDSCs, have been shown to correlate with poor response to anti-CTLA-4 therapy in multiple studies in melanoma patients [87–90]. The ratio of neutrophil to lymphocytes was associated with decreased PFS and OS after ICB treatment [91–94].

The role of circulating Treg cells is, however, controversial. Higher baseline frequency of Treg has been linked to disease control after ICB [84]. In two studies in patients with advanced melanoma treated with ipilimumab, one reported that an on-therapy increase in frequency of circulating Treg at week 6 was associated with improved PFS [95]. In contrast, another study reported that a decrease in frequency of circulating Treg at a later timepoint of week 12 was associated with disease control upon ICB [96]. It is possible to speculate that an initial increase followed by decrease in Treg might be a sign of clinical response to immunotherapy.

It is therefore important to study the dynamic changes of various peripheral immune subsets at defined time points after immunotherapy, for a more accurate comparison. Such studies are currently lacking in HCC.

2.2.5. Other Extratumoral Biomarkers

Other noncellular biomarkers in the blood include lactate dehydrogenase (LDH), an enzyme that is released by rapidly growing tumors and associated with large tumor burden, tumor hypoxia, angiogenesis, and worse prognosis [116,117]. High baseline serum LDH levels are associated with worse outcomes with ICB [89,97,98,100]. Dynamic changes of LDH levels while on treatment could also predict outcomes. On-treatment reduction in LDH levels was associated with better response in patients with advanced melanoma treated with ipilimumab [96,99]. As serum LDH level has been used as a biomarker in predicting response to TACE [118] and sorafenib [119] in HCC, its role in predicting responses to ICB would be of interest.

As described earlier, increased circulating exosomal PD-L1 during early stages of anti-PD-1 therapy positively correlated with clinical response in melanoma patients [47]. Apart from this, circulating cell-free DNA (cfDNA) carrying tumor-related genetic and epigenetic alterations have been shown to be related to cancer development, progression, and resistance to therapy [120]. This makes cfDNA an easily accessible biomarker to predict tumor response to therapy, which is potentially not affected by intratumoral heterogeneity [121,122]. In fact, specific mutations or TMB can be detected from circulating cfDNA and that have been shown to associate with responses to ICB [101,102]. One particular study found that hypermutated circulating tumor DNA was associated with clinical outcome in 69 cancer patients, including three HCC patients, treated with a variety of ICBs [101].

Lastly, the gut microbiome analyzed from the feces also seem to play an important role in determining response to ICB. In fact, the role of microbiota in human health and disease, particularly in cancer, has been increasingly appreciated [123,124]. Interestingly, different strains of microbiome have been found to be associated with response to ICB in four major reports on baseline fecal sample analysis from melanoma [104–106] and other cancer types [107]. Of note, transferring the response-associated gut microbiota to germ-free or antibiotic-treated mice could induce ICB response, making fecal transfer an area of intense research interest at present. Other than the specific microbiota strain, the general increase in microbiota diversity [106] and the ratio of response-associated to resistance-associated microbiome [103] were also associated with better response to ICB. It remains to be determined if such a microbiome is related to response to ICB in HCC patients.

2.3. IrAEs and Its Association with Outcomes of ICB in HCC

Data from several key clinical trials using ICB in HCC patients showed that 15%–45% of the patients may experience grade 3 or greater treatment-related AEs, most of which being irAEs (see Table 1). Association between incidence of irAEs and clinical outcomes with ICB are conflicting. Overall irAEs have been found to be associated with better clinical outcomes in both melanoma and NSCLC patients treated with nivolumab [125,126]. Some studies, however, reported no association between irAEs and clinical outcome in selected malignancies [127,128]. Interestingly, a retrospective study of patients with various nonmelanoma cancers who received anti-PD-1 therapy demonstrated that only low-grade irAEs were associated with better responses in these patients [127]. Some studies even suggested that cancer-specific irAEs may be important in determining response to immunotherapy. The association of vitiligo to better responses in melanoma patients [129] and thyroid toxicity with better outcomes in NSCLC patients [130] with ICB are two such examples. A recent study on 114 HCC patients treated with mixed ICB reported a correlation of irAEs with higher DCR, median PFS and OS [131]. A future study involving a larger number of HCC patients with better defined immunotherapy regimens would be necessary to have a more conclusive assessment.

Several studies, with the majority of them in ipilimumab-treated melanoma patients, have reported various predictive biomarkers for irAEs, such as level of circulating IL-6, autoantibodies, blood-cell counts, T-cell repertoire, and gut microbiome [132]. For instance, the level of baseline circulating IL6 and being female are associated with higher incidences of irAEs in ipilimumab-treated advanced-melanoma patients [133]. A retrospective review of 167 patients with various solid tumor types treated with nivolumab or pembrolizumab suggested that patients with higher baseline lymphocyte counts have a greater risk for irAEs [134]. Another study on a group of 101 Japanese melanoma patients treated with nivolumab showed that the increase in total white-blood-cell count and decrease in relative lymphocyte count at the point of or just prior to irAEs were associated with lung and gastrointestinal irAEs [135]. Putting these two studies together, a higher baseline levels of lymphocytes predispose the patients to irAEs and the decrease of lymphocytes prior to or during the event of irAEs could indicate relocation or recruitment to the site of toxicities. The pretreatment or on-therapy level of autoantibodies, which is a known factor for autoimmune diseases, has also been implicated as a predictive biomarker for the development of irAEs in various cancer types with ICB [136–139]. There remains no study evaluating the association of irAE response with ICB in HCC. It would be interesting to note whether liver-specific toxicities would be related to response to immunotherapy in HCC.

2.4. Current Landscape and Rationale of Combination Immunotherapy in HCC

A combination of four major factors are needed to achieve an effective and sustained immune response: (1) release of tumor-specific antigens to induce T-cell response; (2) adequate generation of tumor-specific cytotoxic T cells with effective trafficking into TME; (3) appropriate TME remodelling strategies; and (4) overcoming exhaustion pathways which inevitably follows after the local immune activation (Figure 1). We next provide the rationale for various combination therapies currently pursued in HCC by ascribing to these factors. A list of current combination therapy involving ICB in HCC are listed in Table 3.

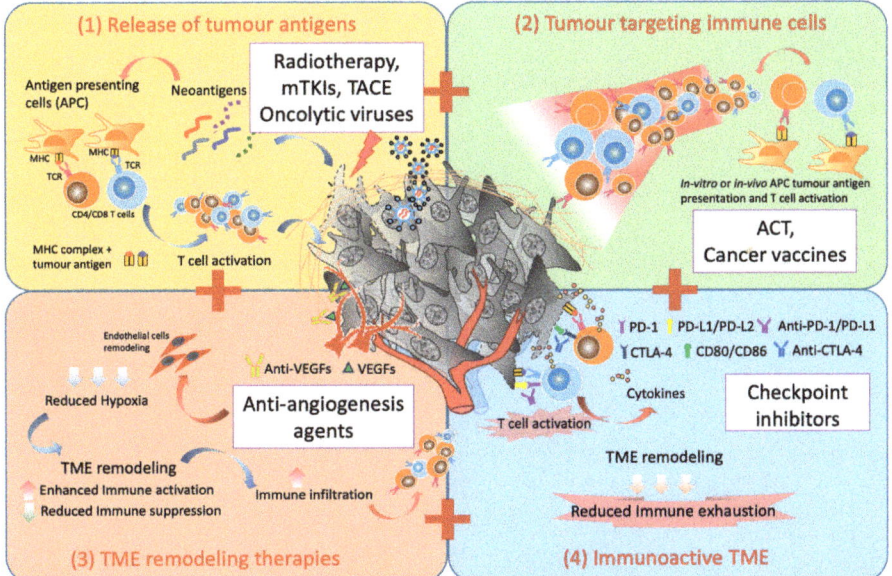

Figure 1. Combination strategies for immunotherapy in HCC. There are four key elements for successful immunotherapeutic strategies: (**1**) the release of tumor antigen to prime the tumor-antigen-specific T-cell response, i.e., the use of radiotherapy, multitargeted tyrokine kinase inhibitors (mTKIs), TACE, or oncolytic viruses that can induce immunogenic cell death; (**2**) the increase in the frequency of tumor-specific cytotoxic T cells which could home into the TME, i.e., by adoptive cell therapy (ACT) or cancer vaccines; (**3**) the TME remodelling strategies such as normalization of the blood to reduce the hypoxic and immunosuppressive microenvironment, i.e., with anti-antiangiogenesis agents; and (**4**) the blocking of the exhaustion pathways which inevitably follows after the local immune activation to reinvigorate the antitumor immune response, i.e., checkpoint-inhibitors.

Table 3. Immune checkpoint combination therapy clinical trials in HCC.

Study Name	Phase	Target	Treatments	Estimated Enrolment ^	ORR (%)	DCR (%)	PFS (Median, mo)	OS (Median, mo)	Adverse Effect † > Grade 3
			ICB-ICB Combination trials						
NCT01658878 (CheckMate040) * [140]	I/II	PD-1 + CTLA-4	Nivolumab + Ipilimumab	148	31%	49%	NR	40% (24-mo)	37%
NCT03298451 (HIMALAYA) [141]	III	PD-L1 + CTLA-4	Durvalumab versus Durvalumab + Tremelimumab vs. Sorafenib	1310	T.B.A.	T.B.A.	T.B.A.	T.B.A.	T.B.A.
NCT03680508	II	PD-1 + TIM3	TSR-042 + TSR-022	42	Not recruiting yet	T.B.A.	T.B.A.	T.B.A.	TBA
			ICB-others Combination trials						
NCT02519348 [25]	I/II	PD-L1 alone or CTLA4 alone or PD-L1 + CTLA-4or PD-L1 + VEGF	Durvalumabor Tremelimumab or Durvalumab + Tremelimumab or Durvalumab + Bevacizumab	545	15% (6/40 patients)	57.5% #	NR	NR	20% (8/40 patients)
NCT03434379 (IMBrave150) [142]	III	PD-L1 + VEGF	Atezolizumab + Bevacizumab vs. Sorafenib	501 (336 vs. 165)	27% vs. 55%	74% vs. 55%	6.8 vs. 4.3	NE vs. 13.2	57% vs. 55%
NCT02715531 (Arm A) [22]	Ib	PD-L1 + VEGF	Atezolizumab + Bevacizumab	104	36%	71%	7.3	17.1	27%
NCT02715531 (Arm F) [22]	Ib	PD-L1 + VEGF	Atezolizumab + Bevacizumab vs. Atezolizumab	60 / 59	20% / 17%	67% / 49%	5.6 / 3.4	NR	37% / 14%
NCT03755791 (COSMIC-312) [143]	III	PD-L1 + mTKIs	Atezolizumab + Cabozantinib vs. Sorafenib vs. Cabozantinib	740	T.B.A.	T.B.A.	T.B.A.	T.B.A.	T.B.A.
NCT03006926 (Keynote 524) [144]	Ib	PD-1 + mTKIs	Pembrolizumab + Lenvatinib	30	36.7%	90%	9.7 (TTP)	14.6	73%
NCT03713593 (LEAP-002) [145]	III	PD-1 + mTKIs	Pembrolizumab + Lenvatinib vs. Lenvatinib	750	TBA	TBA	TBA	TBA	TBA
NCT03289533 [146]	I	PD-L1 + mTKIs	Avelumab + Axitinib	22	13.6%	68.2%	5.5	12.7	72.7%
NCT03092895 [147]	II	PD-1 + FOLFOX4 or GEMOX	SHR-1210 + FOLFOX4 or GEMOX	34 (HCC patients)	26.5%	79.4%	5.5	NR	85.3%
NCT03071094	I/II	PD-1 + oncolytic virus	Nivolumab + Pexa-Vec	30	TBA	TBA	TBA	TBA	TBA

^, most updated from clinicaltrials.gov as of August 2019; n, number of patients; ORR, overall response rate; DCR, disease control rate; PFS, progression-free survival; OS, overall survival; mo, months; †, treatment-related adverse effects. *, Divided to three arms: Arm A: Nivolumab 1 mg/kg + Ipilimumab 3 mg/kg Q3W (4 doses); Arm B: Nivolumab 3 mg/kg + Ipilimumab 1 mg/kg Q3W (four doses), each followed by Nivolumab 240 mg Q2W, or Arm C: Nivolumab 3 mg/kg Q2W + Ipilimumab 1 mg/kg Q6W; NR, not reported; T.B.A., To be announced; #, CR+ PR + SD > 16 weeks; mTKIs, multitargeted tyrosine kinase inhibitors; NE, non-estimable; TTP, time to progression.

2.4.1. ICB and ICB Combination

It is known that anti-PD-1 and anti-CTLA-4 antibodies have differences in their underlying functional mechanisms [13,148]. For instance, anti-PD-1 ICB was thought to act primarily at the interface of T cells and tumor cells within the local tumor microenvironment, while anti-CTL4 ICB was shown to be able to act more upstream at the phase of T cells priming at the lymph nodes [13,148]. Hence, this combination was based on its potential synergistic antitumor activity [149]. A combination of nivolumab and Ipilimumab, which was first evaluated in a phase III trial for patients with advanced melanoma, demonstrated superior outcomes in terms of both progression-free survival and median survival compared to monotherapy with nivolumab or ipilimumab alone [17]. This provided impetus for other solid tumors, including HCC. The third arm of CheckMate 040 evaluated combination nivolumab and ipilimumab in 148 sorafenib-treated patients. Subjects were randomized to three arms: [A] NIVO 1 mg/kg + IPI 3 mg/kg Q3W (4 doses) or [B] NIVO 3 mg/kg + IPI 1 mg/kg Q3W (four doses), each followed by NIVO 240 mg Q2W, or [C] NIVO 3 mg/kg Q2W + IPI 1 mg/kg Q6W. The overall response rate was 31%, with seven complete responses (see Table 3). The 24-month OS rate was 40%, with 37% of patients having a grade 3–4 irAEs (most common all-grade adverse events were pruritus and rash) [140] and 5% having grade 3–4 adverse events, leading to discontinuation. Encouraging results prompted the commencement of CA209-9DW, a phase 3 trial comparing combination ipilimumab and nivolumab against sorafenib or lenvatinib in treatment naïve advanced HCC.

Combination of durvalumab (anti-PD-L1 antibody) and tremelimumab (anti-CTLA4 antibody) (NCT02519348) is currently being evaluated in a Phase I/II study. Preliminary results based only on 40 patients showed a modest ORR of 15% [25] (see Table 3). A large multicenter phase III trial of durvalumab and tremelimumab as first-line treatment in patients with unresectable HCC: HIMALAYA study (NCT03298451) [141] with estimated enrolment of 1310 patients is currently ongoing.

2.4.2. ICB and Anti-Angiogenesis Agent

Angiogenesis, one of the hallmarks of cancer, leads to leaky vasculature, hypoxia, and activation of multiple immunosuppressive pathways in TME as a consequence of rapid tumor growth [150–152]. An anti-angiogenic agent aims to normalize the intratumoral vasculature, hence restoring the equilibrium toward a less protumoral or less immunosuppressive TME [153,154]. The role of vascular endothelial growth factor (VEGF) in driving tumor angiogenesis has made it an attractive therapeutic target. Bevacizumab, a humanized monoclonal antibody against VEGF, has gained FDA approval for many advanced malignancies [155]. The multiple roles of VEGF in reprogramming the tumor microenvironment have been discussed in depth previously [154]. Chiefly, VEGF plays an important role in immunosuppressive regulatory T cells' (Treg) recruitment into the tumor. VEGF inhibition is purported to enhance local antitumor immunity by reducing accumulation of Treg [156]. It was also previously shown that anti-angiogenic agents can increase infiltration of adoptively transferred T cells into a tumor [157]. In a recent study using murine models of HCC, it was shown that this combination therapy reprogrammed the TME by increasing cytotoxic CD8 T cell, while reducing Treg infiltration in HCC tissue and shifting the M1/M2 macrophages ratio in favor of antitumoral TME [158]. A randomized study evaluating atezolizumab (anti-PD-L1 therapy) as monotherapy vs. the combination of atezolizumab + bevacizumab (anti-VEGF therapy; Arm F), as well as single-arm atezolizumab + bevacizumab (Arm A) from a Phase 1b GO30140 study, was conducted in advanced HCC patients and suggested superiority of combination therapy [22]. Concurrently, the outcome from IMbrave150 (NCT03434379) a Phase III, open-label, multicenter, randomized study evaluating combination atezolizumab and bevacizumab versus sorafenib in patients with locally advanced or metastatic and/or unresectable HCC was recently announced [142]. This study met its co-primary endpoints of demonstrating statistically significant and clinically meaningful improvements in both PFS and OS in favor of combination atezolizumab and bevacizumab [142]. With increasing appreciation of immune-modulatory properties of targeted therapies, future combinations of immunotherapy and targeted therapy based on strong rationale and well-studied mechanism of actions would be paramount for drug development.

2.4.3. ICB and Multitargeted Tyrosine Kinase Inhibitors (mTKIs)

Sorafenib, an oral multitargeted tyrosine kinase inhibitor (mTKIs), has been the only systemic therapy for treatment of advanced HCC following the successful SHARP trial in 2008 [16]. Targets of Sorafenib include VEGFR, PDGFR, and RAF kinases, hence exerting antitumor effects through anti-angiogenesis, antiproliferation, and pro-apoptosis [159]. The impact of mTKIs on the TME has also been discussed before [160–162]. Most studies demonstrated the immunomodulating properties of mTKIs, such as reduction of MDSCs and Treg [163–166], enhancing T and NK cells tumor infiltration and activation [167,168], and boosting antitumor immune response. Studies have also discussed the immuno-modulatory properties of mTKIs which could synergize with immunotherapy [169,170]. Furthermore, tumor-cell death induced by mTKIs could serve as a source of tumor antigens that could then activate the specific T cells capable of more cell killing (see Figure 1). Besides that, angiogenesis is one of the common targets for these mTKIs, as well. Two large randomized studies in front-line systemic therapy employs this strategy. Combination atezolizumab (anti-PD-L1 therapy) + cabozantinib (mTKIs) in the COSMIC-312 trial (NCT03755791) [143] and combination of pembrolizumab (anti-PD-1 therapy) and lenvatinib (mTKIs) in the LEAP-002 trial (NCT03713593 or Keynote 524) are currently enrolling [144,145]. Twenty-two systemic treatment-naïve HCC patients were treated with combination avelumab (anti-PD-L1 therapy) and axitinib (mTKIs) with an ORR of 13.6% and median progression-free survival (mPFS) of 5.5 months (see Table 3) [146]. However, toxicities of this combination might be a concern. Grade 3/4 treatment-related adverse events were reported to be 72.7%. Eleven (50%) patients encountered grade 3/4 hypertension, and 22.7% experienced grade 3/4 palmar–plantar erythrodysesthesia (PPE) [146].

2.4.4. Other ICB Combinations

Release of tumor antigen upon tumor-cell killing by chemotherapy, radiotherapy, or transarterial-chemoembolization (TACE) [171,172] further enhances immunogenic cell death. This provides the rationale for combination strategies with ICB (Figure 1). Potential immunogenic cell death induced by oxaliplatin-based chemotherapy containing FOLFOX4 (infusional fluorouracil, leucovorin and oxaliplatin) or GEMOX (gemcitabine and oxaliplatin) provides rationale for an ongoing phase II study in combination with camrelizumab (an anti-PD-1 antibody) in advanced HCC and biliary tract cancer [147]. A number of clinical studies evaluating combination radiotherapy with ICB are in progress [173]. One study in HCC patients treated with external beam RT (EBRT) showed an increase in soluble PD-L1 level post-treatment [174]. Another study, using selective internal radiotherapy (SIRT) in HCC patients, reported enhanced immune cell activation and recruitment, particularly ones that express checkpoint molecule PD-1 [175]. Both studies suggest that combination radiotherapy with ICB could be synergistic. Other locoregional therapies, like transarterial chemoembolization (TACE), have also been explored in combination with immunotherapy. For instance, a multicenter pilot study evaluating the safety of combination of nivolumab with drug-eluting bead-TACE (deb-TACE) in patients with HCC is currently underway [176]. Another study evaluating the safety and efficacy of combination treatment with pembrolizumab and TACE is also ongoing (NCT03397654). Apart from anti-PD-1 therapy, the combination of tremelimumab (anti-CTLA4 therapy) with local therapy (RFA or TACE) has been explored in 32 HCC patients [177].

2.5. Other Immunotherapies and Their Potential as Combination in HCC

2.5.1. Adoptive Cell Therapy (ACT)

T cells engineered to express chimeric antigen receptors (CARs), or autologous T cells expanded and engineered ex vivo with specific targeted tumor antigen(s), have been explored as an immunotherapeutic strategy in cancers, including HCC [178,179]. CAR-T cells directed against GPC-3, CEA, or Mucin 1 are currently being evaluated in early phase trials in various solid tumors, including HCC [180]. Of note, T-cell therapy targeting HCC-specific antigens, such as AFP, has been evaluated previously with disappointing outcomes (NCT03349255). Possible explanation behind this lack of activity could be attributable to low T-cell affinities and high expression of PD-1 T-cell exhaustion markers [181].

Other ACTs such as the use of IL-2-activated and -expanded autologous TILs in vitro have demonstrated improved recurrence-free survival (RFS) after resection in 150 HCC patients [182]. In addition, Cytokine-induced killer (CIK) cells, a heterogeneous cytotoxic immune populations consisting of CD8+ T cells, CD56+ NK cells, and CD3+CD56+ NKT cells, was demonstrated to be safe, with a lower recurrence rate and improved RFS and OS in HCC [183]. NK cell therapy has also been explored for HCC treatment, based on findings that NK cells are dysfunctional in HCC and tumor-infiltration with activated NK cells is associated with superior survival in HCC patients [184,185]. More recently, engineered NK cells or CAR-NK cell therapy with tumor specificity are being explored for various cancer types, including HCC [186].

ACT could enhance the frequency of tumor-specific T cells, however, these tumor antigen-specific T cells would migrate to TME and eventually became exhausted given the immunosuppressive state. Therefore, combination with checkpoint inhibitors could potentially reinvigorate the activity of these T cells (Figure 1). Combination ACT with checkpoint inhibitor is yet to be explored in HCC.

2.5.2. Cancer Vaccines

Cancer vaccines either in the form of peptide, dendritic cell-pulsed with synthetic peptide or RNA vectors based on personalized neoantigens have demonstrated promising outcome in patients with advanced melanoma [187,188]. In contrast to that, cancer vaccines targeting individual tumor-associated antigens (TAAs), such as NY-ESO1, glypican-3 (GPC3), and alpha-fetoprotein (AFP), have met with limited success in HCC [189]. This is most likely due to significant intra and inter-tumor genomic heterogeneity, compounded by a highly immunosuppressive TME. For instance, the AFP vaccine showed limited clinical benefit despite detectable T-cell responses [190,191].

To circumvent this, an ongoing trial evaluating therapeutic cancer vaccine IMA970A, a multi-peptide-based HCC vaccine composed of 16 newly discovered and overexpressed tumor-associated peptides (TUMAPs) identified from resected HCC tissues (clinical trial: NCT03203005) was envisioned. It remains to be determined if such multi-peptide cancer vaccines in HCC will be successful. Given the immunosuppressive internal milieu of HCC, it is likely that combinations with other immunotherapeutic agents will be needed (Figure 1). One Phase Ib/II trial using DSP-7888, a novel WT1 Peptide-Based Vaccine, in combination with nivolumab or pembrolizumab for patients with advanced solid tumors including HCC (NCT03311334), is currently enrolling patients.

2.5.3. Oncolytic Virus Therapy

Oncolytic virus therapy involves the use of native or genetically modified viruses that show selective infection, replication and killing of tumors cells [192,193]. These viruses can also be engineered to express immune-stimulatory genes such as GM-CSF, a cytokine which could enhance antitumor immunity by stimulating antigen-presenting cells and promote the tumour infiltration and maturation of NK cells and T cells [194]. Oncolytic virus therapies have been tested in preclinical and phase I/II clinical trials for HCC [195]. For instance, JX-594, an engineered vaccinia virus with thymidine kinase-deactivated, was well tolerated [196] and demonstrated promising outcome in phase II clinical trial in HCC patients [197]. However, a randomized Phase III trial comparing JX-594 versus sorafenib in patients with advanced HCC (PHOCUS) (NCT02562755) halted enrolment recently due to futility. We believe part of the reason for such failures could be due to the immunosuppressive TME of HCC [40]. It is therefore likely that the success of oncolytic virus could be enhanced in combination with ICB (Figure 1). Indeed, several clinical trials using combination of oncolytic virus and ICB are ongoing, including in advanced HCC [198].

3. Future Perspectives

Challenges remain in identifying HCC patients who could best benefit from immunotherapy. Based on the biomarker studies in other tumor types (see Table 2), the presence of tumor infiltrating T cells, particularly cytotoxic CD8 T cells, predicts for response to immunotherapy. As HCC tumors are enriched with Treg [40] and generally not well infiltrated by immune cells [67,115], strategies to inhibit Treg and enhance T cells infiltration, in combination with ICB, is important. Given the recent success of Phase III trial in HCC, using ICB plus anti-angiogenesis agent (IMbrave150) [142], it is increasingly clear that a combination strategy with clear scientific rationale is necessary. We also need robust biomarkers from longitudinal tumor and blood sampling, as well as multi-omics interrogation to uncover the intrinsic and acquired resistance mechanisms or incidence of irAEs to these treatments. While we acknowledge the potential of combination immunotherapeutic strategies in future, potential enhanced toxicities, given the coexisting liver dysfunction in HCC patients, are also the main concerns to be considered. Further characterization of irAEs in tandem with various combination strategies is of current utmost importance when treating patients.

4. Concluding Remarks

Clinical trials evaluating the use of monotherapy or combination immunotherapeutic agents in HCC are underway. Intensive studies on the mechanisms of actions for evidence-based combination strategies, as well as identification of predictive biomarkers of response and irAEs, are also ongoing. This will result in safer, more effective, and, perhaps, more personalized immunotherapeutic strategies for patients with HCC in the near future.

Author Contributions: D.T., S.P.C. and V.C. contributed in design, drafting, revising and approving the final version of the manuscript.

Funding: This work was supported by the National Medical Research Council (NMRC), Singapore (ref numbers: TCR15Jun006, CIRG16may048, CSAS16Nov006, CSASI17may003 and LCG17MAY003).

Acknowledgments: We would like to thank Thomas Yau from Queen Mary Hospital, Hong Kong, for his insightful discussions and comments.

Conflicts of Interest: The authors declare no conflicts of interest.

References

1. Couzin-Frankel, J. Breakthrough of the year 2013. Cancer immunotherapy. *Science* **2013**, *342*, 1432–1433. [CrossRef] [PubMed]
2. Disis, M.L. Mechanism of action of immunotherapy. *Semin. Oncol.* **2014**, *41*, S3–S13. [CrossRef] [PubMed]
3. Zhang, H.; Chen, J. Current status and future directions of cancer immunotherapy. *J. Cancer* **2018**, *9*, 1773–1781. [CrossRef] [PubMed]
4. Park, Y.J.; Kuen, D.S.; Chung, Y. Future prospects of immune checkpoint blockade in cancer: From response prediction to overcoming resistance. *Exp. Mol. Med.* **2018**, *50*, 109. [CrossRef]
5. Siu, L.L.; Ivy, S.P.; Dixon, E.L.; Gravell, A.E.; Reeves, S.A.; Rosner, G.L. Challenges and Opportunities in Adapting Clinical Trial Design for Immunotherapies. *Clin. Cancer Res.* **2017**, *23*, 4950–4958. [CrossRef]
6. Bray, F.; Ferlay, J.; Soerjomataram, I.; Siegel, R.L.; Torre, L.A.; Jemal, A. Global cancer statistics 2018: GLOBOCAN estimates of incidence and mortality worldwide for 36 cancers in 185 countries. *CA Cancer J. Clin.* **2018**, *68*, 394–424. [CrossRef]
7. Chen, X.P.; Qiu, F.Z.; Wu, Z.D.; Zhang, Z.W.; Huang, Z.Y.; Chen, Y.F. Long-term outcome of resection of large hepatocellular carcinoma. *Br. J. Surg.* **2006**, *93*, 600–606. [CrossRef]
8. Poon, R.T.; Fan, S.T.; Lo, C.M.; Liu, C.L.; Wong, J. Long-term survival and pattern of recurrence after resection of small hepatocellular carcinoma in patients with preserved liver function: Implications for a strategy of salvage transplantation. *Ann. Surg.* **2002**, *235*, 373–382. [CrossRef]
9. Portolani, N.; Coniglio, A.; Ghidoni, S.; Giovanelli, M.; Benetti, A.; Tiberio, G.A.; Giulini, S.M. Early and late recurrence after liver resection for hepatocellular carcinoma: Prognostic and therapeutic implications. *Ann. Surg.* **2006**, *243*, 229–235. [CrossRef]

10. Sanyal, A.J.; Yoon, S.K.; Lencioni, R. The etiology of hepatocellular carcinoma and consequences for treatment. *Oncologist* **2010**, *15*, 14–22. [CrossRef]
11. El-Khoueiry, A.B.; Sangro, B.; Yau, T.; Crocenzi, T.S.; Kudo, M.; Hsu, C.; Kim, T.Y.; Choo, S.P.; Trojan, J.; Welling, T.H.R.; et al. Nivolumab in patients with advanced hepatocellular carcinoma (CheckMate 040): An open-label, non-comparative, phase 1/2 dose escalation and expansion trial. *Lancet* **2017**, *389*, 2492–2502. [CrossRef]
12. Zhu, A.X.; Finn, R.S.; Edeline, J.; Cattan, S.; Ogasawara, S.; Palmer, D.; Verslype, C.; Zagonel, V.; Fartoux, L.; Vogel, A.; et al. Pembrolizumab in patients with advanced hepatocellular carcinoma previously treated with sorafenib (KEYNOTE-224): A non-randomised, open-label phase 2 trial. *Lancet Oncol.* **2018**, *19*, 940–952. [CrossRef]
13. Wei, S.C.; Duffy, C.R.; Allison, J.P. Fundamental Mechanisms of Immune Checkpoint Blockade Therapy. *Cancer Discov.* **2018**, *8*, 1069–1086. [CrossRef] [PubMed]
14. Pardoll, D.M. The blockade of immune checkpoints in cancer immunotherapy. *Nat. Rev. Cancer* **2012**, *12*, 252–264. [CrossRef] [PubMed]
15. Yau, T.; Hsu, C.; Kim, T.Y.; Choo, S.P.; Kang, Y.K.; Hou, M.M.; Numata, K.; Yeo, W.; Chopra, A.; Ikeda, M.; et al. Nivolumab in advanced hepatocellular carcinoma: Sorafenib-experienced Asian cohort analysis. *J. Hepatol.* **2019**, *71*, 543–552. [CrossRef]
16. Llovet, J.M.; Ricci, S.; Mazzaferro, V.; Hilgard, P.; Gane, E.; Blanc, J.F.; de Oliveira, A.C.; Santoro, A.; Raoul, J.L.; Forner, A.; et al. Sorafenib in advanced hepatocellular carcinoma. *N. Engl. J. Med.* **2008**, *359*, 378–390. [CrossRef]
17. Wolchok, J.D.; Chiarion-Sileni, V.; Gonzalez, R.; Rutkowski, P.; Grob, J.J.; Cowey, C.L.; Lao, C.D.; Wagstaff, J.; Schadendorf, D.; Ferrucci, P.F.; et al. Overall Survival with Combined Nivolumab and Ipilimumab in Advanced Melanoma. *N. Engl. J. Med.* **2017**, *377*, 1345–1356. [CrossRef] [PubMed]
18. Motzer, R.J.; Escudier, B.; McDermott, D.F.; George, S.; Hammers, H.J.; Srinivas, S.; Tykodi, S.S.; Sosman, J.A.; Procopio, G.; Plimack, E.R.; et al. Nivolumab versus Everolimus in Advanced Renal-Cell Carcinoma. *N. Engl. J. Med.* **2015**, *373*, 1803–1813. [CrossRef] [PubMed]
19. Finn, R.S.; Ryoo, B.-Y.; Merle, P.; Kudo, M.; Bouattour, M.; Lim, H.-Y.; Breder, V.V.; Edeline, J.; Chao, Y.; Ogasawara, S.; et al. Results of KEYNOTE-240: Phase 3 study of pembrolizumab (Pembro) vs. best supportive care (BSC) for second line therapy in advanced hepatocellular carcinoma (HCC). *J. Clin. Oncol.* **2019**, *37*, 4004. [CrossRef]
20. Yau, T.; Park, J.W.; Finn, R.S.; Cheng, A.; Mathurin, P.; Edeline, J.; Kudo, M.; Han, K.; Harding, J.J.; Merle, P.; et al. CheckMate 459: A Randomized, Multi-Center Phase 3 Study of Nivolumab (NIVO) vs. Sorafenib (SOR) as First-Line (1L) Treatment in Patients (pts) With Advanced Hepatocellular Carcinoma (aHCC). *Ann. Oncol.* **2019**, *30*, v851–v934. [CrossRef]
21. Wainberg, Z.A.; Segal, N.H.; Jaeger, D.; Lee, K.-H.; Marshall, J.; Antonia, S.J.; Butler, M.; Sanborn, R.E.; Nemunaitis, J.J.; Carlson, C.A.; et al. Safety and clinical activity of durvalumab monotherapy in patients with hepatocellular carcinoma (HCC). *J. Clin. Oncol.* **2017**, *35*, 4071. [CrossRef]
22. Lee, M.; Ryoo, B.; Hsu, C.; Numata, K.; Stein, S.; Verret, W.; Hack, S.; Spahn, J.; Liu, B.; Abdullah, H.; et al. Randomised Efficacy and Safety Results for Atezolizumab (Atezo) + Bevacizumab (Bev) in Patients (pts) with Previously Untreated, Unresectable Hepatocellular Carcinoma (HCC). *Ann. Oncol.* **2019**, *30*, v851–v934. [CrossRef]
23. Hodi, F.S.; O'Day, S.J.; McDermott, D.F.; Weber, R.W.; Sosman, J.A.; Haanen, J.B.; Gonzalez, R.; Robert, C.; Schadendorf, D.; Hassel, J.C.; et al. Improved survival with ipilimumab in patients with metastatic melanoma. *N. Engl. J. Med.* **2010**, *363*, 711–723. [CrossRef]
24. Sangro, B.; Gomez-Martin, C.; de la Mata, M.; Inarrairaegui, M.; Garralda, E.; Barrera, P.; Riezu-Boj, J.I.; Larrea, E.; Alfaro, C.; Sarobe, P.; et al. A clinical trial of CTLA-4 blockade with tremelimumab in patients with hepatocellular carcinoma and chronic hepatitis C. *J. Hepatol.* **2013**, *59*, 81–88. [CrossRef]
25. Kelley, R.K.; Abou-Alfa, G.K.; Bendell, J.C.; Kim, T.-Y.; Borad, M.J.; Yong, W.-P.; Morse, M.; Kang, Y.-K.; Rebelatto, M.; Makowsky, M.; et al. Phase I/II study of durvalumab and tremelimumab in patients with unresectable hepatocellular carcinoma (HCC): Phase I safety and efficacy analyses. *J. Clin. Oncol.* **2017**, *35*, 4073. [CrossRef]
26. Buder-Bakhaya, K.; Hassel, J.C. Biomarkers for Clinical Benefit of Immune Checkpoint Inhibitor Treatment-A Review from the Melanoma Perspective and Beyond. *Front. Immunol.* **2018**, *9*, 1474. [CrossRef]

27. Havel, J.J.; Chowell, D.; Chan, T.A. The evolving landscape of biomarkers for checkpoint inhibitor immunotherapy. *Nat. Rev. Cancer* **2019**, *19*, 133–150. [CrossRef]
28. Spencer, K.R.; Wang, J.; Silk, A.W.; Ganesan, S.; Kaufman, H.L.; Mehnert, J.M. Biomarkers for Immunotherapy: Current Developments and Challenges. *Am. Soc. Clin. Oncol. Educ. Book* **2016**, *35*, e493–e503. [CrossRef]
29. Gibney, G.T.; Weiner, L.M.; Atkins, M.B. Predictive biomarkers for checkpoint inhibitor-based immunotherapy. *Lancet Oncol.* **2016**, *17*, e542–e551. [CrossRef]
30. Topalian, S.L.; Taube, J.M.; Anders, R.A.; Pardoll, D.M. Mechanism-driven biomarkers to guide immune checkpoint blockade in cancer therapy. *Nat. Rev. Cancer* **2016**, *16*, 275–287. [CrossRef]
31. Ferris, R.L.; Blumenschein, G., Jr.; Fayette, J.; Guigay, J.; Colevas, A.D.; Licitra, L.; Harrington, K.; Kasper, S.; Vokes, E.E.; Even, C.; et al. Nivolumab for Recurrent Squamous-Cell Carcinoma of the Head and Neck. *N. Engl. J. Med.* **2016**, *375*, 1856–1867. [CrossRef] [PubMed]
32. Borghaei, H.; Paz-Ares, L.; Horn, L.; Spigel, D.R.; Steins, M.; Ready, N.E.; Chow, L.Q.; Vokes, E.E.; Felip, E.; Holgado, E.; et al. Nivolumab versus Docetaxel in Advanced Nonsquamous Non-Small-Cell Lung Cancer. *N. Engl. J. Med.* **2015**, *373*, 1627–1639. [CrossRef] [PubMed]
33. Reck, M.; Rodriguez-Abreu, D.; Robinson, A.G.; Hui, R.; Csoszi, T.; Fulop, A.; Gottfried, M.; Peled, N.; Tafreshi, A.; Cuffe, S.; et al. Pembrolizumab versus Chemotherapy for PD-L1-Positive Non-Small-Cell Lung Cancer. *N. Engl. J. Med.* **2016**, *375*, 1823–1833. [CrossRef] [PubMed]
34. Rosenberg, J.E.; Hoffman-Censits, J.; Powles, T.; van der Heijden, M.S.; Balar, A.V.; Necchi, A.; Dawson, N.; O'Donnell, P.H.; Balmanoukian, A.; Loriot, Y.; et al. Atezolizumab in patients with locally advanced and metastatic urothelial carcinoma who have progressed following treatment with platinum-based chemotherapy: A single-arm, multicentre, phase 2 trial. *Lancet* **2016**, *387*, 1909–1920. [CrossRef]
35. Topalian, S.L.; Hodi, F.S.; Brahmer, J.R.; Gettinger, S.N.; Smith, D.C.; McDermott, D.F.; Powderly, J.D.; Carvajal, R.D.; Sosman, J.A.; Atkins, M.B.; et al. Safety, activity, and immune correlates of anti-PD-1 antibody in cancer. *N. Engl. J. Med.* **2012**, *366*, 2443–2454. [CrossRef]
36. Hanna, G.J.; Lizotte, P.; Cavanaugh, M.; Kuo, F.C.; Shivdasani, P.; Frieden, A.; Chau, N.G.; Schoenfeld, J.D.; Lorch, J.H.; Uppaluri, R.; et al. Frameshift events predict anti-PD-1/L1 response in head and neck cancer. *JCI Insight* **2018**, *3*. [CrossRef]
37. Carbone, D.P.; Reck, M.; Paz-Ares, L.; Creelan, B.; Horn, L.; Steins, M.; Felip, E.; van den Heuvel, M.M.; Ciuleanu, T.E.; Badin, F.; et al. First-Line Nivolumab in Stage IV or Recurrent Non-Small-Cell Lung Cancer. *N. Engl. J. Med.* **2017**, *376*, 2415–2426. [CrossRef]
38. Gao, Q.; Wang, X.Y.; Qiu, S.J.; Yamato, I.; Sho, M.; Nakajima, Y.; Zhou, J.; Li, B.Z.; Shi, Y.H.; Xiao, Y.S.; et al. Overexpression of PD-L1 significantly associates with tumor aggressiveness and postoperative recurrence in human hepatocellular carcinoma. *Clin. Cancer Res.* **2009**, *15*, 971–979. [CrossRef]
39. Kuang, D.M.; Zhao, Q.; Peng, C.; Xu, J.; Zhang, J.P.; Wu, C.; Zheng, L. Activated monocytes in peritumoral stroma of hepatocellular carcinoma foster immune privilege and disease progression through PD-L1. *J. Exp. Med.* **2009**, *206*, 1327–1337. [CrossRef]
40. Chew, V.; Lai, L.; Pan, L.; Lim, C.J.; Li, J.; Ong, R.; Chua, C.; Leong, J.Y.; Lim, K.H.; Toh, H.C.; et al. Delineation of an immunosuppressive gradient in hepatocellular carcinoma using high-dimensional proteomic and transcriptomic analyses. *Proc. Natl. Acad. Sci. USA* **2017**, *114*, E5900–E5909. [CrossRef]
41. Kim, H.D.; Song, G.W.; Park, S.; Jung, M.K.; Kim, M.H.; Kang, H.J.; Yoo, C.; Yi, K.; Kim, K.H.; Eo, S.; et al. Association Between Expression Level of PD1 by Tumor-Infiltrating CD8(+) T Cells and Features of Hepatocellular Carcinoma. *Gastroenterology* **2018**, *155*, 1936–1950. [CrossRef] [PubMed]
42. Pinato, D.J.; Mauri, F.A.; Spina, P.; Cain, O.; Siddique, A.; Goldin, R.; Victor, S.; Pizio, C.; Akarca, A.U.; Boldorini, R.L.; et al. Clinical implications of heterogeneity in PD-L1 immunohistochemical detection in hepatocellular carcinoma: The Blueprint-HCC study. *Br. J. Cancer* **2019**, *120*, 1033–1036. [CrossRef] [PubMed]
43. Nishino, M.; Ramaiya, N.H.; Hatabu, H.; Hodi, F.S. Monitoring immune-checkpoint blockade: Response evaluation and biomarker development. *Nat. Rev. Clin. Oncol.* **2017**, *14*, 655–668. [CrossRef] [PubMed]
44. Tang, H.; Liang, Y.; Anders, R.A.; Taube, J.M.; Qiu, X.; Mulgaonkar, A.; Liu, X.; Harrington, S.M.; Guo, J.; Xin, Y.; et al. PD-L1 on host cells is essential for PD-L1 blockade-mediated tumor regression. *J. Clin. Investig.* **2018**, *128*, 580–588. [CrossRef] [PubMed]
45. Herbst, R.S.; Soria, J.C.; Kowanetz, M.; Fine, G.D.; Hamid, O.; Gordon, M.S.; Sosman, J.A.; McDermott, D.F.; Powderly, J.D.; Gettinger, S.N.; et al. Predictive correlates of response to the anti-PD-L1 antibody MPDL3280A in cancer patients. *Nature* **2014**, *515*, 563–567. [CrossRef] [PubMed]

46. Mariathasan, S.; Turley, S.J.; Nickles, D.; Castiglioni, A.; Yuen, K.; Wang, Y.; Kadel, E.E., III; Koeppen, H.; Astarita, J.L.; Cubas, R.; et al. TGFbeta attenuates tumour response to PD-L1 blockade by contributing to exclusion of T cells. *Nature* **2018**, *554*, 544–548. [CrossRef]
47. Chen, G.; Huang, A.C.; Zhang, W.; Zhang, G.; Wu, M.; Xu, W.; Yu, Z.; Yang, J.; Wang, B.; Sun, H.; et al. Exosomal PD-L1 contributes to immunosuppression and is associated with anti-PD-1 response. *Nature* **2018**, *560*, 382–386. [CrossRef]
48. Poggio, M.; Hu, T.; Pai, C.C.; Chu, B.; Belair, C.D.; Chang, A.; Montabana, E.; Lang, U.E.; Fu, Q.; Fong, L.; et al. Suppression of Exosomal PD-L1 Induces Systemic Anti-tumor Immunity and Memory. *Cell* **2019**, *177*, 414–427 e413. [CrossRef]
49. Xue, R.; Li, R.; Guo, H.; Guo, L.; Su, Z.; Ni, X.; Qi, L.; Zhang, T.; Li, Q.; Zhang, Z.; et al. Variable Intra-Tumor Genomic Heterogeneity of Multiple Lesions in Patients with Hepatocellular Carcinoma. *Gastroenterology* **2016**, *150*, 998–1008. [CrossRef]
50. Zhai, W.; Lim, T.K.; Zhang, T.; Phang, S.T.; Tiang, Z.; Guan, P.; Ng, M.H.; Lim, J.Q.; Yao, F.; Li, Z.; et al. The spatial organization of intra-tumour heterogeneity and evolutionary trajectories of metastases in hepatocellular carcinoma. *Nat. Commun.* **2017**, *8*, 4565. [CrossRef]
51. Wang, Q.; Lin, W.; Tang, X.; Li, S.; Guo, L.; Lin, Y.; Kwok, H.F. The Roles of microRNAs in Regulating the Expression of PD-1/PD-L1 Immune Checkpoint. *Int. J. Mol. Sci.* **2017**, *18*, 2540. [CrossRef] [PubMed]
52. Wang, Y.; Wang, H.; Yao, H.; Li, C.; Fang, J.Y.; Xu, J. Regulation of PD-L1: Emerging Routes for Targeting Tumor Immune Evasion. *Front. Pharmacol.* **2018**, *9*, 536. [CrossRef] [PubMed]
53. Wei, R.; Guo, L.; Wang, Q.; Miao, J.; Kwok, H.F.; Lin, Y. Targeting PD-L1 Protein: Translation, Modification and Transport. *Curr. Protein Pept. Sci.* **2019**, *20*, 82–91. [CrossRef] [PubMed]
54. Burr, M.L.; Sparbier, C.E.; Chan, Y.C.; Williamson, J.C.; Woods, K.; Beavis, P.A.; Lam, E.Y.N.; Henderson, M.A.; Bell, C.C.; Stolzenburg, S.; et al. CMTM6 maintains the expression of PD-L1 and regulates anti-tumour immunity. *Nature* **2017**, *549*, 101–105. [CrossRef] [PubMed]
55. Garcia-Diaz, A.; Shin, D.S.; Moreno, B.H.; Saco, J.; Escuin-Ordinas, H.; Rodriguez, G.A.; Zaretsky, J.M.; Sun, L.; Hugo, W.; Wang, X.; et al. Interferon Receptor Signaling Pathways Regulating PD-L1 and PD-L2 Expression. *Cell Rep.* **2017**, *19*, 1189–1201. [CrossRef] [PubMed]
56. Abiko, K.; Matsumura, N.; Hamanishi, J.; Horikawa, N.; Murakami, R.; Yamaguchi, K.; Yoshioka, Y.; Baba, T.; Konishi, I.; Mandai, M. IFN-gamma from lymphocytes induces PD-L1 expression and promotes progression of ovarian cancer. *Br. J. Cancer* **2015**, *112*, 1501–1509. [CrossRef] [PubMed]
57. Ayers, M.; Lunceford, J.; Nebozhyn, M.; Murphy, E.; Loboda, A.; Kaufman, D.R.; Albright, A.; Cheng, J.D.; Kang, S.P.; Shankaran, V.; et al. IFN-gamma-related mRNA profile predicts clinical response to PD-1 blockade. *J. Clin. Investig.* **2017**, *127*, 2930–2940. [CrossRef]
58. Karachaliou, N.; Gonzalez-Cao, M.; Crespo, G.; Drozdowskyj, A.; Aldeguer, E.; Gimenez-Capitan, A.; Teixido, C.; Molina-Vila, M.A.; Viteri, S.; De Los Llanos Gil, M.; et al. Interferon gamma, an important marker of response to immune checkpoint blockade in non-small cell lung cancer and melanoma patients. *Ther. Adv. Med. Oncol.* **2018**, *10*, 1758834017749748. [CrossRef]
59. Goodman, A.M.; Kato, S.; Bazhenova, L.; Patel, S.P.; Frampton, G.M.; Miller, V.; Stephens, P.J.; Daniels, G.A.; Kurzrock, R. Tumor Mutational Burden as an Independent Predictor of Response to Immunotherapy in Diverse Cancers. *Mol. Cancer Ther.* **2017**, *16*, 2598–2608. [CrossRef]
60. Hellmann, M.D.; Ciuleanu, T.E.; Pluzanski, A.; Lee, J.S.; Otterson, G.A.; Audigier-Valette, C.; Minenza, E.; Linardou, H.; Burgers, S.; Salman, P.; et al. Nivolumab plus Ipilimumab in Lung Cancer with a High Tumor Mutational Burden. *N. Engl. J. Med.* **2018**, *378*, 2093–2104. [CrossRef]
61. Rizvi, N.A.; Hellmann, M.D.; Snyder, A.; Kvistborg, P.; Makarov, V.; Havel, J.J.; Lee, W.; Yuan, J.; Wong, P.; Ho, T.S.; et al. Cancer immunology. Mutational landscape determines sensitivity to PD-1 blockade in non-small cell lung cancer. *Science* **2015**, *348*, 124–128. [CrossRef] [PubMed]
62. Snyder, A.; Makarov, V.; Merghoub, T.; Yuan, J.; Zaretsky, J.M.; Desrichard, A.; Walsh, L.A.; Postow, M.A.; Wong, P.; Ho, T.S.; et al. Genetic basis for clinical response to CTLA-4 blockade in melanoma. *N. Engl. J. Med.* **2014**, *371*, 2189–2199. [CrossRef] [PubMed]
63. Yarchoan, M.; Hopkins, A.; Jaffee, E.M. Tumor Mutational Burden and Response Rate to PD-1 Inhibition. *N. Engl. J. Med.* **2017**, *377*, 2500–2501. [CrossRef] [PubMed]

64. Ang, C.; Klempner, S.J.; Ali, S.M.; Madison, R.; Ross, J.S.; Severson, E.A.; Fabrizio, D.; Goodman, A.; Kurzrock, R.; Suh, J.; et al. Prevalence of established and emerging biomarkers of immune checkpoint inhibitor response in advanced hepatocellular carcinoma. *Oncotarget* **2019**, *10*, 4018–4025. [CrossRef] [PubMed]
65. Ma, L.; Hernandez, M.O.; Zhao, Y.; Mehta, M.; Tran, B.; Kelly, M.; Rae, Z.; Hernandez, J.M.; Davis, J.L.; Martin, S.P.; et al. Tumor Cell Biodiversity Drives Microenvironmental Reprogramming in Liver Cancer. *Cancer Cell* **2019**, *36*, 418–430 e416. [CrossRef]
66. Spranger, S.; Bao, R.; Gajewski, T.F. Melanoma-intrinsic beta-catenin signalling prevents anti-tumour immunity. *Nature* **2015**, *523*, 231–235. [CrossRef]
67. Sia, D.; Jiao, Y.; Martinez-Quetglas, I.; Kuchuk, O.; Villacorta-Martin, C.; Castro de Moura, M.; Putra, J.; Camprecios, G.; Bassaganyas, L.; Akers, N.; et al. Identification of an Immune-specific Class of Hepatocellular Carcinoma, Based on Molecular Features. *Gastroenterology* **2017**, *153*, 812–826. [CrossRef]
68. Harding, J.J.; Nandakumar, S.; Armenia, J.; Khalil, D.N.; Albano, M.; Ly, M.; Shia, J.; Hechtman, J.F.; Kundra, R.; El Dika, I.; et al. Prospective Genotyping of Hepatocellular Carcinoma: Clinical Implications of Next-Generation Sequencing for Matching Patients to Targeted and Immune Therapies. *Clin. Cancer Res.* **2019**, *25*, 2116–2126. [CrossRef]
69. Tumeh, P.C.; Harview, C.L.; Yearley, J.H.; Shintaku, I.P.; Taylor, E.J.; Robert, L.; Chmielowski, B.; Spasic, M.; Henry, G.; Ciobanu, V.; et al. PD-1 blockade induces responses by inhibiting adaptive immune resistance. *Nature* **2014**, *515*, 568–571. [CrossRef]
70. Chen, P.L.; Roh, W.; Reuben, A.; Cooper, Z.A.; Spencer, C.N.; Prieto, P.A.; Miller, J.P.; Bassett, R.L.; Gopalakrishnan, V.; Wani, K.; et al. Analysis of Immune Signatures in Longitudinal Tumor Samples Yields Insight into Biomarkers of Response and Mechanisms of Resistance to Immune Checkpoint Blockade. *Cancer Discov.* **2016**, *6*, 827–837. [CrossRef]
71. Haratani, K.; Hayashi, H.; Tanaka, T.; Kaneda, H.; Togashi, Y.; Sakai, K.; Hayashi, K.; Tomida, S.; Chiba, Y.; Yonesaka, K.; et al. Tumor immune microenvironment and nivolumab efficacy in EGFR mutation-positive non-small-cell lung cancer based on T790M status after disease progression during EGFR-TKI treatment. *Ann. Oncol.* **2017**, *28*, 1532–1539. [CrossRef] [PubMed]
72. Kaseb, A.O.; Vence, L.; Blando, J.; Yadav, S.S.; Ikoma, N.; Pestana, R.C.; Vauthey, J.N.; Allison, J.P.; Sharma, P. Immunologic Correlates of Pathologic Complete Response to Preoperative Immunotherapy in Hepatocellular Carcinoma. *Cancer Immunol. Res.* **2019**, *7*, 1390–1395. [CrossRef] [PubMed]
73. Riaz, N.; Havel, J.J.; Makarov, V.; Desrichard, A.; Urba, W.J.; Sims, J.S.; Hodi, F.S.; Martin-Algarra, S.; Mandal, R.; Sharfman, W.H.; et al. Tumor and Microenvironment Evolution during Immunotherapy with Nivolumab. *Cell* **2017**, *171*, 934–949. [CrossRef] [PubMed]
74. Forde, P.M.; Chaft, J.E.; Smith, K.N.; Anagnostou, V.; Cottrell, T.R.; Hellmann, M.D.; Zahurak, M.; Yang, S.C.; Jones, D.R.; Broderick, S.; et al. Neoadjuvant PD-1 Blockade in Resectable Lung Cancer. *N. Engl. J. Med.* **2018**, *378*, 1976–1986. [CrossRef]
75. Inoue, H.; Park, J.H.; Kiyotani, K.; Zewde, M.; Miyashita, A.; Jinnin, M.; Kiniwa, Y.; Okuyama, R.; Tanaka, R.; Fujisawa, Y.; et al. Intratumoral expression levels of PD-L1, GZMA, and HLA-A along with oligoclonal T cell expansion associate with response to nivolumab in metastatic melanoma. *Oncoimmunology* **2016**, *5*, e1204507. [CrossRef]
76. Chowell, D.; Morris, L.G.T.; Grigg, C.M.; Weber, J.K.; Samstein, R.M.; Makarov, V.; Kuo, F.; Kendall, S.M.; Requena, D.; Riaz, N.; et al. Patient HLA class I genotype influences cancer response to checkpoint blockade immunotherapy. *Science* **2018**, *359*, 582–587. [CrossRef]
77. Thommen, D.S.; Koelzer, V.H.; Herzig, P.; Roller, A.; Trefny, M.; Dimeloe, S.; Kiialainen, A.; Hanhart, J.; Schill, C.; Hess, C.; et al. A transcriptionally and functionally distinct PD-1(+) CD8(+) T cell pool with predictive potential in non-small-cell lung cancer treated with PD-1 blockade. *Nat. Med.* **2018**, *24*, 994–1004. [CrossRef]
78. Sade-Feldman, M.; Yizhak, K.; Bjorgaard, S.L.; Ray, J.P.; de Boer, C.G.; Jenkins, R.W.; Lieb, D.J.; Chen, J.H.; Frederick, D.T.; Barzily-Rokni, M.; et al. Defining T Cell States Associated with Response to Checkpoint Immunotherapy in Melanoma. *Cell* **2018**, *175*, 998–1013 e1020. [CrossRef]
79. Charoentong, P.; Finotello, F.; Angelova, M.; Mayer, C.; Efremova, M.; Rieder, D.; Hackl, H.; Trajanoski, Z. Pan-cancer Immunogenomic Analyses Reveal Genotype-Immunophenotype Relationships and Predictors of Response to Checkpoint Blockade. *Cell Rep.* **2017**, *18*, 248–262. [CrossRef]

80. Kamada, T.; Togashi, Y.; Tay, C.; Ha, D.; Sasaki, A.; Nakamura, Y.; Sato, E.; Fukuoka, S.; Tada, Y.; Tanaka, A.; et al. PD-1(+) regulatory T cells amplified by PD-1 blockade promote hyperprogression of cancer. *Proc. Natl. Acad. Sci. USA* **2019**, *116*, 9999–10008. [CrossRef]
81. Neubert, N.J.; Schmittnaegel, M.; Bordry, N.; Nassiri, S.; Wald, N.; Martignier, C.; Tille, L.; Homicsko, K.; Damsky, W.; Maby-El Hajjami, H.; et al. T cell-induced CSF1 promotes melanoma resistance to PD1 blockade. *Sci. Transl. Med.* **2018**, *10*. [CrossRef] [PubMed]
82. Postow, M.A.; Manuel, M.; Wong, P.; Yuan, J.; Dong, Z.; Liu, C.; Perez, S.; Tanneau, I.; Noel, M.; Courtier, A.; et al. Peripheral T cell receptor diversity is associated with clinical outcomes following ipilimumab treatment in metastatic melanoma. *J. Immunother. Cancer* **2015**, *3*, 23. [CrossRef] [PubMed]
83. Hopkins, A.C.; Yarchoan, M.; Durham, J.N.; Yusko, E.C.; Rytlewski, J.A.; Robins, H.S.; Laheru, D.A.; Le, D.T.; Lutz, E.R.; Jaffee, E.M. T cell receptor repertoire features associated with survival in immunotherapy-treated pancreatic ductal adenocarcinoma. *JCI Insight* **2018**, *3*. [CrossRef] [PubMed]
84. Cha, E.; Klinger, M.; Hou, Y.; Cummings, C.; Ribas, A.; Faham, M.; Fong, L. Improved survival with T cell clonotype stability after anti-CTLA-4 treatment in cancer patients. *Sci. Transl. Med.* **2014**, *6*, 238ra270. [CrossRef]
85. Huang, A.C.; Postow, M.A.; Orlowski, R.J.; Mick, R.; Bengsch, B.; Manne, S.; Xu, W.; Harmon, S.; Giles, J.R.; Wenz, B.; et al. T-cell invigoration to tumour burden ratio associated with anti-PD-1 response. *Nature* **2017**, *545*, 60–65. [CrossRef] [PubMed]
86. Krieg, C.; Nowicka, M.; Guglietta, S.; Schindler, S.; Hartmann, F.J.; Weber, L.M.; Dummer, R.; Robinson, M.D.; Levesque, M.P.; Becher, B. High-dimensional single-cell analysis predicts response to anti-PD-1 immunotherapy. *Nat. Med.* **2018**, *24*, 144–153. [CrossRef]
87. Meyer, C.; Cagnon, L.; Costa-Nunes, C.M.; Baumgaertner, P.; Montandon, N.; Leyvraz, L.; Michielin, O.; Romano, E.; Speiser, D.E. Frequencies of circulating MDSC correlate with clinical outcome of melanoma patients treated with ipilimumab. *Cancer Immunol. Immunother.* **2014**, *63*, 247–257. [CrossRef]
88. Gebhardt, C.; Sevko, A.; Jiang, H.; Lichtenberger, R.; Reith, M.; Tarnanidis, K.; Holland-Letz, T.; Umansky, L.; Beckhove, P.; Sucker, A.; et al. Myeloid Cells and Related Chronic Inflammatory Factors as Novel Predictive Markers in Melanoma Treatment with Ipilimumab. *Clin. Cancer Res.* **2015**, *21*, 5453–5459. [CrossRef]
89. Martens, A.; Wistuba-Hamprecht, K.; Geukes Foppen, M.; Yuan, J.; Postow, M.A.; Wong, P.; Romano, E.; Khammari, A.; Dreno, B.; Capone, M.; et al. Baseline Peripheral Blood Biomarkers Associated with Clinical Outcome of Advanced Melanoma Patients Treated with Ipilimumab. *Clin. Cancer Res.* **2016**, *22*, 2908–2918. [CrossRef]
90. Sade-Feldman, M.; Kanterman, J.; Klieger, Y.; Ish-Shalom, E.; Olga, M.; Saragovi, A.; Shtainberg, H.; Lotem, M.; Baniyash, M. Clinical Significance of Circulating CD33+CD11b+HLA-DR- Myeloid Cells in Patients with Stage IV Melanoma Treated with Ipilimumab. *Clin. Cancer Res.* **2016**, *22*, 5661–5672. [CrossRef]
91. Bilen, M.A.; Dutcher, G.M.A.; Liu, Y.; Ravindranathan, D.; Kissick, H.T.; Carthon, B.C.; Kucuk, O.; Harris, W.B.; Master, V.A. Association Between Pretreatment Neutrophil-to-Lymphocyte Ratio and Outcome of Patients With Metastatic Renal-Cell Carcinoma Treated With Nivolumab. *Clin. Genitourin. Cancer* **2018**, *16*, e563–e575. [CrossRef]
92. Jiang, T.; Qiao, M.; Zhao, C.; Li, X.; Gao, G.; Su, C.; Ren, S.; Zhou, C. Pretreatment neutrophil-to-lymphocyte ratio is associated with outcome of advanced-stage cancer patients treated with immunotherapy: A meta-analysis. *Cancer Immunol. Immunother.* **2018**, *67*, 713–727. [CrossRef]
93. Bagley, S.J.; Kothari, S.; Aggarwal, C.; Bauml, J.M.; Alley, E.W.; Evans, T.L.; Kosteva, J.A.; Ciunci, C.A.; Gabriel, P.E.; Thompson, J.C.; et al. Pretreatment neutrophil-to-lymphocyte ratio as a marker of outcomes in nivolumab-treated patients with advanced non-small-cell lung cancer. *Lung Cancer* **2017**, *106*, 1–7. [CrossRef]
94. Ferrucci, P.F.; Gandini, S.; Battaglia, A.; Alfieri, S.; Di Giacomo, A.M.; Giannarelli, D.; Cappellini, G.C.; De Galitiis, F.; Marchetti, P.; Amato, G.; et al. Baseline neutrophil-to-lymphocyte ratio is associated with outcome of ipilimumab-treated metastatic melanoma patients. *Br. J. Cancer* **2015**, *112*, 1904–1910. [CrossRef]
95. Tarhini, A.A.; Edington, H.; Butterfield, L.H.; Lin, Y.; Shuai, Y.; Tawbi, H.; Sander, C.; Yin, Y.; Holtzman, M.; Johnson, J.; et al. Immune monitoring of the circulation and the tumor microenvironment in patients with regionally advanced melanoma receiving neoadjuvant ipilimumab. *PLoS ONE* **2014**, *9*, e87705. [CrossRef]

96. Simeone, E.; Gentilcore, G.; Giannarelli, D.; Grimaldi, A.M.; Caraco, C.; Curvietto, M.; Esposito, A.; Paone, M.; Palla, M.; Cavalcanti, E.; et al. Immunological and biological changes during ipilimumab treatment and their potential correlation with clinical response and survival in patients with advanced melanoma. *Cancer Immunol. Immunother.* **2014**, *63*, 675–683. [CrossRef]
97. Mezquita, L.; Auclin, E.; Ferrara, R.; Charrier, M.; Remon, J.; Planchard, D.; Ponce, S.; Ares, L.P.; Leroy, L.; Audigier-Valette, C.; et al. Association of the Lung Immune Prognostic Index with Immune Checkpoint Inhibitor Outcomes in Patients with Advanced Non-Small Cell Lung Cancer. *JAMA Oncol.* **2018**, *4*, 351–357. [CrossRef]
98. Kelderman, S.; Heemskerk, B.; van Tinteren, H.; van den Brom, R.R.; Hospers, G.A.; van den Eertwegh, A.J.; Kapiteijn, E.W.; de Groot, J.W.; Soetekouw, P.; Jansen, R.L.; et al. Lactate dehydrogenase as a selection criterion for ipilimumab treatment in metastatic melanoma. *Cancer Immunol. Immunother.* **2014**, *63*, 449–458. [CrossRef]
99. Dick, J.; Lang, N.; Slynko, A.; Kopp-Schneider, A.; Schulz, C.; Dimitrakopoulou-Strauss, A.; Enk, A.H.; Hassel, J.C. Use of LDH and autoimmune side effects to predict response to ipilimumab treatment. *Immunotherapy* **2016**, *8*, 1033–1044. [CrossRef]
100. Weide, B.; Martens, A.; Hassel, J.C.; Berking, C.; Postow, M.A.; Bisschop, K.; Simeone, E.; Mangana, J.; Schilling, B.; Di Giacomo, A.M.; et al. Baseline Biomarkers for Outcome of Melanoma Patients Treated with Pembrolizumab. *Clin. Cancer Res.* **2016**, *22*, 5487–5496. [CrossRef]
101. Khagi, Y.; Goodman, A.M.; Daniels, G.A.; Patel, S.P.; Sacco, A.G.; Randall, J.M.; Bazhenova, L.A.; Kurzrock, R. Hypermutated Circulating Tumor DNA: Correlation with Response to Checkpoint Inhibitor-Based Immunotherapy. *Clin. Cancer Res.* **2017**, *23*, 5729–5736. [CrossRef] [PubMed]
102. Gandara, D.R.; Paul, S.M.; Kowanetz, M.; Schleifman, E.; Zou, W.; Li, Y.; Rittmeyer, A.; Fehrenbacher, L.; Otto, G.; Malboeuf, C.; et al. Blood-based tumor mutational burden as a predictor of clinical benefit in non-small-cell lung cancer patients treated with atezolizumab. *Nat. Med.* **2018**, *24*, 1441–1448. [CrossRef] [PubMed]
103. Zitvogel, L.; Ma, Y.; Raoult, D.; Kroemer, G.; Gajewski, T.F. The microbiome in cancer immunotherapy: Diagnostic tools and therapeutic strategies. *Science* **2018**, *359*, 1366–1370. [CrossRef] [PubMed]
104. Matson, V.; Fessler, J.; Bao, R.; Chongsuwat, T.; Zha, Y.; Alegre, M.L.; Luke, J.J.; Gajewski, T.F. The commensal microbiome is associated with anti-PD-1 efficacy in metastatic melanoma patients. *Science* **2018**, *359*, 104–108. [CrossRef]
105. Chaput, N.; Lepage, P.; Coutzac, C.; Soularue, E.; Le Roux, K.; Monot, C.; Boselli, L.; Routier, E.; Cassard, L.; Collins, M.; et al. Baseline gut microbiota predicts clinical response and colitis in metastatic melanoma patients treated with ipilimumab. *Ann. Oncol.* **2017**, *28*, 1368–1379. [CrossRef]
106. Gopalakrishnan, V.; Spencer, C.N.; Nezi, L.; Reuben, A.; Andrews, M.C.; Karpinets, T.V.; Prieto, P.A.; Vicente, D.; Hoffman, K.; Wei, S.C.; et al. Gut microbiome modulates response to anti-PD-1 immunotherapy in melanoma patients. *Science* **2018**, *359*, 97–103. [CrossRef]
107. Routy, B.; Le Chatelier, E.; Derosa, L.; Duong, C.P.M.; Alou, M.T.; Daillere, R.; Fluckiger, A.; Messaoudene, M.; Rauber, C.; Roberti, M.P.; et al. Gut microbiome influences efficacy of PD-1-based immunotherapy against epithelial tumors. *Science* **2018**, *359*, 91–97. [CrossRef]
108. Spranger, S.; Spaapen, R.M.; Zha, Y.; Williams, J.; Meng, Y.; Ha, T.T.; Gajewski, T.F. Up-regulation of PD-L1, IDO, and T(regs) in the melanoma tumor microenvironment is driven by CD8(+) T cells. *Sci. Transl. Med.* **2013**, *5*, 200ra116. [CrossRef]
109. Le, D.T.; Durham, J.N.; Smith, K.N.; Wang, H.; Bartlett, B.R.; Aulakh, L.K.; Lu, S.; Kemberling, H.; Wilt, C.; Luber, B.S.; et al. Mismatch repair deficiency predicts response of solid tumors to PD-1 blockade. *Science* **2017**, *357*, 409–413. [CrossRef]
110. Galon, J.; Costes, A.; Sanchez-Cabo, F.; Kirilovsky, A.; Mlecnik, B.; Lagorce-Pages, V.; Tosolini, M.; Camus, M.; Berger, A.; Wind, P.; et al. Type, density, and location of immune cells within human colorectal tumors predict clinical outcome. *Science* **2006**, *313*, 1960–1964. [CrossRef]
111. Pages, F.; Berger, A.; Camus, M.; Sanchez-Cabo, F.; Costes, A.; Molidor, R.; Mlecnik, B.; Kirilovsky, A.; Nilsson, M.; Damotte, D.; et al. Effector memory T cells, early metastasis, and survival in colorectal cancer. *N. Engl. J. Med.* **2005**, *353*, 2654–2666. [CrossRef]

112. Sato, E.; Olson, S.H.; Ahn, J.; Bundy, B.; Nishikawa, H.; Qian, F.; Jungbluth, A.A.; Frosina, D.; Gnjatic, S.; Ambrosone, C.; et al. Intraepithelial CD8+ tumor-infiltrating lymphocytes and a high CD8+/regulatory T cell ratio are associated with favorable prognosis in ovarian cancer. *Proc. Natl. Acad. Sci. USA* **2005**, *102*, 18538–18543. [CrossRef]
113. Chew, V.; Tow, C.; Teo, M.; Wong, H.L.; Chan, J.; Gehring, A.; Loh, M.; Bolze, A.; Quek, R.; Lee, V.K.; et al. Inflammatory tumour microenvironment is associated with superior survival in hepatocellular carcinoma patients. *J. Hepatol.* **2010**, *52*, 370–379. [CrossRef]
114. Gabrielson, A.; Wu, Y.; Wang, H.; Jiang, J.; Kallakury, B.; Gatalica, Z.; Reddy, S.; Kleiner, D.; Fishbein, T.; Johnson, L.; et al. Intratumoral CD3 and CD8 T-cell Densities Associated with Relapse-Free Survival in HCC. *Cancer Immunol. Res.* **2016**, *4*, 419–430. [CrossRef]
115. Kurebayashi, Y.; Ojima, H.; Tsujikawa, H.; Kubota, N.; Maehara, J.; Abe, Y.; Kitago, M.; Shinoda, M.; Kitagawa, Y.; Sakamoto, M. Landscape of immune microenvironment in hepatocellular carcinoma and its additional impact on histological and molecular classification. *Hepatology* **2018**, *68*, 1025–1041. [CrossRef]
116. Tas, F.; Aykan, F.; Alici, S.; Kaytan, E.; Aydiner, A.; Topuz, E. Prognostic factors in pancreatic carcinoma: Serum LDH levels predict survival in metastatic disease. *Am. J. Clin. Oncol.* **2001**, *24*, 547–550. [CrossRef]
117. Miao, P.; Sheng, S.; Sun, X.; Liu, J.; Huang, G. Lactate dehydrogenase A in cancer: A promising target for diagnosis and therapy. *IUBMB Life* **2013**, *65*, 904–910. [CrossRef]
118. Scartozzi, M.; Faloppi, L.; Bianconi, M.; Giampieri, R.; Maccaroni, E.; Bittoni, A.; Del Prete, M.; Loretelli, C.; Belvederesi, L.; Svegliati Baroni, G.; et al. The role of LDH serum levels in predicting global outcome in HCC patients undergoing TACE: Implications for clinical management. *PLoS ONE* **2012**, *7*, e32653. [CrossRef]
119. Faloppi, L.; Scartozzi, M.; Bianconi, M.; Svegliati Baroni, G.; Toniutto, P.; Giampieri, R.; Del Prete, M.; De Minicis, S.; Bitetto, D.; Loretelli, C.; et al. The role of LDH serum levels in predicting global outcome in HCC patients treated with sorafenib: Implications for clinical management. *BMC Cancer* **2014**, *14*, 110. [CrossRef]
120. Schwarzenbach, H.; Hoon, D.S.; Pantel, K. Cell-free nucleic acids as biomarkers in cancer patients. *Nat. Rev. Cancer* **2011**, *11*, 426–437. [CrossRef]
121. Haber, D.A.; Velculescu, V.E. Blood-based analyses of cancer: Circulating tumor cells and circulating tumor DNA. *Cancer Discov.* **2014**, *4*, 650–661. [CrossRef]
122. Parikh, A.R.; Leshchiner, I.; Elagina, L.; Goyal, L.; Levovitz, C.; Siravegna, G.; Livitz, D.; Rhrissorrakrai, K.; Martin, E.E.; Van Seventer, E.E.; et al. Liquid versus tissue biopsy for detecting acquired resistance and tumor heterogeneity in gastrointestinal cancers. *Nat. Med.* **2019**, *25*, 1415–1421. [CrossRef]
123. Garrett, W.S. Cancer and the microbiota. *Science* **2015**, *348*, 80–86. [CrossRef]
124. Zitvogel, L.; Ayyoub, M.; Routy, B.; Kroemer, G. Microbiome and Anticancer Immunosurveillance. *Cell* **2016**, *165*, 276–287. [CrossRef]
125. Freeman-Keller, M.; Kim, Y.; Cronin, H.; Richards, A.; Gibney, G.; Weber, J.S. Nivolumab in Resected and Unresectable Metastatic Melanoma: Characteristics of Immune-Related Adverse Events and Association with Outcomes. *Clin. Cancer Res.* **2016**, *22*, 886–894. [CrossRef]
126. Teraoka, S.; Fujimoto, D.; Morimoto, T.; Kawachi, H.; Ito, M.; Sato, Y.; Nagata, K.; Nakagawa, A.; Otsuka, K.; Uehara, K.; et al. Early Immune-Related Adverse Events and Association with Outcome in Advanced Non-Small Cell Lung Cancer Patients Treated with Nivolumab: A Prospective Cohort Study. *J. Thorac. Oncol.* **2017**, *12*, 1798–1805. [CrossRef]
127. Judd, J.; Zibelman, M.; Handorf, E.; O'Neill, J.; Ramamurthy, C.; Bentota, S.; Doyle, J.; Uzzo, R.G.; Bauman, J.; Borghaei, H.; et al. Immune-Related Adverse Events as a Biomarker in Non-Melanoma Patients Treated with Programmed Cell Death 1 Inhibitors. *Oncologist* **2017**, *22*, 1232–1237. [CrossRef]
128. Horvat, T.Z.; Adel, N.G.; Dang, T.O.; Momtaz, P.; Postow, M.A.; Callahan, M.K.; Carvajal, R.D.; Dickson, M.A.; D'Angelo, S.P.; Woo, K.M.; et al. Immune-Related Adverse Events, Need for Systemic Immunosuppression, and Effects on Survival and Time to Treatment Failure in Patients with Melanoma Treated with Ipilimumab at Memorial Sloan Kettering Cancer Center. *J. Clin. Oncol.* **2015**, *33*, 3193–3198. [CrossRef]
129. Teulings, H.E.; Limpens, J.; Jansen, S.N.; Zwinderman, A.H.; Reitsma, J.B.; Spuls, P.I.; Luiten, R.M. Vitiligo-like depigmentation in patients with stage III-IV melanoma receiving immunotherapy and its association with survival: A systematic review and meta-analysis. *J. Clin. Oncol.* **2015**, *33*, 773–781. [CrossRef]

130. Kim, H.I.; Kim, M.; Lee, S.H.; Park, S.Y.; Kim, Y.N.; Kim, H.; Jeon, M.J.; Kim, T.Y.; Kim, S.W.; Kim, W.B.; et al. Development of thyroid dysfunction is associated with clinical response to PD-1 blockade treatment in patients with advanced non-small cell lung cancer. *Oncoimmunology* **2017**, *7*, e1375642. [CrossRef]
131. Wong, L.; Ang, A.; Ng, K.; Tan, S.H.; Choo, S.P.; Tai, D.; Lee, J. Association between immune-related adverse events and efficacy of immune checkpoint inhibitors in patients with advanced hepatocellular carcinoma. *Ann. Oncol.* **2019**, *30*. [CrossRef]
132. Nakamura, Y. Biomarkers for Immune Checkpoint Inhibitor-Mediated Tumor Response and Adverse Events. *Front. Med. (Lausanne)* **2019**, *6*, 119. [CrossRef] [PubMed]
133. Valpione, S.; Pasquali, S.; Campana, L.G.; Piccin, L.; Mocellin, S.; Pigozzo, J.; Chiarion-Sileni, V. Sex and interleukin-6 are prognostic factors for autoimmune toxicity following treatment with anti-CTLA4 blockade. *J. Transl. Med.* **2018**, *16*, 94. [CrossRef] [PubMed]
134. Diehl, A.; Yarchoan, M.; Hopkins, A.; Jaffee, E.; Grossman, S.A. Relationships between lymphocyte counts and treatment-related toxicities and clinical responses in patients with solid tumors treated with PD-1 checkpoint inhibitors. *Oncotarget* **2017**, *8*, 114268–114280. [CrossRef]
135. Fujisawa, Y.; Yoshino, K.; Otsuka, A.; Funakoshi, T.; Fujimura, T.; Yamamoto, Y.; Hata, H.; Gosho, M.; Tanaka, R.; Yamaguchi, K.; et al. Fluctuations in routine blood count might signal severe immune-related adverse events in melanoma patients treated with nivolumab. *J. Dermatol. Sci.* **2017**, *88*, 225–231. [CrossRef]
136. de Moel, E.C.; Rozeman, E.A.; Kapiteijn, E.H.; Verdegaal, E.M.E.; Grummels, A.; Bakker, J.A.; Huizinga, T.W.J.; Haanen, J.B.; Toes, R.E.M.; van der Woude, D. Autoantibody Development under Treatment with Immune-Checkpoint Inhibitors. *Cancer Immunol. Res.* **2019**, *7*, 6–11. [CrossRef]
137. Kobayashi, T.; Iwama, S.; Yasuda, Y.; Okada, N.; Tsunekawa, T.; Onoue, T.; Takagi, H.; Hagiwara, D.; Ito, Y.; Morishita, Y.; et al. Patients with Antithyroid Antibodies Are Prone to Develop Destructive Thyroiditis by Nivolumab: A Prospective Study. *J. Endocr. Soc.* **2018**, *2*, 241–251. [CrossRef]
138. Stamatouli, A.M.; Quandt, Z.; Perdigoto, A.L.; Clark, P.L.; Kluger, H.; Weiss, S.A.; Gettinger, S.; Sznol, M.; Young, A.; Rushakoff, R.; et al. Collateral Damage: Insulin-Dependent Diabetes Induced with Checkpoint Inhibitors. *Diabetes* **2018**, *67*, 1471–1480. [CrossRef]
139. Cooling, L.L.; Sherbeck, J.; Mowers, J.C.; Hugan, S.L. Development of red blood cell autoantibodies following treatment with checkpoint inhibitors: A new class of anti-neoplastic, immunotherapeutic agents associated with immune dysregulation. *Immunohematology* **2017**, *33*, 15–21.
140. Yau, T.; Kang, Y.-K.; Kim, T.-Y.; El-Khoueiry, A.B.; Santoro, A.; Sangro, B.; Melero, I.; Kudo, M.; Hou, M.-M.; Matilla, A.; et al. Nivolumab (NIVO) + ipilimumab (IPI) combination therapy in patients (pts) with advanced hepatocellular carcinoma (aHCC): Results from CheckMate 040. *J. Clin. Oncol.* **2019**, *37*, 4012. [CrossRef]
141. Abou-Alfa, G.K.; Chan, S.L.; Furuse, J.; Galle, P.R.; Kelley, R.K.; Qin, S.; Armstrong, J.; Darilay, A.; Vlahovic, G.; Negro, A.; et al. A randomized, multicenter phase 3 study of durvalumab (D) and tremelimumab (T) as first-line treatment in patients with unresectable hepatocellular carcinoma (HCC): HIMALAYA study. *J. Clin. Oncol.* **2018**, *36*, TPS4144. [CrossRef]
142. Cheng, A.-L.; Qin, S.; Ikeda, M.; Galle, P.; Ducreux, M.; Zhu, A.; Kim, T.-Y.; Kudo, M.; Breder, V.; Merle, P.; et al. Abstract LBA3—IMbrave150: Efficacy and safety results from a ph III study evaluating atezolizumab (atezo) + bevacizumab (bev) vs. sorafenib (Sor) as first treatment (tx) for patients (pts) with unresectable hepatocellular carcinoma (HCC). *Ann. Oncol.* **2019**, *30*. [CrossRef]
143. Kelley, R.K.; Cheng, A.-L.; Braiteh, F.S.; Park, J.-W.; Benzaghou, F.; Milwee, S.; Borgman, A.; El-Khoueiry, A.B.; Kayali, Z.K.; Zhu, A.X.; et al. Phase 3 (COSMIC-312) study of cabozantinib (C) in combination with atezolizumab (A) versus sorafenib (S) in patients (pts) with advanced hepatocellular carcinoma (aHCC) who have not received previous systemic anticancer therapy. *J. Clin. Oncol.* **2019**, *37*, TPS4157. [CrossRef]
144. Ikeda, M.; Sung, M.W.; Kudo, M.; Kobayashi, M.; Baron, A.D.; Finn, R.S.; Kaneko, S.; Zhu, A.X.; Kubota, T.; Kralijevic, S.; et al. Abstract CT061: A Phase Ib trial of lenvatinib (LEN) plus pembrolizumab (PEMBRO) in unresectable hepatocellular carcinoma (uHCC): Updated results. *Cancer Res.* **2019**, *79*, CT061. [CrossRef]
145. Llovet, J.M.; Kudo, M.; Cheng, A.-L.; Finn, R.S.; Galle, P.R.; Kaneko, S.; Meyer, T.; Qin, S.; Dutcus, C.E.; Chen, E.; et al. Lenvatinib (len) plus pembrolizumab (pembro) for the first-line treatment of patients (pts) with advanced hepatocellular carcinoma (HCC): Phase 3 LEAP-002 study. *J. Clin. Oncol.* **2019**, *37*, TPS4152. [CrossRef]

146. Kudo, M.; Motomura, K.; Wada, Y.; Inaba, Y.; Sakamoto, Y.; Kurosaki, M.; Umeyama, Y.; Kamei, Y.; Yoshimitsu, J.; Fujii, Y.; et al. First-line avelumab + axitinib in patients with advanced hepatocellular carcinoma: Results from a phase 1b trial (VEGF Liver 100). *J. Clin. Oncol.* **2019**, *37*, 4072. [CrossRef]
147. Qin, S.; Chen, Z.; Liu, Y.; Xiong, J.; Ren, Z.; Meng, Z.; Gu, S.; Wang, L.; Zou, J. A phase II study of anti–PD-1 antibody camrelizumab plus FOLFOX4 or GEMOX systemic chemotherapy as first-line therapy for advanced hepatocellular carcinoma or biliary tract cancer. *J. Clin. Oncol.* **2019**, *37*, 4074. [CrossRef]
148. Buchbinder, E.I.; Desai, A. CTLA-4 and PD-1 Pathways: Similarities, Differences, and Implications of Their Inhibition. *Am. J. Clin. Oncol.* **2016**, *39*, 98–106. [CrossRef]
149. Granier, C.; De Guillebon, E.; Blanc, C.; Roussel, H.; Badoual, C.; Colin, E.; Saldmann, A.; Gey, A.; Oudard, S.; Tartour, E. Mechanisms of action and rationale for the use of checkpoint inhibitors in cancer. *ESMO Open* **2017**, *2*, e000213. [CrossRef]
150. Jain, R.K. Antiangiogenesis strategies revisited: From starving tumors to alleviating hypoxia. *Cancer Cell* **2014**, *26*, 605–622. [CrossRef]
151. Hanahan, D.; Weinberg, R.A. Hallmarks of cancer: The next generation. *Cell* **2011**, *144*, 646–674. [CrossRef] [PubMed]
152. Motz, G.T.; Coukos, G. The parallel lives of angiogenesis and immunosuppression: Cancer and other tales. *Nat. Rev. Immunol.* **2011**, *11*, 702–711. [CrossRef] [PubMed]
153. Goel, S.; Wong, A.H.; Jain, R.K. Vascular normalization as a therapeutic strategy for malignant and nonmalignant disease. *Cold Spring Harb. Perspect. Med.* **2012**, *2*, a006486. [CrossRef] [PubMed]
154. Datta, M.; Coussens, L.M.; Nishikawa, H.; Hodi, F.S.; Jain, R.K. Reprogramming the Tumor Microenvironment to Improve Immunotherapy: Emerging Strategies and Combination Therapies. *Am. Soc. Clin. Oncol. Educ. Book* **2019**, *39*, 165–174. [CrossRef]
155. Ferrara, N.; Hillan, K.J.; Gerber, H.P.; Novotny, W. Discovery and development of bevacizumab, an anti-VEGF antibody for treating cancer. *Nat. Rev. Drug Discov.* **2004**, *3*, 391–400. [CrossRef]
156. Hansen, W.; Hutzler, M.; Abel, S.; Alter, C.; Stockmann, C.; Kliche, S.; Albert, J.; Sparwasser, T.; Sakaguchi, S.; Westendorf, A.M.; et al. Neuropilin 1 deficiency on CD4+Foxp3+ regulatory T cells impairs mouse melanoma growth. *J. Exp. Med.* **2012**, *209*, 2001–2016. [CrossRef]
157. Shrimali, R.K.; Yu, Z.; Theoret, M.R.; Chinnasamy, D.; Restifo, N.P.; Rosenberg, S.A. Antiangiogenic agents can increase lymphocyte infiltration into tumor and enhance the effectiveness of adoptive immunotherapy of cancer. *Cancer Res.* **2010**, *70*, 6171–6180. [CrossRef]
158. Shigeta, K.; Datta, M.; Hato, T.; Kitahara, S.; Chen, I.X.; Matsui, A.; Kikuchi, H.; Mamessier, E.; Aoki, S.; Ramjiawan, R.R.; et al. Dual Programmed Death Receptor-1 and Vascular Endothelial Growth Factor Receptor-2 Blockade Promotes Vascular Normalization and Enhances Antitumor Immune Responses in Hepatocellular Carcinoma. *Hepatology* **2019**. [CrossRef]
159. Wilhelm, S.M.; Adnane, L.; Newell, P.; Villanueva, A.; Llovet, J.M.; Lynch, M. Preclinical overview of sorafenib, a multikinase inhibitor that targets both Raf and VEGF and PDGF receptor tyrosine kinase signaling. *Mol. Cancer Ther.* **2008**, *7*, 3129–3140. [CrossRef]
160. Tan, H.Y.; Wang, N.; Lam, W.; Guo, W.; Feng, Y.; Cheng, Y.C. Targeting tumour microenvironment by tyrosine kinase inhibitor. *Mol. Cancer* **2018**, *17*, 43. [CrossRef]
161. Gurule, N.J.; Heasley, L.E. Linking tyrosine kinase inhibitor-mediated inflammation with normal epithelial cell homeostasis and tumor therapeutic responses. *Cancer Drug Resist.* **2018**, *1*, 118–125. [CrossRef] [PubMed]
162. Galluzzi, L.; Buque, A.; Kepp, O.; Zitvogel, L.; Kroemer, G. Immunological Effects of Conventional Chemotherapy and Targeted Anticancer Agents. *Cancer Cell* **2015**, *28*, 690–714. [CrossRef] [PubMed]
163. Ko, J.S.; Rayman, P.; Ireland, J.; Swaidani, S.; Li, G.; Bunting, K.D.; Rini, B.; Finke, J.H.; Cohen, P.A. Direct and differential suppression of myeloid-derived suppressor cell subsets by sunitinib is compartmentally constrained. *Cancer Res.* **2010**, *70*, 3526–3536. [CrossRef] [PubMed]
164. Xin, H.; Zhang, C.; Herrmann, A.; Du, Y.; Figlin, R.; Yu, H. Sunitinib inhibition of Stat3 induces renal cell carcinoma tumor cell apoptosis and reduces immunosuppressive cells. *Cancer Res.* **2009**, *69*, 2506–2513. [CrossRef] [PubMed]
165. Ozao-Choy, J.; Ma, G.; Kao, J.; Wang, G.X.; Meseck, M.; Sung, M.; Schwartz, M.; Divino, C.M.; Pan, P.Y.; Chen, S.H. The novel role of tyrosine kinase inhibitor in the reversal of immune suppression and modulation of tumor microenvironment for immune-based cancer therapies. *Cancer Res.* **2009**, *69*, 2514–2522. [CrossRef]

166. Wang, Q.; Yu, T.; Yuan, Y.; Zhuang, H.; Wang, Z.; Liu, X.; Feng, M. Sorafenib reduces hepatic infiltrated regulatory T cells in hepatocellular carcinoma patients by suppressing TGF-beta signal. *J. Surg. Oncol.* **2013**, *107*, 422–427. [CrossRef]
167. Mustjoki, S.; Ekblom, M.; Arstila, T.P.; Dybedal, I.; Epling-Burnette, P.K.; Guilhot, F.; Hjorth-Hansen, H.; Hoglund, M.; Kovanen, P.; Laurinolli, T.; et al. Clonal expansion of T/NK-cells during tyrosine kinase inhibitor dasatinib therapy. *Leukemia* **2009**, *23*, 1398–1405. [CrossRef]
168. Rusakiewicz, S.; Semeraro, M.; Sarabi, M.; Desbois, M.; Locher, C.; Mendez, R.; Vimond, N.; Concha, A.; Garrido, F.; Isambert, N.; et al. Immune infiltrates are prognostic factors in localized gastrointestinal stromal tumors. *Cancer Res.* **2013**, *73*, 3499–3510. [CrossRef]
169. Kwilas, A.R.; Donahue, R.N.; Tsang, K.Y.; Hodge, J.W. Immune consequences of tyrosine kinase inhibitors that synergize with cancer immunotherapy. *Cancer Cell Microenviron.* **2015**, *2*. [CrossRef]
170. Vanneman, M.; Dranoff, G. Combining immunotherapy and targeted therapies in cancer treatment. *Nat. Rev. Cancer* **2012**, *12*, 237–251. [CrossRef]
171. Golden, E.B.; Apetoh, L. Radiotherapy and immunogenic cell death. *Semin. Radiat. Oncol.* **2015**, *25*, 11–17. [CrossRef] [PubMed]
172. Kroemer, G.; Galluzzi, L.; Kepp, O.; Zitvogel, L. Immunogenic cell death in cancer therapy. *Annu. Rev. Immunol.* **2013**, *31*, 51–72. [CrossRef] [PubMed]
173. Choi, C.; Yoo, G.S.; Cho, W.K.; Park, H.C. Optimizing radiotherapy with immune checkpoint blockade in hepatocellular carcinoma. *World J. Gastroenterol.* **2019**, *25*, 2416–2429. [CrossRef] [PubMed]
174. Kim, H.J.; Park, S.; Kim, K.J.; Seong, J. Clinical significance of soluble programmed cell death ligand-1 (sPD-L1) in hepatocellular carcinoma patients treated with radiotherapy. *Radiother. Oncol.* **2018**, *129*, 130–135. [CrossRef] [PubMed]
175. Chew, V.; Lee, Y.H.; Pan, L.; Nasir, N.J.M.; Lim, C.J.; Chua, C.; Lai, L.; Hazirah, S.N.; Lim, T.K.H.; Goh, B.K.P.; et al. Immune activation underlies a sustained clinical response to Yttrium-90 radioembolisation in hepatocellular carcinoma. *Gut* **2019**, *68*, 335–346. [CrossRef] [PubMed]
176. Harding, J.J.; Erinjeri, J.P.; Tan, B.R.; Reiss, K.A.; Mody, K.; Khalil, D.; Yarmohammadi, H.; Nadolski, G.; Giardina, J.D.; Capanu, M.; et al. A multicenter pilot study of nivolumab (NIVO) with drug eluting bead transarterial chemoembolization (deb-TACE) in patients (pts) with liver limited hepatocellular carcinoma (HCC). *J. Clin. Oncol.* **2018**, *36*, TPS4146. [CrossRef]
177. Duffy, A.G.; Ulahannan, S.V.; Makorova-Rusher, O.; Rahma, O.; Wedemeyer, H.; Pratt, D.; Davis, J.L.; Hughes, M.S.; Heller, T.; ElGindi, M.; et al. Tremelimumab in combination with ablation in patients with advanced hepatocellular carcinoma. *J. Hepatol.* **2017**, *66*, 545–551. [CrossRef]
178. Dudley, M.E.; Rosenberg, S.A. Adoptive-cell-transfer therapy for the treatment of patients with cancer. *Nat. Rev. Cancer* **2003**, *3*, 666–675. [CrossRef]
179. Zhang, R.; Zhang, Z.; Liu, Z.; Wei, D.; Wu, X.; Bian, H.; Chen, Z. Adoptive cell transfer therapy for hepatocellular carcinoma. *Front. Med.* **2019**, *13*, 3–11. [CrossRef]
180. Hoseini, S.S.; Cheung, N.V. Immunotherapy of hepatocellular carcinoma using chimeric antigen receptors and bispecific antibodies. *Cancer Lett.* **2017**, *399*, 44–52. [CrossRef]
181. Gehring, A.J.; Ho, Z.Z.; Tan, A.T.; Aung, M.O.; Lee, K.H.; Tan, K.C.; Lim, S.G.; Bertoletti, A. Profile of tumor antigen-specific CD8 T cells in patients with hepatitis B virus-related hepatocellular carcinoma. *Gastroenterology* **2009**, *137*, 682–690. [CrossRef] [PubMed]
182. Takayama, T.; Sekine, T.; Makuuchi, M.; Yamasaki, S.; Kosuge, T.; Yamamoto, J.; Shimada, K.; Sakamoto, M.; Hirohashi, S.; Ohashi, Y.; et al. Adoptive immunotherapy to lower postsurgical recurrence rates of hepatocellular carcinoma: A randomised trial. *Lancet* **2000**, *356*, 802–807. [CrossRef]
183. Xu, L.; Wang, J.; Kim, Y.; Shuang, Z.Y.; Zhang, Y.J.; Lao, X.M.; Li, Y.Q.; Chen, M.S.; Pawlik, T.M.; Xia, J.C.; et al. A randomized controlled trial on patients with or without adjuvant autologous cytokine-induced killer cells after curative resection for hepatocellular carcinoma. *Oncoimmunology* **2016**, *5*, e1083671. [CrossRef] [PubMed]
184. Chew, V.; Chen, J.; Lee, D.; Loh, E.; Lee, J.; Lim, K.H.; Weber, A.; Slankamenac, K.; Poon, R.T.; Yang, H.; et al. Chemokine-driven lymphocyte infiltration: An early intratumoural event determining long-term survival in resectable hepatocellular carcinoma. *Gut* **2012**, *61*, 427–438. [CrossRef]
185. Liu, P.; Chen, L.; Zhang, H. Natural Killer Cells in Liver Disease and Hepatocellular Carcinoma and the NK Cell-Based Immunotherapy. *J. Immunol. Res.* **2018**, *2018*, 1206737. [CrossRef]

186. Daher, M.; Rezvani, K. Next generation natural killer cells for cancer immunotherapy: The promise of genetic engineering. *Curr. Opin. Immunol.* **2018**, *51*, 146–153. [CrossRef]
187. Ott, P.A.; Hu, Z.; Keskin, D.B.; Shukla, S.A.; Sun, J.; Bozym, D.J.; Zhang, W.; Luoma, A.; Giobbie-Hurder, A.; Peter, L.; et al. An immunogenic personal neoantigen vaccine for patients with melanoma. *Nature* **2017**, *547*, 217–221. [CrossRef]
188. Sahin, U.; Derhovanessian, E.; Miller, M.; Kloke, B.P.; Simon, P.; Lower, M.; Bukur, V.; Tadmor, A.D.; Luxemburger, U.; Schrors, B.; et al. Personalized RNA mutanome vaccines mobilize poly-specific therapeutic immunity against cancer. *Nature* **2017**, *547*, 222–226. [CrossRef]
189. Buonaguro, L.; Petrizzo, A.; Tagliamonte, M.; Tornesello, M.L.; Buonaguro, F.M. Challenges in cancer vaccine development for hepatocellular carcinoma. *J. Hepatol.* **2013**, *59*, 897–903. [CrossRef]
190. Butterfield, L.H.; Ribas, A.; Dissette, V.B.; Lee, Y.; Yang, J.Q.; De la Rocha, P.; Duran, S.D.; Hernandez, J.; Seja, E.; Potter, D.M.; et al. A phase I/II trial testing immunization of hepatocellular carcinoma patients with dendritic cells pulsed with four alpha-fetoprotein peptides. *Clin. Cancer Res.* **2006**, *12*, 2817–2825. [CrossRef]
191. Wang, X.; Wang, Q. Alpha-Fetoprotein and Hepatocellular Carcinoma Immunity. *Can. J. Gastroenterol. Hepatol.* **2018**, *2018*, 9049252. [CrossRef] [PubMed]
192. Howells, A.; Marelli, G.; Lemoine, N.R.; Wang, Y. Oncolytic Viruses-Interaction of Virus and Tumor Cells in the Battle to Eliminate Cancer. *Front. Oncol.* **2017**, *7*, 195. [CrossRef] [PubMed]
193. Kaufman, H.L.; Kohlhapp, F.J.; Zloza, A. Oncolytic viruses: A new class of immunotherapy drugs. *Nat. Rev. Drug Discov.* **2015**, *14*, 642–662. [CrossRef] [PubMed]
194. Jhawar, S.R.; Thandoni, A.; Bommareddy, P.K.; Hassan, S.; Kohlhapp, F.J.; Goyal, S.; Schenkel, J.M.; Silk, A.W.; Zloza, A. Oncolytic Viruses-Natural and Genetically Engineered Cancer Immunotherapies. *Front. Oncol.* **2017**, *7*, 202. [CrossRef]
195. Yoo, S.Y.; Badrinath, N.; Woo, H.Y.; Heo, J. Oncolytic Virus-Based Immunotherapies for Hepatocellular Carcinoma. *Mediat. Inflamm.* **2017**, *2017*, 5198798. [CrossRef]
196. Park, B.H.; Hwang, T.; Liu, T.C.; Sze, D.Y.; Kim, J.S.; Kwon, H.C.; Oh, S.Y.; Han, S.Y.; Yoon, J.H.; Hong, S.H.; et al. Use of a targeted oncolytic poxvirus, JX-594, in patients with refractory primary or metastatic liver cancer: A phase I trial. *Lancet Oncol.* **2008**, *9*, 533–542. [CrossRef]
197. Heo, J.; Reid, T.; Ruo, L.; Breitbach, C.J.; Rose, S.; Bloomston, M.; Cho, M.; Lim, H.Y.; Chung, H.C.; Kim, C.W.; et al. Randomized dose-finding clinical trial of oncolytic immunotherapeutic vaccinia JX-594 in liver cancer. *Nat. Med.* **2013**, *19*, 329–336. [CrossRef]
198. LaRocca, C.J.; Warner, S.G. Oncolytic viruses and checkpoint inhibitors: Combination therapy in clinical trials. *Clin. Transl. Med.* **2018**, *7*, 35. [CrossRef]

© 2019 by the authors. Licensee MDPI, Basel, Switzerland. This article is an open access article distributed under the terms and conditions of the Creative Commons Attribution (CC BY) license (http://creativecommons.org/licenses/by/4.0/).

Review

Overview of Immune Checkpoint Inhibitors Therapy for Hepatocellular Carcinoma, and The ITA.LI.CA Cohort Derived Estimate of Amenability Rate to Immune Checkpoint Inhibitors in Clinical Practice

Edoardo G. Giannini [1,*], Andrea Aglitti [2], Mauro Borzio [3], Martina Gambato [4], Maria Guarino [5], Massimo Iavarone [6], Quirino Lai [7], Giovanni Battista Levi Sandri [8], Fabio Melandro [9], Filomena Morisco [5], Francesca Romana Ponziani [10], Maria Rendina [11], Francesco Paolo Russo [12], Rodolfo Sacco [13], Mauro Viganò [14], Alessandro Vitale [15], Franco Trevisani [16] and on behalf of the Associazione Italiana per lo Studio del Fegato (AISF) HCC Special Interest Group [17,†]

1. Gastroenterology Unit, Department of Internal Medicine, Università di Genova, IRCCS (Istituto di Ricovero e Cura a Carattere Scientifico)-Ospedale Policlinico San Martino, 16132 Genoa, Italy
2. Department of Medicine and Surgery, Internal Medicine and Hepatology Unit, University of Salerno, 84084 Fisciano, Italy; andreaaglitti@gmail.com
3. Unità Operativa Complessa (UOC) Gastroenterologia ed Endoscopia Digestiva, ASST (Azienda Socio Sanitaria Territoriale) Melegnano Martesana, 20063 Milan, Italy; mauro.borzio@gmail.com
4. Multivisceral Transplant Unit, Department of Surgery, Oncology and Gastroenterology, Padua University Hospital, 35124 Padua, Italy; martina.gambato@gmail.com
5. Gastroenterology Unit, Department of Clinical Medicine and Surgery, University of Naples Federico II, 80138 Naples, Italy; maria.guarino86@gmail.com (M.G.); filomena.morisco@unina.it (F.M.)
6. CRC "A. M. and A. Migliavacca" Center for Liver Disease, Division of Gastroenterology and Hepatology, Fondazione IRCCS Cà Granda Ospedale Maggiore Policlinico, Università degli Studi di Milano, 20122 Milan, Italy; massimo.iavarone@gmail.com
7. Liver Transplantation Program, Sapienza University, 00185 Rome, Italy; lai.quirino@libero.it
8. Department of Surgery, Sant'Eugenio Hospital, 00144 Roma, Italy; gblevisandri@gmail.com
9. Dipartimento Assistenziale Integrato di Chirurgia Generale, Unità Operativa Complessa Epatica e Trapianto Fegato, Azienda Ospedaliera Universitaria Pisana, 56126 Pisa, Italy; fabmelan@yahoo.it
10. Internal Medicine, Gastroenterology and Hepatology, Fondazione Policlinico Universitario A. Gemelli IRCCS, 00168 Rome, Italy; francesca.ponziani@yahoo.it
11. UOC Gastroenterologia Universitaria, Dipartimento Emergenza e trapianti di organo, Azienda Policlinico-Universita' di Bari, 70124 Bari, Italy; mariarendina@virgilio.it
12. Gastroenterology and Multivisceral Transplant Unit, Padua University Hospital, 35124 Padua, Italy; francescopaolo.russo@unipd.it
13. UOC Gastroenterologia ed Endoscopia Digestiva, Azienda Ospedaliera Universitaria "Ospedali Riuniti", 71122 Foggia, Italy; saccorodolfo@hotmail.com
14. Division of Hepatology, Ospedale San Giuseppe, University of Milan, 20122 Milan, Italy; mvigano72@gmail.com
15. UOC di Chirurgia Epatobiliare, Dipartimento di Scienze Chirurgiche Oncologiche e Gastroenterologiche, Azienda Università di Padova, 35124 Padua, Italy; alessandro.vitale.10@gmail.com
16. Dipartimento di Scienze Mediche e Chirurgiche Alma Mater Studiorum, Università di Bologna, 40126 Bologna, Italy; franco.trevisani@unibo.it
17. HCC Special Interest Group, Associazione Italiana per lo Studio del Fegato (AISF), 00199 Roma, Italy
* Correspondence: egiannini@unige.it; Tel.: +39-010-353-7950; Fax: +39-010-353-8638
† Members of the AISF HCC Special Interest Group: Aliberti C., Baccarani U., Bhoori S., Brancaccio G., Burra P., Cabibbo G., Casadei Gardini A., Carrai P., Cillo U., Conti F., Cucchetti A., D'Ambrosio R., Dell'Unto C., Dematthaeis N., Di Costanzo G.G., Di Sandro S., Foschi F.G., Fucilli F., Galati G., Gasbarrini A., Giuliante F., Ghinolfi D., Grieco A., Gruttaduria S., Kostandini A., Lenci I., Losito F., Lupo L.G., Manzia T.M., Mazzocato S., Mescoli C., Miele L., Muley M., Nicolini D., Persico M., Pompili M., Pravisani R., Rapaccini G.L., Renzulli

M., Rossi M., Rreka E., Sangiovanni A., Sessa A., Simonetti N., Sposito C., Tortora R., Viganò L., Villa E., Vincenzi V., Violi P.

Received: 8 October 2019; Accepted: 24 October 2019; Published: 30 October 2019

Abstract: Despite progress in our understanding of the biology of hepatocellular carcinoma (HCC), this tumour remains difficult-to-cure for several reasons, starting from the particular disease environment where it arises—advanced chronic liver disease—to its heterogeneous clinical and biological behaviour. The advent, and good results, of immunotherapy for cancer called for the evaluation of its potential application also in HCC, where there is evidence of intra-hepatic immune response activation. Several studies advanced our knowledge of immune checkpoints expression in HCC, thus suggesting that immune checkpoint blockade may have a strong rationale even in the treatment of HCC. According to this background, initial studies with tremelimumab, a cytotoxic T-lymphocyte-associated protein 4 (CTLA-4) inhibitor, and nivolumab, a programmed cell death protein 1 (PD-1) antibody, showed promising results, and further studies exploring the effects of other immune checkpoint inhibitors, alone or with other drugs, are currently underway. However, we are still far from the identification of the correct setting, and sequence, where these drugs might be used in clinical practice, and their actual applicability in real-life is unknown. This review focuses on HCC immunobiology and on the potential of immune checkpoint blockade therapy for this tumour, with a critical evaluation of the available trials on immune checkpoint blocking antibodies treatment for HCC. Moreover, it assesses the potential applicability of immune checkpoint inhibitors in the real-life setting, by analysing a large, multicentre cohort of Italian patients with HCC.

Keywords: check-point inhibitors; liver disease; immunotherapy; outcome

1. Introduction

Hepatocellular carcinoma (HCC) is the sixth most common cause of cancer, and ranks fourth among the causes of cancer-related death [1]. Major risk factors for HCC include chronic infection with the hepatitis C (HCV) and B (HBV) viruses, heavy alcohol drinking, and aflatoxins B1 exposure, depending on geographical epidemiology. In recent years, Non-Alcoholic Fatty Liver Disease (NAFLD), the hepatological aspect of the metabolic syndrome, has been recognised as a relevant cause of advanced chronic liver disease, and the fastest growing cause of cirrhosis and HCC in Western countries [2]. Although mixed data exist about the exact magnitude of HCC risk in patients with NAFLD, and different epidemiological and methodological confounders must be taken into account, in a recent retrospective cohort study involving 130 facilities in the United States Veterans Administration health service, Kanwal et al. found that the risk of HCC was higher in NAFLD patients than in the general population, with a 5- and 10-year cumulative incidence rate of HCC of 0.8 and 1.7 per 1000 patients in NAFLD patients as compared to 0.09 and 0.18 per 1000 patients in controls [3].

HCC represents a unique and peculiar neoplastic setting, as in up to 80% of cases it arises on the background of cirrhosis and chronic inflammation, which is now considered an important factor involved in cancer progression [4]. Indeed, liver cirrhosis is a recognised model of local chronic inflammation driven by infiltrating immune cells and resident liver cells like Kupffer cells, dendritic cells, liver sinusoidal cells and hepatic stellate cells. Chronic inflammation initiates tissue remodelling and determines an oxidative microenvironment, triggering DNA damage and genomic aberrations that eventually culminate in neoplasia, and as a fact it is recognised that cirrhosis and chronic inflammation act as a favourable preneoplastic setting [4]. Although precise molecular links between inflammation and HCC have not yet been fully elucidated, most data rely on the activation of the tumour necrosis factor-nuclear factor-κB axis, transcription target STAT3 and janus kinases activation as procarcinogenetic in the liver, while another player recently identified in this field is the inflammasome, a multiprotein complex and sensor of cellular damage [5,6]. Thus, as in no other neoplasia, the development of HCC is a multi-event process involving a series of genetic mutations (pr3RB, β-catenin, chromatin and transcription modulation) and epigenetic events such as hystone acetylation/deacetylation leading to a dysregulation of various genes, which may also represent putative therapeutic targets [7,8].

In the past fifteen years, advances in molecular and tumour biology significantly modified the paradigm of cancer treatment, moving from a histopathological basis to targeting specific molecular patterns. This review focuses on HCC immunobiology and the rationale for immune checkpoint blockade in these patients, while a specific discussion has been dedicated to a critical evaluation of the available trials on immune checkpoint inhibitors, alone or with other therapies, for HCC. Lastly, we assessed the potential applicability of immune checkpoint inhibitors to the real-life setting analysing a large cohort of Italian patients with HCC.

2. Cancer Immunotherapy

The principle of tumour immune surveillance presumes that most pre-malignant and early malignant cells can be eliminated (or controlled) by the immune system [9]. However, a critical feature of advanced tumours compared to early malignant lesions is their ability to escape adaptive immune response. During malignant transformation, tumour-associated antigens generated by gene mutations are created and recognised by the immune system, and adaptive tumour antigen-specific T-cell responses are generated, leading to cancer-cell elimination [10]. Therefore, to survive, growing tumours must adapt to their immunological environment by either turning off immune responses, and/or creating a local microenvironment that inhibits immune cell tumouricidal activity.

In normal circumstances, T-cells with a different T-cell receptor (TCR) repertoire circulate in the body patrolling for evidence of foreign peptides presented on the surface of cells due to infection or cancer development. The identification of tumoural antigen by T-cell determines an activation, with clonal proliferation/expansion, and a cytolytic response. On the other hand, the immune system plays a critical role in promoting tumour progression. This dual role by which the immune system can suppress and/or promote cancer growth is termed "cancer immunoediting" and consists of three phases: elimination, equilibrium, and escape [11].

In cancer immunotherapy, agents such as interferon, interleukins, vaccines and oncolytic viruses are used to enhance the immune system activation to attack tumoural cells through natural mechanisms. In particular, this goal can be achieved with several drug classes: checkpoint inhibitors, lymphocyte-promoting cytokines, engineered T-cells such as Chimeric Antigen Receptor T-cell (CAR-T) and TCR T-cells, agonistic antibodies against co-stimulatory receptors, and cancer vaccines [12]. The efficacy of cancer immunotherapies has been demonstrated, determining the rapid integration of these treatments into clinical practice. Moreover, one of the most attractive features of many cancer immunotherapies is that they target malignant cells and spare normal tissues from the damage often seen with radiation and chemotherapy that contributes to patient morbidity and mortality [13]. These properties of immunotherapy have supported the rapid inclusion of such a treatment into clinical practice. Currently, antibodies targeting cytotoxic T-lymphocyte-associated protein 4 (CTLA-4) (tremelimumab and ipilimumab) and the programmed cell death protein 1 (PD-1) or its ligand PD-L1 (nivolumab, pembrolizumab, atezolizumab, and durvalumab) have been approved for different types of solid tumours and acute lymphoblastic leukaemia.

However, despite continue advances in the field of cancer immunotherapy, several problems remain unsolved, including the inability to predict treatment efficacy, the need for additional biomarkers able to guide treatment, the development of cancer resistance immunotherapies, the lack of clinical study designs optimised to determine efficacy, and the high cost of treatment [14]. Moreover, due to the limited results in terms of efficacy and the narrow therapeutic index of some of these drugs, the adoption of a personalised pharmacogenetic approach would represent a turning point to improve results [15]. Even though all these findings are particularly relevant in HCC tissue, the limited efficacy of systemic therapies in HCC patients, and their poor tolerability to anticancer drugs, prompted the exploration of the potential of immunotherapy even in this setting, where immunotherapy is expected to play a pivotal role in the near future.

3. Rationale of Immune Checkpoint Blockade in Hepatocellular Carcinoma

In the last decade, many basic science advancements and discoveries related to tumour biology have been achieved by transcriptomic, genomic and epigenomic studies [16,17]. However, in the case of HCC, they poorly translated into clinical practice and only suboptimal results have been obtained in clinical trials testing many drugs in the last decade. As a result, although targeted systemic therapies for HCC provided some clinical benefits, the improvement in patient outcome remains modest, and HCC remains a difficult-to-cure tumour for various reasons: firstly, 70–80% of cases occur in the context of liver cirrhosis; secondly, intra-tumour morphologic and genetic heterogeneity make difficult our understanding of liver cancer, and may determine the resistance to targeted therapies; and, thirdly, either drivers or passengers mutations can be present in the tumour, making an effective molecularly-targeted therapy quite difficult [16–18]. This can explain why, despite good rationale and promising Phase II data, drug development in Phase III trials failed in many instances.

Currently, the standard-of-care for first-line treatment for advanced HCC is represented by two multikinase inhibitors (sorafenib and lenvatinib), and in patients who fail first-line treatment with these drugs, the second-line treatment is again represented by multikinase inhibitors (regorafenib and cabozantinib) [19–22]. The survival benefit obtained with multikinase inhibitors over the best supportive care is limited, and their tolerability is generally poor, indicating the urgent need for more efficacious and better tolerated therapeutic approaches.

One of the alternative strategies against the tumour relies on the modulation of the already existing immune response through the enhancement of activators and the block of inhibitors. T-cell exhaustion, defined as an impaired T-cell capacity to secrete cytokines and proliferate, with overexpression of immune checkpoint receptors (e.g., PD-1, CTLA4, and lymphocyte-activating 3) has been observed in certain types of cancer, including HCC [23]. Immune inhibitory receptors and ligands play a major role in induction and maintenance of HCC immune tolerance [24–26]. In particular, CTLA-4 is essential for the activation of helper CD4+ T-cells and the priming phase of the immune response. Upon binding of its ligands, CTLA-4 decreases T-cell activation following antigen presentation. CTLA-4 also plays a major role in the function of regulatory T-cells (Treg), a subset of CD4+ T-cells that inhibit the immune response. Moreover, CTLA-4 expression on CD14+ dendritic cells was associated with IL-10 and indoleamine-2,3-dioxygenase (IDO)-mediated inhibition of T-cell proliferation and induction of T-cell apoptosis [26]. In HCC patients, high CTLA-4 expression on Tregs in peripheral blood has been reported in association with a decrease in cytolytic granzyme B production by CD8+ T-cells [27]. Another immune checkpoint pathway is the one regulated by PD-1 receptor. PD-1 is a key factor in the effector phase of the immune-response, and is expressed by activated T- and B-cells and other cell types such as in the skin and in the lung: upon binding to its ligands (PD-L1 and PD-L2), PD-1 inhibits T-cell activation and proliferation [28]. The increased expression of PD-1 has been reported on CD8+ T-cells in patients with HCC, as well as an increase in tumour infiltrating and circulating PD-1+CD8+ T-cells associated with disease progression after curative hepatic resection [24,29]. In addition to the upregulation of PD-1 on T-cells, its ligand PD-L1 is highly expressed on peritumoural stroma cells as well as cancer cells, promoting a PD-L1/PD-1 pathway-driven inhibition of anti-tumour T-cell responses [29–31].

From a clinical standpoint, there is evidence of a role played by an activated immune-response in HCC behaviour: (i) the infiltration of T-cells in the tumour is correlated to neoplastic recurrence after liver transplantation; (ii) the presence of different immune cells infiltrating the tumour have been correlated to patients' survival; and (iii) different immune-subtypes of the tumour microenvironment are variously associated with histological and molecular classification of HCC—with potential prognostic implications—and the presence of exhausted T-cells was associated with poorer patient survival [32–36].

The longer experience accrued with immunotherapy for other tumours is essential to guide clinicians in the HCC landscape. It is known that expression of tumour-infiltrating lymphocytes, features of inflammatory cells (PD-1 and PD-L1 expression), percentage of mutations in tumour cells and gene expression profiles correlate with the activity and efficacy of these drugs against several tumour types [37–40]. In selected neoplasms, tumour mutational burden measured by targeted next-generation sequencing panels or by whole-exome sequencing, may predict clinical response to immunotherapy [41,42]. Tumours with high rate mutations present highly immunogenic antigens and more immune infiltration and they are more suitable to be managed with immune checkpoint inhibitors. Conversely, tumours with lower mutational burden present less immunogenic antigens and lower immune infiltration and, therefore, they are better candidates to other therapies [43,44]. Recently, Samstein et al. analysed the genomic data (targeted next generation sequencing) of patients with several tumour types (but not HCC) treated with immunotherapy or other therapies. Among all patients, higher somatic tumour mutational burden (highest 20% in each histology) was associated with better overall survival, but the tumour mutational burden cut-points associated with improved survival varied markedly among tumour types, indicating that there may not be one universal definition of high tumour mutational burden [45]. Moreover, Sia et al. focused their attention on HCC and its microenvironment (interactions among tumour cells, immune cells, and other immunomodulators present in the microenvironment) showing that 25% of HCC have markers of an inflammatory response, with high expression levels of PD-L1, markers of cytolytic activity, and fewer chromosomal aberrations [46]. The authors called this group of tumours the "immune class", and subdivided this class in two subtypes, characterised by active or exhausted immune response, the latter representing the ideal one to receive immunotherapy. Conversely, Harding et al. reported that HCC "cold" tumours (with Wnt/CTNNB1 mutations) are refractory to immune checkpoint inhibitors [47].

Better characterisation and understanding of increased immune checkpoints expression provide the rationale for the use of immune checkpoint blocking antibodies in HCC treatment. Figure 1 reports a schematic representation of the potential factors involved in immune system paralysis in HCC patients, and the potential pathways of action of various drugs. Binding the targeted molecules, the immune checkpoint inhibitors block the signalling, putting the immune response on hold, and allowing cytotoxic T-cells to strike tumour cells. Many Phase III trials testing the efficacy of monoclonal antibodies that target this pathway in HCC patients are ongoing, but the encouraging results reported in Phase I investigations has spurred the approval by FDA of immunotherapy even for this cancer [48,49]. Indeed, this therapy is the most interesting approach proposed according to the new discoveries in HCC biology, and especially the knowledge that liver has developed intrinsic tolerogenic mechanisms within the innate and adaptive immune system as a result of its constant exposure to antigens from portal-venous blood [50]. To date, all immune checkpoint targeted therapies for HCC consist of monoclonal antibodies developed for a specific immune target. Although several immune checkpoint blocking agents were identified in preclinical models, the majority of clinically tested therapies rely on antibodies targeting PD-1, PD-L1 and CTLA-4 molecules. The first small Phase II clinical trial using an immune checkpoint inhibitor, tremelimumab (a CTLA-4 blocking monoclonal antibody), targeted patients with HCC and chronic HCV infection, including a significant proportion (42.9%) of patients in Child-Pugh stage B [51]. A notable disease control rate (76.4%) was observed and the safety profile was acceptable. In a second small pilot trial, tremelimumab was combined with (incomplete) tumour ablation using locoregional therapies with the aim to synergise the effects by inducing immunogenic tumour cell death [52]. In this study, all aetiologies patients were included, liver function was preserved in the most patients, and 26.3% achieved a confirmed partial response. This study represented a proof of concept that immunotherapy in combination with tumour ablation is a potential way to treat patients with advanced HCC, and leads to the accumulation of intratumoural CD8+ T-cells. In fact, in tumour biopsies performed at six weeks, a clear increase in CD8+ T-cells occurred in patients showing a clinical response.

In patients with advanced HCC, PD-1 antibodies (nivolumab and pembrolizumab) have shown promising efficacy in therapy-naïve, as well as pre-treated patients. However, only 10–20% of them showed an objective and durable response. Therefore, combination schedules including different immune-therapies, (e.g., PD-1/PD-L1 and CTLA-4 antibodies) or the combination of immunotherapy and small molecules, or bifunctional antibodies are likely needed to improve response rates.

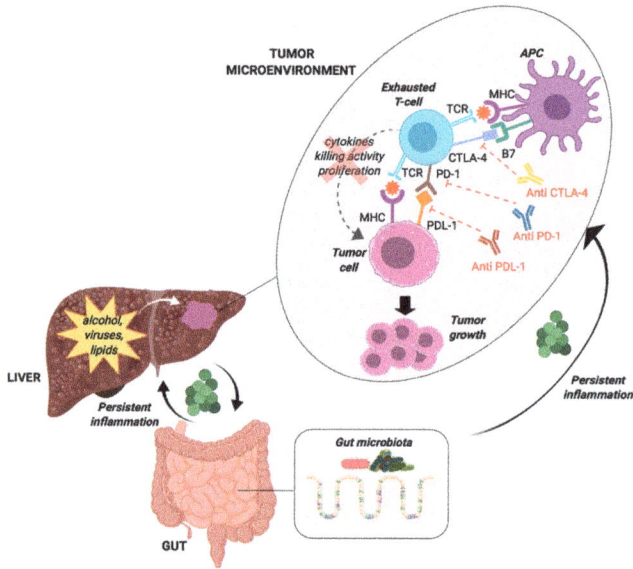

Figure 1. Mechanisms of immune system paralysis in patients with hepatocellular carcinoma (HCC). Inflammatory damage triggered from various factors (alcohol, hepatitis viruses, lipid accumulation, etc.) and from the gut microbiota is involved in the pathogenesis of HCC both directly and indirectly, through T-cells exhaustion. Exhausted T-cells express inhibitory receptor proteins and have a diminished capacity to produce cytokines, proliferate and kill cells. Indeed, antigen presenting cells (APC) and tumour cells express inhibitory molecules such as programmed cell death ligand 1 (PDL-1) and B7 that interact with the surface antigens programmed cell death 1 (PD-1) and cytotoxic T-lymphocyte-associated protein 4 (CTLA-4) on T-lymphocytes, inhibiting the downstream signalling caused by the T-cell receptor (TCR)/ major histocompatibility complex (MHC) interaction with tumour antigens thus favouring tumour growth.

4. Strategies for Patients Selection

4.1. T-Cell Exhaustion

To select patients who are likely to clinically benefit from immune checkpoint inhibitors and to establish optimal strategies, a better understanding of T-cell exhaustion in the HCC microenvironment is crucial. It is known that pro-inflammatory cytokines such as IL-1β, IL-6, IL-8, IL-12, IL-18, and IFN-γ have been shown to enhance T-cell response, while anti-inflammatory ones, e.g., TGF-β and IL-10, promote T-cell exhaustion and infiltration in tumours [53].

The exhaustion profile of tumour-infiltrating CD8+ T-cells in HCC patients needs to be characterised in detail regarding heterogeneous subsets of exhausted T-cells. A recent study suggested that combination blockade of immune checkpoint receptors additionally restores the functions of tumour-infiltrating T-cells from HCC patients, although the identification of HCC patients eligible for a combined approach remains unclear [23]. Interestingly, Kim et al. investigated the heterogeneity of exhausted tumour-infiltrating CD8+ T-cells and the relationship with clinical features of HCC, focussing on the different molecular and cellular characteristics of the tumour-infiltrating CD8+ T-cell subpopulations, distinguished by differential PD-1 expression. They demonstrated that HCCs with a discrete population of PD-1-high CD8+ T-cells might be more susceptible to combined immune checkpoint blockade–based therapies [54]. Recently, Feun et al. performed a correlative study to investigate the correlation between circulating biomarkers and response to pembrolizumab. They found that the mean plasma TGF-β levels in responders were lower than in non-responders, and that a TGF-β level ≥ 200 pg/mL was an indicator of poor response to treatment. Furthermore, low baseline plasma levels of TGF-β were significantly associated with improved overall survival and progression free survival after treatment with pembrolizumab. These results support a study showing that TGF-β signalling diminishes tumour response to PD-1/PD-L1 blockade by excluding CD8-positive effector T cells from the tumour parenchyma [55,56].

4.2. The Gut Microbiota

The gut microbiota is a well-known modulator of the immune response and is able to mediate the response to immunological treatments, as shown in patients with melanoma, renal tumour and non-small cell lung cancer [57–59]. Recent studies also provided evidence that the gut microbiota is linked with the pathogenesis of HCC. In animal models of HCC, the correlation between circulating levels of inflammatory mediators and lipopolysaccharides (LPS) and the number and size of tumours suggests an interplay between the outgrowth of harmful bacteria, such as Gram-negative ones, and tumourigenesis [60–62]. Administration of antibiotics and probiotics or blocking the expression of toll-like receptor-4 (the LPS receptor) not only inhibits tumour cells proliferation but also reduces the infiltration of macrophages and the expression of tumour necrosis factor (TNF)-alpha and IL-6 in the liver tissue (Figure 1) [62,63].

In cirrhotic patients with HCC and non-alcoholic fatty liver disease (NAFLD), an altered gut microbiota profile, consisting in the reduction of beneficial and anti-inflammatory bacteria such as *Akkermansia* and *Bifidobacterium* and the increase of harmful ones such as *Enterobacteriaceae* and *Ruminococcus*, was associated with a pronounced intestinal inflammation that, in association with the increased intestinal permeability typical of cirrhotic patients, led to a systemic inflammatory response [64]. In these patients, an increase in circulating activated monocytes and monocytic myeloid-derived suppressor cells expressing PD-1 and PD-L1 points out that the persistence of an inflammatory stimulation derived from the gut eventually results in the paralysis of the immune system, favouring the process of hepatocarcinogenesis [65].

Based on these data, it is conceivable that the gut microbiota is implicated in the pathogenesis of HCC through immunostimulating and immunosuppressive mechanisms. Consequently, it can be expected that the response to immunotherapy might be modulated by the microbiota composition of HCC patients. The identification of a microbial signature associated with the response to immunotherapy could allow implementing modulation strategies, such as faecal microbial transplantation or the use of prebiotics, probiotics or postbiotics to personalise the therapeutic approach and maximise its effectiveness. This is an exciting and important field of future research aimed at improving the results of immunotherapy in HCC patients.

5. Outcome of Current Studies on Immunotherapy in Patients with HCC

5.1. Efficacy

Checkpoint inhibitor-based treatments will be, in the near future, an important enrichment of the therapeutic armamentarium against HCC, and probably not only as first/second line approach to advanced stage tumours as a single or combined systemic therapy, but also in early and intermediate stages in combination with surgery and locoregional treatments. The first drugs of this class tested in HCC were tremelimumab, a CTLA-4 inhibitor, and nivolumab, a PD-1 antibody [48,51]. Until now, various other drugs have been tested: CTLA-4 antibodies ipilimumab, pembrolizumab, spartalizumab, tislelizumab and camrelizumab with a strong PD-1 inhibitory activity, and PD-L1 antibodies durvalumab, avelumab and atezolizumab [65]. The ongoing clinical trials exploring immune checkpoint inhibitors alone, or in combination with other drugs or with local therapies are summarised in Table 1.

Table 1. Ongoing clinical trials exploring immune checkpoint inhibitors: alone, in combination with other drugs or with local therapies.

NCT	Phase	Drug	Procedure	Line of Treatment	Primary End-Point	Estimated Study Completion Date	Company Conducting the Trial
NCT03298451	III	Tremelilumab (+Durvalumab) vs Sorafenib	-	1	OS	06/2021	Astra Zeneca
NCT02576509	III	Nivolumab vs Sorafenib	-	1	OS	07/2020	BMS
NCT03412773	III	Tislelizumab vs Sorafenib	-	1	OS	05/2022	BeiGene
NCT03062358	III	Pembrolizumab vs placebo	-	2	OS	01/2022	MSD
NCT02702401	III	Pembrolizumab vs placebo	-	2	OS, PFS	06/2020	MSD
NCT02702414	II	Pembrolizumab	-	1-2	ORR	05/2021	MSD
NCT02519348	II	Tremelilumab (+Durvalumab)	-	2	Safety, DLI	04/2021	MedImmune LLC
NCT03163992	II	Pembrolizumab	-	2	ORR	12/2020	Samsung Medical Center
NCT02658019	II	Pembrolizumab	-	>2	DCR, Safety	11/2020	Lynn Feun
NCT03389126	II	Avelumab	-	>2	ORR	03/2020	Seoul National University Hospital
NCT03419897	II	Tislelizumab	-	>2	ORR	09/2021	BeiGene
NCT03033446	II	Nivolumab	SIRT	Any	ORR	12/2019	National Cancer Centre, Singapore
NCT03572582	II	Nivolumab	TACE	1	ORR	09/2022	AIO-Studien-gGmbH
NCT03380130	II	Nivolumab	SIRT	1	Safety	10/2019	Clinica Universidad de Navarra, Universidad de Navarra
NCT03638141	II	Tremelilumab (+Durvalumab)	debTACE	1	ORR	11/2020	Sidney Kimmel Comprehensive Cancer Center at Johns Hopkins
NCT02821754	II	Tremelilumab	Local ablation	1	PFS	04/2021	National Cancer Institute (NCI)
NCT03630640	II	Nivolumab	Electroporation	1	RFS	09/2020	Assistance Publique—Hôpitaux de Paris
NCT03482102	II	Tremelilumab (+Durvalumab)	BRT	2	ORR	10/2025	Massachusetts General Hospital
NCT03316872	II	Pembrolizumab	SBRT	2	ORR	04/2022	University Health Network, Toronto
NCT01658878	IB/II	Nivolumab vs Sorafenib	-	1	ORR	12/2019	BMS
NCT02423343	IB/II	Nivolumab + Galunisertib	-	2	MTD, Safety	12/2019	Eli Lilly and Company
NCT01658878	IB/II	Nivolumab + Ipilimumab	-	>2	ORR	12/2019	BMS
NCT02940496	I/II	Pembrolizumab	TACE	2	Biomarkers	12/2019	M.D. Anderson Cancer Center
NCT03397654	IB	Pembrolizumab	TACE	Any	Safety	12/2020	Imperial College London
NCT02837029	I	Nivolumab	SIRT	Any	MTD	07/2020	Northwestern University
NCT03099564	I	Pembrolizumab	SIRT	1	PFS	01/2020	Autumn McRee, MD
NCT03143270	I	Nivolumab	debTACE	1	Safety	04/2020	Memorial Sloan Kettering Cancer Center
NCT03203304	I	Nivolumab/Ipilimumab	SBRT	1	Safety	08/2020	University of Chicago
NCT01853618	I	Tremelilumab	Local Ablation	1	Safety	12/2020	National Cancer Institute (NCI)

NCT, number of clinical trial (Clinicaltrials.gov); SIRT, selective intra-arterial radiation treatment; MTD, maximum tolerated dose; ORR, overall response rate; PFS, progression free survival; TACE, transarterial chemoembolisation; debTACE, drug eluting beads transarterial chemoembolisation; SBRT, stereotactic body radiation therapy; RFS, recurrence free survival.

5.2. Results of Monotherapy with Checkpoint Inhibitors in HCC

5.2.1. Tremelimumab

In a Phase II open-label, multicentre clinical trial, Sangro et al. treated with this drug, at a dose of 15 mg/kg IV every 90 days until tumour progression or severe toxicity, 21 patients with HCV-related HCC (57% with an advanced stage and 76% naïve to sorafenib) [51]. Objective response and disease control rate were 76.4% and 17.6%, respectively. Median time-to-progression (TTP) was 6.48 months (95% CI: 3.95–9.14). No toxicities requiring systemic steroid treatment were recorded. These initial results on safety profile and antitumour activity in patients with advanced HCC supported subsequent studies.

5.2.2. Nivolumab

A Phase I/II trial open-label, non-comparative, dose escalation and expansion trial (CheckMate 040) for the anti-PD-1 antibody nivolumab against HCC was completed [47]. In this trial, patients naïve to sorafenib, sorafenib intolerant or sorafenib refractory were treated with nivolumab at dose of 0.1–10 mg/kg once every two weeks (dose-escalating cohort) or at a dose of 3 mg/kg once every two weeks (expansion cohort). In a total of 262 patients, nivolumab 3 mg/kg showed, in the dose-expansion phase and in the dose-escalation phase, a manageable safety profile and an objective response rate of 20% (95% CI 15–26) vs. 15% (95% CI 6–28), and an overall survival at nine months of 74% vs. 66%. Despite these favourable data, preliminary results of a Phase III trial of nivolumab vs. sorafenib in first-line treatment (NCT02576509, CheckMate-459) showed that the study did not meet its primary end-point of overall survival [Hazard Ratio = 0.85 (95% CI: 0.72–1.02); $p = 0.0752$] [66,67].

5.2.3. Pembrolizumab

This anti-PD-1 antibody is being developed primarily as a second-line treatment. In a non-randomised, multicentre, open-label Phase II trial (KEYNOTE-224, NCT02702414), pembrolizumab (200 mg intravenously every three weeks for about two years or until disease progression, unacceptable toxicity, patient withdrawal, or investigator decision), was administered, at three-week intervals, to 104 Child–Pugh class A sorafenib-refractory or sorafenib-intolerant patients. The interim results of this trial showed an objective response in 18 patients (17%; 95% CI 11–26) and a median survival of 12.9 months. The best overall responses were one (1%) complete and 17 (16%) partial responses; meanwhile, 46 patients (44%) had stable disease, 34 (33%) had progressive disease, and 6 patients (6%) who did not have a post-baseline assessment were considered not to be assessable. Treatment-related adverse events occurred in 76 patients (73%), which were serious in 16 (15%). Immune-mediated hepatitis occurred in three (3%) patients, but there were no reported cases of viral flares. According to the trial, pembrolizumab was effective and tolerable in patients with advanced HCC who had previously been treated with sorafenib and that the drug might be a treatment option for these patients [68].

In the global Phase III trial allocating patients with advanced HCC who were previously treated with systemic therapy to pembrolizumab or best supportive care (KEYNOTE-240, NCT02702401), pembrolizumab improved overall survival (Hazard Ratio: 0.78; one sided $p = 0.0238$) and progression-free survival (Hazard Ratio: 0.78; one sided $p = 0.0209$), although these differences did not meet significance per the prespecified statistical plan [69]. In the second ongoing, double-blind, randomised Phase III trial (KEYNOTE-394, NCT03062358), pembrolizumab is being tested against placebo in Asian patients with advanced HCC who previously received systemic therapy, having as primary endpoints progression-free and overall survival.

5.2.4. Camrelizumab

A Phase II/III trials is ongoing with this anti-PD-1 antibody in China, enrolling patients with failure or intolerance to prior systemic treatment. Two-hundred seventeen patients were randomised (1:1) to camrelizumab 3 mg/kg iv for q2w ($n = 109$) or q3w ($n = 108$). Interim results showed an objective response rate of 13.8% (95% CI 9.5–19.1) (30/217) and six-month overall survival rate of 74.7%. Median time to response was two months (range: 1.7–6.2). Of the 30 responses, 22 were ongoing, and median duration of response was not reached. Disease control rate was 44.7% (95% CI 38.0–51.6), median time to progression was 2.6 months (95% CI 2.0–3.3), and median progression-free survival was 2.1 months (95% CI 2.0–3.2). The most common treatment-related adverse events were reactive cutaneous capillary endothelial proliferation (66.8%, all grade ≤ 2), increased aspartate (24.4%) or alanine aminotransferase (23.0%), and proteinuria (23.0%). Camrelizumab showed high objective response rate, durable response and acceptable toxicities in Chinese pretreated advanced HCC patients [70].

5.2.5. Tislelizumab

A Phase I trial recruiting including 61 patients with various solid cancers (including HCC) showed safety profile (NCT02407990). In a Phase III trial started in December 2017, patients with HCC were allocated to tislelizumab (200 mg iv for q3w) or sorafenib (400 mg bid) as first-line treatment. The primary endpoint is overall survival and this trial is designed to consider the non-inferiority of tislelizumab compared to sorafenib. The study opened to accrual in December 2017 and is currently recruiting patients; approximately 640 patients will be recruited from approximately 100 sites globally [71].

5.2.6. Durvalumab

A Phase I/II trial of durvalumab monotherapy for solid cancers, including a cohort of 30 patients with HCC, was completed (NCT01693562). A 10% objective response rate and a median survival time of 13.2 months were observed [72].

5.3. Combination of Two Immune Checkpoint Inhibitors

As previously reported, the anti-PD-1/PD-L1 and anti-CTLA-4 antibodies are expected to be promising agents in HCC immunotherapy not only as single agents, but also by combined with agents that have different targets. Therefore, several clinical trials evaluating the simultaneous blockade of multiple immune checkpoints are currently ongoing (Table 1). The high efficacy of combination therapy has already been shown in other solid tumours. For instance, the inhibition of the PD-1/PD-L1 pathway alone might not activate tumour immunity as expected if the required CD8+ T cells are not adequately represented in the tumour microenvironment. However, simultaneous inhibition of the B7-CTLA-4 pathway by an anti-CTLA-4 antibody may increase the number of activated CD8+ T cells in lymph nodes, followed by an increase in the number of activated CD8+ T cells infiltrating into tumoural tissues, thereby enhancing their antitumour effects. In addition, anti-CTLA-4 antibody therapy may be effective against regulatory T cells in the cancer immunosuppressive microenvironment.

5.3.1. Durvalumab plus Tremelimumab

This combination, tested in a Phase I/II study in 40 patients, reported a 15% objective response rate, demonstrating that combined therapy is more effective than durvalamab alone [73]. This combination also showed manageable safety profile. Currently, a Phase III is ongoing to compare different regimens as a first-line treatment; the four arms consist of durvalumab monotherapy, two types of durvalumab plus tremelimumab combination therapies (regimens 1 and 2) and sorafenib monotherapy (NCT03298451) [74].

5.3.2. Nivolumab plus Ipilimumab

A sub-cohort of the CHECKMATE-040 study is evaluating the combination of nivolumab plus ipilimumab in sorafenib-treated patients (NCT01658878). Preliminary results showed an objective response rate of 31%, with a median duration of response of 17 months and a median overall survival that varies between 12 and 23 months according to the different treatment schedules applied in the three arms of the study [75]. There are two other Phase II clinical studies evaluating this combination regimen: one of these studies is comparing, in USA, nivolumab monotherapy with nivolumab plus ipilimumab (NCT03222076), while the second study is evaluating, in Taiwan, the combination therapy alone (NCT03510871).

5.4. Combination of Immune Checkpoint Inhibitors with Molecular-Targeted Agents

The therapeutic outcomes of immune checkpoint inhibitors with molecular target therapies has demonstrated to be superior to those of monotherapy in other solid tumours. Therefore, even for HCC, therapies involving an immune checkpoint inhibitor plus a molecular targeted agent was suggested as a promising strategy in recent years. In particular, interstitial cells (Kupffer cells, dendritic cells, liver endothelial cells, and liver stellate cells) and immunosuppressive cytokines (e.g., IL-10 or TGF-β) may contribute to the immunosuppressive environment of HCC, and the PD-1/PD-L1 pathway plays an important role in the development of the immunosuppressive microenvironment in HCC. Combining a molecular targeted agent and an immune checkpoint inhibitor is expected to improve this immunosuppressive microenvironment.

5.4.1. Atezolizumab plus Bevacizumab

A Phase III randomised controlled trial of atezolizumab plus bevacizumab versus sorafenib as a first-line treatment was started and is currently ongoing to confirm the results of the Phase Ib trial [76]. Preliminary results of this Phase III study (NCT03434379) have recently been released, and the combination of atezolizumab (1200 mg on day 1 of each 21-day cycle, intravenously) plus bevacizumab (15 mg/kg on day 1 of each 21-day cycle, intravenously) met both co-primary end-points of improvement in overall and progression-free survivals as compared with sorafenib (400 mg twice per day, on days 1–21 of each 21-day cycle), although survival figures have not yet been communicated [77].

5.4.2. Pembrolizumab plus Lenvatinib

A Phase I trial for this therapy is also underway in patients with HCC. According to preliminary results, 46% of patients had either partial response or stable disease in the mRECIST criteria among the patients who had been evaluated [78].

5.4.3. Other Combinations

Currently there are several early stage clinical studies considering various combination of PD-1 inhibitors and targeted agents for HCC, without available data for the moment. They include: nivolumab plus lenvatinib (NCT03418922), nivolumab plus cabozantinib (NCT03299946), nivolumab plus bevacizumab (NCT03382886), pembrolizumab plus regorafenib (NCT03347292), pembrolizumab plus sorafenib (NCT03211416), and PDR001 (spartalizumab) plus sorafenib (NCT02988440).

5.5. Immune Checkpoint Inhibitors as Neo-Adjuvant or Adjuvant Therapy, or in Combination with Local Treatments

Despite significant improvements in the treatment of early HCC, curative therapies remain associated with high recurrence rates ($\approx$70% at 5 years), and adjuvant therapies able to curb this figure currently represent an unmet need. In both settings of surgery and locoregional treatment, treatment-induced liberation of tumour-associated antigens has previously been demonstrated, thus providing a strong rationale for a combined treatment with immunostimulating agents, as previously shown for other solid tumours [79]. Thus, several studies have been recently initiated in HCC in order to evaluate the safety and efficacy of adjuvant treatments in patients who are at high risk of recurrence after curative hepatic resection or ablation. As an example, a study is currently recruiting patients to test nivolumab against placebo in the adjuvant setting following resection or local ablation (NCT03383458). Similarly, the MK-3475-937/KEYNOTE-937 trial with pembrolizumab is also undergoing in the neoadjuvant setting (NCT03867084). Phase II trials are also evaluating tremelimumab in a similar setting.

Similar to ablation, chemoembolisation has been shown to be associated with enhanced tumour-associated antigens spread together with an increase of vascular endothelial growth factor. In this regard, at least one study on transarterial chemoembolisation plus nivolumab is undergoing (NCT03143270). In this setting, a more complex approach has recently been proposed by the combination of chemoembolisation with both an immune checkpoint inhibitor and a molecular-target agent with an anti-VEGF effect: the LEAP-01 study (combination of chemoembolisation with pembrolizumab and lenvatinib, NCT03713593) and the EMERALD-1 study (combination of chemoembolisation with durvalumab and bevacizumab, NCT03778957). Lastly, transarterial radioembolisation promotes radiation-induced tumour damage similar to that induced by stereotactic radiation therapy: several early studies (Phase I and II) by combining this emergent locoregional approaches to immune checkpoint inhibitors are going to start recruitment (NCT02837029, NCT03033446, NCT03099564, and NCT03380130).

6. Liver Involvement in Immune-Related-Adverse Events

Compared to tyrosine-kinase inhibitors as sorafenib and lenvatinib, immunotherapy has significant differences in terms of both toxicity and response. Checkpoint inhibitors are generally better tolerated than tyrosine-kinase inhibitors, although some patients may rarely experience serious, immune-related adverse events involving different organs and systems, such as endocrine glands, the skin, the gastrointestinal tract, the brain and the liver itself [80]. Acute hepatitis is rare, occurring in 4–9% of patients considering all grades of liver injury, and in 3.5% for grade 3 or 4 hepatitis [81,82]. No predictors of checkpoint inhibitors toxicity and immune-related adverse events have been clearly demonstrated. However, the presence of baseline sarcopenia, a family history of autoimmune diseases, tumour infiltration and liver metastases, previous viral infections (such as HIV or hepatitis) and the concomitant use of drugs with autoimmune mechanism of toxicity (anti-arrhythmics, antibiotics, anticonvulsants or antipsychotics) have been suggested to be potential predictors of severe treatment-related toxicity [83,84]. Histological features of the immune-related hepatitis are still little known, due to its rarity and the uncommon utilisation of liver biopsy. A recent French study showed a different histological pattern between patients receiving anti PD-1/PD-L1 and anti-CTLA-4 agents. Anti-CTLA-4-associated injury is typically a granulomatous hepatitis with severe globular necrotic and inflammatory activity and lymphocyte T CD8 cells activation, while the histological pattern of liver damage associated with use of anti-PD-1/PD-L1 agents is more heterogeneous, showing a spotty and confluent necrosis and mild-to-moderate lobular and periportal inflammatory activity, involving both CD4 and CD8 lymphocytes in equal proportions [85]. Finally, three cases of checkpoint inhibitors-induced hepatotoxicity were characterised by biliary injury, and in one patient receiving pembrolizumab for metastatic melanoma a vanishing bile ducts syndrome has been described [86]. According to guidelines, a grade 2 transaminase or bilirubin elevation should prompt the interruption of checkpoint inhibitors therapy and transaminase/bilirubin should be checked twice weekly [80]. A grade 2 elevation lasting for more than two weeks, in the absence of any other cause of liver damage, should be approached with steroids [1 mg/kg/day (methyl)prednisolone or equivalent]. Upon improvement, immunotherapy can be resumed after steroids tapering. Conversely, in the case of worsening, steroids should be increased to 2 mg/kg/day, with permanent discontinuation of checkpoint inhibitors. In the case of grade 3 or 4 transaminase/bilirubin increase, checkpoint inhibitors should be permanently discontinued, steroids must be started (1–2 mg/kg/day) and, if needed, mycophenolate mofetil should be added. In steroid and mycophenolate refractory cases consultation with the hepatologist and liver biopsy are recommended [80].

7. Assessment of Treatment Response: The iRECIST Criteria

The response of HCC to immunotherapy appears as low as with tyrosine kinase inhibitors in terms of objective response rates but with longer durability, a finding that appears completely new for this tumour. In particular, the concept of radiological response likely needs a different approach from the one we have used to define treatment response with tyrosine kinase inhibitors: the RECIST and mRECIST criteria will have to be paralleled by a new system specifically designed for these drugs (i.e., the immune-related response criteria, iRECIST) [87].

Indeed, the peculiar tumour response observed with immunotherapy raised questions about the appropriateness of the conventional classification of tumour response, i.e., objective response and disease progression. The RECIST working group has recently developed consensus guidelines for the use of a common language in cancer immunotherapy trials, to ensure consistent design and data collection [87]. The need of a different modality to consider radiological response with checkpoint inhibitors has been raised in early trials in melanoma, when investigators described for the first time a unique response pattern, termed *pseudoprogression*: the disease behaviour met the criteria for disease progression based on RECIST criteria but later patients had marked and durable responses. Thereafter, following a long process of revision of different trials, the major innovation of iRECIST is the concept of resetting the bar if RECIST progression is followed by tumour shrinkage at the subsequent assessment [87]. This evolutive pattern has been defined "unconfirmed progression" (iUPD): if progression is not confirmed, but the tumour shrinks (compared with baseline), which meets the criteria of complete response, partial response or stable disease, then the bar is reset so that iUPD needs to occur again (compared with nadir values) and then be confirmed (by further growth) at the next assessment for confirmed progression (iCPD) to be assigned. Other aspects of lesion assessment are unique to iRECIST. If a new lesion is identified (thus meeting the criteria for iUPD) and the patient is clinically stable, treatment should be continued. Progressive disease is confirmed (iCPD) in the new lesion category if the next imaging assessment, done at 4–8 weeks after iUPD, confirms additional new lesions or a further increase in new lesion size from iUPD (sum of measures increase in new lesion target ≥5 mm, any increase for new lesion non-target).

8. Rationale Underlying the Use of the ITA.LI.CA Database to Assess Real-Life Applicability of Checkpoint Inhibitors

With the intent to explore the potential use in clinical practice of two checkpoint inhibitors, nivolumab and pembrolizumab, in HCC patients, we developed some scenarios applying to the Italian Liver Cancer (ITA.LI.CA) database the inclusion criteria adopted by the checkpoint inhibitors studies [48,68]. The ITA.LI.CA database is a large, multi-centre database including patients with newly diagnosed or recurrent HCC approaches managed in a large number of Italian centres with different levels of expertise (secondary and tertiary referral centres) [88]. This database, due to its heterogeneity in terms of tumour stage, underlying liver disease severity, and therapeutic approach, provides a reliable insight into the characteristics of HCC in a Western population, and allows predicting figures of the potential utilisation of these drugs in real-life clinical practice [88,89].

To define the "real world" scenario where these drugs could be used either as up-front or in second-line treatment, we used the selection criteria reported by El-Khoueiry et al. for nivolumab and those proposed by Zhu et al. for pembrolizumab [48,68]. This analysis was mainly aimed at providing the clinicians with a tentative foresight of the proportion of eligible patients and field of applicability of the checkpoint inhibitors in the real-life clinical practice.

8.1. First-Line Scenario

To construct the first-line scenarios, we adopted three patient-removal steps: (i) firstly, removal based on the period and type (naive vs. recurrent) of HCC diagnosis; (ii) secondly, removal based on missing data for at least one of the parameters used to identify potential candidates to checkpoint inhibitors; and (iii) lastly, removal based on the selection criteria used for the investigated drug.

As far as nivolumab is concerned, we identified 27 different selection parameters to build the first-line scenario and specifically we considered not amenable to nivolumab patients with one or more of the 27 conditions reported in Table 2, which also reports the number of patients excluded by each step. Thus, we firstly removed patients with recurrent HCC ($n = 4453$) and those with a tumour diagnosis before 2008 ($n = 3144$), and then those in whom data regarding one or more of the 27 selected criteria were missing ($n = 1403$). The remaining 2483 patients with a first diagnosis of HCC over the period 2008–2016 formed the cohort where we tested the amenability to immunotherapy. Among them, 525 patients (21.1%) met the criteria for nivolumab treatment. According to the year of HCC diagnosis, the proportion of potentially treatable cases ranged from 18.3% to 30.3% (Figure 2A), with a median eligibility rate of 20.1% (19.9–20.3% interquartile range).

Considering the eligibility to first-line pembrolizumab, we adopted 30 selection parameters to build the first-line scenario (Table 2). The first two steps were identical to the nivolumab scenario: of the 2483 patients selected by these steps only 268 (10.8%) patients were considered eligible to receive pembrolizumab. Over time, the proportion of patients eligible to pembrolizumab ranged from 9.4% to 21.2% (Figure 2B), with a median eligibility rate of 10.6% (10.2–11.1% interquartile range).

Table 2. Potential use of nivolumab and pembrolizumab as first-line therapy in HCC patients according to the ITA.LI.CA database.

ITA.LI.CA Database	Number of HCCs = 11,483 (including recurrences)	
(A) First-step removal	1. HCC diagnosis before 01/01/2008 = 3144	
	2. HCC recurrence = 4453	
	Number of patients = 3886 (01/01/2008-31/12/2016)	
(B) Second step removal	Missing data = 1403	
	Examined population = 2483 (100.0%)	
	Nivolumab	Pembrolizumab
(C) Third step removal	1. Child-Pugh > B7 = 601 2. ECOG PST > 1 = 343 3. ECOG PST = 1, BCLC C, resected or RFA/PEI, MC-IN = 86 4. BCLC 0-A resected = 99 5. BCLC 0-A RFA/PEI = 238 6. BCLC B resected = 55 7. Transplantation = 55 8. TACE with CR/PR/SD = 577 9. PBC = 18 10. Autoimmune hepatitis = 5 11. Active HBV + HCV = 12 12. Active HBV + HDV = 12 13. Autoimmune diseases = 34 14. Active alcohol abuse = 323 15. Brain metastases = 2 16. Story of encephalopathy = 155 17. Severe ascites = 380 18. Malignancies previous 3 years = 27 19. HIV = 22 20. Leucocytes < 2000/mcL = 63 21. PLT < 60,000/mcL = 299 22. Hb < 9 g/dL = 107 23. GFR < 40 mL/min = 147 24. Total bilirubin > 3.0 mg/dL = 214 25. AST/ALT > 5× = 123 26. Albumin < 2.8 g/dL = 226 27. INR > 2.3 = 34	1. Child-Pugh > B7 = 601 2. ECOG PST > 1 = 343 3. ECOG PST = 1, BCLC C, resected or RFA/PEI, MC-IN = 86 4. BCLC 0-A resected = 99 5. BCLC 0-A RFA/PEI = 238 6. BCLC B resected = 55 7. Transplantation = 55 8. TACE with CR/PR/SD = 577 9. PBC = 18 10. Autoimmune hepatitis = 5 11. Active HBV = 95 12. Double infection HBV/HCV = 36 13. Autoimmune diseases = 34 14. Active alcohol abuse = 323 15. Brain metastases = 2 16. Story of encephalopathy = 155 17. Clinically apparent ascites = 1009 18. Malignancies previous 5 years = 43 19. HIV = 22 20. Leucocytes < 1200/mcL = 23 21. PLT < 60,000/mcL = 299 22. Hb < 8 g/dL = 33 23. sCr > 1.5 mg/dL = 121 24. GFR < 60 mL/min if sCr < 1.5 mg/dL = 502 25. Total bilirubin > 2.0 mg/dL = 440 26. AST/ALT > 5× = 123 27. Albumin < 3.0 mg/dL = 414 28. INR > 1.5× = 60 29. Variceal bleeding < 6 months = 103 30. Main branch PVT/IVC thrombosis = 187
	Final population = 525/2483 (21.1%)	Final population = 268/2483 (10.8%)

Abbreviations: ITA.LI.CA, Italian Liver Cancer; HCC, hepatocellular cancer; ECOG, Eastern Cooperative Oncology Group; PST, performance status; BCLC, Barcelona Clinic Liver Cancer; RFA, radio-frequency ablation; PEI, percutaneous ethanol injection; MC, Milan Criteria; TACE, trans-arterial chemoembolisation; CR, complete response; PR, partial response; SD, stable disease; PBC, primitive biliary cholangitis; HBV, hepatitis B virus; HCV, hepatitis C virus; HDV, hepatitis D virus; HIV, human immunodeficiency virus; PLT, platelets; Hb, hemoglobin; GFR, glomerular filtration rate; sCr, serum creatinine; AST, aspartate transaminases; ALT, alanine transaminases; INR, international normalised ratio; PVT, portal vein thrombosis; IVC, inferior vena cava.

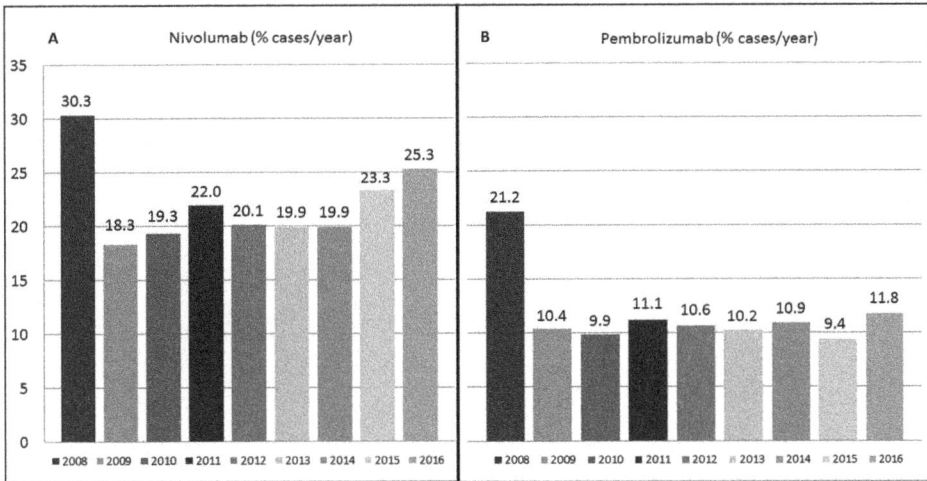

Figure 2. Proportion of patients within the Italian Liver Cancer cohort meeting the criteria for: first-line nivolumab treatment (**A**); and second-line nivolumab treatment (**B**).

8.2. Second-Line Scenario

To build second-line scenarios, we followed the removal steps reported in Table 2. First, we removed the cases diagnoses before 2008 ($n = 3144$) and those with a naive HCC as well as those with ≥2 two recurrences ($n = 6485$). The removal of the 1413 cases with missing data selected 441 patients for nivolumab and 266 patients for pembrolizumab with a first recurrence of HCC after any type of first-line treatment during the period 2008–2016. According to the 27 criteria for nivolumab, only 24 patients (5.4%) resulted eligible for second-line treatment. The proportion of potentially treatable patients ranged from 0% to 10% across the years, with a median of 4.8% (2.9–6.4 interquartile range) (Figure 3A). Likewise, after removing patients with missing data for the 30 variables ($n = 1588$) used for pembrolizumab, only 266 patients with HCC recurrence after any first-line treatment in the period 2008–2016 were selected. Of them, 26 (9.8%) were considered eligible for pembrolizumab treatment, and their proportion ranged over time from 0% to 12.9%, with a median eligibility rate of 8.0% (6.5–10.3 interquartile range) (Figure 3B).

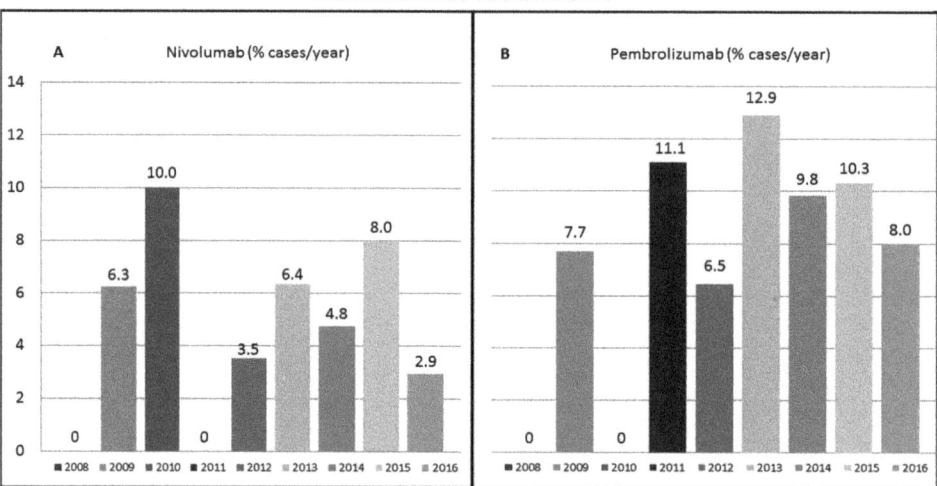

Figure 3. Proportion of patients within the Italian Liver Cancer cohort meeting the criteria for: first-line pembrolizumab treatment (**A**); or second-line pembrolizumab treatment (**B**).

9. Conclusions

All the described encouraging results are enriching the scenario of HCC treatment, with a trend to expand the use of immune checkpoint inhibitors, alone or in combination with other molecules, for advanced stage HCCs, as adjuvant therapy after curative approaches in patients with a high risk of disease recurrence, or in combination with transcatheter arterial chemoembolisation in those carrying an intermediate stage HCC.

Nevertheless, despite the expectancy related to ongoing studies, the application of immune checkpoint inhibitors in patients with HCC may still not fulfil the unmet needs of these patients, since as many as 30–40% of them do not respond to these agents, and we have shown by analysing the large ITA.LI.CA database—despite the limitations related to the retrospective nature of the analysis—that in the real-life clinical practice the eligibility rate to immune checkpoint inhibitors is approximately 10–20% in the first-line, and less than 10% in the second-line treatment. The mechanisms of primary resistance to immunotherapy are largely unknown, but combination strategies may overcome this limit, considering that HCC-induced immune tolerance in the setting of a tolerogenic liver environment and chronic inflammation is associated with multiple immunosuppressive mechanisms. Thus, dual or triple combinations of immune targeting agents, associated with inhibitory checkpoint blockage as a backbone of therapy, might be the most promising strategies. Moreover, in this context, it is necessary to identify easily accessible biomarkers to predict tumour response and help us in selecting optimal candidates to immunotherapy. How we will select and monitor these therapies, and use them safely in different groups of patients is not yet clear, as the field is limited by the lack of either tissue or circulating biomarkers to guide clinical decision-making. Additional studies are warranted to identify how many patients (among the whole HCC population, and also among those who undergo this therapy) will actually benefit from immune checkpoint inhibitors treatment, and to assess its cost-effectiveness in this complex disease.

Funding: This research received no external funding.

Conflicts of Interest: A.A., M.B., M.G. (Martina Gambato), M.G. (Maria Guarino), Q.L., G.B.L.S., F.M. (Fabio Melandro), F.R.P., M.R., F.P.R., and R.S.: The authors declare no conflict of interest. E.G.G.: Bayer, Bristol-Myers Squibb (advisory board, lecturing fees). M.V.: Gilead Sciences, Merck Sharp & Dohme (MSD), Fujirebio (speaking and teaching). M.I.: Bayer: Gilead Science, Janssen, BTG Corporate, Abbvie (speaking and teaching); BTG Corporate (consultancy). F.M. (Filomena Morisco): Abbvie, Bristol-Myers Squibb, Gilead Science, Janssen, Merck Sharp & Dohme (research grants, lecturing fees, advisory boards, scientific consultancy). A.V.: Bayer (advisory board). F.T.: Bayer, Alfasigma, Sirtex, Bristol-Myers Squibb (advisory board, consulting, conferences, research grants).

References

1. Bray, F.; Ferlay, J.; Soerjomataram, I.; Siegel, R.L.; Torre, L.A.; Jemal, A. Global cancer statistics 2018: GLOBOCAN estimates of incidence and mortality worldwide for 36 cancers in 185 countries. *CA Cancer J. Clin.* **2018**, *68*, 394–424. [CrossRef] [PubMed]
2. White, D.L.; Kanwal, F.; Serag, H.B.E. Association between nonalcoholic fatty liver disease and risk for hepatocellular cancer, based on systematic review. *Clin. Gastroenterol. Hepatol.* **2012**, *10*, 1342–1359. [CrossRef] [PubMed]
3. Kanwal, F.; Kramer, J.R.; Mapakshi, S.; Natarajan, Y.; Chayanupatkul, M.; Richardson, P.A.; Li, L.; Desiderio, R.; Thrift, A.P.; Asch, S.M.; et al. Risk of hepatocellular cancer in patients with non-alcoholic fatty liver disease. *Gastroenterology* **2018**, *155*, 1828–1837. [CrossRef] [PubMed]
4. Hanahan, D.; Weinberg, R.A. Chronic inflammation due to the various underlying etiologies is a mainstay of HCC development. *Cell* **2011**, *144*, 646–674. [CrossRef] [PubMed]
5. Liu, P.; Kimmoun, E.; Legrand, A.; Sauvanet, A.; Degott, C.; Lardeux, B.; Bernuau, D. Activation of NF-kappa B, AP-1 and STAT transcription factors is a frequent and early event in human hepatocellular carcinomas. *J. Hepatol.* **2002**, *37*, 63–71. [CrossRef]
6. He, G.; Yu, G.Y.; Temkin, V.; Ogata, H.; Kuntzen, C.; Sakurai, T.; Sieghart, W.; Peck-Radosavljevic, M.; Leffert, H.L.; Karin, M. Hepatocyte IKKβ/NF-κB inhibits tumor promotion and progression by preventing oxidative stress-driven STAT3 activation. *Cancer Cell* **2010**, *17*, 286–297. [CrossRef]
7. Kanda, T.; Goto, T.; Hirotsu, Y.; Moriyama, M.; Omata, M. Molecular mechanisms driving progression of liver cirrhosis towards hepatocellular carcinoma in chronic hepatitis B and C infections: A review. *Int. J. Mol. Sci.* **2019**, *20*, 1358. [CrossRef]
8. Bitzer, M.; Horger, M.; Giannini, E.G.; Ganten, T.M.; Wörns, M.A.; Siveke, J.T.; Dollinger, M.M.; Gerken, G.; Scheulen, M.E.; Wege, H.; et al. Resminostat plus sorafenib as second-line therapy of advanced hepatocellular carcinoma—The SHELTER study. *J. Hepatol.* **2016**, *65*, 280–288. [CrossRef]
9. Klener, P., Jr.; Otahal, P.; Lateckova, L.; Klener, P. Immunotherapy approaches in cancer treatment. *Curr. Pharm. Biotechnol.* **2015**, *16*, 771–781. [CrossRef]
10. Ventola, C.L. Cancer Immunotherapy, Part 1: Current Strategies and Agents. *Pharm. Ther.* **2017**, *42*, 375–383.
11. Schreiber, R.D.; Old, L.J.; Smyth, M.J. Cancer immunoediting: Integrating immunity's roles in cancer suppression and promotion. *Science* **2011**, *331*, 1565–1570. [CrossRef] [PubMed]
12. Riley, R.S.; June, C.H.; Langer, R.; Mitchell, M.J. Delivery technologies for cancer immunotherapy. *Nat. Rev. Drug Discov.* **2019**, *18*, 175–195. [CrossRef] [PubMed]
13. Alatrash, G.; Jakher, H.; Stafford, P.D.; Mittendorf, E.A. Cancer immunotherapies, their safety and toxicity. *Expert Opin. Drug Saf.* **2013**, *12*, 631–645. [CrossRef] [PubMed]
14. Yang, Y. Cancer immunotherapy: Harnessing the immune system to battle cancer. *J. Clin. Investig.* **2015**, *125*, 3335–3337. [CrossRef]
15. Dhanasekaran, R.; Nault, J.C.; Roberts, L.R.; Zucman-Rossi, J. Genomic medicine and implications for hepatocellular carcinoma prevention and therapy. *Gastroenterology* **2019**, *156*, 492–509. [CrossRef]
16. Llovet, J.M.; Montal, R.; Sia, D.; Finn, R.S. Molecular therapies and precision medicine for hepatocellular carcinoma. *Nat. Rev. Clin. Oncol.* **2018**, *15*, 599–616. [CrossRef]
17. Zucman-Rossi, J.; Villanueva, A.; Nault, J.C.; Llovet, J.M. Genetic landscape and biomarkers of hepatocellular carcinoma. *Gastroenterology* **2015**, *149*, 1226–1239. [CrossRef]
18. Nault, J.C.; Villanueva, A. Intratumor molecular and phenotypic diversity in hepatocellular carcinoma. *Clin. Cancer Res.* **2015**, *21*, 1786–1788. [CrossRef]
19. Llovet, J.M.; Ricci, S.; Mazzaferro, V.; Hilgard, P.; Gane, E.; Blanc, J.F.; de Oliveira, A.C.; Santoro, A.; Raoul, J.L.; Forner, A.; et al. Sorafenib in advanced hepatocellular carcinoma. *N. Engl. J. Med.* **2008**, *359*, 378–390. [CrossRef]

20. Kudo, M.; Finn, R.S.; Qin, S.; Han, K.H.; Ikeda, K.; Piscaglia, F.; Baron, A.; Park, J.W.; Han, G.; Jassem, J.; et al. Lenvatinib versus sorafenib in first-line treatment of patients with unresectable hepatocellular carcinoma: A randomised phase 3 non-inferiority trial. *Lancet* **2018**, *391*, 1163–1173. [CrossRef]
21. Bruix, J.; Qin, S.; Merle, P.; Granito, A.; Huang, Y.H.; Bodoky, G.; Pracht, M.; Yokosuka, O.; Rosmorduc, O.; Breder, V.; et al. RESORCE investigators. Regorafenib for patients with hepatocellular carcinoma who progressed on sorafenib treatment (RESORCE): A randomised, double-blind, placebo-controlled, phase 3 trial. *Lancet* **2017**, *389*, 56–66. [CrossRef]
22. Abou-Alfa, G.K.; Meyer, T.; Cheng, A.L.; El-Khoueiry, A.B.; Rimassa, L.; Ryoo, B.Y.; Cicin, I.; Merle, P.; Chen, Y.; Park, J.W.; et al. Cabozantinib in patients with advanced and progressing hepatocellular carcinoma. *N. Engl. J. Med.* **2018**, *379*, 54–63. [CrossRef] [PubMed]
23. Zhou, G.; Sprengers, D.; Boor, P.P.C.; Doukas, M.; Schutz, H.; Mancham, S.; Pedroza-Gonzalez, A.; Polak, W.G.; de Jonge, J.; Gaspersz, M.; et al. Antibodies against immune checkpoint molecules restore functions of tumor-infiltrating T cells in hepatocellular carcinomas. *Gastroenterology* **2017**, *153*, 1107–1119. [CrossRef] [PubMed]
24. Shi, F.; Shi, M.; Zeng, Z.; Qi, R.Z.; Liu, Z.W.; Zhang, J.Y.; Yang, Y.P.; Tien, P.; Wang, F.S. PD-1 and PD-L1 upregulation promotes CD8$^+$ Tcell apoptosis and postoperative recurrence in hepatocellular carcinoma patients. *Int. J. Cancer* **2011**, *128*, 887–896. [CrossRef]
25. Kuang, D.M.; Zhao, Q.; Peng, C.; Xu, J.; Zhang, J.P.; Wu, C.; Zheng, L. Activated monocytes in peritumoral stroma of hepatocellular carcinoma foster immune privilege and disease progression through PD-L1. *J. Exp. Med.* **2009**, *206*, 1327–1337. [CrossRef] [PubMed]
26. Han, Y.; Chen, Z.; Yang, Y.; Jiang, Z.; Gu, Y.; Liu, Y.; Lin, C.; Pan, Z.; Yu, Y.; Jiang, M.; et al. Human CD14$^+$ CTLA-4$^+$ regulatory dendritic cells suppress T-cell response by cytotoxic T-lymphocyte antigen-4-dependent IL-10 and indoleamine-2,3-dioxygenase production in hepatocellular carcinoma. *Hepatology* **2014**, *59*, 567–579. [CrossRef] [PubMed]
27. Kalathil, S.; Lugade, A.A.; Miller, A.; Iyer, R.; Thanavala, Y. Higher frequencies of GARP$^+$ CTLA-4$^+$ Foxp3$^+$ T regulatory cells and myeloid-derived suppressor cells in hepatocellular carcinoma patients are associated with impaired T-cell functionality. *Cancer Res.* **2013**, *73*, 2435–2444. [CrossRef]
28. Wu, X.; Gu, Z.; Chen, Y.; Chen, B.; Chen, W.; Weng, L.; Liu, X. Application of PD-1 blockade in cancer immunotherapy. *Comput. Struct. Biotechnol. J.* **2019**, *17*, 661–674. [CrossRef]
29. Wu, K.; Kryczek, I.; Chen, L.; Zou, W.; Welling, T.H. Kupffer cell suppression of CD8$^+$ T cells in human hepatocellular carcinoma is mediated by B7–H1/programmed death-1 interactions. *Cancer Res.* **2009**, *69*, 8067–8075. [CrossRef]
30. Wang, B.J.; Bao, J.J.; Wang, J.Z.; Wang, Y.; Jiang, M.; Xing, M.Y.; Zhang, W.G.; Qi, J.Y.; Roggendorf, M.; Lu, M.J.; et al. Immunostaining of PD-1/PD-Ls in liver tissues of patients with hepatitis and hepatocellular carcinoma. *World J. Gastroenterol.* **2011**, *17*, 3322–3329. [CrossRef]
31. Gao, Q.; Wang, X.Y.; Qiu, S.J.; Yamato, I.; Sho, M.; Nakajima, Y.; Zhou, J.; Li, B.Z.; Shi, Y.H.; Xiao, Y.S.; et al. Overexpression of PD-L1 significantly associates with tumor aggressiveness and postoperative recurrence in human hepatocellular carcinoma. *Clin. Cancer Res.* **2009**, *15*, 971–979. [CrossRef] [PubMed]
32. Unitt, E.; Marshall, A.; Gelson, W.; Rushbrook, S.M.; Davies, S.; Vowler, S.L.; Morris, L.S.; Coleman, N.; Alexander, G.J. Tumour lymphocytic infiltrate and recurrence of hepatocellular carcinoma following liver transplantation. *J. Hepatol.* **2006**, *45*, 246–253. [CrossRef] [PubMed]
33. Wada, Y.; Nakashima, O.; Kutami, R.; Yamamoto, O.; Kojiro, M. Clinicopathological study on hepatocellular carcinoma with lymphocytic infiltration. *Hepatology* **1998**, *27*, 407–414. [CrossRef] [PubMed]
34. Flecken, T.; Schmidt, N.; Hild, S.; Gostick, E.; Drognitz, O.; Zeiser, R.; Schemmer, P.; Bruns, H.; Eiermann, T.; Price, D.A.; et al. Immunodominance and functional alterations of tumor-associated antigen-specific CD8$^+$ T-cell responses in hepatocellular carcinoma. *Hepatology* **2014**, *59*, 1415–1426. [CrossRef] [PubMed]
35. Kurebayashi, Y.; Ojima, H.; Tsujikawa, H.; Kubota, N.; Maehara, J.; Abe, Y.; Kitago, M.; Shinoda, M.; Kitagawa, Y.; Sakamoto, M. Landscape of immune microenvironment in hepatocellular carcinoma and its additional impact on histological and molecular classification. *Hepatology* **2018**, *68*, 1025–1041. [CrossRef] [PubMed]

36. Zheng, C.; Zheng, L.; Yoo, J.K.; Guo, H.; Zhang, Y.; Guo, X.; Kang, B.; Hu, R.; Huang, J.Y.; Zhang, Q.; et al. Landscape of infiltrating T cells in liver cancer revealed by single-cell sequencing. *Cell* **2017**, *169*, 1342–1356. [CrossRef] [PubMed]
37. Topalian, S.L.; Hodi, F.S.; Brahmer, J.R.; Gettinger, S.N.; Smith, D.C.; McDermott, D.F.; Powderly, J.D.; Carvajal, R.D.; Sosman, J.A.; Atkins, M.B.; et al. Safety, activity, and immune correlates of anti-PD-1 antibody in cancer. *N. Engl. J. Med.* **2012**, *366*, 2443–2454. [CrossRef]
38. Bald, T.; Landsberg, J.; Lopez-Ramos, D.; Renn, M.; Glodde, N.; Jansen, P.; Gaffal, E.; Steitz, J.; Tolba, R.; Kalinke, U.; et al. Immune cell-poor melanomas benefit from PD-1 blockade after targeted type I IFN activation. *Cancer Discov.* **2014**, *4*, 674–687. [CrossRef]
39. Postow, M.A.; Callahan, M.K.; Wolchok, J.D. Immune checkpoint blockade in cancer therapy. *J. Clin. Oncol.* **2015**, *33*, 1974–1982. [CrossRef]
40. O'Donnell, J.S.; Long, G.V.; Scolyer, R.A.; Teng, M.W.; Smyth, M.J. Resistance to PD1/PDL1 checkpoint inhibition. *Cancer Treat. Rev.* **2017**, *52*, 71–81. [CrossRef]
41. Zehir, A.; Benayed, R.; Shah, R.H.; Syed, A.; Middha, S.; Kim, H.R.; Srinivasan, P.; Gao, J.; Chakravarty, D.; Devlin, S.M.; et al. Mutational landscape of metastatic cancer revealed from prospective clinical sequencing of 10,000 patients. *Nat. Med.* **2017**, *23*, 703–713. [CrossRef] [PubMed]
42. Chalmers, Z.R.; Connelly, C.F.; Fabrizio, D.; Gay, L.; Ali, S.M.; Ennis, R.; Schrock, A.; Campbell, B.; Shlien, A.; Chmielecki, J.; et al. Analysis of 100,000 human cancer genomes reveals the landscape of tumor mutational burden. *Genome Med.* **2017**, *9*, 34. [CrossRef] [PubMed]
43. Yarchoan, M.; Hopkins, A.; Jaffee, E.M. Tumor mutational burden and response rate to PD-1 inhibition. *N. Engl. J. Med.* **2017**, *377*, 2500–2501. [CrossRef] [PubMed]
44. Goodman, A.M.; Kato, S.; Bazhenova, L.; Patel, S.P.; Frampton, G.M.; Miller, V.; Stephens, P.J.; Daniels, G.A.; Kurzrock, R. Tumor mutational burden as an independent predictor of response to immunotherapy in diverse cancers. *Mol. Cancer Ther.* **2017**, *16*, 2598–2608. [CrossRef] [PubMed]
45. Samstein, R.M.; Lee, C.H.; Shoushtari, A.N.; Hellmann, M.D.; Shen, R.; Janjigian, Y.Y.; Barron, D.A.; Zehir, A.; Jordan, E.J.; Omuro, A.; et al. Tumor mutational load predicts survival after immunotherapy across multiple cancer types. *Nat. Genet.* **2019**, *51*, 202–206. [CrossRef] [PubMed]
46. Sia, D.; Jiao, Y.; Martinez-Quetglas, I.; Kuchuk, O.; Villacorta-Martin, C.; de Moura, M.C.; Putra, J.; Camprecios, G.; Bassaganyas, L.; Akers, N.; et al. Identification of an immune-specific class of hepatocellular carcinoma, based on molecular features. *Gastroenterology* **2017**, *153*, 812–826. [CrossRef] [PubMed]
47. Harding, J.J.; Zhu, A.X.; Bauer, T.M.; Choueiri, T.K.; Drilon, A.; Voss, M.H.; Fuchs, C.S.; Abou-Alfa, G.K.; Wijayawardana, S.R.; Wang, X.A.; et al. A Phase 1b/2 study of ramucirumab in combination with emibetuzumab in patients with advanced cancer. *Clin. Cancer Res.* **2019**, *25*, 2116–2126. [CrossRef]
48. El-Khoueiry, A.B.; Sangro, B.; Yau, T.; Crocenzi, T.S.; Kudo, M.; Hsu, C.; Kim, T.Y.; Choo, S.P.; Trojan, J.; Meyer, T.; et al. Nivolumab in patients with advanced hepatocellular carcinoma (CheckMate 040): An open-label, non-comparative, phase 1/2 dose escalation and expansion trial. *Lancet* **2017**, *389*, 2492–2502. [CrossRef]
49. Sangro, B.; Crocenzi, T.S.; Welling, T.H.; Iñarrairaegui, M.; Prieto, J.; Fuertes, C.; Feely, W.; Anderson, J.; Grasela, D.M.; Wigginton, J.M.; et al. Phase I dose escalation study of nivolumab (anti-PD-1; BMS-936558; ONO-4538) in patients (pts) with advanced hepatocellular carcinoma (HCC) with or without chronic viral hepatitis. *J. Clin. Oncol.* **2013**, *31* (Suppl. 15). [CrossRef]
50. Knolle, P.A.; Thimme, R. Hepatic immune regulation and its involvement in viral hepatitis infection. *Gastroenterology* **2014**, *146*, 1193–1207. [CrossRef]
51. Sangro, B.; Gomez-Martin, C.; de la Mata, M.; Iñarrairaegui, M.; Garralda, E.; Barrera, P.; Riezu-Boj, J.I.; Larrea, E.; Alfaro, C.; Sarobe, P.; et al. A clinical trial of CTLA-4 blockade with tremelimumab in patients with hepatocellular carcinoma and chronic hepatitis C. *J. Hepatol.* **2013**, *59*, 81–88. [CrossRef] [PubMed]
52. Duffy, A.G.; Ulahannan, S.V.; Makorova-Rusher, O.; Rahma, O.; Wedemeyer, H.; Pratt, D.; Davis, J.L.; Hughes, M.S.; Heller, T.; ElGindi, M.; et al. Tremelimumab in combination with ablation in patients with advanced hepatocellular carcinoma. *J. Hepatol.* **2017**, *66*, 545–551. [CrossRef] [PubMed]
53. Yi, J.S.; Cox, M.A.; Zajac, A.J. T-cell exhaustion: Characteristics, causes and conversion. *Immunology* **2010**, *129*, 474–481. [CrossRef] [PubMed]

54. Kim, H.D.; Song, G.W.; Park, S.; Jung, M.K.; Kim, M.H.; Kang, H.J.; Yoo, C.; Yi, K.; Kim, K.H.; Eo, S.; et al. Association between expression level of PD1 by tumor-infiltrating CD8$^+$ T cells and features of hepatocellular carcinoma. *Gastroenterology* **2018**, *155*, 1936–1950.
55. Feun, L.G.; Li, Y.Y.; Wu, C.; Wangpaichitr, M.; Jones, P.D.; Richman, S.P.; Madrazo, B.; Kwon, D.; Garcia-Buitrago, M.; Martin, P.; et al. Phase 2 study of pembrolizumab and circulating biomarkers to predict anticancer response in advanced, unresectable hepatocellular carcinoma. *Cancer* **2019**, *125*, 3603–3614. [CrossRef]
56. Mariathasan, S.; Turley, S.J.; Nickles, D.; Castiglioni, A.; Yuen, K.; Wang, Y.; Koeppen, H.; Astarita, J.L.; Cubas, R.; Jhunjhunwala, S.; et al. TGFβ attenuates tumour response to PD-L1 blockade by contributing to exclusion of T cells. *Nature* **2018**, *554*, 544–548. [CrossRef]
57. Blander, J.M.; Longman, R.S.; Iliev, I.D.; Sonnenberg, G.F.; Artis, D. Regulation of inflammation by microbiota interactions with the host. *Nat. Immunol.* **2017**, *18*, 851–860. [CrossRef]
58. Sivan, A.; Corrales, L.; Hubert, N.; Williams, J.B.; Aquino-Michaels, K.; Earley, Z.M.; Benyamin, F.W.; Lei, Y.M.; Jabri, B.; Alegre, M.L.; et al. Commensal *Bifidobacterium* promotes antitumor immunity and facilitates anti-PD-L1 efficacy. *Science* **2015**, *350*, 1084–1089. [CrossRef]
59. Routy, B.; le Chatelier, E.; Derosa, L.; Duong, C.P.M.; Alou, M.T.; Daillère, R.; Fluckiger, A.; Messaoudene, M.; Rauber, C.; Roberti, M.P.; et al. Gut microbiome influences efficacy of PD-1-based immunotherapy against epithelial tumors. *Science* **2018**, *359*, 91–97. [CrossRef]
60. Li, J.; Sung, C.Y.; Lee, N.; Ni, Y.; Pihlajamäki, J.; Panagiotou, G.; El-Nezami, H. Probiotics modulated gut microbiota suppresses hepatocellular carcinoma growth in mice. *Proc. Natl. Acad. Sci. USA* **2016**, *113*, E1306–E1315. [CrossRef]
61. Yu, L.X.; Yan, H.X.; Liu, Q.; Yang, W.; Wu, H.P.; Dong, W.; Tang, L.; Lin, Y.; He, Y.Q.; Zou, S.S.; et al. Endotoxin accumulation prevents carcinogen-induced apoptosis and promotes liver tumorigenesis in rodents. *Hepatology* **2010**, *52*, 1322–1333. [CrossRef] [PubMed]
62. Dapito, D.H.; Mencin, A.; Gwak, G.Y.; Pradere, J.P.; Jang, M.K.; Mederacke, I.; Caviglia, J.M.; Khiabanian, H.; Adeyemi, A.; Bataller, R.; et al. Promotion of hepatocellular carcinoma by the intestinal microbiota and TLR4. *Cancer Cell* **2012**, *21*, 504–516. [CrossRef] [PubMed]
63. Zhang, H.L.; Yu, L.X.; Yang, W.; Tang, L.; Lin, Y.; Wu, H.; Zhai, B.; Tan, Y.X.; Shan, L.; Liu, Q.; et al. Profound impact of gut homeostasis on chemically-induced pro-tumorigenic inflammation and hepatocarcinogenesis in rats. *J. Hepatol.* **2012**, *57*, 803–812. [CrossRef] [PubMed]
64. Ponziani, F.R.; Bhoori, S.; Castelli, C.; Putignani, L.; Rivoltini, L.; del Chierico, F.; Sanguinetti, M.; Morelli, D.; Paroni Sterbini, F.; Petito, V.; et al. Hepatocellular carcinoma is associated with gut microbiota profile and inflammation in nonalcoholic fatty liver disease. *Hepatology* **2019**, *69*, 107–120. [CrossRef]
65. Okusaka, T.; Ikeda, M. Immunotherapy for hepatocellular carcinoma: Current status and future perspectives. *ESMO Open* **2018**, *3* (Suppl. 1), e000455. [CrossRef]
66. Sangro, B.; Park, J.W.; Cruz, C.M.D.; Anderson, J.; Lang, L.; Neely, J.; Shaw, J.W.; Cheng, A.L. A randomized, multicenter, phase 3 study of nivolumab vs. sorafenib as first-line treatment in patients (pts) with advanced hepatocellular carcinoma (HCC): CheckMate-459. *J. Clin. Oncol.* **2016**, *34*, TPS4147. [CrossRef]
67. Bristol-Myers Squibb Announces Results from CheckMate-459 Study Evaluating Opdivo (Nivolumab) as a First-Line Treatment for Patients with Unresectable Hepatocellular Carcinoma. Available online: https://news.bms.com/press-release/bmy/bristol-myers-squibb-announces-results-checkmate-459-study-evaluating-opdivo-nivol (accessed on 13 August 2019).
68. Zhu, A.X.; Finn, R.S.; Edeline, J.; Cattan, S.; Ogasawara, S.; Palmer, D.; Verslype, C.; Zagonel, V.; Fartoux, L.; Vogel, A.; et al. Pembrolizumab in patients with advanced hepatocellular carcinoma previously treated with sorafenib (KEYNOTE-224): A non-randomised, open-label phase 2 trial. *Lancet Oncol.* **2018**, *19*, 940–952. [CrossRef]
69. Finn, R.S.; Chan, S.L.; Zhu, A.X.; Knox, J.J.; Cheng, A.L.; Siegel, A.B.; Bautista, O.; Kudo, M. Phase 3, randomized study of pembrolizumab (pembro) vs best supportive care (BSC) for second-line advanced hepatocellular carcinoma (HCC): KEYNOTE-240. *J. Clin. Oncol.* **2017**, *35* (Suppl. 15), TPS4143. [CrossRef]

70. Qin, S.K.; Ren, Z.G.; Meng, Z.Q.; Chen, Z.D.; Chai, X.L.; Xiong, J.P.; Bai, Y.X.; Yang, L.; Zhu, H.; Fang, W.J.; et al. A randomized multicentered phase II study to evaluate SHR-1210 (PD-1 antibody) in subjects with advanced hepatocellular carcinoma (HCC) who failed or intolerable to prior systemic treatment. *Ann. Oncol.* **2018**, *29* (Suppl. 8). [CrossRef]
71. Qin, S.; Finn, R.S.; Kudo, M.; Meyer, T.; Vogel, A.; Ducreux, M.; Macarulla, T.M.; Tomasello, G.; Boisserie, F.; Hou, J.; et al. Rationale 301 study: Tislelizumab versus sorafenib as first-line treatment for unresectable hepatocellular carcinoma. *Future Oncol.* **2019**, *15*. [CrossRef]
72. Wainberg, Z.A.; Segal, N.H.; Jaeger, D.; Lee, K.H.; Marshall, J.; Joseph, A.S.; Butler, M.; Sandborn, R.E.; Nemunaitis, J.J.; Carlson, C.A.; et al. Safety and clinical activity of durvalumab monotherapy in patients with hepatocellular carcinoma (HCC). *J. Clin. Oncol.* **2017**, *35* (Suppl. 15), 4071. [CrossRef]
73. Kelley, R.K.; Abou-Alfa, G.K.; Bendell, J.C.; Kim, T.Y.; Borad, M.J.; Yong, W.P.; Morse, M.; Kang, Y.K.; Rebelatto, M.; Makowsky, M.; et al. Phase I/II study of durvalumab and tremelimumab in patients with unresectable hepatocellular carcinoma (HCC): Phase I safety and efficacy analyses. *J. Clin. Oncol.* **2017**, *35* (Suppl. 15), 4073. [CrossRef]
74. Abou-Alfa, G.K.; Chan, S.L.; Furuse, J.; Galle, P.R.; Kelley, R.K.; Qin, A.; Armstrong, J.; Darilay, A.; Vlahovic, G.; Negro, A.; et al. A randomized, multi center phase 3 study of durvalumab (D) and tremelimumab (T) as first-line treatment in patients with unresectable hepatocellular carcinoma (HCC): HIMALAYA study. *J. Clin. Oncol.* **2018**, *36* (Suppl. 15), TPS4144. [CrossRef]
75. Yau, T.; Kang, Y.K.; Kim, T.Y.; El-Khoueiry, A.B.; Santoro, A.; Sangro, B.; Melero, I.; Kudo, M.; Hou, M.M.; Matilla, A.; et al. Nivolumab (NIVO) + ipilimumab (IPI) combination therapy in patients (pts) with advanced hepatocellular carcinoma (aHCC): Results from CheckMate 040. *J. Clin. Oncol.* **2019**, *37* (Suppl. 15), 4012. [CrossRef]
76. Finn, R.S.; Ducreux, M.; Qin, S.; Galle, P.R.; Zhu, A.X.; Ikeda, M.; Kim, T.Y.; Xu, D.Z.; Verret, W.; Liu, J.; et al. IMbrave150: A randomized phase III study of atezolizumab plus bevacizumab vs sorafenib in locally advanced or metastatic hepatocellular carcinoma. *J. Clin. Oncol.* **2018**, *36* (Suppl. 15), TPS4141. [CrossRef]
77. Roche's Tecentriq in Combination with Avastin Increased Overall Survival and Progression-Free Survival in People with Unresectable Hepatocellular Carcinoma. Available online: https://www.roche.com/media/releases/med-cor-2019-10-21.htm (accessed on 28 October 2019).
78. Ikeda, M.; Sung, M.W.; Kudo, M.; Kobayashi, M.; Baron, A.D.; Finn, R.S.; Kanek, S.; Zhu, A.X.; Kubota, T.; Kraljevic, S.; et al. A phase 1b trial of lenvatinib (LEN) plus pembrolizumab (PEM) in patients (pts) with unresectable hepatocellular carcinoma (uHCC). *J. Clin. Oncol.* **2018**, *36* (Suppl. 15), 4076. [CrossRef]
79. Weber, J.; Mandala, M.; del Vecchio, M.; Gogas, H.J.; Arance, A.M.; Cowey, C.L.; Dalle, S.; Schenker, M.; Chiarion-Sileni, V.; Marquez-Rodas, I.; et al. Adjuvant nivolumab versus ipilimumab in resected Stage III or IV melanoma. *N. Engl. J. Med.* **2017**, *377*, 1824–1835. [CrossRef]
80. Haanen, J.B.A.G.; Carbonnel, F.; Robert, C.; Kerr, K.M.; Peters, S.; Larkin, J.; Jordan, K. ESMO Guidelines Committee Management of toxicities from immunotherapy: ESMO Clinical Practice Guidelines for diagnosis, treatment and follow-up. *Ann. Oncol.* **2018**, *29*, iv264–iv266. [CrossRef]
81. Larkin, J.; Chiarion-Sileni, V.; Gonzalez, R.; Grob, J.J.; Cowey, C.L.; Lao, C.D.; Schadendorf, D.; Dummer, R.; Smylie, M.; Rutkowski, P.; et al. Combined nivolumab and ipilimumab or monotherapy in untreated melanoma. *N. Engl. J. Med.* **2015**, *373*, 23–34. [CrossRef]
82. O'Day, S.J.; Maio, M.; Chiarion-Sileni, V.; Gajewski, T.F.; Pehamberger, H.; Bondarenko, I.N.; Queirolo, P.; Lundgren, L.; Mikhailov, S.; Roman, L.; et al. Efficacy and safety of ipilimumab monotherapy in patients with pretreated advanced melanoma: A multicenter single-arm phase II study. *Ann. Oncol.* **2010**, *21*, 1712–1717. [CrossRef]
83. Daly, L.E.; Power, D.G.; O'Reilly, Á.; Donnellan, P.; Cushen, S.J.; O'Sullivan, K.; Twomey, M.; Woodlock, D.P.; Redmond, H.P.; Ryan, A.M. The impact of body composition parameters on ipilimumab toxicity and survival in patients with metastatic melanoma. *Br. J. Cancer* **2017**, *116*, 310–317. [CrossRef] [PubMed]
84. Hopkins, A.M.; Rowland, A.; Kichenadasse, G.; Wiese, M.D.; Gurney, H.; McKinnon, R.A.; Karapetis, C.S.; Sorich, M.J. Predicting response and toxicity to immune checkpoint inhibitors using routinely available blood and clinical markers. *Br. J. Cancer* **2017**, *117*, 913–920. [CrossRef] [PubMed]

85. De Martin, E.; Michot, J.M.; Papouin, B.; Champiat, S.; Mateus, C.; Lambotte, O.; Roche, B.; Antonini, T.M.; Coilly, A.; Laghouati, S.; et al. Characterization of liver injury induced by cancer immunotherapy using immune checkpoint inhibitors. *J. Hepatol.* **2018**, *68*, 1181–1190. [CrossRef] [PubMed]
86. Doherty, G.J.; Duckworth, A.M.; Davies, S.E.; Mells, G.F.; Brais, R.; Harden, S.V.; Parkinson, C.A.; Corrie, P.G. Severe steroid-resistant anti-PD1 T-cell checkpoint inhibitor-induced hepatotoxicity driven by biliary injury. *ESMO Open* **2017**, *2*, e000268. [CrossRef] [PubMed]
87. Seymour, L.; Bogaerts, J.; Perrone, A.; Ford, R.; Schwartz, L.H.; Mandrekar, S.; Lin, N.U.; Litière, S.; Dancey, J.; Chen, A.; et al. iRECIST: Guidelines for response criteria for use in trials testing immunotherapeutics. *Lancet Oncol.* **2017**, *18*, e143–e152. [CrossRef]
88. Giannini, E.G.; Bucci, L.; Garuti, F.; Brunacci, M.; Lenzi, B.; Valente, M.; Caturelli, E.; Cabibbo, G.; Piscaglia, F.; Virdone, R.; et al. Patients with advanced hepatocellular carcinoma need a personalized management: A lesson from clinical practice. *Hepatology* **2018**, *67*, 1784–1796. [CrossRef]
89. Bucci, L.; Garuti, F.; Lenzi, B.; Pecorelli, A.; Farinati, F.; Giannini, E.G.; Granito, A.; Ciccarese, F.; Rapaccini, G.L.; di Marco, M.; et al. The evolutionary scenario of hepatocellular carcinoma in Italy: An update. *Liver Int.* **2017**, *37*, 259–270. [CrossRef]

© 2019 by the authors. Licensee MDPI, Basel, Switzerland. This article is an open access article distributed under the terms and conditions of the Creative Commons Attribution (CC BY) license (http://creativecommons.org/licenses/by/4.0/).

Review

Predictive Factors for Response to PD-1/PD-L1 Checkpoint Inhibition in the Field of Hepatocellular Carcinoma: Current Status and Challenges

Zuzana Macek Jilkova [1,2,3,*], Caroline Aspord [1,2,4] and Thomas Decaens [1,2,3,*]

1. Université Grenoble Alpes, 38000 Grenoble, France; caroline.aspord@efs.sante.fr
2. Institute for Advanced Biosciences, Research Center UGA/Inserm U 1209/CNRS 5309, 38700 La Tronche, France
3. Service d'hépato-gastroentérologie, Pôle Digidune, CHU Grenoble Alpes, 38700 La Tronche, France
4. Etablissement Français du Sang Auvergne-Rhône-Alpes, R&D-Laboratory, 38701 Grenoble, France
* Correspondence: zuzana.mjilkova@gmail.com or ZMacekjilkova@chu-grenoble.fr (Z.M.J.); tdecaens@chu-grenoble.fr (T.D.); Tel.: +33-4-7654-9433 (Z.M.J.); +33-4-7676-5441 (T.D.); Fax: +33-4-7676-5179 (T.D.)

Received: 27 August 2019; Accepted: 10 October 2019; Published: 14 October 2019

Abstract: Immunotherapies targeting immune checkpoints are fast-developing therapeutic approaches adopted for several tumor types that trigger unprecedented rates of durable clinical responses. Immune checkpoint programmed cell death protein 1 (PD-1), expressed primarily by T cells, and programmed cell death ligand 1 (PD-L1), expressed mainly by tumor cells, macrophages, and dendritic cells, are molecules that impede immune function, thereby allowing tumor cells to proliferate, grow and spread. PD-1/PD-L1 checkpoint inhibitors have emerged as a promising treatment strategy of hepatocellular carcinoma (HCC). However, only a minority of HCC patients benefit from this therapy. To find a niche for immune checkpoint inhibition in HCC patients, future strategies might require predictive factor-based patient selection, to identify patients who are likely to respond to the said therapy and combination strategies in order to enhance anti-tumor efficacy and clinical success. This review provides an overview of the most recent data pertaining to predictive factors for response to PD-1/PD-L1 checkpoint inhibition in the field of HCC.

Keywords: PD-1; PD-L1; hepatocellular carcinoma; predictive factors; immunotherapy; immune checkpoint inhibition

1. PD-1/PD-L1/PD-L2: A Physiological Immune Checkpoint Axis Exploited by Cancer Cells and Viruses to Escape Immunity

Programmed cell death protein 1 (PD-1) was discovered in 1992 by the group led by Tasuku Honjo [1], who received the 2018 Nobel Prize in Physiology or Medicine for this discovery. Honjo and his group of researchers described PD-1 antigen expression on the surface of stimulated mouse T and B lymphocytes [2] and showed the importance of PD-1 activation during the late phase of immune responses, involvement in the effector phase, memory response, and chronic infections in peripheral tissues. This pathway displays a physiologic role in maintaining self-tolerance and dampening immune responses to immune reactions. Programmed cell death ligand 1 (PD-L1) was identified as PD-1 ligand by Honjo's group in 2000, as a receptor expressed by antigen-presenting cells, primarily in the heart, lungs, kidney, and placenta [3]. In 2001, the second ligand for PD-1, i.e., PD-L2, was described, and the expression of PD-1 ligands on tumor cell lines was demonstrated [4]. This report suggested, for the first time, that blocking the PD-1 pathway might enhance anti-tumor immunity.

At present, it is known that PD-L1 is expressed in non-lymphoid and lymphoid tissues, whereas PD-L2 expression is more restricted. PD-L1 expression is upregulated upon activation in hematopoietic cells, especially antigen-presenting cells such as dendritic cells and macrophages. Most importantly, PD-L1 is expressed in different tumor cells and in virus-infected cells, and upon ligation with PD-1, it directly inhibits T-cell proliferation and T-cell effector functions such as IFN-gamma production and cytotoxic activity against the target cells [5].

2. PD-1/PD-L1 Pathway in Hepatocellular Carcinoma

A recent study based on tumor samples with advanced solid tumors and melanoma depicted hepatocellular carcinoma (HCC) as a tumor type with low/moderate immunogenicity [6], which may explain the lower rate of response of HCC patients to immune checkpoint blockers compared to melanoma patients.

HCC is commonly developed on the background of chronic liver disease (chronic hepatitis B virus (HBV) or hepatitis C virus (HCV) infection, metabolic disorders, or chronic alcohol consumption), which promotes an immunosuppressive status of liver and T-cell exhaustion [7,8]. During tumor development and growth, the effective anti-tumor immune surveillance in the liver microenvironment is impaired, and immune checkpoints, especially the PD-1/PD-L1 signaling pathway, are greatly involved in the said process [9]. In patients with HCC, the expression of PD-1 was increased in $CD8^+$ T cells [10], and the high frequency of both circulating and tumor-infiltrating PD-1^+ $CD8^+$ T cells was associated with progression following curative hepatic resection in patients who were never treated via immunotherapy [11]. Furthermore, high PD-1 expression on tumor-infiltrating lymphocytes and the correlation between an exhausted phenotype and impaired effector function have been observed in HCC patients [12,13]. The expression of PD-L1 in HCC cells inhibits function of T cells in the liver tumor microenvironment. Not surprisingly, high PD-L1 expression on tumor cells was determined as a predictor of recurrence for HCC patients [14]. Analyses of the samples obtained from HCC resection depicted higher expression of PD-L1, in addition to its association with tumor aggressiveness [15] and poor prognosis [16] in patients who were never treated via immunotherapy.

Blocking the interaction between PD-1 and PD-L1 leads to impressive and long-lasting anti-tumor responses in a subset of patients with many tumor types. PD-1 and PD-L1 blockades largely showed similar efficacy, though the objective response rates were 5% higher with PD-1 blockade than with PD-L1 blockade in non-small-cell lung carcinoma [17]. Agents targeting PD-1/PD-L1 have initiated a revolution also in HCC treatment as recently reviewed elsewhere [18,19].

In September 2017, anti-PD-1 antibody nivolumab (Opdivo) was approved for use by the Food and Drug Administration (FDA) for second-line treatment in sorafenib-pretreated patients with advanced HCC, based on the data derived from a dose-escalation and dose-expansion phase trial within the CheckMate-040 multi-cohort trial [20] (Table 1). The clinical activity of nivolumab was investigated in four sub-groups of advanced HCC, namely (i) sorafenib untreated or intolerant without viral hepatitis; (ii) sorafenib progressors without viral hepatitis; and (iii) HBV infected; or (iv) HCV infected HCC patients. The objective response rate was 20% in patients treated with stable dose of nivolumab and 15% in the dose-escalation phase, without differences according to the underlying liver disease [20].

Similarly, the efficacy of anti-PD-1 inhibitor pembrolizumab (Keytruda) was investigated in a phase 2 study for second-line treatment in advanced HCC patients following sorafenib failure. The study confirmed an objective response rate of 17% [21]. Thus, the FDA approved pembrolizumab for the treatment of HCC patients who have been previously treated with sorafenib in November 2018.

Table 1. Results obtained from clinical trials of programmed cell death protein 1 (PD-1) inhibitors in hepatocellular carcinoma (HCC). Administration every two weeks (Q2W) and every three weeks (Q3W).

Agent (Clinical Trial)	Dose	Objective Response Rate	Partial Response	Complete Response	Reference
Nivolumab (CheckMate 040)	Escalation 0.1–10 mg/kg (Q2W)	15%	4/48 (8.3%)	2/48 (4.2%)	[20]
Nivolumab (CheckMate 040)	Expansion 3 mg/kg (Q2W)	20%	39/214 (18.2%)	3/214 (1.4%)	[20]
Pembrolizumab (KEYNOTE-224)	200 mg (Q3W)	17%	17/104 (16%)	1/104 (1%)	[21]

Despite the improvement of clinical outcomes in a subset of patients, anti-PD-1/PD-L1 blockers are still inefficient in 80% of HCC patients. Further, they are costly and cause many severe side effects [22]. There is an urgent need to define predictive factors of response to spare patients from toxicity in the absence of clinical benefits. However, none of the current trials select HCC patients according to the potential predictive factors of tumor response.

3. Predictive Biomarkers of Response to PD-1/PD-L1 Blockade in Order to Better Select Patients and Guide Therapeutic Choices

To date, very little has been described about predictive biomarkers of response to PD-1/PD-L1 blockade in HCC. Therefore, in this study, we will present predictive biomarkers highlighted in other tumor types, which could be relevant in the HCC field, in addition to recent data available for the said field. Furthermore, predictive markers of response to PD-1/PD-L1 blockade will be divided into three subsections: (i) liver tissue and tumor side factors; (ii) circulating prognostic factors; and (iii) host factors (Figure 1).

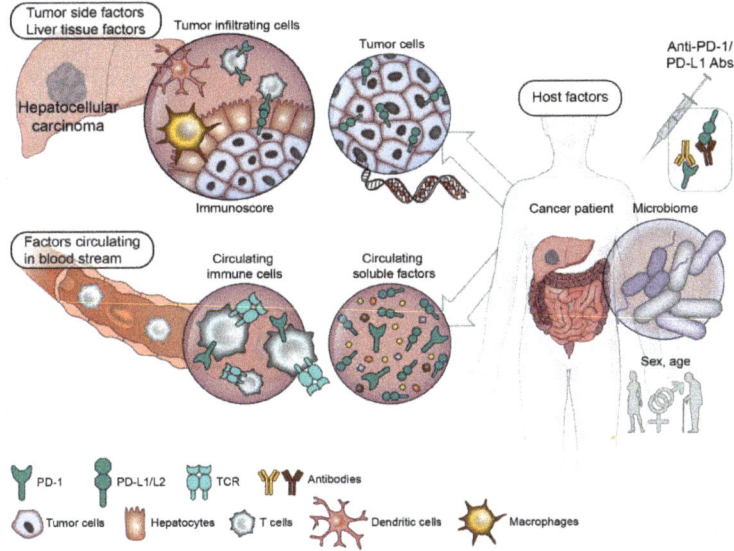

Figure 1. Overview of the predictive factors for PD-1/programmed cell death ligand 1 (PD-L1) blockade: potential factors to explore in HCC.

3.1. Liver Tissue and Tumor Side Factors

3.1.1. Immunological Biomarkers

As a logical extension of our knowledge concerning the PD-1/PD-L1 pathway, the first candidate biomarkers to be explored for PD-1/PD-L1 checkpoint inhibition were immunological. Theoretically, PD-1/PD-L1 blockade should work in patients positive for PD-1 and/or PD-L1. However, we are still unravelling the complexities of the PD-1/PD-L1 interaction between tumor and different immune cell populations.

PD-L1 Expression by Tumor Cells and Immune Infiltrate

In 2012, it has been reported that among 17 patients with PD-L1 negative tumors, none of them responded to anti-PD-1 therapy, whereas among the 25 patients with PD-L1 positive tumors 9 presented an objective response [23]. PD-L1 expression by tumor cells was intensively studied as possible predictive biomarker for ascertaining the efficacy of anti-PD-1/PD-L1 therapy. High PD-L1 expression prior to anti-PD-1/PD-L1 therapy was later demonstrated to be associated with improved objective response rate and survival in patients with non-small-cell lung cancer [24], melanoma [25] and head and neck squamous cell carcinoma [26].

However, in patients with advanced HCC, baseline expression of PD-L1 on tumor cell did not have an impact on the objective response rates to anti-PD-1 therapy [20]. In fact, as a part of the CheckMate 040 clinical trial, tumor biopsies collected at the baseline were retrospectively assessed for PD-L1 status. Membrane expression of PD-L1 on at least 1% of the tumor cells was observed in 20% patients at the baseline and majority of patients had PD-L1 expression on less than 1% of the tumor cells. Response to therapy was observed in 26% patients with PD-L1 expression on at least 1% of the tumor cells and in 19% patients with PD-L1 on less than 1% of the tumor cells. Thus, a fraction of PD-L1-negative HCC patients showed objective clinical responses, demonstrating no significant difference compared to PD-L1 positive patients. This was recently confirmed by a study where the response to anti-PD-1 did not correlate with PD-L1 tumor staining in advanced HCC [27]. However, archival tissue samples were used and the number of evaluable patients in this study was very limited ($n = 10$).

Interestingly, PD-L1 expression on immune cells may be more predictive of anti-PD-1/PD-L1 response than PD-L1 expression on tumor cells in certain tumor types, such as bladder cancer or breast cancer [28]. This might be the case for HCC as well. In fact, the relevance of PD-L1 expression on immune cells versus that of tumor cells has been revealed by comparing three tumor models with varying sensitivity to PD-1/PD-L1 blockade. Juneja et al. demonstrated that the relative contribution of tumor-derived versus host-derived PD-L1 is context-dependent and that both these PD-L1 expressions play a role in tumor microenvironment [29]. In view of the fact that PD-L1 expression on immune cells is critical for inhibiting anti-tumor immunity, PD-L1 expression within the tumor, but not necessarily on tumor cells, may be sufficient for an anti-PD-1/PD-L1 response, as reviewed recently [30]. Thus, PD-L1 expression on immune cells should be included in the list of potential markers of response to PD-1/PD-L1 inhibition.

However, several unsolved problems remain regarding the interpretation of PD-L1 expression, such as the cut-off value to define positivity and the temporal and spatial heterogeneity of PD-L1 expression. First, the lack of standardized analyses and methods makes it difficult to compare results from individual studies, in order to reach robust overall consensus [31]. In fact, most studies evaluated PD-L1 status as the percentage of tumor cells positive for cell-surface and/or membranous PD-L1 staining. However, variable cut-off values have been used to identify positivity of PD-L1 [32]. Moreover, to detect PD-L1 staining, different types and clones of anti-PD-L1 antibodies are currently on the market. Some anti-PD-L1 antibodies result in a mixture of both membranous and cytoplasmic staining of tumor cells, which obscures the interpretation of results and affects the accuracy of the analysis [33]. Three clones of anti-PD-L1 recombinant monoclonal antibodies (Clone 28-8, 73-10, and SP142) have been approved by the FDA as complementary diagnostics for PD-1/PD-L1 checkpoint inhibitors. However, using these antibodies, some differences in detecting PD-L1 staining have still been observed [34]. Recently, five anti-PD-L1 antibody clones were used to stain HCC samples [35], showing very high diversity that impacts the reliability and reproducibility of PD-L1 assays. In addition, PD-L1 expression is inducible and can change over the course of the disease and/or during treatment [36]. Thus, the lack of standardization renders interpretation across clinical trials highly difficult.

Features of Intratumoral Lymphoid Infiltrates

The potential of the adaptive immune system to control or eradicate tumors has been clearly demonstrated. The immune contexture, defined by the type, location, density, and functional orientation of the tumor-infiltrating immune cells (in particular $CD8^+$ cytotoxic T cells), allows one to predict the clinical outcome [37–39], especially in HCC [40]. Moreover, the score of immune system is a critical prognosis factor in cancer patients, and immune checkpoint blockers impact this parameter. Four different types of tumor microenvironments have been proposed by combining PD-L1 expression and T-cell density. This stratification allows one to better predict the immunotherapeutic strategy best suited to target each type [41]. Notably, different classes of HCC have been identified based on the genomic profiling of the concerned tumor microenvironment [42]. One of them, called the "immune class" (present in about 25% patients), is more susceptible to therapeutic agents blocking regulatory pathways in T cells and is characterized by markers of adaptive immune responses as well as exhausted immune responses. Therefore, it is evident that the immune contexture in HCC is critical to predict clinical outcomes following PD-1/PD-L1 checkpoint inhibition.

In addition, an IFN-γ-related signature was associated with the clinical benefits of anti-PD-1 treatment across nine different cancer cohorts [43]. In fact, the signature established from the tumor tissue at the baseline contained IFNγ-responsive genes related to antigen presentation, cytotoxic activity, chemokine expression, and adaptive immune resistance. In parallel, a resistance signature to PD-1 blockade has been identified in melanoma patients, involving high expression of the genes involved in cell adhesion, regulation of mesenchymal transition, angiogenesis, matrix remodeling, and wound healing [44]. These studies highlighted the complex biology and importance of the pre-existing tumor immune microenvironment with regard to its ability or inability to respond to PD-1/PD-L1 checkpoint inhibition.

The impact of tumor infiltration of $CD8^+$ T-cell on the survival of cancer patients has been the most well-studied topic. A meta-analysis summarized that in majority of articles published, $CD8^+$ immune cell infiltrates were associated with good prognosis in a wide variety of solid tumor types [45,46], and also associated with improved responses to chemotherapy and immunotherapy [47]. In addition, the number of tumor infiltrating lymphocytes expressing PD-1 was shown to be predictive of the clinical response following PD-1 blockade [48]. Similarly, it has been reported that tumor response to PD-1 blockade requires pre-existing $CD8^+$ T cells that are negatively regulated by PD-1/PD-L1-mediated adaptive immune resistance [49]. Particularly, the PD-1^{high} T cells seem to be very important as this subset demonstrates higher capacity for tumor recognition and markedly different profile compared to PD-1^{int} cells in patients with non-small-cell lung cancer, where the frequency of PD-1^{high} cells strongly predicted the response and survival of patients [50]. Similarly, a clinical study performed on melanoma patients showed that PD-1^{high} expression before treatment was correlated to the response to PD-1 blockade [51].

Recently, we demonstrated that the responders to anti-PD-1/PD-L1 therapy had high baseline frequency of PD-1^{high} $CD8^+$ T cells in tumor tissue, as determined by extensive phenotypic flow cytometry analyses of fresh biopsies obtained from advanced HCC patients before start of anti-PD-1/PD-L1 therapy [52]. This is in accordance with the observations of a recent study that investigated $CD8^+$ T cells isolated from HCC tissue and showed in vitro that tumors with high proportions of PD-1^{high} $CD8^+$ T cells are more susceptible to PD-1 blockade [13]. Similarly, high numbers of PD-1⁺ intratumoral lymphocytes predict survival benefit of cytokine-induced killer cells for HCC patients [53].

The main problem regarding the interpretation of PD-1 expression on CD8⁺ T cells is connected to the complexity of the methods needed for analyses. Simple immunohistochemistry is unable to distinguish PD-1⁺ CD8⁺ T cells since a combination of several antibodies is necessary to characterize these cells. For instance, the majority of NK cells express CD8 receptors, and their frequency is very high in the liver [54,55]. However, the CD56^{bright} subpopulation of NK cells that is present at high frequency in the liver [56] do not express PD-1 [57]. Thus, NK cells should be excluded from immunohistochemical analysis to allow correctly quantify the frequency of PD-1⁺ cells in the CD8⁺ T cell population. Moreover, tumor heterogeneity and sampling variability are inherent limitations when using liver biopsies. Due to the invasiveness of tissue sampling, only one of multiple lesions is usually selected for liver biopsy. Thus, a tissue sample might not necessarily reflect the entire picture of HCC. Additionally, both PD-1 and PD-L1 expression levels can change over time, as can the distribution of CD8⁺ T cells. Therefore, to develop clear predictive factors, specific time restrictions need to be defined, for instance, the requirement of analyzing tissue biopsies obtained at a maximum of three months prior to the start of the treatment.

In addition to tumor immune infiltrates, it is important to take into consideration that the prognostic factor for the response to PD-1/PD-L1 could also come from the non-tumoral tissue. Especially in HCC, as demonstrated previously, microarrays from surrounding non-tumoral liver tissues can predict overall survival after curative treatment of HCC, rather than the analyses obtained from tumor tissues [58]. Moreover, the frequency of infiltrated lymphocytes is much higher in a non-tumoral liver compared to a tumor area [59].

3.1.2. Mutations of Tumor Cells and Microsatellite Instability

Tumor mutational burden (TMB) is a measure of the total number of mutations per coding area of a tumor genome. Tumors with higher levels of TMB are believed to express more neoantigens that may allow for a more robust immune anti-tumor response and therefore, potentially, a better response to immunotherapy. Certainly, high TMB and neoantigen load have been noted to predict the response to immunotherapies, including anti-PD-1 therapy (higher objective response rate and/or prolonged survival) in melanoma, non-small-cell lung carcinoma [23], and across diverse tumors [60]. When compared to other tumor types, HCC is described by an above-average TMB with frequent formation of neoantigens [61], expected to have a good response to PD-1/PD-L1 blockage. Nevertheless, TMB is a rough marker because a mutation could or could not be immunogenic. Currently, bioinformatics tools are available to better predict the immunogenicity of mutations [62]. Recently, next-generation sequencing recognized Wnt/CTNNB1 mutations, typical for the immune-excluded tumor class, as possible biomarkers predicting resistance to immune checkpoint inhibitors in patients with advanced HCC [63]. However, this type of sequencing is complex and costly, therefore difficult for routine clinical use. Microsatellite instability (MSI) is a phenotype of hyper-mutations arising from mismatch-repair deficiency (dMMR), that is the first predictive biomarker for anti-PD-1 blockage approved by the FDA [64]. To be more precise, in May 2017, the FDA granted accelerated approval to pembrolizumab for pediatric and adult patients suffering from unresectable or metastatic MSI or dMMR solid tumors that have progressed following first-line treatment, in addition to the standard of care. Previously, MSI-high tumors were observed to display upregulation of multiple immune checkpoints, including PD-1, thus making PD-1/PD-L1 blockade a rational treatment approach. In an expanded study of advanced dMMR cancers across 12 different tumor types, objective radiographic responses were observed in 53% of patients, while complete responses were achieved in 21% of patients across 12 different tumor types [65]. However, in HCC, MSI seems to be a rare event [66].

3.2. Circulating Prognostic Factors

Circulating markers possess the advantage of being suitable for sampling over the course of treatment period and may be quickly established and accessible for clinical practice.

3.2.1. Circulating Immune Cells

The predictive value of circulating markers has been evaluated in melanoma patients treated with pembrolizumab. Moreover, high relative eosinophil and lymphocyte count were associated with favorable overall survival [67]. In another study, high relative eosinophils and basophils, low absolute monocytes, and a low neutrophil-to-lymphocyte ratio served as significant independent variables for favorable overall survival of patients with advanced melanoma [68]. Additionally, T-cell receptor (TCR) diversity could be a critical determinant of the clinical outcome regarding PD-1/PD-L1 checkpoint inhibition. A high pre-treatment clonality of TCR (indicative of a repertoire that is not diverse) was associated with poor clinical outcomes in patients with urothelial cancer treated with anti-PD-L1 [69].

In HCC, the expression of immune checkpoint molecules, such as PD-1, Tim-3, and Lag-3, in the tumor tissue may be partially reflected on the circulating immune cells [12,13,52], and immunomonitoring conducted at the circulating level has the potential to highlight prognosis factors of clinical evolution and distinguish responders from non-responders.

3.2.2. Circulating Soluble Factors

Recently, Feun et al. published a study where high baseline plasma levels of anti-inflammatory cytokine TGF-β were significantly correlated with poor outcomes after anti-PD-1 treatment in patients with advanced, unresectable HCC [27]. This promising finding is based on small cohort of patients and requires a larger number of patients for confirmation.

A baseline protein signature of patients that were PD-1 resistant as analyzed via mass spectrometry, was characterized by complement, acute phase, and wound healing molecules in metastatic melanoma patients receiving PD-1 blocking antibody [70]. Soluble immune checkpoints may also serve as potent biomarkers of response to PD-1/PD-L1 checkpoint inhibition, as it has been shown that elevated pre-treatment levels of soluble PD-L1 were associated with a progression in melanoma patients treated with PD-1 blockade [71]. In a cohort of HCC patients, the high level of the soluble PD-L1 was correlated with a poor outcome [72] but the association between soluble immune checkpoints and the response to PD-1/PD-L1 needs to be further investigated.

Extracellular vesicles such as exosomes and microvesicles are actively released from various cells, including cancer cells, and carry bioactive molecules that influence the immune system. A recent study from Chen et al. indicated that the circulating exosomal PD-L1 may reflect the states of anti-tumor immunity in melanoma patients as responders to anti-PD-1 were characterized by the increase in circulating exosomal PD-L1 during early stages of treatment [73]. However, the application of exosomal PD-L1 as a possible predictor for anti-PD-1 therapy remains controversial. HCC-derived exosomes and their potential as biomarkers were recently reviewed elsewhere [74].

3.3. Host Factors

3.3.1. Sex and Age

Recently, Conforti et al. provided evidence for the fact that the benefit of immune checkpoint inhibitors might be sex dependent [75]. In their meta-analysis of 20 randomized controlled trials testing anti-PD-1 and anti-CTLA-4, the authors observed a significantly higher overall survival benefit for men than women. This could be predictable if we consider that on an average, women mount stronger immune responses than men, and this immune response is hypothesized to lower their risk of cancer-related mortality [76,77]. Men are at almost two-times higher risk of mortality from most cancers, including HCC [78], compared to women. This male-biased mortality reflects differences not only in behavioral and biological factors but also in the immune system that is less active in men, including less effective anti-tumor immune responses [76,77]. Therefore, men's immune system might be easier to activate via immunotherapies targeting immune cells. Sex hormones influence innate and adaptive immune responses [76,77,79] and directly regulate the expression as well as function of PD-1 and PD-L1 [80,81]. A retrospective analysis found that the female sex and the age <65 years are associated with lower objective response rates to anti-PD-1 therapy compared to the male sex [82]. Thus, as far as sex hormones are concerned, it should be noted that aging is associated with the loss of sex hormones in both men and women. Thus, sex-dependent differences might disappear in part with age. In addition, the immune system is less active in older patients. As anti-PD-1/PD-L1 antibodies are therapies that should restore a lost anti-tumor immunity [83], older patients may benefit more from this treatment. Recently, Kugel et al. reported that patients aged over 60 had better response to anti-PD-1 therapy, and the likelihood of response increased with age [84]. Nevertheless, further studies are required in this regard to provide clues pertaining to the effectiveness of immunotherapy according to one's sex and age among HCC patients.

3.3.2. Influence of the Gut Microbiome

Recent evidence suggests that modulation of the gut microbiome may affect responses to immunotherapy. In fact, a significant association was observed between commensal microbial composition and clinical response to PD-1/PD-L1 therapy in melanoma patients [85]. Moreover, extensive work on the biology of the gut–liver axis has assisted in better understanding the relationship between the said microbiome and HCC [86]. For instance, in patients suffering from cirrhosis and fatty liver, the gut microbiota profile and systemic inflammation were significantly correlated and linked to HCC development [87]. A recent review summarized the knowledge about the modulatory effect of gut microbiota on immune system leading to chronic inflammation and HCC development [88]. Additionally, Zheng et al. reported the characteristics of the gut microbiome during anti-PD-1 immunotherapy in HCC, by metagenomic sequencing of periodic fecal samples [89]. Authors observed that fecal samples from patients responding to immunotherapy ($n = 3$) showed higher taxa richness and more gene counts compared to non-responders ($n = 5$), suggesting for the first time that gut microbiome may affect the response to anti-PD-1/PD-L1 immunotherapy in patients with HCC. Thus, the role of the gut microbiome in response to PD-1/PD-L1 checkpoint inhibition in HCC patients needs to be further investigated.

4. Conclusions

Although PD-1/PD-L1 checkpoint inhibition has improved the response rate for HCC, such treatments help only a minority of patients at present. A major focus involves determining the reason immunotherapies succeed or fail, in addition to the way they can be improved further. Predictive biomarkers are necessary to identify HCC patients with a greater likelihood of response, thereby guiding clinical decision-making for first-line and second-line therapies. However, even the most promising predictors of response to anti-PD-1/PD-L1 therapy in HCC, low baseline plasma levels of TGF-β or high frequency of intratumoral $CD8^+$ or PD-1^{high} $CD8^+$ T cells, need to be verified using a larger number of patients in a prospective trial. Thus, in order to propose a clinical decision-making algorithm in HCC based on such biomarkers, extensive translation research is currently required.

Author Contributions: Z.M.J. drafted the manuscript; Z.M.J., C.A. and T.D. contributed to literature search, figure design and the final drafting of the manuscript.

Funding: This work is funded by la Ligue Nationale Contre le Cancer under Grant 2016-R16145CC, Le comité de Haute-Savoie de La Ligue Contre le Cancer (CD74).

Conflicts of Interest: The authors declare no conflict of interest.

References

1. Ishida, Y.; Agata, Y.; Shibahara, K.; Honjo, T. Induced expression of PD-1, a novel member of the immunoglobulin gene superfamily, upon programmed cell death. *EMBO J.* **1992**, *11*, 3887–3895. [CrossRef] [PubMed]
2. Agata, Y.; Kawasaki, A.; Nishimura, H.; Ishida, Y.; Tsubata, T.; Yagita, H.; Honjo, T. Expression of the PD-1 antigen on the surface of stimulated mouse T and B lymphocytes. *Int. Immunol.* **1996**, *8*, 765–772. [CrossRef] [PubMed]
3. Freeman, G.J.; Long, A.J.; Iwai, Y.; Bourque, K.; Chernova, T.; Nishimura, H.; Fitz, L.J.; Malenkovich, N.; Okazaki, T.; Byrne, M.C.; et al. Engagement of the PD-1 immunoinhibitory receptor by a novel B7 family member leads to negative regulation of lymphocyte activation. *J. Exp. Med.* **2000**, *192*, 1027–1034. [CrossRef] [PubMed]
4. Latchman, Y.; Wood, C.R.; Chernova, T.; Chaudhary, D.; Borde, M.; Chernova, I.; Iwai, Y.; Long, A.J.; Brown, J.A.; Nunes, R.; et al. PD-L2 is a second ligand for PD-1 and inhibits T cell activation. *Nat. Immunol.* **2001**, *2*, 261–268. [CrossRef] [PubMed]
5. Iwai, Y.; Hamanishi, J.; Chamoto, K.; Honjo, T. Cancer immunotherapies targeting the PD-1 signaling pathway. *J. Biomed. Sci.* **2017**, *24*, 26. [CrossRef]
6. Cristescu, R.; Mogg, R.; Ayers, M.; Albright, A.; Murphy, E.; Yearley, J.; Sher, X.; Liu, X.Q.; Lu, H.; Nebozhyn, M.; et al. Pan-tumor genomic biomarkers for PD-1 checkpoint blockade-based immunotherapy. *Science* **2018**, *362*. [CrossRef]
7. Ringelhan, M.; Pfister, D.; O'Connor, T.; Pikarsky, E.; Heikenwalder, M. The immunology of hepatocellular carcinoma. *Nat. Immunol.* **2018**, *19*, 222–232. [CrossRef]
8. Elsegood, C.L.; Tirnitz-Parker, J.E.; Olynyk, J.K.; Yeoh, G.C. Immune checkpoint inhibition: Prospects for prevention and therapy of hepatocellular carcinoma. *Clin. Transl. Immunol.* **2017**, *6*, e161. [CrossRef]
9. Shrestha, R.; Prithviraj, P.; Anaka, M.; Bridle, K.R.; Crawford, D.H.G.; Dhungel, B.; Steel, J.C.; Jayachandran, A. Monitoring Immune Checkpoint Regulators as Predictive Biomarkers in Hepatocellular Carcinoma. *Front. Oncol.* **2018**, *8*, 269. [CrossRef]
10. Wang, B.J.; Bao, J.J.; Wang, J.Z.; Wang, Y.; Jiang, M.; Xing, M.Y.; Zhang, W.G.; Qi, J.Y.; Roggendorf, M.; Lu, M.J.; et al. Immunostaining of PD-1/PD-Ls in liver tissues of patients with hepatitis and hepatocellular carcinoma. *World J. Gastroenterol. WJG* **2011**, *17*, 3322–3329. [CrossRef]
11. Shi, F.; Shi, M.; Zeng, Z.; Qi, R.Z.; Liu, Z.W.; Zhang, J.Y.; Yang, Y.P.; Tien, P.; Wang, F.S. PD-1 and PD-L1 upregulation promotes CD8(+) T-cell apoptosis and postoperative recurrence in hepatocellular carcinoma patients. *Int. J. Cancer* **2011**, *128*, 887–896. [CrossRef] [PubMed]
12. Zhou, G.; Sprengers, D.; Boor, P.P.C.; Doukas, M.; Schutz, H.; Mancham, S.; Pedroza-Gonzalez, A.; Polak, W.G.; de Jonge, J.; Gaspersz, M.; et al. Antibodies Against Immune Checkpoint Molecules Restore Functions of Tumor-Infiltrating T Cells in Hepatocellular Carcinomas. *Gastroenterology* **2017**, *153*, 1107–1119.e1110. [CrossRef] [PubMed]
13. Kim, H.D.; Song, G.W.; Park, S.; Jung, M.K.; Kim, M.H.; Kang, H.J.; Yoo, C.; Yi, K.; Kim, K.H.; Eo, S.; et al. Association Between Expression Level of PD1 by Tumor-Infiltrating CD8(+) T Cells and Features of Hepatocellular Carcinoma. *Gastroenterology* **2018**. [CrossRef] [PubMed]
14. Gao, Q.; Wang, X.Y.; Qiu, S.J.; Yamato, I.; Sho, M.; Nakajima, Y.; Zhou, J.; Li, B.Z.; Shi, Y.H.; Xiao, Y.S.; et al. Overexpression of PD-L1 significantly associates with tumor aggressiveness and postoperative recurrence in human hepatocellular carcinoma. *Clin. Cancer Res. Off. J. Am. Assoc. Cancer Res.* **2009**, *15*, 971–979. [CrossRef] [PubMed]
15. Calderaro, J.; Rousseau, B.; Amaddeo, G.; Mercey, M.; Charpy, C.; Costentin, C.; Luciani, A.; Zafrani, E.S.; Laurent, A.; Azoulay, D.; et al. Programmed death ligand 1 expression in hepatocellular carcinoma: Relationship With clinical and pathological features. *Hepatology* **2016**, *64*, 2038–2046. [CrossRef] [PubMed]
16. Jung, H.I.; Jeong, D.; Ji, S.; Ahn, T.S.; Bae, S.H.; Chin, S.; Chung, J.C.; Kim, H.C.; Lee, M.S.; Baek, M.J. Overexpression of PD-L1 and PD-L2 Is Associated with Poor Prognosis in Patients with Hepatocellular Carcinoma. *Cancer Res. Treat. Off. J. Korean Cancer Assoc.* **2017**, *49*, 246–254. [CrossRef]
17. Xu-Monette, Z.Y.; Zhang, M.; Li, J.; Young, K.H. PD-1/PD-L1 Blockade: Have We Found the Key to Unleash the Antitumor Immune Response? *Front. Immunol.* **2017**, *8*, 1597. [CrossRef]

18. Nishida, N.; Kudo, M. Immune checkpoint blockade for the treatment of human hepatocellular carcinoma. *Hepatol. Res. Off. J. Jpn. Soc. Hepatol.* **2018**, *48*, 622–634. [CrossRef]
19. Kudo, M. Targeted and immune therapies for hepatocellular carcinoma: Predictions for 2019 and beyond. *World J. Gastroenterol. WJG* **2019**, *25*, 789–807. [CrossRef]
20. El-Khoueiry, A.B.; Sangro, B.; Yau, T.; Crocenzi, T.S.; Kudo, M.; Hsu, C.; Kim, T.Y.; Choo, S.P.; Trojan, J.; Welling, T.H.R.; et al. Nivolumab in patients with advanced hepatocellular carcinoma (CheckMate 040): An open-label, non-comparative, phase 1/2 dose escalation and expansion trial. *Lancet* **2017**, *389*, 2492–2502. [CrossRef]
21. Zhu, A.X.; Finn, R.S.; Edeline, J.; Cattan, S.; Ogasawara, S.; Palmer, D.; Verslype, C.; Zagonel, V.; Fartoux, L.; Vogel, A.; et al. Pembrolizumab in patients with advanced hepatocellular carcinoma previously treated with sorafenib (KEYNOTE-224): A non-randomised, open-label phase 2 trial. *Lancet Oncol.* **2018**, *19*, 940–952. [CrossRef]
22. Nishida, N.; Kudo, M. Liver damage related to immune checkpoint inhibitors. *Hepatol. Int.* **2019**, *13*, 248–252. [CrossRef] [PubMed]
23. Topalian, S.L.; Hodi, F.S.; Brahmer, J.R.; Gettinger, S.N.; Smith, D.C.; McDermott, D.F.; Powderly, J.D.; Carvajal, R.D.; Sosman, J.A.; Atkins, M.B.; et al. Safety, activity, and immune correlates of anti-PD-1 antibody in cancer. *N. Engl. J. Med.* **2012**, *366*, 2443–2454. [CrossRef] [PubMed]
24. Herbst, R.S.; Soria, J.C.; Kowanetz, M.; Fine, G.D.; Hamid, O.; Gordon, M.S.; Sosman, J.A.; McDermott, D.F.; Powderly, J.D.; Gettinger, S.N.; et al. Predictive correlates of response to the anti-PD-L1 antibody MPDL3280A in cancer patients. *Nature* **2014**, *515*, 563–567. [CrossRef] [PubMed]
25. Daud, A.I.; Wolchok, J.D.; Robert, C.; Hwu, W.J.; Weber, J.S.; Ribas, A.; Hodi, F.S.; Joshua, A.M.; Kefford, R.; Hersey, P.; et al. Programmed Death-Ligand 1 Expression and Response to the Anti-Programmed Death 1 Antibody Pembrolizumab in Melanoma. *J. Clin. Oncol. Off. J. Am. Soc. Clin. Oncol.* **2016**, *34*, 4102–4109. [CrossRef] [PubMed]
26. Gandini, S.; Massi, D.; Mandala, M. PD-L1 expression in cancer patients receiving anti PD-1/PD-L1 antibodies: A systematic review and meta-analysis. *Crit. Rev. Oncol. /Hematol.* **2016**, *100*, 88–98. [CrossRef] [PubMed]
27. Feun, L.G.; Li, Y.Y.; Wu, C.; Wangpaichitr, M.; Jones, P.D.; Richman, S.P.; Madrazo, B.; Kwon, D.; Garcia-Buitrago, M.; Martin, P.; et al. Phase 2 study of pembrolizumab and circulating biomarkers to predict anticancer response in advanced, unresectable hepatocellular carcinoma. *Cancer* **2019**. [CrossRef]
28. Powles, T.; Eder, J.P.; Fine, G.D.; Braiteh, F.S.; Loriot, Y.; Cruz, C.; Bellmunt, J.; Burris, H.A.; Petrylak, D.P.; Teng, S.L.; et al. MPDL3280A (anti-PD-L1) treatment leads to clinical activity in metastatic bladder cancer. *Nature* **2014**, *515*, 558–562. [CrossRef]
29. Juneja, V.R.; McGuire, K.A.; Manguso, R.T.; LaFleur, M.W.; Collins, N.; Haining, W.N.; Freeman, G.J.; Sharpe, A.H. PD-L1 on tumor cells is sufficient for immune evasion in immunogenic tumors and inhibits CD8 T cell cytotoxicity. *J. Exp. Med.* **2017**, *214*, 895–904. [CrossRef]
30. Kleinovink, J.W.; van Hall, T.; Ossendorp, F.; Fransen, M.F. PD-L1 immune suppression in cancer: Tumor cells or host cells? *Oncoimmunology* **2017**, *6*, e1325982. [CrossRef]
31. Yi, M.; Jiao, D.; Xu, H.; Liu, Q.; Zhao, W.; Han, X.; Wu, K. Biomarkers for predicting efficacy of PD-1/PD-L1 inhibitors. *Mol. Cancer* **2018**, *17*, 129. [CrossRef] [PubMed]
32. Udall, M.; Rizzo, M.; Kenny, J.; Doherty, J.; Dahm, S.; Robbins, P.; Faulkner, E. PD-L1 diagnostic tests: A systematic literature review of scoring algorithms and test-validation metrics. *Diagn. Pathol.* **2018**, *13*, 12. [CrossRef] [PubMed]
33. Mahoney, K.M.; Sun, H.; Liao, X.; Hua, P.; Callea, M.; Greenfield, E.A.; Hodi, F.S.; Sharpe, A.H.; Signoretti, S.; Rodig, S.J.; et al. PD-L1 Antibodies to Its Cytoplasmic Domain Most Clearly Delineate Cell Membranes in Immunohistochemical Staining of Tumor Cells. *Cancer Immunol. Res.* **2015**, *3*, 1308–1315. [CrossRef] [PubMed]
34. Yu, H.; Batenchuk, C.; Badzio, A.; Boyle, T.A.; Czapiewski, P.; Chan, D.C.; Lu, X.; Gao, D.; Ellison, K.; Kowalewski, A.A.; et al. PD-L1 Expression by Two Complementary Diagnostic Assays and mRNA In Situ Hybridization in Small Cell Lung Cancer. *J. Thorac. Oncol. Off. Publ. Int. Assoc. Study Lung Cancer* **2017**, *12*, 110–120. [CrossRef] [PubMed]

35. Pinato, D.J.; Mauri, F.A.; Spina, P.; Cain, O.; Siddique, A.; Goldin, R.; Victor, S.; Pizio, C.; Akarca, A.U.; Boldorini, R.L.; et al. Clinical implications of heterogeneity in PD-L1 immunohistochemical detection in hepatocellular carcinoma: The Blueprint-HCC study. *Br. J. Cancer* **2019**. [CrossRef] [PubMed]
36. Vilain, R.E.; Menzies, A.M.; Wilmott, J.S.; Kakavand, H.; Madore, J.; Guminski, A.; Liniker, E.; Kong, B.Y.; Cooper, A.J.; Howle, J.R.; et al. Dynamic Changes in PD-L1 Expression and Immune Infiltrates Early During Treatment Predict Response to PD-1 Blockade in Melanoma. *Clin. Cancer Res. Off. J. Am. Assoc. Cancer Res.* **2017**, *23*, 5024–5033. [CrossRef]
37. Fridman, W.H.; Pages, F.; Sautes-Fridman, C.; Galon, J. The immune contexture in human tumours: Impact on clinical outcome. *Nat. Rev. Cancer* **2012**, *12*, 298–306. [CrossRef]
38. Angell, H.; Galon, J. From the immune contexture to the Immunoscore: The role of prognostic and predictive immune markers in cancer. *Curr. Opin. Immunol.* **2013**, *25*, 261–267. [CrossRef]
39. Pages, F.; Mlecnik, B.; Marliot, F.; Bindea, G.; Ou, F.S.; Bifulco, C.; Lugli, A.; Zlobec, I.; Rau, T.T.; Berger, M.D.; et al. International validation of the consensus Immunoscore for the classification of colon cancer: A prognostic and accuracy study. *Lancet* **2018**, *391*, 2128–2139. [CrossRef]
40. Foerster, F.; Hess, M.; Gerhold-Ay, A.; Marquardt, J.U.; Becker, D.; Galle, P.R.; Schuppan, D.; Binder, H.; Bockamp, E. The immune contexture of hepatocellular carcinoma predicts clinical outcome. *Sci. Rep.* **2018**, *8*, 5351. [CrossRef]
41. Teng, M.W.; Ngiow, S.F.; Ribas, A.; Smyth, M.J. Classifying Cancers Based on T-cell Infiltration and PD-L1. *Cancer Res.* **2015**, *75*, 2139–2145. [CrossRef] [PubMed]
42. Sia, D.; Jiao, Y.; Martinez-Quetglas, I.; Kuchuk, O.; Villacorta-Martin, C.; Castro de Moura, M.; Putra, J.; Camprecios, G.; Bassaganyas, L.; Akers, N.; et al. Identification of an Immune-specific Class of Hepatocellular Carcinoma, Based on Molecular Features. *Gastroenterology* **2017**, *153*, 812–826. [CrossRef] [PubMed]
43. Ayers, M.; Lunceford, J.; Nebozhyn, M.; Murphy, E.; Loboda, A.; Kaufman, D.R.; Albright, A.; Cheng, J.D.; Kang, S.P.; Shankaran, V.; et al. IFN-gamma-related mRNA profile predicts clinical response to PD-1 blockade. *J. Clin. Investig.* **2017**, *127*, 2930–2940. [CrossRef] [PubMed]
44. Hugo, W.; Zaretsky, J.M.; Sun, L.; Song, C.; Moreno, B.H.; Hu-Lieskovan, S.; Berent-Maoz, B.; Pang, J.; Chmielowski, B.; Cherry, G.; et al. Genomic and Transcriptomic Features of Response to Anti-PD-1 Therapy in Metastatic Melanoma. *Cell* **2016**, *165*, 35–44. [CrossRef] [PubMed]
45. Gooden, M.J.; de Bock, G.H.; Leffers, N.; Daemen, T.; Nijman, H.W. The prognostic influence of tumour-infiltrating lymphocytes in cancer: A systematic review with meta-analysis. *Br. J. Cancer* **2011**, *105*, 93–103. [CrossRef]
46. Zeng, D.Q.; Yu, Y.F.; Ou, Q.Y.; Li, X.Y.; Zhong, R.Z.; Xie, C.M.; Hu, Q.G. Prognostic and predictive value of tumor-infiltrating lymphocytes for clinical therapeutic research in patients with non-small cell lung cancer. *Oncotarget* **2016**, *7*, 13765–13781. [CrossRef]
47. Danilova, L.; Wang, H.; Sunshine, J.; Kaunitz, G.J.; Cottrell, T.R.; Xu, H.; Esandrio, J.; Anders, R.A.; Cope, L.; Pardoll, D.M.; et al. Association of PD-1/PD-L axis expression with cytolytic activity, mutational load, and prognosis in melanoma and other solid tumors. *Proc. Natl. Acad. Sci. USA* **2016**, *113*, E7769–E7777. [CrossRef]
48. Taube, J.M.; Klein, A.; Brahmer, J.R.; Xu, H.; Pan, X.; Kim, J.H.; Chen, L.; Pardoll, D.M.; Topalian, S.L.; Anders, R.A. Association of PD-1, PD-1 ligands, and other features of the tumor immune microenvironment with response to anti-PD-1 therapy. *Clin. Cancer Res. Off. J. Am. Assoc. Cancer Res.* **2014**, *20*, 5064–5074. [CrossRef]
49. Tumeh, P.C.; Harview, C.L.; Yearley, J.H.; Shintaku, I.P.; Taylor, E.J.; Robert, L.; Chmielowski, B.; Spasic, M.; Henry, G.; Ciobanu, V.; et al. PD-1 blockade induces responses by inhibiting adaptive immune resistance. *Nature* **2014**, *515*, 568–571. [CrossRef]
50. Thommen, D.S.; Koelzer, V.H.; Herzig, P.; Roller, A.; Trefny, M.; Dimeloe, S.; Kiialainen, A.; Hanhart, J.; Schill, C.; Hess, C.; et al. A transcriptionally and functionally distinct PD-1(+) CD8(+) T cell pool with predictive potential in non-small-cell lung cancer treated with PD-1 blockade. *Nat. Med.* **2018**, *24*, 994–1004. [CrossRef]
51. Daud, A.I.; Loo, K.; Pauli, M.L.; Sanchez-Rodriguez, R.; Sandoval, P.M.; Taravati, K.; Tsai, K.; Nosrati, A.; Nardo, L.; Alvarado, M.D.; et al. Tumor immune profiling predicts response to anti-PD-1 therapy in human melanoma. *J. Clin. Investig.* **2016**, *126*, 3447–3452. [CrossRef] [PubMed]

52. Macek Jilkova, Z.; Aspord, C.; Kurma, K.; Granon, A.; Sengel, C.; Sturm, N.; Marche, P.N.; Decaens, T. Immunologic Features of Patients With Advanced Hepatocellular Carcinoma Before and During Sorafenib or Anti-programmed Death-1/Programmed Death-L1 Treatment. *Clin. Transl. Gastroenterol.* **2019**. [CrossRef] [PubMed]
53. Chang, B.; Shen, L.; Wang, K.; Jin, J.; Huang, T.; Chen, Q.; Li, W.; Wu, P. High number of PD-1 positive intratumoural lymphocytes predicts survival benefit of cytokine-induced killer cells for hepatocellular carcinoma patients. *Liver Int. Off. J. Int. Assoc. Study Liver* **2018**, *38*, 1449–1458. [CrossRef] [PubMed]
54. Macek Jilkova, Z.; Afzal, S.; Marche, H.; Decaens, T.; Sturm, N.; Jouvin-Marche, E.; Huard, B.; Marche, P.N. Progression of fibrosis in patients with chronic viral hepatitis is associated with IL-17(+) neutrophils. *Liver Int. Off. J. Int. Assoc. Study Liver* **2016**, *36*, 1116–1124. [CrossRef] [PubMed]
55. Fugier, E.; Marche, H.; Thelu, M.A.; Macek Jilkova, Z.; Van Campenhout, N.; Dufeu-Duchesne, T.; Leroy, V.; Zarski, J.P.; Sturm, N.; Marche, P.N.; et al. Functions of liver natural killer cells are dependent on the severity of liver inflammation and fibrosis in chronic hepatitis C. *PLoS ONE* **2014**, *9*, e95614. [CrossRef] [PubMed]
56. Mikulak, J.; Bruni, E.; Oriolo, F.; Di Vito, C.; Mavilio, D. Hepatic Natural Killer Cells: Organ-Specific Sentinels of Liver Immune Homeostasis and Physiopathology. *Front. Immunol.* **2019**, *10*. [CrossRef] [PubMed]
57. Pesce, S.; Greppi, M.; Grossi, F.; Del Zotto, G.; Moretta, L.; Sivori, S.; Genova, C.; Marcenaro, E. PD/1-PD-Ls Checkpoint: Insight on the Potential Role of NK Cells. *Front. Immunol.* **2019**, *10*, 1242. [CrossRef] [PubMed]
58. Hoshida, Y.; Villanueva, A.; Kobayashi, M.; Peix, J.; Chiang, D.Y.; Camargo, A.; Gupta, S.; Moore, J.; Wrobel, M.J.; Lerner, J.; et al. Gene expression in fixed tissues and outcome in hepatocellular carcinoma. *N. Engl. J. Med.* **2008**, *359*, 1995–2004. [CrossRef]
59. Ramzan, M.; Sturm, N.; Decaens, T.; Bioulac-Sage, P.; Bancel, B.; Merle, P.; Tran Van Nhieu, J.; Slama, R.; Letoublon, C.; Zarski, J.P.; et al. Liver-infiltrating CD8(+) lymphocytes as prognostic factor for tumour recurrence in hepatitis C virus-related hepatocellular carcinoma. *Liver Int. Off. J. Int. Assoc. Study Liver* **2016**, *36*, 434–444. [CrossRef]
60. Goodman, A.M.; Kato, S.; Bazhenova, L.; Patel, S.P.; Frampton, G.M.; Miller, V.; Stephens, P.J.; Daniels, G.A.; Kurzrock, R. Tumor Mutational Burden as an Independent Predictor of Response to Immunotherapy in Diverse Cancers. *Mol. Cancer Ther.* **2017**, *16*, 2598–2608. [CrossRef]
61. Khalil, D.N.; Smith, E.L.; Brentjens, R.J.; Wolchok, J.D. The future of cancer treatment: Immunomodulation, CARs and combination immunotherapy. *Nat. Rev. Clin. Oncol.* **2016**, *13*, 273–290. [CrossRef] [PubMed]
62. Liu, X.S.; Mardis, E.R. Applications of Immunogenomics to Cancer. *Cell* **2017**, *168*, 600–612. [CrossRef]
63. Harding, J.J.; Nandakumar, S.; Armenia, J.; Khalil, D.N.; Albano, M.; Ly, M.; Shia, J.; Hechtman, J.F.; Kundra, R.; El Dika, I.; et al. Prospective Genotyping of Hepatocellular Carcinoma: Clinical Implications of Next-Generation Sequencing for Matching Patients to Targeted and Immune Therapies. *Clin. Cancer Res. Off. J. Am. Assoc. Cancer Res.* **2019**, *25*, 2116–2126. [CrossRef]
64. Boyiadzis, M.M.; Kirkwood, J.M.; Marshall, J.L.; Pritchard, C.C.; Azad, N.S.; Gulley, J.L. Significance and implications of FDA approval of pembrolizumab for biomarker-defined disease. *J. Immunother. Cancer* **2018**, *6*, 35. [CrossRef]
65. Le, D.T.; Durham, J.N.; Smith, K.N.; Wang, H.; Bartlett, B.R.; Aulakh, L.K.; Lu, S.; Kemberling, H.; Wilt, C.; Luber, B.S.; et al. Mismatch repair deficiency predicts response of solid tumors to PD-1 blockade. *Science* **2017**, *357*, 409–413. [CrossRef]
66. Goumard, C.; Desbois-Mouthon, C.; Wendum, D.; Calmel, C.; Merabtene, F.; Scatton, O.; Praz, F. Low Levels of Microsatellite Instability at Simple Repeated Sequences Commonly Occur in Human Hepatocellular Carcinoma. *Cancer Genom. Proteom.* **2017**, *14*, 329–339. [CrossRef]
67. Weide, B.; Martens, A.; Hassel, J.C.; Berking, C.; Postow, M.A.; Bisschop, K.; Simeone, E.; Mangana, J.; Schilling, B.; Di Giacomo, A.M.; et al. Baseline Biomarkers for Outcome of Melanoma Patients Treated with Pembrolizumab. *Clin. Cancer Res. Off. J. Am. Assoc. Cancer Res.* **2016**, *22*, 5487–5496. [CrossRef] [PubMed]
68. Rosner, S.; Kwong, E.; Shoushtari, A.N.; Friedman, C.F.; Betof, A.S.; Brady, M.S.; Coit, D.G.; Callahan, M.K.; Wolchok, J.D.; Chapman, P.B.; et al. Peripheral blood clinical laboratory variables associated with outcomes following combination nivolumab and ipilimumab immunotherapy in melanoma. *Cancer Med.* **2018**, *7*, 690–697. [CrossRef]
69. Snyder, A.; Nathanson, T.; Funt, S.A.; Ahuja, A.; Buros Novik, J.; Hellmann, M.D.; Chang, E.; Aksoy, B.A.; Al-Ahmadie, H.; Yusko, E.; et al. Contribution of systemic and somatic factors to clinical response

and resistance to PD-L1 blockade in urothelial cancer: An exploratory multi-omic analysis. *PLoS Med.* **2017**, *14*, e1002309. [CrossRef]
70. Weber, J.S.; Sznol, M.; Sullivan, R.J.; Blackmon, S.; Boland, G.; Kluger, H.M.; Halaban, R.; Bacchiocchi, A.; Ascierto, P.A.; Capone, M.; et al. A Serum Protein Signature Associated with Outcome after Anti-PD-1 Therapy in Metastatic Melanoma. *Cancer Immunol. Res.* **2018**, *6*, 79–86. [CrossRef]
71. Zhou, J.; Mahoney, K.M.; Giobbie-Hurder, A.; Zhao, F.; Lee, S.; Liao, X.; Rodig, S.; Li, J.; Wu, X.; Butterfield, L.H.; et al. Soluble PD-L1 as a Biomarker in Malignant Melanoma Treated with Checkpoint Blockade. *Cancer Immunol. Res.* **2017**, *5*, 480–492. [CrossRef] [PubMed]
72. Finkelmeier, F.; Canli, O.; Tal, A.; Pleli, T.; Trojan, J.; Schmidt, M.; Kronenberger, B.; Zeuzem, S.; Piiper, A.; Greten, F.R.; et al. High levels of the soluble programmed death-ligand (sPD-L1) identify hepatocellular carcinoma patients with a poor prognosis. *Eur. J. Cancer* **2016**, *59*, 152–159. [CrossRef] [PubMed]
73. Chen, G.; Huang, A.C.; Zhang, W.; Zhang, G.; Wu, M.; Xu, W.; Yu, Z.; Yang, J.; Wang, B.; Sun, H.; et al. Exosomal PD-L1 contributes to immunosuppression and is associated with anti-PD-1 response. *Nature* **2018**, *560*, 382–386. [CrossRef] [PubMed]
74. Sasaki, R.; Kanda, T.; Yokosuka, O.; Kato, N.; Matsuoka, S.; Moriyama, M. Exosomes and Hepatocellular Carcinoma: From Bench to Bedside. *Int. J. Mol. Sci.* **2019**, *20*, 1406. [CrossRef] [PubMed]
75. Conforti, F.; Pala, L.; Bagnardi, V.; De Pas, T.; Martinetti, M.; Viale, G.; Gelber, R.D.; Goldhirsch, A. Cancer immunotherapy efficacy and patients' sex: A systematic review and meta-analysis. *Lancet Oncol.* **2018**, *19*, 737–746. [CrossRef]
76. Klein, S.L.; Flanagan, K.L. Sex differences in immune responses. *Nat. Rev. Immunol.* **2016**. [CrossRef]
77. Pennell, L.M.; Galligan, C.L.; Fish, E.N. Sex affects immunity. *J. Autoimmun.* **2012**, *38*, J282–J291. [CrossRef]
78. Li, Y.; Xu, A.; Jia, S.; Huang, J. Recent advances in the molecular mechanism of sex disparity in hepatocellular carcinoma. *Oncol. Lett.* **2019**, *17*, 4222–4228. [CrossRef]
79. Macek Jilkova, Z.; Decaens, T.; Marlu, A.; Marche, H.; Jouvin-Marche, E.; Marche, P.N. Sex Differences in Spontaneous Degranulation Activity of Intrahepatic Natural Killer Cells during Chronic Hepatitis B: Association with Estradiol Levels. *Mediat. Inflamm.* **2017**, *2017*, 3214917. [CrossRef]
80. Wang, C.; Dehghani, B.; Li, Y.; Kaler, L.J.; Proctor, T.; Vandenbark, A.A.; Offner, H. Membrane estrogen receptor regulates experimental autoimmune encephalomyelitis through up-regulation of programmed death 1. *J. Immunol.* **2009**, *182*, 3294–3303. [CrossRef]
81. Polanczyk, M.J.; Hopke, C.; Vandenbark, A.A.; Offner, H. Estrogen-mediated immunomodulation involves reduced activation of effector T cells, potentiation of Treg cells, and enhanced expression of the PD-1 costimulatory pathway. *J. Neurosci. Res.* **2006**, *84*, 370–378. [CrossRef] [PubMed]
82. Nosrati, A.; Tsai, K.K.; Goldinger, S.M.; Tumeh, P.; Grimes, B.; Loo, K.; Algazi, A.P.; Nguyen-Kim, T.D.L.; Levesque, M.; Dummer, R.; et al. Evaluation of clinicopathological factors in PD-1 response: Derivation and validation of a prediction scale for response to PD-1 monotherapy. *Br. J. Cancer* **2017**, *116*, 1141–1147. [CrossRef]
83. Sanmamed, M.F.; Chen, L. A Paradigm Shift in Cancer Immunotherapy: From Enhancement to Normalization. *Cell* **2018**, *175*, 313–326. [CrossRef]
84. Kugel, C.H., 3rd; Douglass, S.M.; Webster, M.R.; Kaur, A.; Liu, Q.; Yin, X.; Weiss, S.A.; Darvishian, F.; Al-Rohil, R.N.; Ndoye, A.; et al. Age Correlates with Response to Anti-PD1, Reflecting Age-Related Differences in Intratumoral Effector and Regulatory T-Cell Populations. *Clin. Cancer Res. Off. J. Am. Assoc. Cancer Res.* **2018**. [CrossRef] [PubMed]
85. Matson, V.; Fessler, J.; Bao, R.; Chongsuwat, T.; Zha, Y.; Alegre, M.L.; Luke, J.J.; Gajewski, T.F. The commensal microbiome is associated with anti-PD-1 efficacy in metastatic melanoma patients. *Science* **2018**, *359*, 104–108. [CrossRef] [PubMed]
86. Tripathi, A.; Debelius, J.; Brenner, D.A.; Karin, M.; Loomba, R.; Schnabl, B.; Knight, R. The gut-liver axis and the intersection with the microbiome. *Nat. Rev. Gastroenterol. Hepatol.* **2018**, *15*, 397–411. [CrossRef] [PubMed]
87. Ponziani, F.R.; Bhoori, S.; Castelli, C.; Putignani, L.; Rivoltini, L.; Del Chierico, F.; Sanguinetti, M.; Morelli, D.; Paroni Sterbini, F.; Petito, V.; et al. Hepatocellular carcinoma is associated with gut microbiota profile and inflammation in nonalcoholic fatty liver disease. *Hepatology* **2018**. [CrossRef] [PubMed]

88. Ponziani, F.R.; Nicoletti, A.; Gasbarrini, A.; Pompili, M. Diagnostic and therapeutic potential of the gut microbiota in patients with early hepatocellular carcinoma. *Ther. Adv. Med Oncol.* **2019**, *11*, 1758835919848184. [CrossRef] [PubMed]
89. Zheng, Y.; Wang, T.; Tu, X.; Huang, Y.; Zhang, H.; Tan, D.; Jiang, W.; Cai, S.; Zhao, P.; Song, R.; et al. Gut microbiome affects the response to anti-PD-1 immunotherapy in patients with hepatocellular carcinoma. *J. Immunother. Cancer* **2019**, *7*, 193. [CrossRef] [PubMed]

© 2019 by the authors. Licensee MDPI, Basel, Switzerland. This article is an open access article distributed under the terms and conditions of the Creative Commons Attribution (CC BY) license (http://creativecommons.org/licenses/by/4.0/).

Article

Etiology-Specific Analysis of Hepatocellular Carcinoma Transcriptome Reveals Genetic Dysregulation in Pathways Implicated in Immunotherapy Efficacy

Wei Tse Li [1,†], Angela E. Zou [1,†], Christine O. Honda [1,†], Hao Zheng [1,†], Xiao Qi Wang [2], Tatiana Kisseleva [1], Eric Y. Chang [3] and Weg M. Ongkeko [1,*]

1. Department of Surgery, University of California, San Diego, CA 92093, USA
2. Department of Surgery, The University of Hong Kong, 21 Sassoon Road, Pokfulam, Hong Kong, China
3. Department of Radiology, California and Radiology Service, VA San Diego Healthcare System, University of California, San Diego, CA 92093, USA
* Correspondence: wongkeko@ucsd.edu; Tel.: +858-552-8585 (ext. 7165)
† Authors contributed equally.

Received: 23 July 2019; Accepted: 8 August 2019; Published: 30 August 2019

Abstract: Immunotherapy has emerged in recent years as arguably the most effective treatment for advanced hepatocellular carcinoma (HCC), but the failure of a large percentage of patients to respond to immunotherapy remains as the ultimate obstacle to successful treatment. Etiology-associated dysregulation of immune-associated (IA) genes may be central to the development of this differential clinical response. We identified immune-associated genes potentially dysregulated by alcohol or viral hepatitis B in HCC and validated alcohol-induced dysregulations in vitro while using large-scale RNA-sequencing data from The Cancer Genome Atlas (TCGA). Thirty-four clinically relevant dysregulated IA genes were identified. We profiled the correlation of all genomic alterations in HCC patients to IA gene expression while using the information theory-based algorithm REVEALER to investigate the molecular mechanism for their dysregulation and explore the possibility of genome-based patient stratification. We also studied gene expression regulators and identified multiple microRNAs that were implicated in HCC pathogenesis that can potentially regulate these IA genes' expression. Our study identified potential key pathways, including the IL-7 signaling pathway and TNFRSF4 (OX40)- NF-κB pathway, to target in immunotherapy treatments and presents microRNAs as promising therapeutic targets for dysregulated IA genes because of their extensive regulatory roles in the cancer immune landscape.

Keywords: cancer immunotherapy; TCGA; mutations; copy number variations; microRNAs

1. Introduction

Hepatocellular carcinoma (HCC) is the most prevalent class of liver cancer and the second leading cause of cancer-related mortality around the world [1]. Late diagnosis of HCC is common because of current limitations in diagnostic methods. Patients with late stages of HCC have five-year survival rates of less than 16% [2]. The most effective standard treatments of HCC, including liver transplantation, ablation, or surgical resection, are only recommended for early stages of HCC and they have high rates of recurrence [3]. Systemic treatments, such as chemotherapy, which are commonly used in other cancers, are relatively ineffective in HCC because of resistance to therapeutic agents and poor metabolism in cirrhotic livers, which contribute to the development of about 90% of HCC cases [4].

The extremely limited treatment options for advanced stage HCC patients led to great interest in recent advancements in cancer immunotherapy.

Most previous immunotherapy studies in HCC, including cytokine or antigen-based therapies, have failed to achieve adequate anti-cancer effects [5]. However, recent interests in immunotherapy were stimulated by the success of oncolytic viral gene therapy using JX-594 [6], the use of anti-Glypican-3 (anti-GPC3) antibodies to neutralize GPC3 antigens present on 80% of HCC cells [7], and CTLA-4 blockade clinical trials [8]. Unwanted immunogenicity and limited response rates remain significant challenges despite the variety of treatment strategies under investigation [5].

An understanding of the mechanisms through which gene regulation leads to evasion of tumor cells from immune recognition is important for the continual advancement of immunotherapy. microRNAs (miRNAs) are non-coding RNAs 18–25 nucleotides long that regulate critical cellular processes, such as development, division, and differentiation through mRNA silencing. miRNA dysregulation has been consistently documented across a large body of studies in human cancers since the recognition of their regulatory significance. miRNAs have also been documented to modulate the development of cells in both the innate and adaptive immune system, as well as regulate the release of cytokines and other proteins that induce key immune processes, such as inflammation [9].

In this study, we first identified immune associated (IA) genes that were dysregulated in alcohol-related and hepatitis B-related HCC and then evaluated their potential regulation by miRNAs identified to be dysregulated in HCC in our previous study [10]. Alcohol consumption and viral hepatitis infection are established independent risk factors for the development of HCC, and documented synergism exists between them [11]. A gene qualifies as IA in our study if it is involved in a pathway that regulates immune processes in either the innate or adaptive immune system. The relationship between dysregulation of IA genes and HCC development was explored through statistical correlations with patient survival, clinical variables, and the expression of commonly mutated genes in HCC. Notably, we identified the dysregulation of several genes that were reported to contribute to immunotherapy success or failure in melanoma. Further functional analysis of IA genes we identified, along with their association to etiology and genomic alterations, may reveal unique immune status stratifications of HCC patients, which can be targeted in immunotherapy to improve the clinical response rate. Finally, we illustrate that regulatory miRNAs of key IA genes in HCC may be therapeutically targeted as a novel treatment strategy or as a complement to existing immunotherapy treatments.

2. Results

2.1. Identification of Dysregulated Immune-Associated Genes in Etiology-Specific HCC and Correlation with Patient Survival

We downloaded liver HCC transcriptome data from The Cancer Genome Atlas (TCGA) database for a total of 371 patients and the adjacent normal samples from 48 of these patients. The patients were divided into four cohorts: alcohol drinkers with hepatitis B ($n = 30$), drinkers without hepatitis B ($n = 34$), nondrinkers with hepatitis B ($n = V109$), and nondrinkers without hepatitis B ($n = 101$). The adjacent normal samples were divided into two cohorts: samples from drinkers and samples from nondrinkers. The patient's hepatitis infection status was not taken into consideration in normal cohorts due to evidence that the transformation of normal cells into HCC cells occurs at the same time as the integration of hepatitis B viral DNA into the host cell genome [12]; therefore, adjacent normal cells most likely do not have viral DNA. We expected our characterization of the landscape of IA gene dysregulation in tumor samples to also include genes that were expressed in immune cells of the tumor microenvironment due to the limited purity of TCGA tumor samples [13].

A total of six differential expression analyses were performed to examine IA genes dysregulated in HCC cases with different etiologies (Figure 1a and Table 1). The expressions of differentially expressed genes identified were then correlated with patient survival data that were obtained from TCGA while using the Cox proportional hazards regression ($p < 0.05$, Figure 2a). Thirty-two survival-correlated IA genes were identified from the five differential expression analyses when comparing tumor vs.

normal samples (Figure 1b). The probable etiology cause of gene dysregulation can be deduced from examining the overlaps and exclusions of differentially expressed genes that were found across different comparisons.

Table 1. Differential expression analysis results for survival-correlated immune-associated genes.

HCC Drinkers without HBV vs. Nondrinker Normals				HCC Drinkers without HBV vs. Drinker Normals			
Gene Name	Fold Change	FDR	p-Value	Gene Name	Fold Change	FDR	p-Value
APOB	−2.16	9.0×10^{-13}	7.0×10^{-10}	APLN	16.65	1.7×10^{-6}	3.5×10^{-4}
CAMK4	−4.12	7.2×10^{-16}	3.1×10^{-13}	CAMK4	−5.90	2.6×10^{-8}	1.5×10^{-6}
CAPG	3.73	4.0×10^{-10}	5.5×10^{-7}	CLEC1B	−22.88	6.8×10^{-5}	3.7×10^{-2}
CBX8	2.73	1.9×10^{-15}	9.0×10^{-13}	DNASE1L3	−6.63	1.4×10^{-5}	5.2×10^{-3}
CCL14	−2.93	2.2×10^{-7}	5.4×10^{-4}	SOCS2	−7.10	2.0×10^{-9}	6.9×10^{-8}
CD226	−3.03	8.9×10^{-9}	1.6×10^{-5}	UBE2S	5.93	1.4×10^{-5}	5.2×10^{-3}
CKLF	2.71	1.1×10^{-11}	1.1×10^{-8}	VIPR1	−12.77	5.2×10^{-12}	6.3×10^{-11}
CLEC1B	−26.29	7.3×10^{-12}	6.9×10^{-9}				
CYP2C9	−3.34	7.8×10^{-8}	1.7×10^{-4}				
DNASE1L3	−6.62	2.2×10^{-13}	1.6×10^{-10}				
IGKV4-1	−5.33	1.7×10^{-6}	4.9×10^{-3}				
IMPDH1	2.08	3.6×10^{-6}	1.2×10^{-2}				
IQGAP2	−1.97	1.3×10^{-9}	1.9×10^{-6}				
KITLG	2.14	4.6×10^{-6}	1.5×10^{-2}				
KLRD1	−2.54	7.3×10^{-7}	2.0×10^{-3}				
LPCAT1	3.07	1.2×10^{-10}	1.4×10^{-7}				
MSC	6.00	2.1×10^{-10}	2.6×10^{-7}				
NDRG2	−2.46	1.2×10^{-9}	1.7×10^{-6}				
PGF	2.06	5.2×10^{-6}	1.7×10^{-2}				
RAB24	2.31	9.6×10^{-15}	5.1×10^{-12}				
SEC61G	2.14	5.1×10^{-9}	8.7×10^{-6}				
SOCS2	−6.45	5.3×10^{-19}	1.3×10^{-16}				
SPP1	25.97	8.7×10^{-22}	1.3×10^{-19}				
TAGAP	−2.49	9.6×10^{-6}	3.3×10^{-2}				
UBE2S	5.14	2.2×10^{-21}	3.7×10^{-19}				
VIPR1	−10.80	6.2×10^{-27}	4.0×10^{-25}				
HCC Drinkers with HBV vs. Drinker Normals				HCC Drinkers vs. HCC Nondrinkers			
Gene Name	Fold Change	FDR	p-Value	Gene Name	Fold Change	FDR	p-Value
APLN	18.60	2.6×10^{-6}	5.0×10^{-4}	CCL18	1.72	1.0×10^{-2}	2.6×10^{-2}
CAMK4	−3.83	4.4×10^{-5}	1.7×10^{-2}	CKLF	1.36	3.5×10^{-3}	5.1×10^{-3}
CLEC1B	−28.26	7.9×10^{-8}	5.2×10^{-6}	MMP9	1.88	5.3×10^{-3}	1.0×10^{-2}
DNASE1L3	−5.45	9.6×10^{-6}	2.6×10^{-3}				
NDRG2	−2.97	3.9×10^{-7}	4.2×10^{-5}				
UBE2S	4.85	1.0×10^{-4}	4.8×10^{-2}				
VIPR1	−8.65	2.1×10^{-7}	1.9×10^{-5}				
HCC Drinkers with HBV vs. Nondrinker Normals				HCC Nondrinkers with HBV vs. Nondrinker Normals			
Gene Name	Fold Change	FDR	p-Value	Gene Name	Fold Change	FDR	p-Value
CAMK4	−2.71	5.7×10^{-9}	8.0×10^{-6}	APLN	13.78	1.2×10^{-33}	1.6×10^{-31}
CAPG	3.47	3.1×10^{-9}	4.2×10^{-6}	CAMK4	−2.89	9.0×10^{-13}	1.3×10^{-9}
CBX8	2.23	3.6×10^{-12}	2.5×10^{-9}	CAPG	3.30	1.1×10^{-9}	2.7×10^{-6}
CCL14	−3.01	2.8×10^{-9}	3.7×10^{-6}	CBX8	2.14	7.7×10^{-14}	9.9×10^{-11}
CD226	−2.39	1.0×10^{-6}	2.4×10^{-3}	CCL14	−3.76	4.7×10^{-20}	2.7×10^{-17}

Table 1. Cont.

Gene	Val	p1	p2	Gene	Val	p1	p2
CKLF	3.26	6.9×10^{-14}	3.3×10^{-11}	CD226	−2.82	1.1×10^{-11}	1.9×10^{-8}
CLEC1B	−32.63	7.6×10^{-20}	1.1×10^{-17}	CLEC1B	−76.69	7.5×10^{-45}	3.9×10^{-43}
CYP2C9	−3.38	2.7×10^{-7}	5.7×10^{-4}	CTHRC1	13.81	3.5×10^{-19}	2.2×10^{-16}
DNASE1L3	−5.40	4.4×10^{-14}	2.0×10^{-11}	CYP2C9	−3.88	3.3×10^{-12}	5.3×10^{-9}
DUSP10	−2.15	7.5×10^{-8}	1.4×10^{-4}	DNASE1L3	−7.82	1.7×10^{-37}	1.6×10^{-35}
KITLG	2.97	1.2×10^{-11}	9.7×10^{-9}	DUSP10	−2.35	5.0×10^{-14}	6.2×10^{-11}
KLRD1	−2.73	1.0×10^{-9}	1.2×10^{-6}	FYN	−2.03	1.5×10^{-12}	2.3×10^{-9}
LPCAT1	2.76	2.1×10^{-9}	2.6×10^{-6}	KITLG	2.39	7.4×10^{-8}	2.3×10^{-4}
MSC	7.90	2.4×10^{-11}	2.1×10^{-8}	KLRD1	−2.82	1.3×10^{-11}	2.2×10^{-8}
NDRG2	−2.79	4.8×10^{-19}	8.3×10^{-17}	LPCAT1	2.16	7.6×10^{-8}	2.4×10^{-4}
RAB24	2.26	5.0×10^{-14}	2.3×10^{-11}	MSC	6.65	8.2×10^{-9}	2.2×10^{-5}
SOCS2	−3.70	1.6×10^{-9}	2.0×10^{-6}	MT-RNR2	−2.21	1.5×10^{-11}	2.7×10^{-8}
SPP1	15.71	6.9×10^{-15}	2.7×10^{-12}	NDRG2	−2.81	7.8×10^{-23}	3.2×10^{-20}
UBE2S	4.06	2.2×10^{-17}	5.5×10^{-15}	PGF	3.09	5.0×10^{-8}	1.5×10^{-4}
VIPR1	−7.31	1.2×10^{-16}	3.5×10^{-14}	RAB24	2.19	1.3×10^{-17}	1.0×10^{-14}
				SOCS2	−3.93	3.9×10^{-16}	3.7×10^{-13}
				TNFRSF4	5.61	9.0×10^{-27}	2.3×10^{-24}
				UBE2S	3.67	6.8×10^{-20}	4.0×10^{-17}
				VIPR1	−13.83	9.5×10^{-55}	1.3×10^{-53}

Six IA genes, *APOB*, *IMPDH1*, *SEC61G*, *IQGAP2*, *TAGAP*, and *IGKV4-1*, were dysregulated (differential expression, $p < 0.05$) in only tumor samples of drinkers without hepatitis B virus (HBV) infection as compared to pure normal samples (from nondrinkers). These six genes were not differentially expressed between the tumor samples from drinkers without HBV and normal samples from drinkers, which suggests that they are most likely exclusively dysregulated by alcohol. Four IA genes, *FYN*, *CTHRC1*, *TNFRSF4 (CD134)*, and *MT-RNR2*, are dysregulated in the samples from nondrinkers with HBV when compared to pure normal samples. These genes were not significantly dysregulated in any other comparison, which suggested that they are not genes that are essential to all malignant transformation and they are most likely exclusively dysregulated by HBV. Ten IA genes, *KITLG* (stem cell factor), *KLRD1*, *CCL14*, *CYP2C9*, *CD226*, *CBX8*, *LPCAT1*, *RAB24*, *CAPG*, and *MSC*, are dysregulated in the three comparisons of cancer samples with normal samples (HCC samples from drinkers with HBV infection vs. normal samples, HCC samples from drinkers without HBV vs. normal samples, and HCC samples from nondrinkers with HBV vs. normal samples). Following the above reasoning, we suggest that these genes are most likely dysregulated by both alcohol drinking and HBV infection independently, but they are not essential to malignant transformation. Two IA genes, *SPP1* and *CKLF*, are most likely dysregulated by alcohol, because they are differentially expressed in the same comparisons as those for the six alcohol-associated IA genes above. However, they are additionally dysregulated in tumor samples from patients who are both drinkers and HBV infected as compared to pure normal samples, which suggested a possibility for synergism of alcohol and HBV in dysregulating these genes. The dysregulation of two IA genes, *DUSP10* and *PGF*, seems to be antagonized by the interaction of alcohol and HBV. *DUSP10* is most likely dysregulated by HBV, but it does not appear to be dysregulated in a comparison between tumor samples from drinkers with HBV infection and normal samples from drinkers. *PGF* appears to be independently dysregulated by both alcohol and HBV, but it is not dysregulated in tumor samples from drinkers with HBV infection as compared to pure normal samples or to normal samples from drinkers. Some IA genes are likely to be central to the development of HCC tumor. These genes include *SOCS2*, *APLN*, *NDRG2*, and the five genes found to be dysregulated across all five cancer to normal sample comparisons: *VIPR1*, *CAMK4*, *CLEC1B*, *DNASE1L3*, and *UBE2S*. Figure 1b,c presents a complete summary of these results.

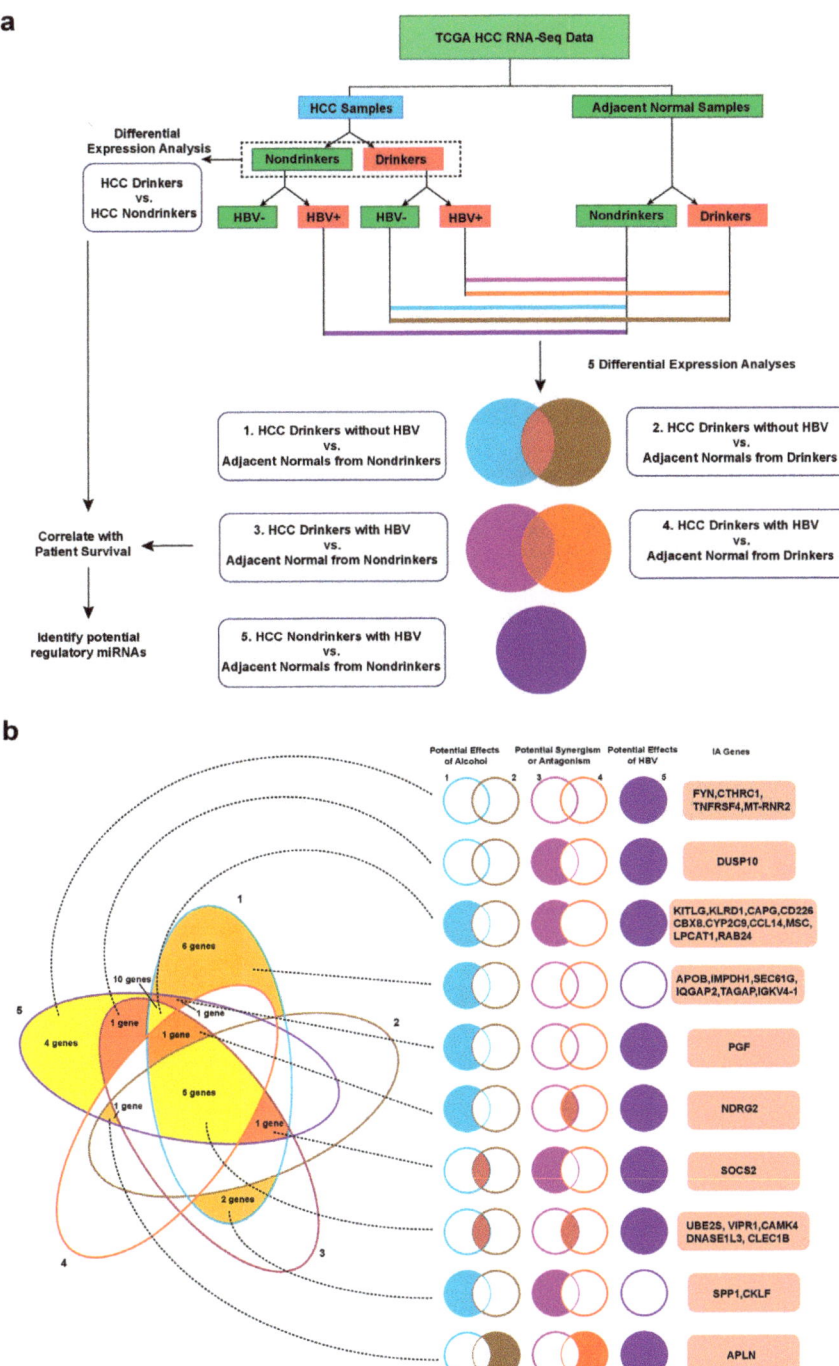

Figure 1. Cont.

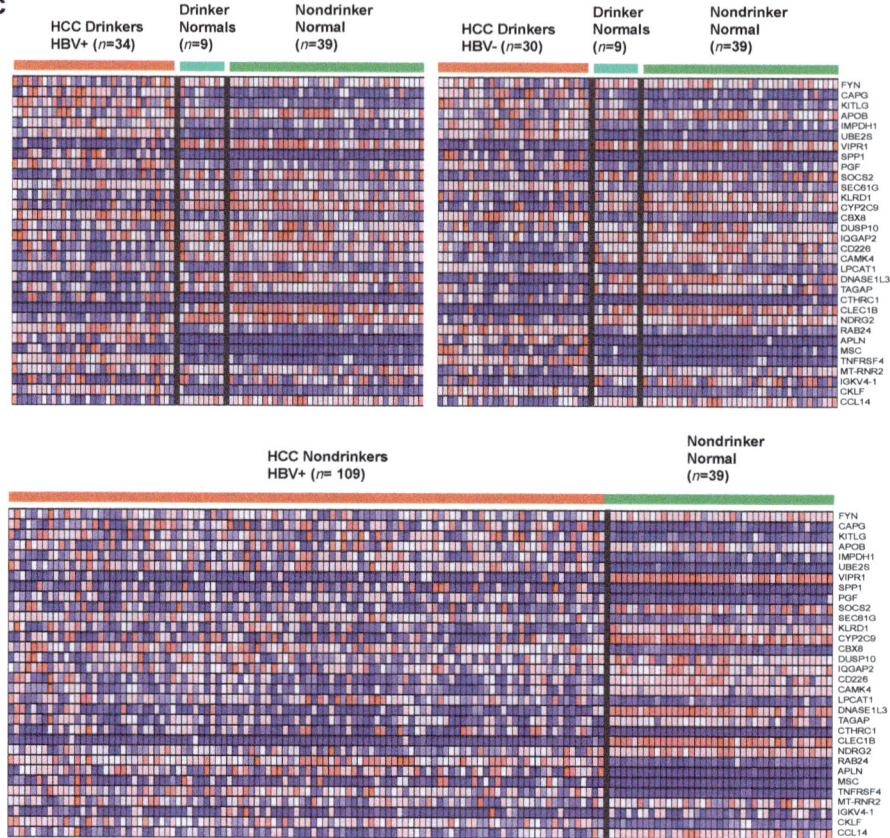

Figure 1. Summary of differential expression analyses results for identification of dysregulated immune-associated genes in hepatocellular carcinoma (HCC). (**a**) Schematic of workflow used to obtain the cohorts for six etiology-specific differential expression analyses is depicted. Each comparison is color-coded, with the color scheme consistent throughout (**a**,**b**). The five differential expression analyses comparing HCC samples to adjacent normal samples were divided into three sets. The first set includes the two comparisons involving samples from HCC drinkers without hepatitis B virus (HBV) and identifies immune associated (IA) genes potentially dysregulated as a result of alcohol consumption. The second set includes the two comparisons involving samples from HCC drinkers with HBV. Genes differentially expressed in this set can be used to examine possible synergism or antagonism between HBV-related HCC and alcohol-related HCC by comparing them to dysregulations identified in other sets of comparisons. The third set compares samples from HCC nondrinkers with HBV to normal samples from nondrinkers and identifies IA genes that were potentially dysregulated by HBV. (**b**) A five-set Venn diagram summarizes the number of IA genes identified as dysregulated in the five comparisons involving normal samples and any overlaps of genes between comparisons. The results are examined in terms of the three sets of comparisons described above. A solid color-filled region indicates the presence of differentially expressed genes for the indicated comparison(s). All IA genes presented correlate with patient survival data. (**c**) Three heatmaps are generated (one for each set of comparisons) for the thirty-two survival-correlated IA genes identified in the five HCC-normal comparisons. Refer to b for the genes differentially expressed in each individual comparison.

2.2. Correlation of IA Gene Expressions with Clinical Variables

To assess the clinical relevance and potential prognostic values of the IA genes that we identified, we correlated their expression to important clinical variables in HCC, including vascular tumor invasion, tumor histological grade, pathological and clinical stages of cancer, and lymphocyte infiltration percentage (Figure 2b). The IA genes, *LPCAT1, NDRG2, SOCS2, CCL14, UBE2S,* and *CYP2C9,* have expression levels that significantly correlate with two or more clinical variables, which suggests their potentially important role in contributing to disease progression.

To understand the relationship between the IA genes, we identified and their extent of involvement in immune-associated processes and summarized the known mechanisms of the functions for these genes in a schematic (Figure 2c).

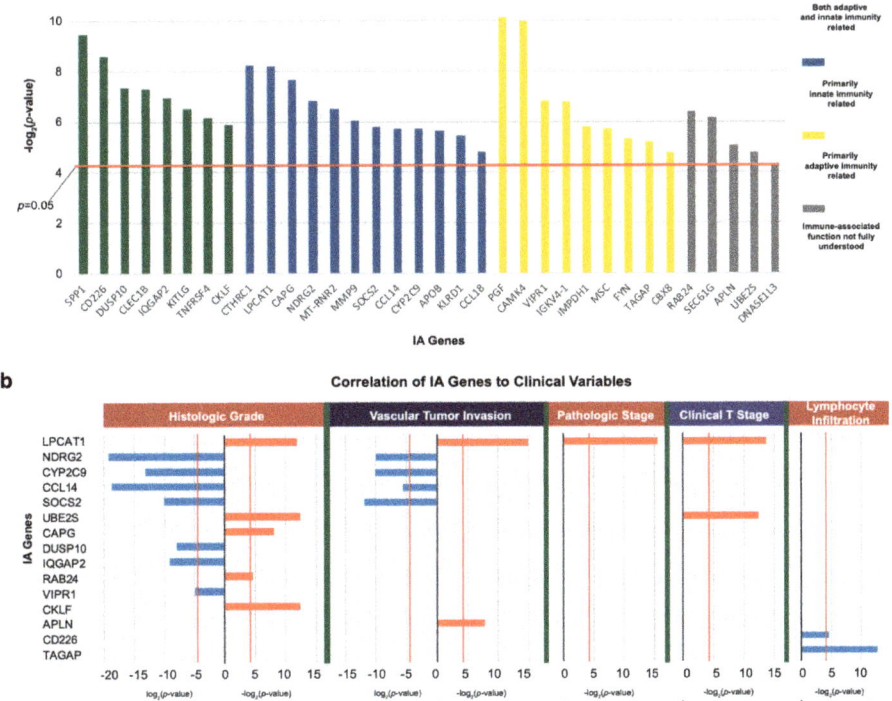

Figure 2. *Cont.*

c

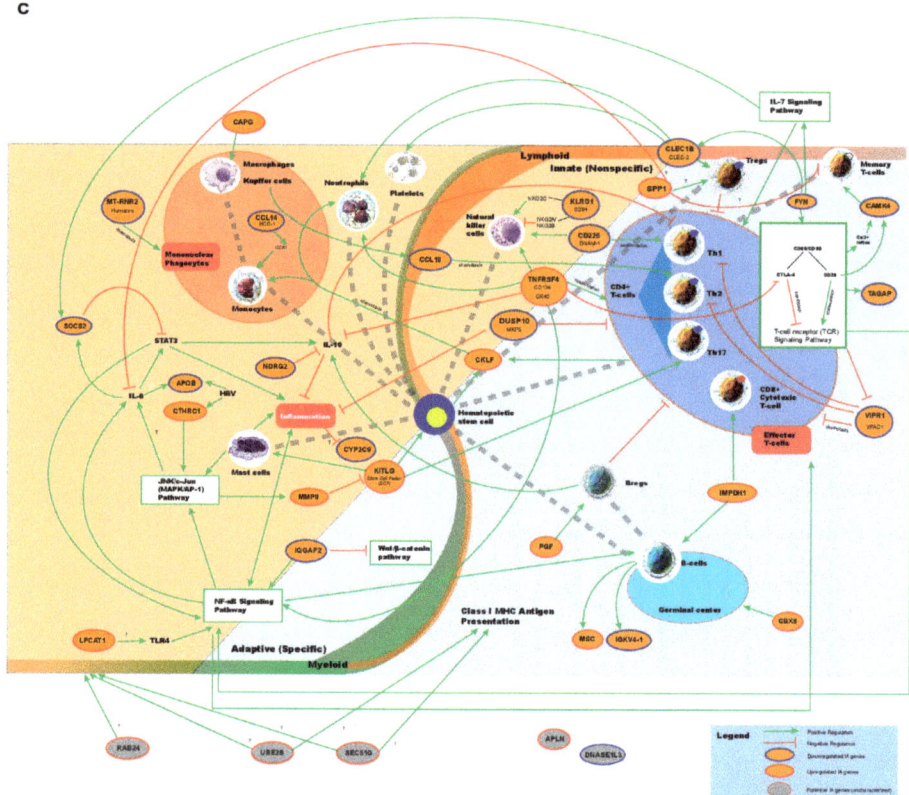

Figure 2. Correlation of IA gene expression with survival and clinical variables. (**a**) Bar graphs plotting the -$\log_2$(p-value) of correlation of IA gene expression with patient survival (Cox regression test, $p < 0.05$). (**b**) Bar graphs plotting the $\pm \log_2$ (p-value) of correlation of IA gene expression with clinical variable (Kruskal-Wallis test, $p < 0.05$). Positive and negative correlation with variables are plotted in opposite directions on the graph. The greater the bars extend from 0, the higher the correlation between the variable and gene expression. (**c**) The interactions between dysregulated IA genes and key immune cells, processes, and pathways are mapped in this schematic. The graphical renderings of immune-cells were obtained from the galleries of Blausen Medical (https://en.wikiversity.org/wiki/WikiJournal_of_Medicine/Medical_gallery_of_Blausen_Medical_2014) and Concepts of Biology (http://philschatz.com/biology-concepts-book/contents/m45542.html).

2.3. Correlation of IA Gene Expressions with Copy Number Variations and Mutation Events Using REVEALER

Copy number variations (CNVs) and mutations are widely recognized as key genomic alterations that drive cancer initiation and progression [14,15]. We used the REVEALER algorithm to systematically correlate all somatic CNVs and mutations that are present in each patient sample to IA gene expression in order to find a set of genomic alterations that are most likely responsible for the dysregulation of each IA gene. Given that genomic alterations initiate and sustain cancer, and that IA gene dysregulation is the likely cause for the sustenance of tumors against immune destruction, the alteration events that highly correlate with dysregulated IA gene expression in multiple patient samples are highly probable to be the cause of such dysregulation. Our REVEALER results illustrate that a large number of IA genes have significant correlation in expression with a set of genomic alterations, and there is minimal overlap between the genomic alterations that are implicated with each IA gene dysregulation

(Figure 3a,b). This result is consistent with the diverse functions of these IA genes and the different IA pathways that they are involved in, as summarized in Figure 2c.

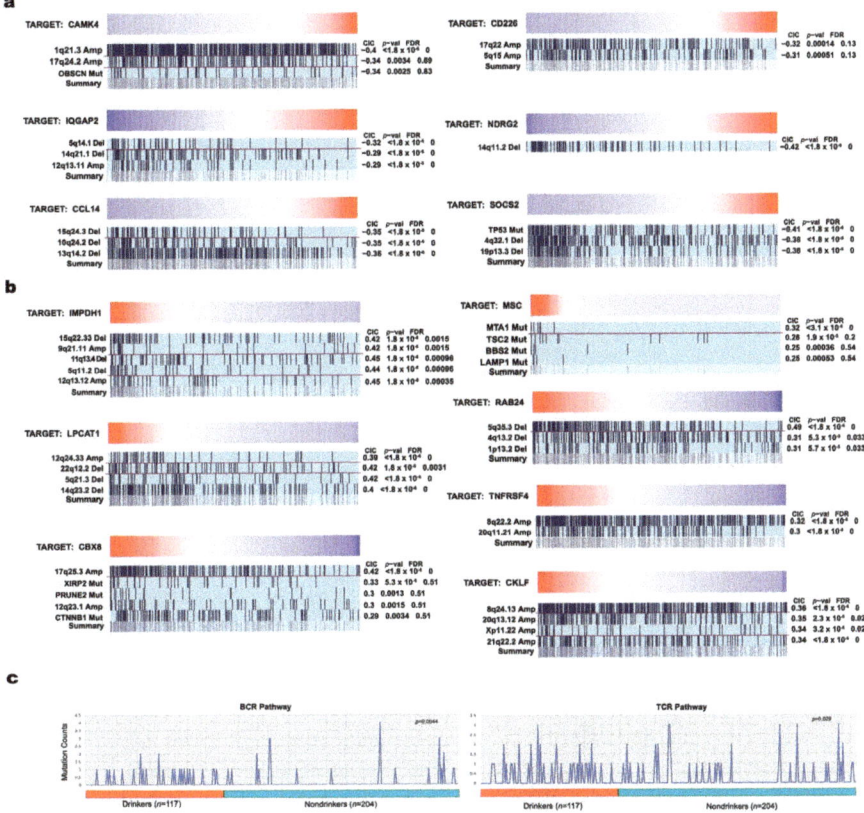

Figure 3. Correlation of IA gene expression to genomic alterations using REVEALER and differential mutation load in key immune-associated pathways. Genomic alterations are analyzed for correlation with low expression in (**a**) downregulated IA genes, while they are analyzed for correlation with high miRNA expression in (**b**) upregulated IA genes. The gradient bar displays the range of the IA gene expression, with the dark red extreme representing the highest expression and the dark blue extreme representing the lowest expression. Each patient sample would be assigned a specific spot along a light blue row based on expression of the IA gene within the sample, and if the indicated genomic alteration is present in a sample, it would be shaded as a dark blue bar. When many shaded bars are clustered towards an extremity of the row, it means many samples with very high or very low expression of the IA gene have the indicated genomic alteration, suggesting that the genomic alteration is significantly correlated with IA gene expression. Genomic alterations negatively correlated with IA gene expression result in negative CIC, while those positively correlated result in positive CIC. Caution should be taken when asserting correlations based on visual inspection because of the nonlinear distribution of expression values in each plot. Significant correlation was determined with CIC of around 0.30 or higher. A red dividing bar in between genomic features signifies a change in iteration. The Summary row combines results from multiple correlations to examine how well the set of genomic alterations identified could collectively account for gene dysregulation. (**c**) Line plots depicting differential mutation counts in B-cell receptor (BCR) and T-cell receptor (TCR) pathways for HCC drinkers versus HCC nondrinkers.

2.4. Identification of Frequently Disrupted Immune-Associated Pathways in HCC through Differential Mutational Load

Genetic mutations that are present in pathways regulating IA functions may reveal key insights into the mechanisms through which HCC evades the anti-cancer immune response. We compared the mutational load within the IA pathways of tumor samples from drinkers to those of tumor samples from nondrinkers to explore the effect of alcohol in dysregulating these pathways. We found that genes in the B cell receptor (BCR) signaling pathway and T cell receptor (TCR) signaling pathway have significantly higher rates of mutations in drinkers as compared to nondrinkers (fisher's exact test—$p < 0.05$, Figure 3c and Table 2).

Table 2. Differential mutation load of selected immune-associated pathways.

Comparison: HCC Drinkers vs. HCC Nondrinkers		
Pathway Name	Odds Ratio	p-Value
B-cell receptor pathway	2.735	0.004
T-cell receptor pathway	1.901	0.029
Classical complement system activation pathway	2.184	0.047

2.5. Identification of Potential miRNAs Involved in Regulation of IA Genes

We used the online target prediction service that was provided by TargetScan–release 7.1 to identify miRNAs that potentially target dysregulated IA genes identified in our study. The website was recently overhauled to incorporate a statistical model that was developed by Agarwal et al., who found their model to be significantly more accurate than other existing computational models for miRNA target prediction [16]. The list of potentially relevant miRNAs was filtered, so that only miRNAs that we identified in our previous study to be dysregulated in HCC in the opposite direction as the dysregulation of IA genes were retained as the candidates for further correlation.

We used gene set enrichment analysis (GSEA) to investigate the correlation of miRNA expression with IA gene expression (Table S1). GSEA also allows for the quantification of the degree of synergistic gene expression suppression by multiple miRNAs (through enrichment score), as well as the ranking of miRNAs that were most closely associated with dysregulated IA genes (through rank metric score). The expression of a large number of miRNAs was found to be negatively enriched in relation to IA gene expression, which demonstrated an inverse relationship between miRNA expression and IA gene expression. Of the 34 dysregulated IA genes that we identified, 11 have 12 or more candidate regulatory miRNAs that potentially contribute to the suppression of gene expression ($p < 0.05$, Figure 4a). Four IA genes were not found to be targeted by any candidate miRNA. While only using miRNAs with core enrichment (as part of the leading-edge subset) and correlation with negative gene expression, we plotted the landscape of the potential interactions between candidate miRNAs and dysregulated IA genes in HCC (Figure 4b and Table S2).

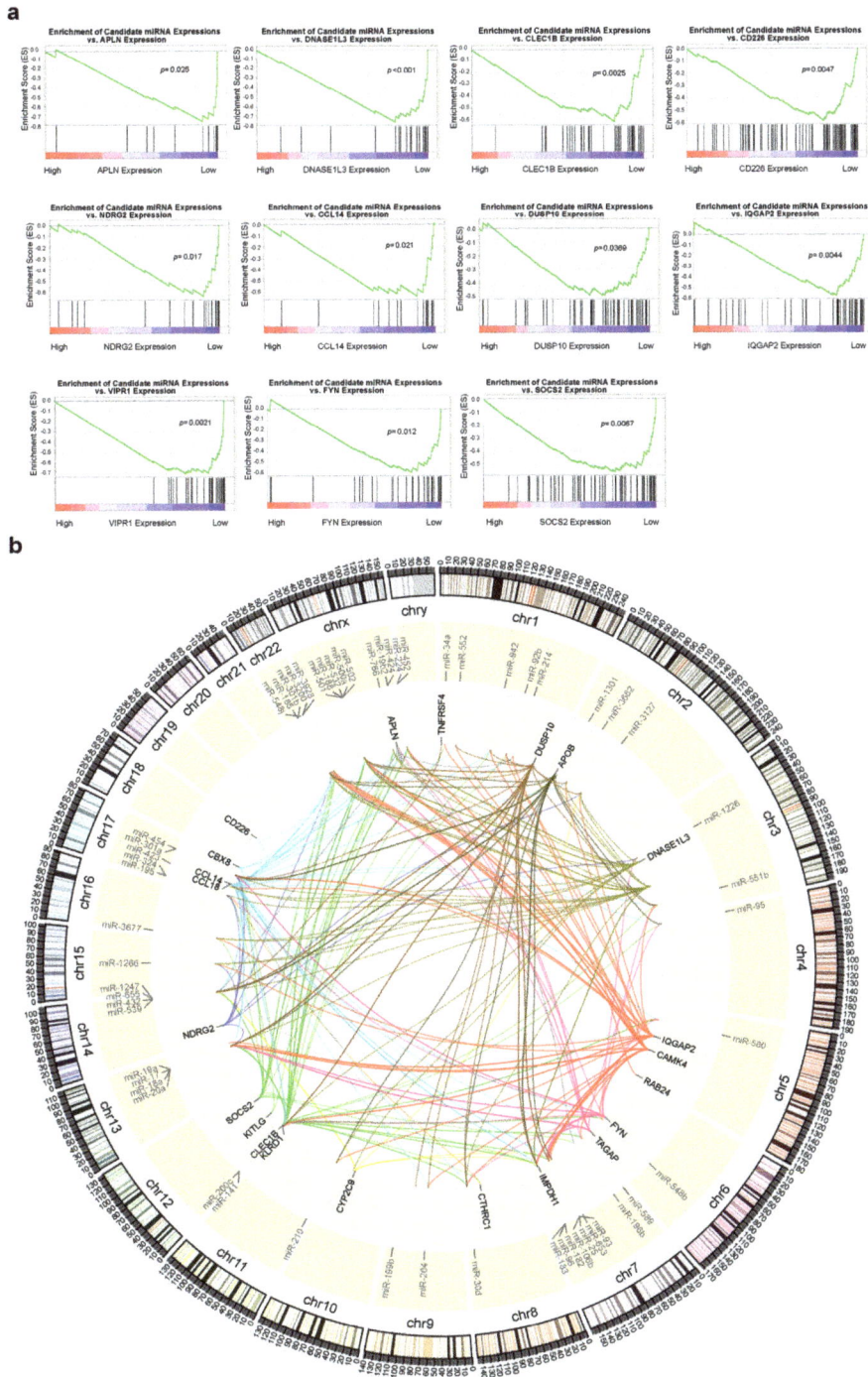

Figure 4. Identification of miRNAs dysregulated in HCC as potential regulators of dysregulated IA genes. (**a**) Gene set enrichment analysis (GSEA) plots ($p < 0.05$) correlate the expression of a set of miRNAs potentially targeting a given IA gene to its gene expression. Negative enrichment suggests

that the expression of the miRNA set depresses the expression of the IA gene. (**b**) A Circos plot depicts the potential interactions between dysregulated miRNAs and dysregulated IA genes with their relative positions in the genome. Only miRNAs with three or more IA genes as potential targets are included in the plot. Potential interactions are defined as interactions forming the leading edge subsets of each GSEA plot. (See also Tables S1 and S2).

2.6. In Vitro Validation of IA Gene Expression in Cell Lines after Alcohol Exposure

Quantitative PCR (qPCR) was used to measure the changes in expression of IA genes in cell lines after treatment with 1% ethanol. Only genes that were previously reported to be expressed in epithelial cells were examined. We found that, for the L-02 cell line, which was derived from normal human fetal hepatocytes, 6 IA genes dysregulated in our analysis are similarly dysregulated after alcohol exposure. The upregulated IA genes—*CKLF*, *KITLG*, and *SEC61G*—in our analysis are upregulated in alcohol-treated L-02 cells, while the downregulated IA genes *CCL14*, *IQGAP2*, *NDRG2*, and *SOCS2* are also downregulated in alcohol-treated L-02 cells (Figure 5a). *NDRG2*, *KITLG*, *SOCS2*, and *SEC61G* are also dysregulated in the HCC cell line MHCC97-L (Figure 5b–e). Additionally, *SOCS2* is downregulated in the HCC cell line HEPG2. We hypothesize that more IA genes are dysregulated in the normal liver cell line, because the HCC cell lines have already been transcriptionally reprogrammed by etiological factors.

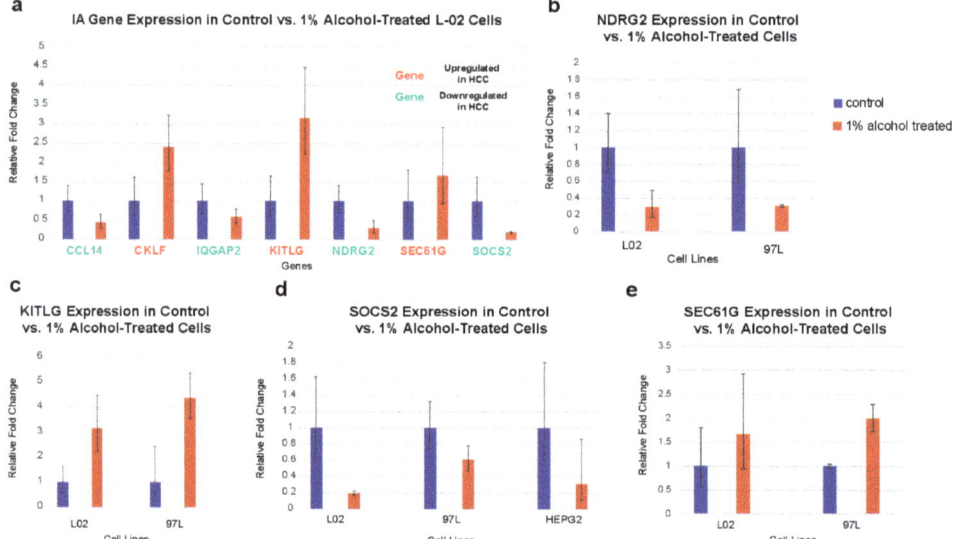

Figure 5. In vitro validation of IA gene dysregulation in liver cell lines by qPCR. (**a**) *CCL14*, *IQGAP2*, *NDRG2*, and *SOCS2* are downregulated, while *CKLF*, *KITLG*, and *SEC61G* are upregulated, after the L-02 cells are exposed to 1% alcohol. (**b**–**e**) *NDRG2* is downregulated, while *KITLG* and *SEC61G* are upregulated, in both MHCC-97L and L-02 cells after 1% alcohol treatment. *SOCS2* is downregulated in the HEPG2 cells, in addition to MHCC-97L and L-02 cells.

3. Discussion

The current treatment options for advanced HCC have shown extremely limited efficacy. The only two drugs that are approved for treating advanced HCC in the United States, sorafenib and regorafenib, only lead to modest increase of the median survival rates of advanced HCC patients [17,18]. The partial response rate of sorafenib is only 2% [19]. Immunotherapy may be the key toward the effective treatment of advanced HCC. A phase I clinical trial with CTLA-4 checkpoint blockade has achieved 17.6% partial response rate in patients with advanced HCC [8]. Another clinical trial using the PD-1 checkpoint inhibitor nivolumab has reported 5% complete response and 18% partial response in advanced HCC [20]. A significant proportion of patients do not respond to immunotherapy treatment despite the promising potential of immunotherapy, particularly checkpoint blockades. When compared to sorafenib's stable disease rate of 58.8%, PD-1 blockade performed worse, with 56% of patients (23 out of 41) in the clinical trial, discontinuing because of progressive disease [20]. The factors leading to the failure of immunotherapy in large fractions of patients remain poorly explored. Understanding differences in the immune status, including functional immune processes and their regulation, of different patients can shed light on the different observed clinical results. We systematically analyzed the gene expression data from TCGA to explore the landscape of dysregulated immune processes in HCC tumors and their surrounding environment. To the best of our knowledge, no previous study has identified the etiologic-specific dysregulation of IA genes or pathways in HCC through large-scale sequencing of RNA expression data, although Sia et al. examined the immune landscape of HCC and Thorsson et. al. examined the pan-cancer immune landscape of the TCGA samples [21,22]. We identified different IA genes that were dysregulated in HCC associated with hepatitis B virus (HBV) infection, attributed to 50% of HCC cases [23], and alcohol intake, which is also a well-established risk factor of HCC [24]. After exploring the relationship between IA gene dysregulation and miRNA dysregulation in HCC, we reveal that miRNA dysregulation might be the key contributor to differences in the immune status of HCC patients and that miRNAs serves as promising candidates for therapeutic intervention or profiling of patient immune status.

We observed several dysregulated genes in HCC that may be key contributors to the potential failure or success of PD-1 blockade, according to a recent study by Manguso et al. describing the effects of certain genes on immunotherapy outcome through loss-of-function screening [25]. Despite the many limitations of that study, including the screening of genes that were only expressed in a single melanoma cell line and the use of murine immune system as model, it was a significant study that sheds light on the many yet unknown factors that influence immunotherapy success. Genes in several pathways that were reported to sensitize tumors or cause resistance to immunotherapy were identified to be dysregulated in our study.

Manguso et al. reported that tumor cells lacking the gene *Jak1* are more resistant to PD-1 blockade [25]. JAK1 is critical to the function of the interferon-gamma signaling pathway and the IL-7 signaling pathway, which implies that reduced activity of these pathways may be responsible for resistance. Our study identified the downregulation of two genes participating in the IL-7 pathway: *FYN*, which codes for the tyrosine protein kinase Fyn and binds to the IL-7 receptor, forming a complex that JAK1 then binds [26]; and, *SOCS2*, the expression of which is induced by IL-7 signaling [27]. *FYN* is most likely downregulated by HBV infection according to our results, while the downregulation of *SOCS2* was observed in four out of five of our comparisons and may be non-etiology specific. The IL-7 signaling pathway is critical to the maintenance and survival of mature T-cells, and the downregulation of both *FYN* and *SOCS2* suggest a downregulation of IL-7 activity [28].

Inactivation of genes that are involved in the induction of the NF-κB pathway was reported to sensitize tumors to PD-1 blockade, which suggested that the activity of such genes may partly explain the failure of PD-1 based immunotherapy in certain patients [25]. We observed *OX40*, which is involved in the activation of canonical NF-κB signaling, to be upregulated in HBV-associated signaling. OX40, also known as TNFRSF4 or CD134, is a co-stimulatory receptor that is expressed on T-cells that binds to OX40L to target NF-κB1 [29]. Therefore, therapeutically antagonizing OX40 through the use of OX40 immunoglobulin may increase the response to PD-1 checkpoint inhibitors. Interestingly, OX40 is also known to increase T-cell viability and inhibit the CTLA-4 checkpoint molecule, so it can be theoretically engaged to complement CTLA-4 checkpoint blockade [30]. Because the wide range of pathways OX40 can potentially activate, further study of its functions in the context of immunotherapy will be useful. If OX40 complements immunotherapy in certain cases and induce resistance to immunotherapy in others, its expression levels in different patients may be a valuable criterion for selecting the appropriate immunotherapy drug to apply.

We also found one gene, *KLRD1* (*CD94*), dysregulated in a direction that potentially sensitizes tumor cells to PD-1 blockade, which may be a mechanism that contributes to the success of PD-1 blockade in certain HCC patients. Manguso et al. reported that the absence of Qa-1b (mice equivalent of HLA-E) binding to NKG2A, which is a inhibitory receptor, on T cells and NK cells increased the effectiveness of PD-1 blockade [25]. In humans, CD94 binds to NKG2A to form the CD94/NKG2A receptor, on which HLA-E then binds, in NK cells [31]. The gene *KLRD1* is likely to be downregulated by both alcohol drinking and HBV infection, which suggested decreased checkpoint activity in these HCC patients. Thus, patients with low *KLRD1* expression may be good candidates for the PD-1 blockade.

The limited results that were obtained by Manguso et al. represent one of the only sources of information elucidating the mechanisms contributing to differences in the effectiveness of immunotherapy. Therefore, we will summarize the rest of our findings in the context of their implications in the HCC immune landscape. The liver is evolutionarily highly tolerant of foreign antigens because of its exposure to blood containing microbial antigens and nutrients flowing from the intestines to the liver through the portal vein [32]. Immunosuppressive mechanisms include the upregulation of immune checkpoint molecules, an increase in the number of regulatory T cells, and the inhibition of natural killer cells [33]. Under this highly immunosuppressive environment, HCC tumor antigens can effectively evade immunity. We identified multiple IA genes dysregulated to increase immunosuppression. *CLEC1* (C-type lectin-like receptor 1B) is downregulated in all cancers as compared to normal tissue in differential expression comparisons, and codes for the protein CLEC-2, which is part of the same pathway as Fyn and is also critical in maintaining lymph node integrity [34,35]. *SPP1*, which is most likely upregulated by alcohol, codes for the protein osteopontin and it is expressed by both tumor cells and myeloid cells to mold an immunosuppressive tumor microenvironment [36]. Other dysregulations leading to immunosuppression include the upregulation of *PGF* and *CTHRC1* and downregulation of *IQGAP2*, *NDRG2*, *CCL14* (*HCC-1*), *CD226* (*DNAM-1*), and *CAMK4*.

On the other hand, HCC is also an inflammation-associated cancer, hence the tumor environment has potential immunogenicity [37]. Inflammation has been recognized as an important factor in malignant transformation and it results in the recruitment of large amounts of immune-associated cells into the tumor environment, which leads to the release of cytokines and growth factors to promote cellular proliferation and regeneration in response to necrosis in the tumor core [38]. We identified a number of IA genes dysregulated to increase immunogenicity. *VIPR1* (vasoactive intestinal polypeptide receptor 1, or *VPAC1*) is consistently downregulated across all cancer to normal sample comparisons. It is expressed on T-cells as a part of the VIP signaling axis, being responsible for suppressing cytokine production and increasing the number of inducible regulatory T-cells [39]. *DUSP10*, which is also known as *MKP5* (mitogen-activated protein kinase phosphatase 5), was identified to be likely downregulated by HBV infection in this study. The downregulation of *DUSP10* was found to increase the levels of pro-inflammatory cytokines and level of T-cell activation [40]. Other dysregulations leading to immunogenicity include the upregulation of *IMPDH1*, *KITLG* (stem cell factor), and *CKLF*.

Correlating the expressions of IA genes we identified with clinical variables allows for us to explore their clinical significance and importance in HCC pathogenesis and progression. Vascular tumor invasion is a strong prognostic factor in HCC and it is arguably the strongest predictor of recurrence after surgical resection or liver transplant [41]. The expressions of six IA genes exhibit significant statistical correlation with vascular tumor invasion. The expressions of several IA genes also correlated with tumor histological grade, pathological stage, and size of primary tumor (clinical T stage). Two IA genes, *CD226* and *TAGAP*, have expressions that directly correlate with lymphocyte infiltration, which suggest that increasing their expression may lead to clinically observable immune activation.

While using qPCR, we were able to validate the dysregulation of *NDRG2*, *KITLG*, *SOCS2*, and *SEC61G* in multiple liver cell lines after exposure to alcohol. In the normal liver cell line, L-02, we observed the dysregulation of *CCL14*, *CKLF*, and *IQGAP2*, in addition to the dysregulation of the genes above. *SEC61G* and *IQGAP2* are exclusively dysregulated by alcohol, while others are dysregulated in both HBV and alcohol-induced HCC, according to our analysis. Several of these genes, including *KITLG*, *CCL14*, and *CKLF*, are cytokines, which possibly provide a mechanism for gene dysregulation in tumor cells to affect the immune phenotype. Additionally, *NDRG2* regulates the release of cytokines in HCC cells, while SEC61G potentially mediates antigen presentation [42,43]. On the other hand, SOCS2 and IQGAP2 relay interactions with factors that are released by immune cells, including various types of cytokines, potentially allowing for cancer cells to be aware of the immune environment [44,45].

In search for potential causal genomic alterations that lead to the dysregulation of these IA genes, we applied the REVEALER algorithm to systematically correlate all mutations and CNVs that are present in HCC to IA gene expression. A diverse set of genomic alterations seems to be implicated in IA gene dysregulation, although we discovered that a number of commonly mutated genes in HCC, including *TP53*, *CTNNB1*, *XIRP2*, and *PRUNE2*, strongly correlated with the dysregulation of certain genes. In addition to common mutations, an increase in the frequency of mutations in IA pathways may offer an explanation to the dysregulation of IA genes. We found that the T-cell receptor pathway and B-cell receptor pathway both have higher mutational load in drinkers with HCC than in nondrinkers with HCC.

We chose the REVEALER correlation method to explore the critical application of our data: the stratification of patients into clinically relevant cohorts that differently respond to immunotherapy drugs, in order to select the most effective treatment option for each patient. REVEALER selects the genomic alteration with the strongest correlations as a seed for subsequent iterations, then the algorithm penalizes the CIC for genomic alterations that are present in patients who also possess the seed alteration, since the dysregulation of IA genes in those patients would be better accounted for by the seed [46]. This innovation allows for the stratification of patient immune status, which we sought to accomplish to some degree through examining the role of etiology in dysregulating IA genes. However, stratification using gene expression has limited effectiveness, because of deficiencies in the RNA sample quality, irreproducibility of expression signatures across different cohorts, and different compositions of cell populations in different samples [47]. An effective stratification method may be developed with genomic alterations when the effects of dysregulated IA genes on immunotherapy outcome, and the genomic alterations that are responsible for their dysregulation, are known. We observed that our REVEALER data adequately demonstrate REVEALER's ability to identify multiple genomic alterations that are associated with each IA gene dysregulation and ensure that as diverse a set of patients as possible possesses these alterations.

We turned to microRNAs (miRNAs) to explore the potential treatment options for a complex stratified patient population. A single miRNA can target multiple mRNAs transcribed from genes in a single network [48]; therefore, therapeutically adjusting the expression of one miRNA can potentially reverse the dysregulation of a large number of genes. For example, the sensitivity of T cell receptors (TCRs) can be effectively increased by miR-181a, which increased interleukin production, stimulated T-cell proliferation, repressed antagonistic phosphatases, and also decreased the number of CTLA-4 molecules [49].

The limitations of previous studies of miRNA-mRNA regulation include an intense focus on validating or predicting individual miRNA-mRNA interaction and the consequent failure to examine complex miRNA-mRNA system interactions. We mapped the association of all dysregulated IA genes in HCC with all potential regulatory miRNAs dysregulated in HCC to assess the wide range of possible interactions that are available for therapeutic targeting.

From our correlations of miRNA dysregulation with IA gene dysregulation, we identified several miRNAs, including miR-106b, miR-17, miR-183, miR-20a, miR-25, miR-301a, miR-30d, miR-532, and miR-93, which can potentially regulate eight to ten different downregulated IA genes. The therapeutic inhibition of the expression of these miRNAs may lead to a dramatic improvement in the immune system's ability to detect tumors. With the exception of miR-532, the upregulation of all the miRNAs that are listed above has been previously described as the prognostic factors of HCC or mechanistically linked to HCC development [50–54]. For example, miR-17 has been reported to regulate the IL-7 signaling pathway through targeting *JAK1* mRNA [55]. The dysregulation of these miRNAs may also be responsible for the HCC tumor evasion of immune processes by dysregulating a large number of IA genes.

4. Materials and Methods

4.1. RNA-Sequencing Datasets and Clinical Data

Level 3 normalized mRNA expression read counts for tumor samples from 371 hepatocellular carcinoma patients and patient clinical data were downloaded on 4 July 2017 from The Cancer Genome Atlas (TCGA) (https://tcga-data.nci.nih.gov/tcga). The mRNA read counts for adjacent solid normal tissue samples of 48 hepatocellular carcinoma patients were also obtained.

4.2. mRNA Differential Expression Analyses

mRNA read count tables were imported into edgeR v3.5 (http://www.bioconductor.org/packages/release/bioc/html/edgeR.html), and lowly expressed mRNAs (counts-per-million < 1 in an amount of samples that ware greater than the size of the smaller cohort of each analysis) were filtered from the analysis. Following TMM (trimmed mean of M-values) normalization, pairwise designs were applied to identify significantly differentially expressed mRNAs in (1) tumor tissue from HCC patients who are drinkers without HBV versus adjacent normal tissue from patients who are nondrinkers, (2) tumor tissue from HCC patients who are drinkers without HBV versus adjacent normal tissue from patients who are drinkers, (3) tumor tissue from HCC patients with HBV who are also drinkers versus normal tissue from patients who are nondrinkers, (4) tumor tissue from HCC patients with HBV who are also drinkers versus normal tissue from patients who are drinkers, (5) tumor tissue from HCC patients with HBV who do not drink versus normal patients from patients who do not drink, and (6) tumor tissue from HCC patients who drink versus tumor tissue from HCC patients who do not drink. Immune-associated genes, from which differentially expressed mRNAs were transcribed, were identified as dysregulated and retained as candidates. Differential expression is defined as $p < 0.05$ and fold change <-2 or >2 in edgeR analysis.

4.3. Association of Candidate Genes' Expressions with Patient Survival and Clinical Variables

Survival analyses were performed while using the Kaplan–Meier Model, with gene expression being designated as a binary variable based on expression above or below the median expression of all the samples. Univariate Cox regression analysis was used to identify candidates that were significantly associated with patient survival ($p < 0.05$). Survival-correlated genes were evaluated for clinical significance. Employing the Kruskal–Wallis test, we investigated gene association with neoplasm histological grade, clinical and pathologic stages, vascular invasion of tumor, and percent lymphocyte infiltration while using clinical data and mRNA expression values (counts-per-million) from HCC patients. In clinical T stage analysis, patients with stages T1a and T1b were grouped into stage T1, and likewise for stages T2, T3, and T4.

4.4. Information-Coefficient Based Correlation of IA Gene Expression with Genomic Alterations

Mutation and copy number variation (CNV) data for the HCC tumors were obtained from mutation and CNV annotation files that were generated by the Broad Institute GDAC Firehose on 28 January 2016. Annotation files were compiled into a binary input file for the program REVEALER (repeated evaluation of variables conditional entropy and redundancy), which was designed to computationally identify a set of specific copy number variations and mutations that were most likely responsible for the change in activity of a target profile [46]. The target profile was defined in our study to be IA gene expression. REVEALER runs multiple iterations of the correlation algorithm, with the genomic feature exhibiting the strongest correlation in each run serving as a seed for the successive run to identify a set of most relevant genomic alterations. We set the maximum number of iterations to three. A seed is a particular mutation or copy number gain or loss event that most likely accounts for the target activity. When given a seed, REVEALER will focus correlation only on patients with altered target activity that was not accounted for by the seed. We set the seed to null for the first iteration. We set the threshold of genomic features to input to features present in less than 75% of all samples.

4.5. Identification of Differential Mutational Load in Immune-Associated Pathways

Immune-associated pathways were manually identified through gene sets described in existing literature. The number of genes with mutations in each immune-associated pathway was tallied for each patient tumor sample, and the Fisher's exact test was performed to identify significant differential mutational load ($p < 0.05$) of each pathway in the samples from alcohol drinkers versus samples from nondrinkers.

4.6. Assessing Potential Involvement of miRNAs in Regulating IA Genes

To identify the possible regulatory miRNAs that were associated with IA genes, we identified a list of miRNAs predicted to bind to each dysregulated mRNA using TargetScan version 7.1 (http://www.targetscan.org/vert_71/) [16]. This list is then filtered to exclude any miRNAs not identified as dysregulated in HCC in our previous study [10], and only miRNAs that were dysregulated in a direction consistent with their regulatory roles of IA genes (i.e., miRNAs that were upregulated if the IA gene was downregulated) were retained as candidates.

The gene set enrichment analysis (GSEA) software was used to characterize the enrichment of miRNA expressions with respect to IA gene expressions [56]. The full set of candidate miRNAs for each IA gene was modeled as a gene set. The continuous expression values of IA genes were used as phenotype labels. The unfiltered expression values of all miRNAs available from TCGA miRNA expression datasets ($n = 1535$) were included in the expression dataset input file. One GSEA plot was produced for each IA gene that was potentially associated with seven or more candidate miRNAs.

4.7. Validation of IA Gene Expression with Quantitative PCR (qPCR)

The cultured cells were treated with 1% ethanol for 24 h. Specifically, 20 µL of pure ethanol was added to a culture plate with 2 mL of media. The plates were sealed following ethanol exposure to keep the ethanol from escaping and to maintain constant ethanol levels. Total cell lysate was collected and mRNA was extracted while using the RNeasy kit (QIAGEN). cDNA was then synthesized from 1.5 µg of total mRNA using reverse transcriptase (RT) (Invitrogen, Carlsbad, CA, USA), as per the manufacturer's instructions. Real-time qPCR was performed by combining 2.5 µL of RT with 22.5 µL of SYBR green (Roche, Basel, Switzerland). The reaction was run while using System 7300 (Applied Biosystems, Foster City, CA, USA) and the results were analyzed by the relative quantity method. Experiments were performed in triplicates with GAPDH expression as the endogenous control. GAPDH was chosen as control, because it was not differentially expressed between the samples from drinkers vs. those from nondrinkers ($p = 0.57$), which suggested that alcohol does not alter GAPDH expression. Primers were custom designed by the authors and created by Eurofin Genomics, Louisville, KY, USA. The following sequences were used:

CCL14 forward: AATACAGCTAAAGTTGGTGGGG
CCL14 reverse: TCAAAGCAGGGAAGCTCCAA
CKLF forward: GGCACTAACTGTGACATCTATGA
CKLF reverse: TCACAAGTGCAAACACAAGCA
IQGAP2 forward: TCAAGTGTAGGAAGGAGTTGTGG
IQGAP2 reverse: CTGGATCTGGGGTGCTATTCC
KITLG forward: TATGTCCCCGGGATGGATGT

KITLG reverse: TTTGGCCTTCCCTTTCTCAGG
NDRG2 forward: GGGACAGGGATGGAAAATGGT
NDRG2 reverse: CCACATGAACCCGCACAAAG
SEC61G forward: TTTAGGTGTCGGTTGGGTAGG
SEC61G reverse: CTCACACCCTCACACTTGTTC
SOCS2 forward: AGAGCCGGAGAGTCTGGTTT
SOCS2 reverse: ATAGCGATCCTTGGCCCTTG

5. Conclusions

Our study demonstrated significant differences in the clinically relevant IA gene dysregulation landscape in HBV-induced and alcohol-induced HCC. We found that several dysregulated IA genes that are associated with pathways reported to contribute to immunotherapy effectiveness or resistance and identified several other dysregulation IA genes that we hope will be examined in the context of immunotherapy outcome in future studies. We correlated IA gene dysregulation to genomic alterations to explore potential methods of stratifying patients into clinically relevant populations because of the diverse expression profiles possible for different patients. Finally, we proposed a novel focus for HCC immunotherapy by examining dysregulated miRNA as the potential targets for therapeutic intervention of a stratified patient population. The presence of large numbers of dysregulated genes in the HCC immune landscape and differences in this dysregulation profile based on HCC etiology precipitate the importance of using regulatory molecules, such as miRNAs, as treatment targets to improve the patient response rate to immunotherapy.

Supplementary Materials: The following are available online at http://www.mdpi.com/2072-6694/11/9/1273/s1, Table S1: GSEA results for list of miRNAs associated with each IA gene, Table S2: List of potential miRNA-mRNA interactions as defined by GSEA leading-edge subsets.

Author Contributions: Conceptualization, W.M.O.; methodology, W.T.L., A.E.Z., and W.M.O.; software, W.T.L. and A.E.Z.; validation, W.T.L., A.E.Z., C.O.H., and H.Z.; formal analysis, W.T.L., A.E.Z., C.O.H., and H.Z.; investigation, W.T.L., A.E.Z., C.O.H., and H.Z.; resources, W.M.O.; data curation, N/A; writing—original draft preparation, W.T.L.; writing—review and editing, X.Q.W., T.K., E.Y.C., and W.M.O.; visualization, W.T.L.; supervision, W.M.O.; project administration, W.M.O.; funding acquisition, E.Y.C. and W.M.O.

Funding: This research was funded by the Academic Senate of the University of California San Diego, grant number RS167R-ONGKEKO.

Conflicts of Interest: The authors declare no conflict of interest.

References

1. Qu, Z.; Yuan, C.-H.; Yin, C.-Q.; Guan, Q.; Chen, H.; Wang, F.-B. Meta-analysis of the prognostic value of abnormally expressed lncRNAs in hepatocellular carcinoma. *OncoTargets Ther.* **2016**, *9*, 5143–5152. [CrossRef] [PubMed]
2. Siegel, R.; Naishadham, D. Cancer statistics, 2013. *CA Cancer J. Clin.* **2013**, *63*, 11–30. [CrossRef] [PubMed]
3. Dufour, J.F.; Greten, T.F.; Raymond, E.; Roskams, T.; De, T.; Ducreux, M.; Mazzaferro, V.; Governing, E. Clinical Practice Guidelines EASL—EORTC Clinical Practice Guidelines: Management of hepatocellular carcinoma European Organisation for Research and Treatment of Cancer. *J. Hepatol.* **2012**, *56*, 908–943. [CrossRef]
4. Asghar, U.; Meyer, T. Are there opportunities for chemotherapy in the treatment of hepatocellular cancer? *J. Hepatol.* **2012**, *56*, 686–695. [CrossRef] [PubMed]
5. Rai, V.; Abdo, J.; Alsuwaidan, A.N.; Agrawal, S.; Sharma, P.; Agrawal, D.K. Cellular and molecular targets for the immunotherapy of hepatocellular carcinoma. *Mol. Cell. Biochem.* **2017**, *437*, 13–36. [CrossRef] [PubMed]
6. Heo, J.; Reid, T.; Ruo, L.; Breitbach, C.J.; Rose, S.; Bloomston, M.; Cho, M.; Lim, H.Y.; Chung, H.C.; Kim, C.W.; et al. Randomized dose-finding clinical trial of oncolytic immunotherapeutic vaccinia JX-594 in liver cancer. *Nat. Med.* **2013**, *19*, 329–336. [CrossRef]

7. Sawada, Y.; Yoshikawa, T.; Fujii, S.; Mitsunaga, S.; Nobuoka, D.; Mizuno, S.; Takahashi, M.; Yamauchi, C.; Endo, I.; Nakatsura, T. Remarkable tumor lysis in a hepatocellular carcinoma patient immediately following glypican-3-derived peptide vaccination. *Hum. Vaccines Immunother.* **2013**, *9*, 1228–1233. [CrossRef]
8. Sangro, B.; Gomez-Martin, C.; de la Mata, M.; Inarrairaegui, M.; Garralda, E.; Barrera, P.; Riezu-Boj, J.I.; Larrea, E.; Alfaro, C.; Sarobe, P.; et al. A clinical trial of CTLA-4 blockade with tremelimumab in patients with hepatocellular carcinoma and chronic hepatitis C. *J. Hepatol.* **2013**, *59*, 81–88. [CrossRef]
9. O'Connell, R.M.; Rao, D.S.; Chaudhuri, A.A.; Baltimore, D. Physiological and pathological roles for microRNAs in the immune system. *Nat. Rev. Immunol.* **2010**, *10*, 111–122. [CrossRef]
10. Zheng, H.; Zou, A.E.; Saad, M.A.; Wang, X.Q.; Kwok, J.G.; Korrapati, A.; Li, P.; Kisseleva, T.; Wang-Rodriguez, J.; Ongkeko, W.M. Alcohol-dysregulated microRNAs in hepatitis B virus-related hepatocellular carcinoma. *PLoS ONE* **2017**, *12*. [CrossRef]
11. Hassan, M.M.; Hwang, L.Y.; Hatten, C.J.; Swaim, M.; Li, D.; Abbruzzese, J.L.; Beasley, P.; Patt, Y.Z. Risk factors for hepatocellular carcinoma: Synergism of alcohol with viral hepatitis and diabetes mellitus. *Hepatology* **2002**, *36*, 1206–1213. [CrossRef] [PubMed]
12. Shafritz, D.A.; Kew, M.C. Identification of integrated hepatitis B virus DNA sequences in human hepatocellular carcinomas. *Hepatology* **1981**, *1*, 1–8. [CrossRef] [PubMed]
13. Aran, D.; Sirota, M.; Butte, A.J. Systematic pan-cancer analysis of tumour purity. *Nat. Commun.* **2015**, *6*, 8971. [CrossRef] [PubMed]
14. Stratton, M.R.; Campbell, P.J.; Futreal, P.A. The cancer genome. *Nature* **2009**, *458*, 719–724. [CrossRef] [PubMed]
15. Zack, T.I.; Schumacher, S.E.; Carter, S.L.; Cherniack, A.D.; Saksena, G.; Tabak, B.; Lawrence, M.S.; Zhang, C.-Z.; Wala, J.; Mermel, C.H.; et al. Pan-cancer patterns of somatic copy number alteration. *Nat. Genet.* **2013**, *45*, 1134–1140. [CrossRef] [PubMed]
16. Agarwal, V.; Bell, G.W.; Nam, J.-W.; Bartel, D.P.; Ameres, S.; Martinez, J.; Schroeder, R.; Anders, G.; Mackowiak, S.; Jens, M.; et al. Predicting effective microRNA target sites in mammalian mRNAs. *Elife* **2015**, *4*, 101–112. [CrossRef] [PubMed]
17. Cheng, A.-L.; Kang, Y.-K.; Chen, Z.; Tsao, C.-J.; Qin, S.; Kim, J.S.; Luo, R.; Feng, J.; Ye, S.; Yang, T.-S.; et al. Efficacy and safety of sorafenib in patients in the Asia-Pacific region with advanced hepatocellular carcinoma: A phase III randomised, double-blind, placebo-controlled trial. *Lancet Oncol.* **2009**, *10*, 25–34. [CrossRef]
18. Bruix, J.; Qin, S.; Merle, P.; Granito, A.; Huang, Y.H.; Bodoky, G.; Pracht, M.; Yokosuka, O.; Rosmorduc, O.; Breder, V.; et al. Regorafenib for patients with hepatocellular carcinoma who progressed on sorafenib treatment (RESORCE): A randomised, double-blind, placebo-controlled, phase 3 trial. *Lancet* **2017**, *389*, 56–66. [CrossRef]
19. Chan, S.L.; Mok, T.; Ma, B.B.Y. Management of hepatocellular carcinoma: Beyond sorafenib. *Curr. Oncol. Rep.* **2012**, *14*, 257–266. [CrossRef]
20. El-Khoueiry, A.B.; Melero, I.; Crocenzi, T.S.; Welling, T.H.; Yau, T.C.; Yeo, W.; Chopra, A.; Grosso, J.; Lang, L.; Anderson, J.; et al. Phase I/II safety and antitumor activity of nivolumab in patients with advanced hepatocellular carcinoma (HCC): CA209-040. *J. Clin. Oncol.* **2015**, *33*, LBA101. [CrossRef]
21. Thorsson, V.; Gibbs, D.L.; Brown, S.D.; Wolf, D.; Bortone, D.S.; Ou Yang, T.-H.; Porta-Pardo, E.; Gao, G.F.; Plaisier, C.L.; Eddy, J.A.; et al. The Immune Landscape of Cancer. *Immunity* **2018**, *48*, 812–830. [CrossRef] [PubMed]
22. Sia, D.; Jiao, Y.; Martinez-Quetglas, I.; Kuchuk, O.; Villacorta-Martin, C.; Castro de Moura, M.; Putra, J.; Camprecios, G.; Bassaganyas, L.; Akers, N.; et al. Identification of an Immune-specific Class of Hepatocellular Carcinoma, Based on Molecular Features. *Gastroenterology* **2017**. [CrossRef] [PubMed]
23. Gurtsevitch, V.E. Human oncogenic viruses: Hepatitis B and hepatitis C viruses and their role in hepatocarcinogenesis. *Biochemistry* **2008**, *73*, 504–513. [CrossRef] [PubMed]
24. Morgan, T.R.; Mandayam, S.; Jamal, M.M. Alcohol and hepatocellular carcinoma. *Gastroenterology* **2004**, *127*, S87–S96. [CrossRef] [PubMed]
25. Manguso, R.T.; Pope, H.W.; Zimmer, M.D.; Brown, F.D.; Yates, K.B.; Miller, B.C.; Collins, N.B.; Bi, K.; LaFleur, M.W.; Juneja, V.R.; et al. In vivo CRISPR screening identifies Ptpn2 as a cancer immunotherapy target. *Nature* **2017**, *547*, 413–418. [CrossRef]
26. Tanner, J.W.; Chen, W.; Young, R.L.; Longmore, G.D.; Shaw, A.S. The conserved box 1 motif of cytokine receptors is required for association with JAK kinases. *J. Biol. Chem.* **1995**, *270*, 6523–6530. [CrossRef]

27. Ghazawi, F.M.; Faller, E.M.; Parmar, P.; El-Salfiti, A.; MacPherson, P.A. Suppressor of cytokine signaling (SOCS) proteins are induced by IL-7 and target surface CD127 protein for degradation in human CD8 T cells. *Cell. Immunol.* **2016**, *306–307*, 41–52. [CrossRef]
28. Carrette, F.; Surh, C.D. IL-7 signaling and CD127 receptor regulation in the control of T cell homeostasis. *Semin. Immunol.* **2012**, *24*, 209–217. [CrossRef]
29. Song, J.; So, T.; Croft, M. Activation of NF-kappaB1 by OX40 contributes to antigen-driven T cell expansion and survival. *J. Immunol.* **2008**, *180*, 7240–7248. [CrossRef]
30. Redmond, W.L.; Linch, S.N.; Kasiewicz, M.J. Combined targeting of costimulatory (OX40) and coinhibitory (CTLA-4) pathways elicits potent effector T cells capable of driving robust antitumor immunity. *Cancer Immunol. Res.* **2014**, *2*, 142–153. [CrossRef]
31. Braud, V.M.; Allan, D.S.J.; O'Callaghan, C.A.; Söderström, K.; D'Andrea, A.; Ogg, G.S.; Lazetic, S.; Young, N.T.; Bell, J.I.; Phillips, J.H.; et al. HLA-E binds to natural killer cell receptors CD94/NKG2A, B and C. *Nature* **1998**, *391*, 795–799. [CrossRef] [PubMed]
32. Longo, V.; Gnoni, A.; Casadei Gardini, A.; Pisconti, S.; Licchetta, A.; Scartozzi, M.; Memeo, R.; Palmieri, V.O.; Aprile, G.; Santini, D.; et al. Immunotherapeutic approaches for hepatocellular carcinoma. *Oncotarget* **2017**, *8*, 33897–33910. [CrossRef] [PubMed]
33. Hato, T.; Goyal, L.; Greten, T.F.; Duda, D.G.; Zhu, A.X. Immune checkpoint blockade in hepatocellular carcinoma: Current progress and future directions. *Hepatology* **2014**, *60*, 1776–1782. [CrossRef] [PubMed]
34. Severin, S.; Pollitt, A.Y.; Navarro-Nunez, L.; Nash, C.A.; Mourao-Sa, D.; Eble, J.A.; Senis, Y.A.; Watson, S.P. Syk-dependent phosphorylation of CLEC-2: A novel mechanism of hem-immunoreceptor tyrosine-based activation motif signaling. *J. Biol. Chem.* **2011**, *286*, 4107–4116. [CrossRef] [PubMed]
35. Bénézech, C.; Nayar, S.; Finney, B.A.; Withers, D.R.; Lowe, K.; Desanti, G.E.; Marriott, C.L.; Watson, S.P.; Caamaño, J.H.; Buckley, C.D.; et al. CLEC-2 is required for development and maintenance of lymph nodes. *Blood* **2014**, *123*, 3200–3207. [CrossRef]
36. Sangaletti, S.; Tripodo, C.; Sandri, S.; Torselli, I.; Vitali, C.; Ratti, C.; Botti, L.; Burocchi, A.; Porcasi, R.; Tomirotti, A.; et al. Osteopontin shapes immunosuppression in the metastatic niche. *Cancer Res.* **2014**, *74*, 4706–4719. [CrossRef] [PubMed]
37. Tarao, K.; Ohkawa, S.; Miyagi, Y.; Morinaga, S.; Ohshige, K.; Yamamoto, N.; Ueno, M.; Kobayashi, S.; Kameda, R.; Tamai, S.; et al. Inflammation in background cirrhosis evokes malignant progression in HCC development from HCV-associated liver cirrhosis. *Scand. J. Gastroenterol.* **2013**, *48*, 729–735. [CrossRef]
38. He, G.; Karin, M. NF-κB and STAT3-key players in liver inflammation and cancer. *Cell Res.* **2011**, *21*, 159–168. [CrossRef]
39. Pozo, D.; Anderson, P.; Gonzalez-Rey, E. Induction of alloantigen-specific human T regulatory cells by vasoactive intestinal peptide. *J. Immunol.* **2009**, *183*, 4346–4359. [CrossRef]
40. Zhang, Y.; Blattman, J.N.; Kennedy, N.J.; Duong, J.; Nguyen, T.; Wang, Y.; Davis, R.J.; Greenberg, P.D.; Flavell, R.A.; Dong, C. Regulation of innate and adaptive immune responses by MAP kinase phosphatase 5. *Nature* **2004**, *430*, 793–797. [CrossRef]
41. Thuluvath, P.J. Vascular Invasion is the Most Important Predictor of Survival in HCC, but How Do We Find It? *J. Clin. Gastroenterol.* **2009**, *43*, 101–102. [CrossRef] [PubMed]
42. Zehner, M.; Marschall, A.L.; Bos, E.; Schloetel, J.G.; Kreer, C.; Fehrenschild, D.; Limmer, A.; Ossendorp, F.; Lang, T.; Koster, A.J.; et al. The Translocon Protein Sec61 Mediates Antigen Transport from Endosomes in the Cytosol for Cross-Presentation to CD8$^+$ T Cells. *Immunity* **2015**. [CrossRef] [PubMed]
43. Hu, W.; Fan, C.; Jiang, P.; Ma, Z.; Yan, X.; Di, S.; Jiang, S.; Li, T.; Cheng, Y.; Yang, Y. Emerging role of N-myc downstream-regulated gene 2 (NDRG2) in cancer. *Oncotarget* **2016**. [CrossRef] [PubMed]
44. Monti-Rocha, R.; Cramer, A.; Leite, P.G.; Antunes, M.M.; Pereira, R.V.S.; Barroso, A.; Queiroz-Junior, C.M.; David, B.A.; Teixeira, M.M.; Menezes, G.B.; et al. SOCS2 is critical for the balancing of immune response and oxidate stress protecting against acetaminophen-induced acute liver injury. *Front. Immunol.* **2019**. [CrossRef] [PubMed]
45. Brisac, C.; Salloum, S.; Yang, V.; Schaefer, E.A.K.; Holmes, J.A.; Chevaliez, S.; Hong, J.; Carlton-Smith, C.; Alatrakchi, N.; Kruger, A.; et al. IQGAP2 is a novel interferon-alpha antiviral effector gene acting non-conventionally through the NF-κB pathway. *J. Hepatol.* **2016**. [CrossRef] [PubMed]

46. Kim, J.W.; Botvinnik, O.B.; Abudayyeh, O.; Birger, C.; Rosenbluh, J.; Shrestha, Y.; Abazeed, M.E.; Hammerman, P.S.; DiCara, D.; Konieczkowski, D.J.; et al. Characterizing genomic alterations in cancer by complementary functional associations. *Nat. Biotechnol.* **2016**, *34*, 539–546. [CrossRef] [PubMed]
47. Raspe, E.; Decraene, C.; Berx, G. Gene expression profiling to dissect the complexity of cancer biology: Pitfalls and promise. *Semin. Cancer Biol.* **2012**, *22*, 250–260. [CrossRef] [PubMed]
48. Garzon, R.; Marcucci, G.; Croce, C.M. Targeting microRNAs in cancer: Rationale, strategies and challenges. *Nat. Rev. Drug Discov.* **2010**, *9*, 775–789. [CrossRef] [PubMed]
49. Li, Q.J.; Chau, J.; Ebert, P.J.R.; Sylvester, G.; Min, H.; Liu, G.; Braich, R.; Manoharan, M.; Soutschek, J.; Skare, P.; et al. miR-181a Is an Intrinsic Modulator of T Cell Sensitivity and Selection. *Cell* **2007**, *129*, 147–161. [CrossRef]
50. Shen, G.; Jia, H.; Tai, Q.; Li, Y.; Chen, D. miR-106b downregulates adenomatous polyposis coli and promotes cell proliferation in human hepatocellular carcinoma. *Carcinogenesis* **2013**, *34*, 211–219. [CrossRef]
51. Zhang, Y.; Zheng, L.; Ding, Y.; Li, Q.; Wang, R.; Liu, T.; Sun, Q.; Yang, H.; Peng, S.; Wang, W.; et al. MiR-20a Induces Cell Radioresistance by Activating the PTEN/PI3K/Akt Signaling Pathway in Hepatocellular Carcinoma. *Int. J. Radiat. Oncol.* **2015**, *92*, 1132–1140. [CrossRef] [PubMed]
52. Ohta, K.; Hoshino, H.; Wang, J.; Ono, S.; Iida, Y.; Hata, K.; Huang, S.K.; Colquhoun, S.; Hoon, D.S.B. MicroRNA-93 activates c-Met/PI3K/Akt pathway activity in hepatocellular carcinoma by directly inhibiting PTEN and CDKN1A. *Oncotarget* **2015**, *6*, 3211–3224. [CrossRef] [PubMed]
53. Zhou, P.; Jiang, W.; Wu, L.; Chang, R.; Wu, K.; Wang, Z. miR-301a Is a Candidate Oncogene that Targets the Homeobox Gene Gax in Human Hepatocellular Carcinoma. *Dig. Dis. Sci.* **2012**, *57*, 1171–1180. [CrossRef] [PubMed]
54. Yao, J.; Liang, L.; Huang, S.; Ding, J.; Tan, N.; Zhao, Y.; Yan, M.; Ge, C.; Zhang, Z.; Chen, T.; et al. MicroRNA-30d promotes tumor invasion and metastasis by targeting Galphai2 in hepatocellular carcinoma. *Hepatology* **2010**, *51*, 846–856. [CrossRef] [PubMed]
55. Katz, G.; Pobezinsky, L.A.; Jeurling, S.; Shinzawa, M.; Van Laethem, F.; Singer, A. T cell receptor stimulation impairs IL-7 receptor signaling by inducing expression of the microRNA miR-17 to target Janus kinase 1. *Sci. Signal.* **2014**, *7*, ra83. [CrossRef] [PubMed]
56. Subramanian, A.; Tamayo, P.; Mootha, V.K.; Mukherjee, S.; Ebert, B.L.; Gillette, M.A.; Paulovich, A.; Pomeroy, S.L.; Golub, T.R.; Lander, E.S.; et al. Gene set enrichment analysis: A knowledge-based approach for interpreting genome-wide expression profiles. *Proc. Natl. Acad. Sci. USA* **2005**, *102*, 15545–15550. [CrossRef] [PubMed]

© 2019 by the authors. Licensee MDPI, Basel, Switzerland. This article is an open access article distributed under the terms and conditions of the Creative Commons Attribution (CC BY) license (http://creativecommons.org/licenses/by/4.0/).

MDPI
St. Alban-Anlage 66
4052 Basel
Switzerland
Tel. +41 61 683 77 34
Fax +41 61 302 89 18
www.mdpi.com

Cancers Editorial Office
E-mail: cancers@mdpi.com
www.mdpi.com/journal/cancers

www.ingramcontent.com/pod-product-compliance
Lightning Source LLC
LaVergne TN
LVHW070509100526
838202LV00014B/1821